T0318737

VOLCANIC AND IGNEOUS PLUMBING SYSTEMS

VOLCANIC AND IGNEOUS PLUMBING SYSTEMS

Understanding Magma Transport, Storage, and Evolution in the Earth's Crust

Edited by

STEFFI BURCHARDT

Uppsala University, Uppsala, Sweden

Elsevier
Radarweg 29, PO Box 211, 1000 AE Amsterdam, Netherlands
The Boulevard, Langford Lane, Kidlington, Oxford OX5 1GB, United Kingdom
50 Hampshire Street, 5th Floor, Cambridge, MA 02139, United States

Copyright © 2018 Elsevier Inc. All rights reserved.

No part of this publication may be reproduced or transmitted in any form or by any means, electronic or mechanical, including photocopying, recording, or any information storage and retrieval system, without permission in writing from the publisher. Details on how to seek permission, further information about the Publisher's permissions policies and our arrangements with organizations such as the Copyright Clearance Center and the Copyright Licensing Agency, can be found at our website: www.elsevier.com/permissions.

This book and the individual contributions contained in it are protected under copyright by the Publisher (other than as may be noted herein).

Notices

Knowledge and best practice in this field are constantly changing. As new research and experience broaden our understanding, changes in research methods, professional practices, or medical treatment may become necessary.

Practitioners and researchers must always rely on their own experience and knowledge in evaluating and using any information, methods, compounds, or experiments described herein. In using such information or methods they should be mindful of their own safety and the safety of others, including parties for whom they have a professional responsibility.

To the fullest extent of the law, neither the Publisher nor the authors, contributors, or editors, assume any liability for any injury and/or damage to persons or property as a matter of products liability, negligence or otherwise, or from any use or operation of any methods, products, instructions, or ideas contained in the material herein.

Library of Congress Cataloging-in-Publication Data
A catalog record for this book is available from the Library of Congress

British Library Cataloguing-in-Publication Data
A catalogue record for this book is available from the British Library

ISBN: 978-0-12-809749-6

For information on all Elsevier publications visit our website at
https://www.elsevier.com/books-and-journals

Publisher: Candice Janco
Acquisition Editor: Marisa LaFleur
Editorial Project Manager: Katerina Zaliva
Production Project Manager: Omer Mukthar
Designer: Victoria Pearson

Typeset by Thomson Digital

CONTENTS

LIST OF CONTRIBUTORS

Talfan Barnie
Nordic Volcanological Center, Institute of Earth Sciences, University of Iceland; Icelandic Meteorological Office, Reykjavik, Iceland

H.S. Bertelsen
Physics of Geological Processes, University of Oslo, Oslo, Norway

Steffi Burchardt
Uppsala University, Uppsala, Sweden

Jim W. Cole
University of Canterbury, Christchurch, New Zealand

A.R. Cruden
Monash University, Melbourne, VIC, Australia

Jonathan R.J. Davidson
University of Canterbury, Christchurch, New Zealand

Audray Delcamp
Vrije Universiteit Brussel, Brussels, Belgium

M. Detienne
Earth and Life Institute, Université catholique de Louvain, Louvain-la-Neuve, Belgium

Katherine J. Dobson
Volcanology, Earth Sciences, Durham University, UK

Vincent Drouin
Nordic Volcanological Center, Institute of Earth Sciences, University of Iceland, Reykjavik, Iceland

Stéphanie Dumont
Instituto Dom Luiz – University of Beira Interior, Covilhã, Portugal

C.H. Eide
University of Bergen, Bergen, Norway

Páll Einarsson
Nordic Volcanological Center, Institute of Earth Sciences, University of Iceland, Reykjavik, Iceland

O. Galland
Physics of Geological Processes, University of Oslo, Oslo, Norway

Halldór Geirsson
Nordic Volcanological Center, Institute of Earth Sciences, University of Iceland, Reykjavik, Iceland

Ronni Grapenthin
New Mexico Tech, Department of Earth and Environmental Science, Socorro, NM, United States

Darren M. Gravley
University of Canterbury, Christchurch, New Zealand

Magnús Guðmundsson
Nordic Volcanological Center, Institute of Earth Sciences, University of Iceland, Reykjavik, Iceland

F. Guldstrand
Physics of Geological Processes, University of Oslo, Oslo, Norway

Þórdís Högnadóttir
Nordic Volcanological Center, Institute of Earth Sciences, University of Iceland, Reykjavik, Iceland

Ø.T. Haug
Physics of Geological Processes, University of Oslo, Oslo, Norway

Elías Rafn Heimisson
Stanford University, Stanford, CA, United States

Ásta Rut Hjartardóttir
Nordic Volcanological Center, Institute of Earth Sciences, University of Iceland, Reykjavik, Iceland

Eoghan P. Holohan
UCD School of Earth Sciences, University College Dublin, Dublin, Ireland

Andy Hooper
COMET, School of Earth and Environment, University of Leeds, Leeds, United Kingdom

Sigrún Hreinsdóttir
GNS Science, Lower Hutt, New Zealand

Kristín Jónsdóttir
Icelandic Meteorological Office, Reykjavik, Iceland

Dougal A. Jerram
Centre for Earth Evolution and Dynamics (CEED), University of Oslo, Norway; DougalEARTH Ltd., Solihull, UK; Queensland University of Technology, Brisbane, Queensland, Australia

Janine L. Kavanagh
University of Liverpool, Liverpool, UK

Ben M. Kennedy
University of Canterbury, Christchurch, New Zealand

Héctor A. Leanza
CONICET, Argentinian Museum of Natural Sciences, Buenos Aires, Argentina

K. Mair
Physics of Geological Processes, University of Oslo, Oslo, Norway

Dan J. Morgan
Institute of Geophysics and Tectonics, University of Leeds, Leeds, UK

Sven Morgan
Iowa State University, Ames, IA, United States

Benedikt G. Ófeigsson
Icelandic Meteorological Office, Reykjavik, Iceland

E.M.R. Paguican
University at Buffalo, The State University of New York, Buffalo, NY, United States;
Earth System Sciences, Vrije Universiteit Brussel, Brussels, Belgium

O. Palma
CONICET Y-TEC and Universidad Nacional de La Plata, La Plata, Argentina

Matthew J. Pankhurst
University of Manchester, Manchester; Rutherford Appleton Laboratories, Didcot; Institute of Geophysics and Tectonics, University of Leeds, Leeds, UK; Instituto Tecnológico y de Energías Renovables, Tenerife, Spain

Michelle Parks
Icelandic Meteorological Office, Reykjavik, Iceland

Rikke Pedersen
Nordic Volcanological Center, Institute of Earth Sciences, University of Iceland, Reykjavik, Iceland

S. Planke
Physics of Geological Processes, University of Oslo, Oslo, Norway

S. Poppe
Vrije Universiteit Brussel, Brussels, Belgium

O. Rabbel
Physics of Geological Processes, University of Oslo, Oslo, Norway

Matthew J. Roberts
Icelandic Meteorological Office, Reykjavik, Iceland

B. Rogers
Physics of Geological Processes, University of Oslo, Oslo, Norway

T. Schmiedel
Physics of Geological Processes, University of Oslo, Oslo, Norway

Freysteinn Sigmundsson
Nordic Volcanological Center, Institute of Earth Sciences, University of Iceland, Reykjavik, Iceland

A. Souche
Physics of Geological Processes, University of Oslo, Oslo, Norway

J.B. Spacapan
YPF, Buenos Aires, Argentina

John Stix
McGill University, Montreal, QC, Canada

Erik Sturkell
University of Gothenburg, Gothenburg, Sweden

Hugh Tuffen
Lancaster Environment Centre, Lancaster University, Lancaster, United Kingdom

Benjamin van Wyk de Vries
Laboratoire Magmas et Volcans, Université Blaise Pascal, Clermont-Ferrand, France

Maximillian van Wyk de Vries
Laboratoire Magmas et Volcans, Université Blaise Pascal, Clermont-Ferrand, France

Kristín Vogfjörð
Icelandic Meteorological Office, Reykjavik, Iceland

Thomas R. Walter
Physics of Earthquakes and Volcanoes, GFZ German Research Centre for Geosciences, Telegrafenberg, Potsdam, Germany

R. F. Weinberg
Monash University, Melbourne, VIC, Australia

CHAPTER 1

Introduction to Volcanic and Igneous Plumbing Systems—Developing a Discipline and Common Concepts

Steffi Burchardt
Uppsala University, Uppsala, Sweden

Contents

1.1 INTRODUCTION

Volcanoes are, in every respect, fascinating: they form breathtaking landscapes and provide billions of people with living space, fertile ground, energy and mineral resources. Volcanoes were most certainly a cradle for the development and evolution of life on Earth, but their vigorous nature has also brought countless species close to or beyond extinction. To understand the inner workings of volcanoes in order to ultimately be able to interpret their behaviour and mitigate volcanic hazards is a motivation that unites all volcanologists. Volcanology is traditionally the study of volcanic eruptions, eruptive products (rocks) and landforms with the ultimate goal to understand volcanic phenomena and predict volcanic eruptions.

However, an essential puzzle piece in understanding volcano behaviour is hidden beneath the volcanic edifice: the volcanic and igneous plumbing system (VIPS; also known as magma plumbing system). The VIPS comprises the network of magma production, storage and transport channels and chambers underlying volcanic regions in all tectonic settings (Fig. 1.1). Where magma is produced, and how it rises and stalls on its way to the Earth's surface controls, to a large extent, when, how, where and for how long a volcano will erupt. Studying VIPS is therefore key to interpret volcano behaviour.

Many active volcanoes are currently studied and monitored with a variety of geological, geochemical, geophysical and geodetic techniques. However, none of these techniques delivers a direct or complete picture of the VIPS, but each sheds light

Volcanic and Igneous Plumbing Systems. http://dx.doi.org/10.1016/B978-0-12-809749-6.00001-7
Copyright © 2018 Elsevier Inc. All rights reserved.

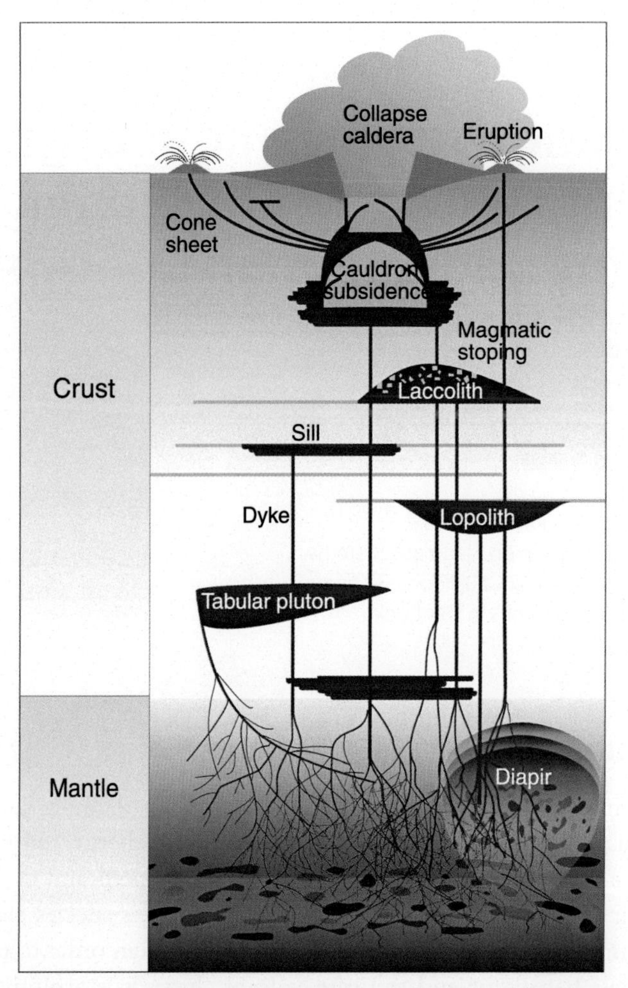

Figure 1.1 ***Schematic sketch of the VIPS, the network of magma transport channels and storage chambers from the Earth's mantle to the surface.*** Different types of magma bodies exist, and their formation and shape depend on, for example, the tectonic setting, the amount of magma available and the properties of the magma and the surrounding rocks.

on some particular characteristics of magma transport and storage beneath volcanoes. In combination, these techniques help to interpret what is going on in the subsurface prior to, during and after a volcanic eruption (Burchardt and Galland, 2016).

This book summarises what we currently know about magma transport and storage in VIPSs, building on various methods from the fields of, for example, structural geology, igneous petrology, geodesy and geophysics. By combining results from multiple disciplines, common concepts emerge and reveal the processes that occur while magma makes its way through the crust. In this chapter, I outline how the study of VIPS relates to other scientific disciplines (Section 1.2) and how it developed historically (Section 1.3). I also summarise how our knowledge on VIPS has evolved (Section 1.4) and give an overview of the chapters of this book (Section 1.5).

1.2 DEVELOPING A SCIENTIFIC DISCIPLINE

The study of VIPSs has traditionally not been its own, defined scientific discipline, which is directly tied to the historical development of the Earth sciences as a whole (Burchardt and Galland, 2016). Instead, VIPS studies remained undefined in the scientific landscape in the grey zone between the fields of volcanology, igneous petrology and geochemistry and pluton studies, all of which focus on magmatism from different perspectives. While volcanology targets near-surface and surface magmatic phenomena in both active and extinct volcanoes, pluton studies look at the structure, timing and development of magmatic intrusions that are now exposed at the Earth's surface but were formed at depth in the geological past. Igneous petrology and geochemistry are closely tied to both volcanology and pluton studies and focus on the formation and evolution of magmatic rocks through detailed analyses and models of their composition. All three of these research fields benefited from methodological advances, such as high-precision seismometry, microscopy and geochronology, and developed in parallel their specific classifications and terminology, sometimes even for the same features (Burchardt and Galland, 2016).

Volcanology and igneous petrology developed into established research fields with dedicated university programs and departments, organizations (e.g. International Association of Volcanology and Chemistry of the Earth's Interior [IAVCEI]), scientific journals (e.g. *Journal of Volcanology and Geothermal Research*, *Journal of Petrology*), conferences (e.g. the annual assembly of the Volcanic and Magmatic Studies Group (VMSG), the Goldschmidt conference), and textbooks. Pluton studies, on the other hand, became an established field of research mainly in the discipline of structural geology, represented by a large community of researchers that use, for example, microstructural analysis, field mapping, petrophysical, geophysical and petrological methods. Pluton studies are a common theme at scientific conferences and are frequently published in specialised journals (e.g. *Journal of Structural Geology*).

VIPS studies draw on methods within all three disciplines and beyond (e.g. geophysics and geodesy) and look at systems between the deep crust and volcanic eruptions. VIPS studies are, to a large degree, rooted in the mapping of eroded volcanic complexes in the late 19th to early 20th century, in particular, in Scotland and Ireland (e.g. Clough et al., 1909; Bailey et al., 1924; Richey and Thomas, 1930) (see Section 1.3). Next to plutonic bodies of various shapes, these complexes hosted numerous magmatic sheet intrusions, as well as volcanic rock assemblages that are closely related with the intrusions. The classification of different intrusion types in subvolcanic complexes and their emplacement mechanisms from then on co-existed with research on plutons, magma and volcanoes, but without being defined as its own field in the past. Unlike volcanology, igneous petrology and pluton studies, researchers working on VIPS did not fund their own organizations, establish their own dedicated scientific journals, or write their own textbooks, nor did VIPS studies become subjects taught as university courses until recently.

Nowadays, VIPS studies are emerging as their own field at the interface between volcanology, igneous petrology and geochemistry and pluton studies. This development is a natural expression of our development of knowledge on VIPSs. In particular, the link between volcanoes and their intrusive roots has become apparent in field studies of extinct volcanoes and volcanic regions that are eroded down to their underlying solidified magma chambers. Moreover, the development and combination of new and precise measurement techniques in active volcanic areas have revealed magma movements and accumulations that fit with the geological record from eroded volcanoes and plutons. Conferences, such as the LASI conference series, as well as conference sessions and workshops on VIPSs are organised. The VIPS commission of the IAVCEI is the first formal association of plumbing system researchers, and this book is the first comprehensive textbook on VIPS for students and researchers.

1.3 THE HISTORY OF RESEARCH ON MAGMATIC SYSTEMS

The study of our planet has been interwoven with the development of human society since humans had to learn how to adjust to natural forces and phenomena and to use natural resources. Magmatic activity has played a key role in building civilisation, for example, by providing raw materials such as copper, on the one hand, but also by bringing humanity on the verge of extinction, on the other hand. Survivors of volcanic eruptions have expressed their experiences in art, literature, mythology and religion throughout human history.

The modern scientific discipline of the Earth sciences started to take shape during the Age of Enlightenment in the late 18th century. During that time, philosophers who studied natural phenomena, the so-called naturalists, engaged in a debate on how the rocks they could see at the Earth's surface formed. When it came to basalt and granite, two schools of thought emerged, whose views were in stark contrast to each other: The Neptunists around the German professor Abraham Gottlob Werner proclaimed that rocks, such as basalt, formed by precipitation from water, pointing to basaltic lava layers that conformably overlie sedimentary rocks. The Plutonists, on the other hand, declared that basalt was composed of originally molten magma. The most prominent advocate of plutonism was James Hutton, a self-taught geologist and a meticulous observer of nature. Hutton proposed that 'subterraneous lava' intruded other rocks to form dykes and sills, as well as granitic rocks (Hutton, 1785). Even other field studies revealed that precipitation of what we today know as igneous rocks was not able to explain the relationship between sedimentary rocks and dyke intrusions. Sir James Hall, a colleague of Hutton's, who melted and cooled volcanic glass and basalt samples to produce rocks with variable crystallinity (Hall, 1800), provided further evidence for a magmatic origin.

The early 19th century was also the era of large expeditions. Alexander von Humboldt and Leopold von Buch, two of Werner's students and initial neptunists, changed their minds when they themselves observed volcanic eruptions and volcanoes in, for example, Italy, the Canary Islands and South America. In the face of increasing evidence that supported plutonism, a magmatic origin of basalt and granite started to become the accepted theory from the 1820s.

In 1841, Giuseppe Mercally established the first volcano observatory, the Vesuvius Observatory, and later became the founding father of seismology. Several large volcanic eruptions subsequently caught the attention of researchers, and their studies shaped modern volcanology. These events include, for example, the eruptions of Tambora in 1815 and of Mount Pelee in 1902. In 1919, the International Association of Volcanology and Chemistry of the Earth's Interior (IAVCEI) was founded, and from 1922 reports on volcanic eruptions and scientific articles were published in IAVCEI's journal *Bullentin Volcanologique* (today *Bulletin of Volcanology*).

The 19th century also saw the rapid development of optical microscopy and its introduction to the study of rocks and minerals (Chisholm, 1911), as well as a number of classifications of igneous rocks and more petrological experiments. Hence, during the second half of the 19th century, igneous petrology had emerged as a scientific discipline.

While early petrologists had started to focus their attention on studying small-scale microscopic features of igneous rock, field geologists were instead concerned with larger-scale phenomena, such as the shape of igneous intrusions and the associated structural features (Young, 2003). Extensive mapping of intrusive rocks revealed that several distinct types of intrusive bodies exist, among them laccoliths (Gilbert, 1877) and batholiths (Suess, 1885). The different shapes and features inspired researchers, such as Gilbert and Daly, to propose mechanisms of magma emplacement (e.g. Gilbert, 1877; Daly, 1903, 1914).

However, the discovery of large granitic intrusions sparked a new controversy, the so-called 'space problem' (e.g. Daly, 1903; Bowen, 1948; Read, 1957; Buddington, 1959). The assumption that enormous volumes of granitic magma intruded into the Earth's crust required equally enormous amounts of space. The space problem has been a rather long-lived controversy that received new fuel from a better understanding of tectonic forces and the beginning quantification of stresses and strains in the field of structural geology (e.g. Roberts, 1970; Hutton, 1988; Paterson and Fowler, 1993). Recently, high-resolution geochronology revealed that granitic plutons formed through addition of small batches of magma (e.g. Glazner et al., 2004), a discovery that partly resolved the space problem. However, some aspects of the controversy, such as the role and efficiency of magmatic stoping, are still discussed (e.g. Glazner and Bartley, 2006; Burchardt et al., 2012, 2016).

During the late 19th and early 20th century, geological mapping of eroded volcanoes provided additional insights into the volcanic plumbing system. Maps of volcanic complexes in the British and Irish Paleogene Igneous Province (BIPIP) revealed circular outlines of ring dykes and cone sheets, adding cone-sheet emplacement and cauldron subsidence to the list of magma emplacement mechanisms (e.g. Clough et al., 1909; Bailey et al., 1924; Richey and Thomas, 1930). The study of subvolcanic structures reached a first peak during the 1960s with the work of George Patrick Leonard Walker and his students in Eastern Iceland and the BIPIP. George Walker was a thorough observer of nature and systematically mapped the eroded volcanoes and surrounding lavas of Eastern Iceland (e.g. Walker, 1960). His legacy is our basic understanding of VIPSs (Sparks, 2009).

Following Hall's (1800) melting experiments, the approach to model processes was introduced even in other fields. For the study of VIPS, the technique of analogue experiments, also called laboratory models, is of particular importance (Galland et al., 2015). While originally invented by, for example, Hall (1815) to qualitatively reproduce natural phenomena in the laboratory, it became increasingly useful when Hubbert (1937) and later Ramberg (1970, 1981) introduced the concept of scaling. Scaling of model dimensions, material properties and process rates allowed us to compare the model with nature. Other modelling techniques that are frequently used to study VIPS are analytical (mathematical) and numerical models. These approaches originate often from fields outside of the Earth sciences, such as fluid mechanics and materials science, and have become increasingly advanced with the development of computational power (Burchardt and Galland, 2016).

During the second half of the 20th century, the scientific discipline of geophysics started to develop a large number of techniques to measure the properties of the Earth's crust, techniques that are nowadays routinely applied to both active volcanoes and eroded plutons and volcanic complexes. Similarly, the discipline of geodesy that maps and measured the Earth surface has contributed to the study of VIPSs, in particular, with techniques that are satellite based (Burchardt and Galland, 2016).

The historical development of the Earth sciences, as a whole, and VIPS studies, in particular, is thus a story of specialisation and diversification, where a deepened understanding of geological concepts went hand in hand with technical developments. The level of understanding of the Earth, in general, and of magma transport and storage, in particular, we have today spans from the scale of atoms to that of our entire planet. We study aspects of the processes and features of VIPS within volcanology, igneous petrology, geochemistry, geodesy, geophysics, structural geology and economic geology. Diversification and specialisation have produced a wealth of data on magma transport and storage, but no individual field or method can deliver a complete picture (cf. Burchardt and Galland, 2016). To grasp the complexity of VIPS, we need to combine our knowledge and perspectives.

1.4 CONFLICTING PERSPECTIVES: OF BALLOONS, BLOBS AND MUSH

Our current knowledge on how magma is transported and stored in the Earth's crust is a product of more than two centuries of research and a combination of observations, measurements, experiments and models. Numerous hypotheses and concepts have been proposed and tested; some of them have survived, but many have been discarded (see also Chapter 12).

A common concept in the study ofVIPS is that magma resides in some sort of chamber or reservoir. When asked to draw a volcano with a magma chamber, most people sketch a cone-shaped mountain (plus optional eruption cloud) with a red, balloon-like circle underneath (cf. Fig. 1.2). In movies and computer games, magma chambers are cavities in the crust, filled with sloshing magma. And even science journalists visualise magma chambers as red-glowing blobs or bulbous bodies beneath supervolcanoes. So how do magma chambers actually look like? Do they resemble balloons, blobs or bulbs? How do they form? Are they full of sloshing, red-glowing magma?

Unfortunately, not even researchers agree on the answers to these questions (cf. Glazner et al., 2016). Volcanologists may see a magma chamber as a shapeless body that produced a certain type of volcanic eruption or deposit. Geodesists think of magma chambers as deformation sources, volume-less points in a homogeneous elastic half-space (see, e.g. Chapter 11). Geophysicists see areas in the subsurface that deviate from the surrounding rock, for example, because of the presence of some amount of liquid (see, e.g. Chapter 10).

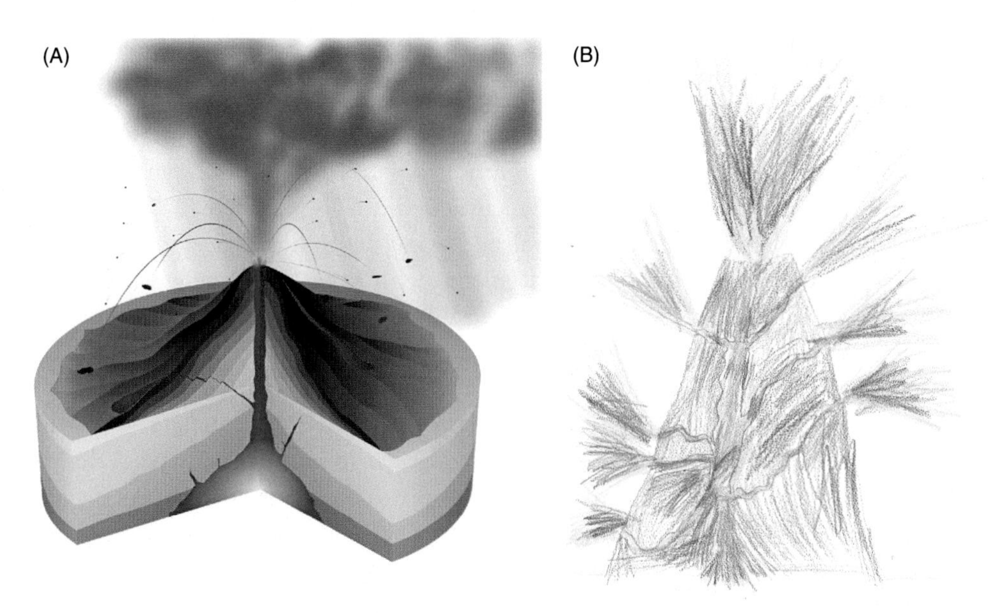

Figure 1.2 (A) Wikipedia's illustration of a magma chamber. *Source: Wikimedia Commons.* (B) Drawing of the VIPS by my 6-year-old niece, C. L. Schmidt.

Seismologists see the fractures produced by moving magma and large areas with high temperatures and melt content through the type and location of earthquakes (see, e.g. Chapter 11). Structural geologists see crystallised intrusions of different shapes and sizes (see, e.g. Chapter 6). Igneous petrologists and geochemists, who prefer the term magma reservoir, see big tanks of melt, crystals and volatiles or diffuse regions filled with a crystal mush (see, e.g. Chapter 8; Lowenstern and Hurvitz, 2008).

While not all of these researchers may actually look at exactly the same thing, most of the fundamental differences in their view of magma chambers reflect the insight they gain using their specific methods (Burchardt and Galland, 2016). Since individual methods do only cover limited aspects of magma transport or storage, we need to accept that we do not see the whole picture when using only one or a few methods. We need to combine the strengths of our approaches and be aware of their weaknesses to eliminate our blind spots. To be able to do this, we need to learn about other disciplines and methods and develop a common terminology to be able to communicate across disciplinary boundaries. This book attempts to provide the base for a holistic study of magma transport and storage in the Earth's crust by summarising our current collective understanding. While all the authors have a background in one of the abovementioned disciplines, they all base their chapters on insights from multiple fields and methods.

1.5 OVERVIEW OF CONTRIBUTIONS IN THIS BOOK

This book follows the way magma takes from the lower and middle crust (Chapter 2) through different types of magma transport channels (Chapters 3 and 4) into different types of magma reservoirs (Chapters 5 and 6). We also take a look at how tectonic forces influence the development and evolution of VIPSs (Chapter 7) and what the magma itself reveals about the processes occurring in VIPSs (Chapter 8). We then explore the destruction and rebuilding of volcanoes and their plumbing systems through edifice collapse (Chapter 9) and caldera collapse (Chapter 10). Finally, we examine how magma transport and storage is expressed through volcanic unrest and eruptions (Chapter 11).

In Chapter 2, we start our journey by examining how magma is segregated and extracted from its source in continental settings. Melt segregation and extraction give rise to a self-organised network of magma transport channels, whereby the volume flux of magma from the source controls the viability of different melt transport mechanisms, such as diapirism, magmatic stoping, and channelised ascent in, for example, dykes. Chapter 2 discussed how the interplay between magma production, extraction, ascent rates and mechanisms, cooling, tectonic stresses, as well as the properties of the surrounding crust influence the formation of magma chambers in the lower and middle crust.

Dykes are the principal magma pathways in the middle and upper Earth's crust and an essential component of VIPSs. They transport magma from its source region into magma reservoirs, and from magma reservoirs to the Earth's surface. Dykes are magma-filled fractures that propagate driven by an internal magma pressure, while interacting with the surrounding rocks. Chapter 3 explores what we can learn from the features of solidified dykes exposed at the surface and the propagation of dykes in numerical and analogue models. Insights from the field and models help to unravel how dykes initiate, propagate and transport magma.

As a volcanic plumbing system evolves, countless dykes and other magmatic sheet intrusions transport magma into the volcanic edifice and to the Earth's surface. Depending on the arrangement of magmatic sheets, we can distinguish different sheet types and sheet-intrusion swarms, which are the focus of Chapter 4. Volcanic rift zones consist of swarms of vertical dykes arranged in the same direction, radial dykes are arranged around a volcano like the spokes of a wheel, and cone sheets form concentric swarms with the shape of an inverted cone. When a magmatic sheet breaches the surface, the eruption can subsequently focus at a conduit, a roughly cylindrical pathway for magma discharge, host-rock erosion, fracturing and eventually intrusion.

While propagating, magmatic sheets may get diverted to intrude along the layering in the host rock, forming sills. Chapter 5 describes the features of sills and the associated host-rock structures as observed in the field and in 3D seismic surveys. Based on these observations, Chapter 5 discusses what we know about how sills initiate and grow as a consequence of the interaction between propagating magma and its host rock. Sills are a significant part of volcanic plumbing system as sites for both transport and storage of magma, but they also play an important role in the maturation of hydrocarbons and the release of greenhouse gases and associated climate change.

In the upper and middle crust, larger magma bodies can grow from the stacking of sheet intrusions and subsequent inflation. Chapter 6 describes the features of a series of multiple sills, laccoliths, and larger magma bodies to derive a conceptual model of magma-chamber emplacement. According to this model, larger magma bodies grow in two stages: (1) lateral growth of an initial magmatic sheet (sill in the upper crust and dyke in the middle crust) and (2) inflation into a laccolith or larger pluton. Both these stages are controlled by Pascal's principle that describes the distribution of magma pressure in relation to external forces, such as the lithostatic load.

On a global scale, magma transport and storage are tightly connected with plate tectonics, whereas on a more local scale, gravitational forces and lithology come into play. Chapter 7 outlines the structure of VIPSs in different tectonic settings, including mid-ocean ridges, continental rift zones, hot spots, oceanic subduction zones, continental collision and transform zones. Although tectonics significantly influences magma storage and transport, magma is a major ingredient in many tectonic processes. Chapter 7 therefore proposes a unified paradigm that explains the interaction of tectonics and VIPSs.

Chapter 8 takes a closer look at magma itself and the processes it experiences during its journey through the VIPS. These processes include crystallisation and the associated separation of crystals from the surrounding melt, magma mixing and mingling, magmatic stoping and contamination and assimilation of the host rock. Since these processes usually occur invisibly hidden within the VIPS, igneous rocks in exposed intrusions or erupted at the Earth's surface contain a record of their history in the form of their crystal cargo. Chapter 8 gives an overview of the information stored in the crystal record and how to retrieve it with state-of-the-art textural and chemical analyses.

During their lifetime, volcanic edifices may collapse in volcanic landslides as a result of instabilities. Chapter 9 gives an overview of the interplay between volcano instability and processes within the VIPS. Volcano instability can be caused by processes in the VIPS, such as shallow magma emplacement in sills or laccoliths, which can lead to edifice collapse and trigger simultaneous eruptions. On the other hand, volcano collapse has a profound effect on magma transport and storage, such as increased melting due to the associated pressure release and affecting the structure of magma transport due to changes in the stress field.

Caldera collapse, and associated eruptions, are among the largest volcanic hazards and can destroy both the volcanic edifice and the underlying plumbing system. Chapter 10 explores the complex VIPS of caldera volcanoes with insights from igneous petrology, geophysics, geodesy and structural geology. Caldera formation requires large amounts of eruptible magma within the VIPS prior to eruption. During caldera collapse, magma can intrude and extrude along ring faults/dykes, which control caldera subsidence. These collapse structures can have a large influence on the re-establishing plumbing system after caldera collapse, as well as on the location and formation of geothermal fields and ore deposits.

In active volcanoes, processes within the VIPS that precede a potential eruption are inaccessibly hidden beneath the Earth's surface, but expressed as volcanic unrest. Chapter 11 summarises the signals of volcanic unrest, how they are measured and how they relate to magma migration. The monitoring of ground deformation in active volcanic areas is a standard method to infer magma movement at depth. Chapter 11 focusses on the different geodetic techniques used to quantify volcano deformation, how the data from these measurements is interpreted employing different types of models and what geodetic monitoring and modelling have contributed to understanding the VIPS of active volcanoes.

In summary, each chapter of this book outlines what we currently know about different aspects of magma transport and storage in the Earth's crust. However, as indicated above, different disciplines and methods have sometimes very different views on this topic, all of them based on actual data. This discrepancy lets us sometimes struggle to unravel what is really happening within the plumbing system of active volcanoes. It is thus important to distinguish between common and accepted understanding and the uncertainties in plumbing system research. Chapter 12 concludes this book with a short overview of some future challenges and directions in VIPS research.

1.6 CONCLUDING REMARKS

VIPS studies are an exciting field of research, emerging as a scientific discipline during a time when the diversification and specialisation of the Earth sciences are transitioning towards integrated and multi-disciplinary approaches. In this time, researchers and students need to have an overview of existing knowledge on magma transport and storage in all the disciplines involved and understand the potential and blind spots of a wide variety of methods. This book is intended to provide such an overview and to inspire students and curious early- and later-career researchers to study magma on its way through the Earth's crust.

REFERENCES

Bailey, E.B., Clough, C.T., Wright, W.B., Richey, J.E., Wilson, G.V., 1924. Tertiary and post-tertiary geology of Mull, Loch Aline, and Oban. Geological Survey of Scotland Memoir. Edinburgh.

Bowen, N.L., 1948. The granite problem and the method of multiple prejudices. Geol. Soc. Am. Mem. 28, 79–90.

Buddington, A., 1959. Granite emplacement with special reference to North America. Geol. Soc. Am. Bull. 70, 671.

Burchardt, S., Galland, O., 2016. Studying Volcanic Plumbing Systems – Multidisciplinary Approaches to a Multifaceted Problem. In: Nemeth, K. (Ed.), Updates in Volcanology – From Volcano Modelling to Volcano Geology. InTech.doi: 10.5772/63959.

Burchardt, S., Tanner, D., Krumbholz, M., 2012. The Slaufrudalur pluton, southeast Iceland—an example of shallow magma emplacement by coupled cauldron subsidence and magmatic stoping. Geol. Soc. Am. Bull. 124 (1–2), 213–227.

Burchardt, S., Troll, V.R., Schmeling, H., Koyi, H., Blythe, L., 2016. Erupted frothy xenoliths may explain lack of country-rock fragments in plutons. Sci. Rep. 6, 34566. doi: 10.1038/srep34566.

Clough, C.T., Maufe, H.B., Bailey, E.B., 1909. The cauldron subsidence of Glen Coe, and associated igneous phenomena. Q. J. Geol. Soc. London 65, 611–678.

Chisholm, H., 1911. Petrology. In: Chisholm, H. (Ed.), Encyclopædia Britannica, 11th ed. Cambridge University Press.

Daly, R.A., 1903. The mechanics of igneous intrusion. Am. J. Sci., 269–298.

Daly, R.A., 1914. Igneous Rocks and Their Origin. McGraw-Hill.

Gilbert, G.K., 1877. Report on the Geology of the Henry Mountains. Government Printing Office, Washington, DC.

Glazner, A.F., Bartley, J.M., 2006. Is stoping a volumetrically significant pluton emplacement process? Geol. Soc. Am. Bull. 118, 1185–1195.

Glazner, A.F., Bartley, J.M., Coleman, D.S., 2016. We need a new definition for "magma". Eos 97. https://doi.org/10.1029/2016EO059741.

Glazner, A.F., Bartley, J.M., Coleman, D.S., Gray, W., Taylor, R.Z., 2004. Are plutons assembled over millions of years by amalgamation from small magma chambers? GSA Today 14 (4/5), 4–12.

Hall, J., 1800. Experiments on whinstone and lava. J. Natural Philos. Chem. Arts 4 (8–18), 56–65.

Hubbert, M.K., 1937. Theory of scale models as applied to the study of geologic structures. Geol. Soc. Am. Bull. 48, 1459–1520.

Hutton, D.H.W., 1988. Granite emplacement mechanisms and tectonic controls: inferences from deformation studies. Trans. R. Soc. Edinburgh 79, 245–255.

Hutton J., 1785. Abstract of a dissertation read in the Royal Society of Edinburgh, upon the Seventh of March, and Fourth of April, MDCCLXXXV, concerning the system of the earth, its duration, and stability.

Lowenstern, J.B., Hurvitz, S., 2008. Monitoring a supervolcano in repose: heat and volatile flux at the Yellowstone Caldera. Elements 4, 35–40.

Paterson, S.R., Fowler, T.K., 1993. Re-examining pluton emplacement processes. J. Struct. Geol. 15, 191–206.

Ramberg, H. (Ed.), 1970. Model Studies in Relation to Intrusion of Plutonic Bodies. Mechanism of Igneous Intrusion. Geol. J. Spec. Iss. 2.

Ramberg, H., 1981. Gravity, deformation and the Earth's crust. Academic Press.

Read, H., 1957. The Granite Controversy. Interscience, New York, 430 pp.

Richey, J.E., Thomas, H.H., 1930. The geology of Ardnamurchan, northwest Mull and Coll. Geological Survey of Scotland Memoirs HMSO, 393 pp.

Roberts, J.L., 1970. The intrusion of magma into brittle rocks. In: Newall, G., Rast, N. (Eds.), Mechanism of Igneous Intrusion. Geol. J. Spec. Iss. 2, 287–338.

Sparks, R.S.J., 2009. The legacy of George Walker to volcanology. In: Thordarson, T., Self, S., Larsen, G., Rowland, S.K., Höskuldsson, A. (Eds.), Studies in Volcanology. The Legacy of George Walker. Special Publications of IAVCEI, London, pp. 1–15.

Suess, E., 1885. Das Antlitz der Erde. 3 v., Vienna, n. p.

Walker, G.P.L., 1960. Zeolite zones and dyke distribution in relation to the structure of eastern Iceland. J. Geol. 68, 515–528.

Young, D.A., 2003. Mind Over Magma: The Story of Igneous Petrology. Princeton University Press, 686 pp.

CHAPTER 2

Mechanisms of Magma Transport and Storage in the Lower and Middle Crust—Magma Segregation, Ascent and Emplacement

A.R. Cruden, R.F. Weinberg
Monash University, Melbourne, VIC, Australia

Contents

2.1 INTRODUCTION

Magmas form by partial melting of the upper mantle and lower crust. This occurs in all tectonic settings, in which rocks are subjected to sufficiently elevated temperatures, decompression or volatile fluxes to initiate melting (Chapter 7; Hildreth and Moorbath, 1988; Brown, 2013). Hence, volcanic and igneous plumbing systems (VIPSs) that transport magmas from deep crustal and upper mantle sources develop in both oceanic and continental lithosphere. In oceanic lithosphere, mid-ocean ridge magma plumbing systems form as a consequence of plate divergence, and oceanic island plumbing systems occur in intraplate settings associated with mantle plumes or hot spots (see Chapter 7). In both cases, the resulting basaltic magmas form by decompression melting

Volcanic and Igneous Plumbing Systems. http://dx.doi.org/10.1016/B978-0-12-809749-6.00002-9
Copyright © 2018 Elsevier Inc. All rights reserved.

of mantle rocks. Magma plumbing systems in continental lithosphere occur in convergent plate margin, rift and intraplate settings. The compositional diversity of felsic to intermediate igneous rocks in continental crust reflects the range of processes and sources involved in their generation (Pitcher, 1979; Huppert and Sparks, 1988; Whitney, 1988; Patiño Douce, 1995; Scaillet et al., 2016). Magmas form in continental settings due to varying degrees of interaction between mantle-derived heat, melts and fluids with crustal source rocks and melts (Brown, 2013).

Given the broad scope and scale of igneous activity on Earth, in this chapter, we focus on VIPSs developed in continental crust at convergent plate margins, with special emphasis on how magmas are extracted from deep crustal source rocks and then transported and emplaced in the lower and middle crust to form plutons and layered intrusions. In the context of crustal-scale VIPSs, we will confine our discussion to the deeper parts of the crust (i.e. $>\sim 5$ km) and refer to Chapters 3–6 for magma transport and emplacement in the shallow crust.

Convergent margins are associated with two distinct styles of magmatism, which are linked to different stages of the Wilson cycle. Igneous activity in continental magmatic arcs is linked to subduction of oceanic lithosphere beneath continental lithosphere, whereby magmas form by: a combination of melting and dehydration of the subducting slab and overlying mantle wedge; intrusion and fractionation of these basaltic melts and related fluids in the lower crust; melting (M) of crustal rocks; and assimilation (A), storage (S) and hybridisation (H) within so-called deep crustal MASH zones (Hildreth and Moorbath, 1988). In this setting, VIPSs tend to be focused in narrow linear belts above the locus of slab dehydration and mantle wedge melting, which migrates across the arc either towards or away from the trench depending on whether the oceanic slab is retreating or advancing, respectively.

The end stages of a Wilson cycle involve closure of an ocean basin and collision between a continental magmatic arc and a continental passive margin or another arc. This involves termination of arc magmatism and extensive deformation and metamorphism of continental crust to form an orogenic belt (see Chapter 7). One consequence of collision-related deformation is crustal thickening, which leads to thermal blanketing of lower crustal rocks (Brown, 2013). The resulting build-up of heat in the lower crust leads to weakening, partial melting and gravitational collapse. Hence, magmatism typically occurs late in the orogenic cycle, 10–40 Ma after collision. Gravitational collapse and/or delamination of the mantle root of the orogen can also lead to the appearance of widespread post-orogenic granites and related intrusive rocks.

Differences in the spatial distribution of intrusive igneous rocks between continental arc settings, in which magmatism is strongly focused in linear belts, versus orogenic belt settings that are characterised by widely distributed plutons partly explain several diverging views of magma plumbing systems that have emerged in the last decade (Fig. 2.1).

A common view of continental magmatic arcs, based on crustal sections observed in the North American Cordillera and the Kohistan Arc, is that they comprise either vertically extensive, closely spaced, or cross-cutting plutons (Fig. 2.1C; Jagoutz and

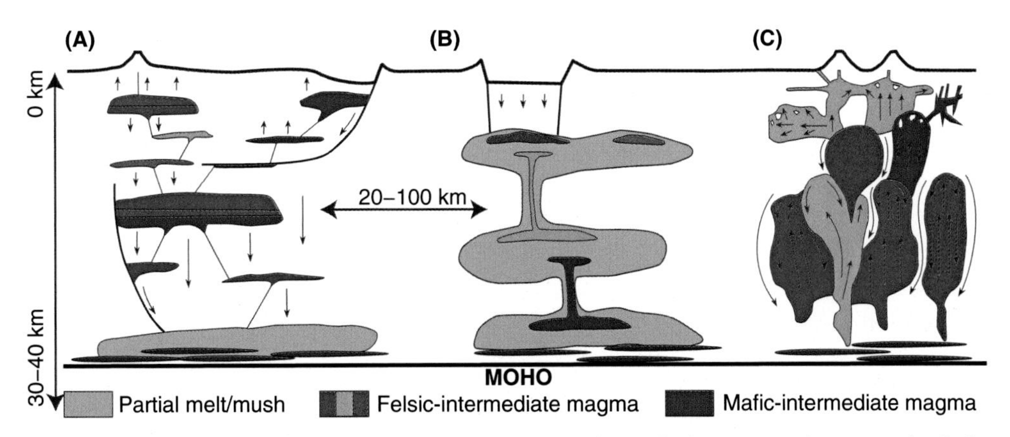

Figure 2.1 ***Spectrum of magma plumbing systems.*** (A) Channelled ascent pathways and tabular intrusions. (B) Magma mush column (after Bachmann and Huber, 2016). (C) Diapirism and downward flow (after Paterson et al., 2011). See text for further discussion.

Schmidt, 2012; Paterson et al., 2011; Saleeby et al., 2003) or an interconnected network of large magma mush bodies (Fig. 2.1B; Lipman and Bachmann, 2015; Bachmann and Huber, 2016; Cashman et al., 2017). In both cases, the mid- to lower crust comprises >50% to ~100% intrusive igneous rock, and mid- to upper crustal magma bodies are envisioned to feed large silicic caldera complexes (Fig. 2.1A; see Chapter 10) or arc volcanoes (Fig. 2.1C).

An alternative view, informed by field and geophysical studies of plutons in orogenic belts and the South American Cordillera, is that plutons occur in interconnected networks of incrementally emplaced tabular intrusions (Fig. 2.1A; Annen et al., 2015; Cruden, 1998; Menand et al., 2011, 2015; see also Chapter 12). These networks can also feed volcanoes and calderas at the surface but the total volume of igneous intrusions in the upper to lower crust may be significantly less than 50%.

In this chapter, we review how magma is extracted from lower crustal source rocks (segregation), how it is transported upwards (ascent) and how space is made for magma in the mid- to upper crust to form plutons and layered mafic intrusions (emplacement). The allied process of magma generation is beyond the scope of this chapter and we refer the reader to many excellent papers on this topic (e.g. Brown, 2013 and references therein).

2.2 PRODUCTION AND REMOVAL OF MAGMAS FROM MIGMATITIC SOURCES

Within continental regions, migmatites are the main sources of crustal magmas (see also Chapter 7). In magmatic arcs, mantle-derived magmas heat up and melt the crust (a process known as anatexis), and interaction between these compositionally distinct magma types leads to hybridisation. Back-arcs in accretionary orogens are particularly important regions of melting of continental rocks, leading to migmatisation of vast fertile turbidite fans and the formation of granite belts (Foster and Goscombe, 2013; Foster and Gray, 2000).

The processes of melting, melt segregation and extraction from crustal sources are responsible for the compositions and volumes of the resulting magmas. Moreover, these processes control the mechanisms and rates of magma transfer to form felsic magma reservoirs (plutons) higher in the crust, their ability to carry solids and how and when magmas fractionate. This section follows the melts from their deep crustal origin as a result of melting reactions and onto how melt is segregated and extracted, and the controls of these processes on the nature of what is ultimately removed from anatectic regions. The section ends with a description of felsic MASH zones in these regions and some remarks on possible controls on melt focusing and self-organisation.

2.2.1 Melting Reactions

The continental crust melts under different pressure–temperature (P–T) conditions depending on the rock composition (*X*) and where H_2O is found. This is because water plays a fundamental role in lowering the melting temperature (solidus) of rocks. When an aqueous fluid is present, water-fluxed or wet melting occurs at low temperatures, involving quartz and plagioclase plus K-feldspar or biotite. In the absence of an aqueous fluid, the break-down of H_2O-bearing minerals (muscovite, biotite, amphibole, typically hornblende) controls melting. These are called dehydration melting reactions, which in order of increasing temperature are: muscovite dehydration, biotite dehydration and amphibole dehydration melting reactions (Fig. 2.2).

The variables controlling melt fertility and composition of the melt generated are therefore: pressure, temperature, H_2O activity, fluid phase volume and source rock composition. It is important to notice that dehydration melting reactions can occur in the presence of minor amounts of added aqueous fluids, which will strongly increase melt

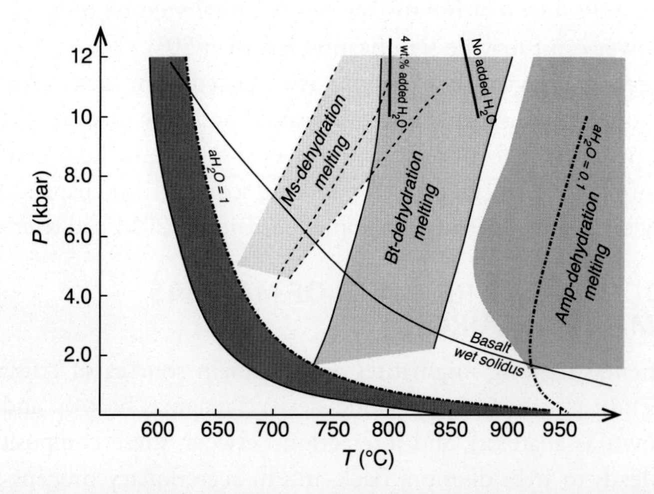

Figure 2.2 ***Summary pressure (P)–temperature (T) phase diagram comparing H_2O-present and -absent melting reactions, each with different slopes and solidus temperature (after Weinberg and Hasalová, 2015).*** Thick solid lines represent similar melting reactions with no added water and with 4 wt.% added water (from Gardien et al., 2000).

fertility of the system without having a strong influence on either the composition of the residue or the melt (Sola et al., 2017).

An important aspect of the melting curves (Fig. 2.2), known as the solidi, is that their slopes indicate the volume change as a result of melting. Curves with negative slopes, such as the water-fluxed melting solidus, indicate an average volume decrease of the system as it changes from solid + H_2O to solid + silicate melt, where H_2O is dissolved in the melt. Conversely, positive slopes indicate an average volume increase.

2.2.2 Segregation and Extraction of Magma from Source Regions

The removal of magma from its source is considered to be a two-step process starting with melt segregation from among the pores into leucosomes (pale-coloured quartzo-felspathic vein material), followed by the rise of a leucosome network that extracts magma (melt plus crystals), thereby draining the source region. While the composition of the pure silicate melt depends on the composition of the source rock and P-T-X-H_2O conditions of anatexis, the nature of segregation and extraction controls the composition of the resulting magma. There are two end members of this process depending on how efficiently melt and residual minerals separate: (a) the seeping out of filtered melt from the source into leucosomes and continued extraction of this clean melt from the source (Sawyer et al., 2011) and (b) melt accumulation within the source, either as segregated irregular pools or layer-parallel melt (leucosomes), eventually leading to complete disaggregation of the rock to form diatexites, which are capable of flowing *en masse* as a magma (Brown, 2013). The former produces leucocratic granites, typically poor in trace elements such as REE and U-Th, which tend to remain within accessory minerals in the source rock. The latter is what White and Chappell (1977) envisaged as the starting point for their restite unmixing process, where *en masse* flow of the mobilised source leads to fractionation, producing granitic magmas carrying different amounts of residual cargo (Fig. 2.3A). The breakdown of the solid (crystal) framework, occurs at relatively high melt fractions, around 50% depending on crystal shapes and sizes, and is known as the solid-to-liquid transition (e.g. Miller et al., 1988).

Most anatectic rock masses will behave somewhere in between these two end members, and magma carrying residual solids will be more or less efficiently filtered during extraction (Fig. 2.3B and C), carrying individual grains, trails of grains, called schlieren, and blocks of migmatites of various sizes, called schollen (Fig. 2.3D and E). There are a number of ways to filter magmas, cleaning melt from their residual solids and even their newly crystallised magmatic grains: (i) mass flow filtering, where the faster flow of the melt in relation to the solids separate the two, (ii) the removal of crystals to form cumulates within the source or in leucosomes by draining the melt (Fig. 2.3B and C) and (iii) choking of conduits (Fig. 2.3E) and filter pressing (Miller et al., 1988; Clarke and Clarke, 1998; Weinberg et al., 2001).

Precisely how each anatectic terrane behaves depends on: (a) the melting reaction, which controls volume change and melt density and (b) the relationship between

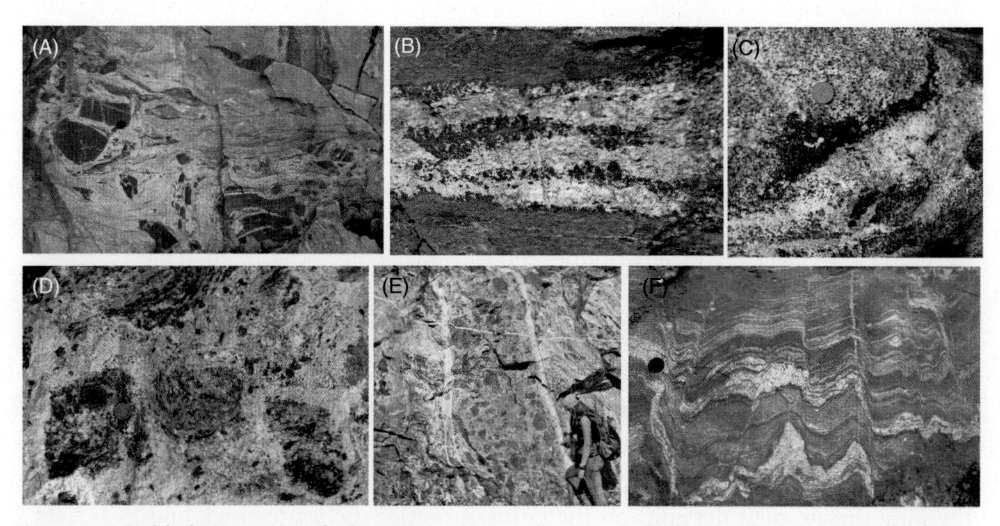

Figure 2.3 ***Field photographs of migmatites.*** (A) Angular to rounded large blocks of amphibolite undergoing minor melting and disaggregation within a schlieren-rich granitic magma (a diatexite). Note the variation in the intensity of the magmatic foliation, with a high-strain zone through the central part of the photograph. Asymmetry of the magmatic foliation and schlieren indicate top-to-the-right shear sense, exaggerated by a cut effect. Cazeca quarry, Camboriú Complex, Santa Catarina state, south Brazil. (B) Hornblende and K-feldspar accumulation in veins in diorite, interpreted to be the result of melt extraction from a leucosome that now preserves a residual assemblage. (C) Peritectic hornblende accumulation in a migmatitic hornblende-biotite-granodiorite. (B) and (C) are from the Karakoram region in NW India (from Reichardt and Weinberg, 2012). (D) Heterogeneous leucocratic granite carrying angular migmatite schollen in different stages of disaggregation and assimilation, broken down to centimetric biotite aggregates. From the Vanov quarry, near the town of Telč, Czech Republic. (E) Vertical dyke with leucocratic margins cutting through migmatites. This dyke is interpreted to represent a paleo-magma pathway linking its source in the Camboriú Complex to pluton (the Itapema granite). The dyke is now clogged by the accumulation of solids in the form of schollen in different states of disaggregation. Interstitial mobile magma was removed towards a more melt-rich and more evolved pluton elsewhere in the system. From Cazeca quarry, Camboriú Complex, Santa Catarina state, south Brazil. (F) Archean migmatitic tonalite gneiss with layer-parallel pegmatite intrusions and regularly spaced (20–50 cm) axial planar leucosomes. Yalgoo Dome, Yilgarn craton, Western Australia (from Weinberg et al., 2015).

the rate of melt production and the rate of melt removal. The latter depends on a number of variables, including: (i) the source of heat and heat flux into the melting region, (ii) the nature of the evolving segregation network and (iii) the magnitude and orientation of external differential stresses and pressure gradients. For example, if the melt source lies in a low-pressure site within the crust, melt will be trapped and accumulate within the source, favouring the disaggregation of the source rock. Conversely, syn-anatectic folding focusses melt migration towards hinge zones of anticlines, efficiently extracting melt from the source rock (Weinberg et al., 2013; Weinberg and Mark, 2008).

2.2.2.1 Melt Segregation

It is clear from the above that the efficiency of melt segregation within the source plays a key role in the nature of the magma that is extracted. Once produced, silica-rich melts, like any fluids, will migrate down existing pressure gradients. Anisotropies, such as foliation, control the shape of melt-filled pore space and subsequent migration to form layer- or foliation-parallel leucosomes, and the extent of this control depends on the orientation of the principal stress axes in relation to the rock anisotropy (Robin, 1979). Melt, being less dense than its solid surroundings, is buoyant giving rise to a pressure gradient that drives it upwards in the crust. Superimposed on buoyancy stresses, there are pressure gradients that originate from tectonic stresses acting on rocks of different strengths that can lead to lateral or even downward melt migration. This leads to magma accumulation in low-pressure sites; boudin necks being a classic example. The speed with which magma is segregated from pore spaces into larger melt pools controls whether or not melt is in chemical and isotopic disequilibrium with the source.

At the microscale, melts are produced as small droplets in amongst reactant minerals. At very small melt fractions, these pools remain isolated. As melt fractions increase, melt pools interconnect. How and when they interconnect depends on the surface tension between melt and crystals, reflected in their dihedral angles. Because these angles are typically $<60°$ for silicate melt and rocks, a stable network will develop for melt fractions of a few percent of the rock (see Holness et al., 2011 for an extensive review of this topic).

In the absence of external differential stresses, melt tends to stay trapped within the pore network, as is commonly recorded in static, partially melted contact aureoles (Holness et al., 2011). Gravity-driven compaction of solids of Newtonian rheology with interstitial melt results in the formation of a melt layer above the compacted solid grains (McKenzie, 1984). This process has been shown to be too slow to account for separation of granitic melts (Brown et al., 1995). More recently, Veveakis et al. (2015) found that a rock with a solid framework of nonlinear, non-Newtonian (power law) rheology and with interstitial melt under compression develops an instability that is absent when the solid is Newtonian. This instability speeds up the process of melt segregation and can give rise to a network of regularly spaced sub-parallel leucosomes that are oriented perpendicular to the maximum compression axis. This may explain the common occurrence of leucosomes parallel to axial planar foliation of folds in migmatites (Fig. 2.3F) (Weinberg et al., 2015).

The first steps in melt migration are generally controlled by local pressure fields. Deformation generates strong local and regional pressure gradients that drives melt migration. The volume and density change in the entire source, which is a function of the melting reaction, is an important variable controlling the first step in melt migration (Weinberg and Hasalová, 2015). A volume increase generates buoyancy, which could lead to a general vertical ascent of the source region as a large diapiric mass. More generally, volume change is capable of modifying regional pressure gradients, leading to fracturing within the source, creating potential magma pathways out of the source region.

When accompanied by deformation, melt segregation is efficient, and typical examples include the axial planar leucosomes described above, magma concentrations in boudin necks, and leucosomes within shear zones (Fig. 2.4A). The melt in rock pore

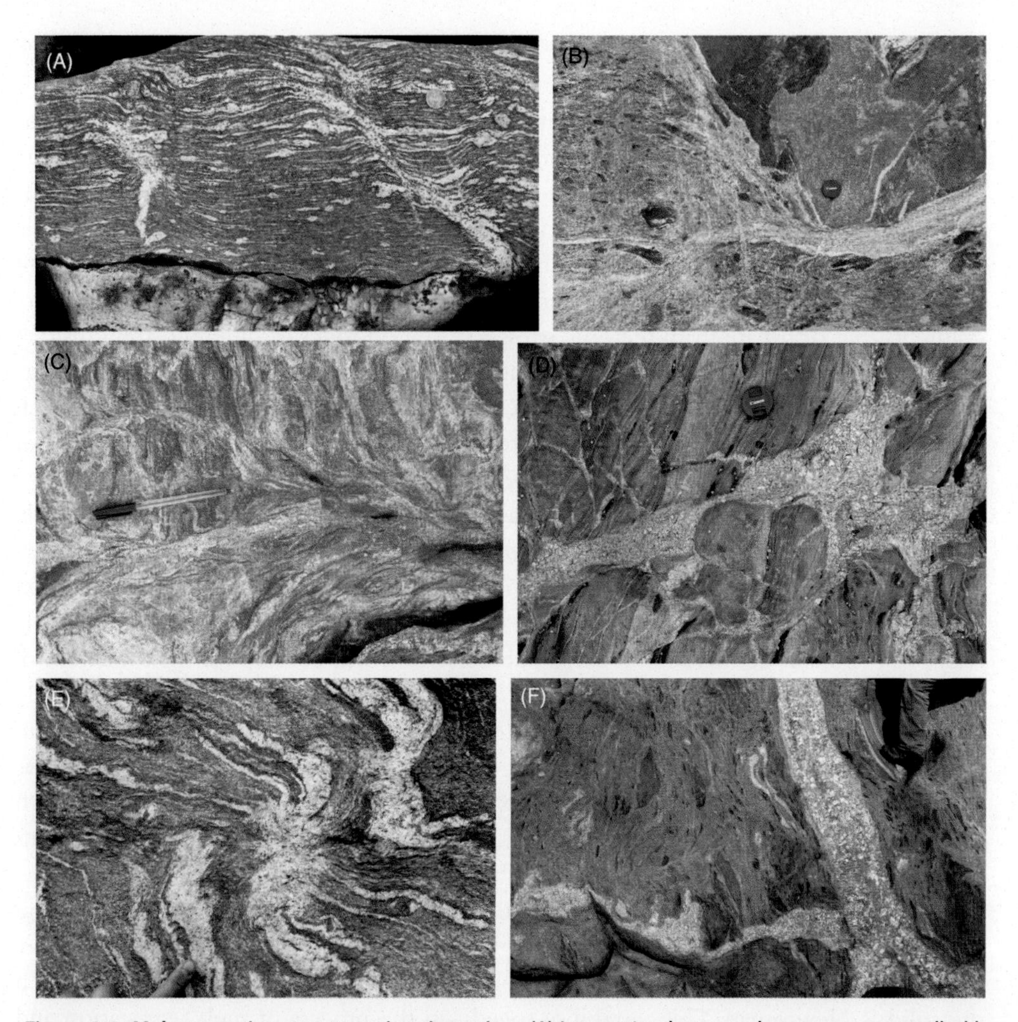

Figure 2.4 ***Melt extraction structures in migmatites.*** (A) Interaction between leucosomes controlled by the foliation anisotropy and seamlessly linked with segregations in shear zones. Miocene anatexis of Paleozoic orthogneiss from the Zanskar Range, NW Indian Himalayas. (B) Magma loaded with schollen being funnelled from left-to-right into an extraction channel where the magma becomes more felsic (dyke on the right). This process of magma channelling and filtering cleans the migmatites of the Camboriú Complex as they are transported towards the Itapema granite pluton, Santa Catarina state, south Brazil. (C) Funnelling of magma from left-to-right into a diffuse extraction channel. Magma in the channels is loaded with disaggregated solids most clearly indicated by their biotite content. (D) A few metres to the right of (C) and in physical continuity with it, the magma is clean of biotite schlieren, richer in K-feldspar and has sharp intrusive contacts with the migmatitic metasedimentary rocks. This is interpreted to represent the intrusive end of the magma paths. In this way, within a few metres, we can follow the evolution from extraction in (C) to intrusion accompanied by fractionation in (D). (C) and (D) are metaturbidites from the Kanmantoo Group, Kangaroo Island, South Australia. (E) Accumulation followed by extraction of melt from leucosomes forming a collapse feature marked by the central leucosome into which foliations converge. Looking down plunge of the lineation. Miocene migmatites with garnet from Sikkim, NE India. (F) Leucocratic granite dykes intruding at different stages during solidification of a diatexite loaded with schollen. Early dykes have been disrupted into irregular felsic streaks within the diatexite. The latest dyke is nearly vertical, with curved margins and irregular widths indicative of a mushy surrounding. Yet another dyke parallel to and close to the base of the photograph has highly irregular margins with protrusions of felsic granite into the diatexite. Metaturbidites from the Kanmantoo Group, Kangaroo Island, South Australia.

spaces and deformation are strongly linked in a feedback process that has not yet been fully explored. Melt weakens the rock mass and therefore accelerates deformation, which is further amplified by melt flow out of rock pore spaces into leucosomes and its removal from the anatectic region. At small melt fractions (<7%), rock strength strongly decreases as melt fraction increases because of a sharp increase in melt interconnectivity (Rosenberg and Handy, 2005). At around 7%, there is a transition where further melt fraction increase has a smaller effect on rock strength. This transitional melt fraction has been called the 'melt connectivity transition' (Rosenberg and Handy, 2005).

2.2.2.2 Melt Extraction: The Early Stages of Migration

Melt extraction from the source is a complex process controlled by the interconnectivity of leucosomes and the development of a pathway out of the source. This process may also be envisaged in terms of two end members: a rapid, catastrophic interconnectivity of the network causing a pulse of magma extraction (Bons and van Milligen, 2001), or the steady, slower establishment of a durable channel network that constantly drains magma from rock pore spaces. In terms of geochemical modelling of igneous processes, the former would be modelled by batch melting with intermittent melt extraction, and the latter by fractional melting.

The nature of this network and the dynamics of magma extraction are fundamental to the evolution of granitic systems. They control the chemical evolution of the melts being produced, their ability to carry solids out of the source, how much magma can be tapped by a propagating dyke, and consequently, how much magma can be fed upwards into plutons. The volume flux of magma from the source controls the viability of different magma transport mechanisms. For example, in the absence of a mature, well-connected magma network in the source, dykes may be unable to successfully drain the source, stalling and freezing before reaching a pluton (see Section 2.2.4). At a small scale, this can be envisaged in Fig. 2.4B–D: the nature and the size of the dyke to the right depend on its drainage network on the left. If a source region is unable to provide a starting dyke with sufficient magma (e.g. Rubin, 1993), the source will remain undrained buoyancy could then favour large-scale bulk ascent of the source through Rayleigh–Taylor instability giving rise to diapirs.

At a larger scale, extraction ultimately controls the spacing between granitic bodies (e.g. Cruden, 2006). In this regard, the hinges of actively folding anticlines play a fundamental role in focusing magma flow. Folding during melting creates pressure gradients, and the upward closure of hinge zones of inclined and upright folds provide focusing points. Magmas segregated into leucosomes parallel to folded layers link-up with the axial planar leucosomes and gradually converge towards the hinge zone. In this way, there is a feedback process, whereby folding creates the conditions for magma extraction, and extraction allows for more efficient folding through magma transfer (Fig. 2.5) (Weinberg and Mark, 2008; Weinberg et al., 2013). Because folds occur at a range of scales, it is possible that large antiforms control the typical length scale of magma extraction from the source (Weinberg et al., 2013; Fig. 2.4F).

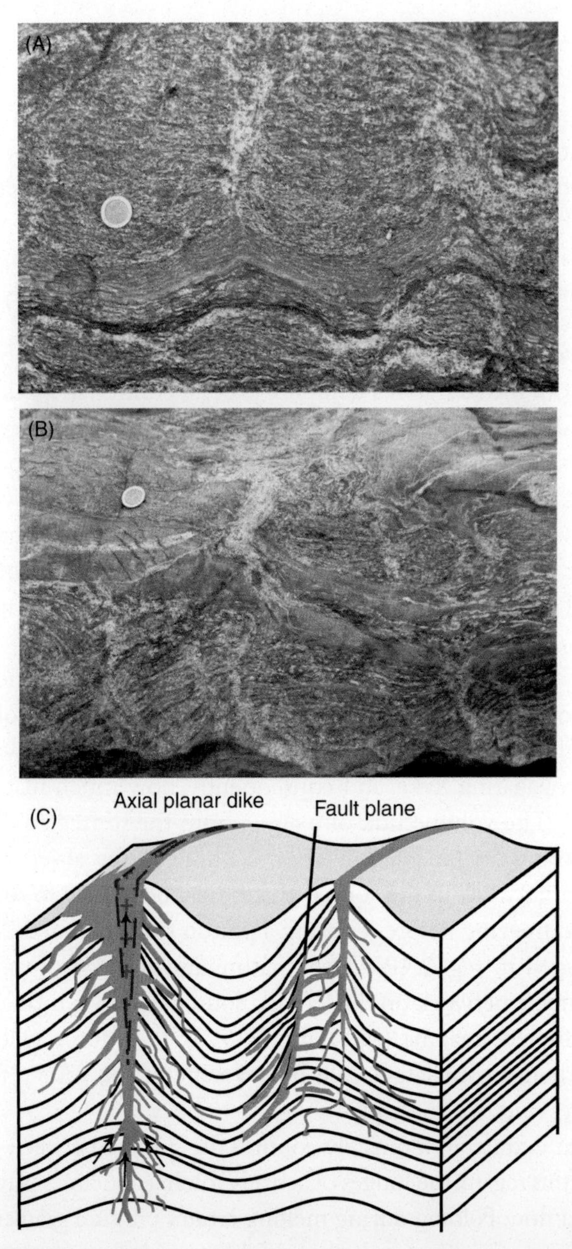

Figure 2.5 ***Magma extraction structures related to folding.*** (A, B) Two field examples of diffuse leucosomes below a folded layer of a less fertile psammite, which become focused above the antiformal hinge zone. The antiformal fold hinge acts to focus flow into a wider dykelet. From Sierra de Quilmes, Famatinian Orogeny, NW Argentina. (C) Leucosome networks developed during open folding of metaturbidites. The combination of layer-parallel and axial planar leucosomes focuses melt migration towards the hinge zone of antiforms where leucosome width increases. Based on outcrop features of the Kanmantoo Group metaturbidites on Kangaroo Island (from Schwindinger and Weinberg, 2017).

2.2.2.3 Features of Magma Loss From Migmatites

When magma is drained from the source, it sometimes leaves recognisable features of this loss of volume. The most typical feature is the presence of relatively wide mafic selvedges around small volumes of leucosome or the presence of highly residual rocks, where the volume of peritectic minerals is incompatible with the small volume of leucosomes, such as cordieritites within a small volume of leucosomes (Finch et al., 2017). More generally, the preservation of anhydrous peritectic minerals in the source is indicative of melt loss. Had the migmatite behaved as a closed system and not lost any melt, the water contained in the melt would fully rehydrate the anhydrous peritectic minerals via retrogression reactions to form minerals such as biotite or muscovite. A more subtle feature of magma loss are collapse structures left by melt removal, which are characterized by the convergence of foliations towards a single central patch of leucosome (Fig. 2.4E) (Bons et al., 2008).

2.2.3 MASH Zones

Hildreth and Moorbath (1988) suggested that mantle magmas in arcs would go through a zone of interaction with crustal melts in the lower part of the crust where processes such as Melting, Assimilation, Storage and Homogenisation would take place. They called this the MASH zone and used it to explain how mantle magmas become hybridised with continental rocks, as evidenced, for example, by the isotopic signatures of arc magmas. More recently, this concept has been expanded to regions lacking mafic magmas, characterised by interaction between different crustal magmas (Fig. 2.6) in felsic MASH zones (Schwindinger and Weinberg, 2017). These zones form and are characterised by a series of reinforcing feedback processes (Figs. 2.4F and 2.6). In essence, felsic MASH zones result from magma migration from deeper parts of an anatectic terrain, transferring heat and H_2O into regions of incipient anatexis triggering further melting. As the melt fraction increases at a given crustal level, it impedes the passage of migrating magmas, trapping any new injections from below and increasing the efficiency of heat and H_2O transfer, causing further melting. At some point, the host rock will disaggregate into a diatexite that mingles and mixes with the intrusive magmas. In this way, the region of anatexis expands and different felsic continental magmas interact and homogenise before proceeding in their migration. If large enough, magmas stored in MASH zones are readily available for migration through the crust by diverse transport mechanisms, such as ductile fracturing, dyking and diapirism (see Section 2.3).

2.2.4 Melt Focusing Mechanisms

The two previous sections convey the idea that the two steps of magma segregation and extraction work together to give rise to a self-organised network of magma-rich leucosomes (Marchildon and Brown, 2003) that may be extracted in different ways from the source. The nature, rates and volumes involved in these processes control everything

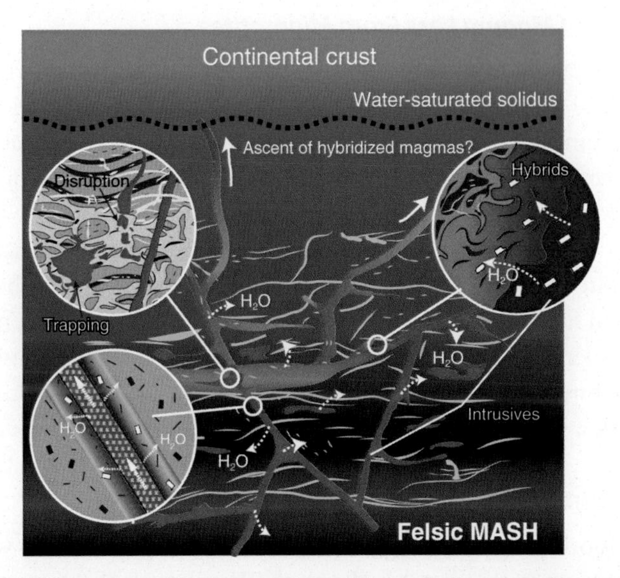

Figure 2.6 ***Felsic MASH zone: granitic magmas intrude into shallower regions undergoing incipient anatexis.*** Intrusions are hot and H_2O-rich, triggering further melting and increasing the in situ melt fraction. As the melt fraction increases, new intrusions are efficiently trapped by the migmatites, increasing heat and H_2O transfer to the surroundings, causing further melting. Intrusive and resident magmas interact, becoming hybrids and in the process remaining solids become increasingly broken down into schollen and schlieren, while early intrusive rocks break up into blocks and blobs in the hybrid rocks (from Schwindinger and Weinberg, 2017).

about the magmas being extracted, including the mechanism of magma migration towards the upper crust. Indeed, felsic magmatic systems may be self-organised from the bottom-up (Cruden, 2006; Brown, 2013), whereby magma segregation and extraction are rate-limiting processes, as well as determinants of the characteristic length scales of the system, such as the spacing between plutons in the mid- to upper crust.

An outstanding question concerns the mechanisms by which flow of partial melt in source regions becomes focused to eventually supply discrete, spaced magma reservoirs in the mid- to upper crust (Figs. 2.5 and 2.7A). That is, how does melt flowing out of the source become self-organised to produce the characteristic spacing of magmatic systems observed higher in the crust, and ultimately of volcanoes? A related question is, how does this melt flow relate to the time scales of intrusive and eruptive processes? Here, we speculate qualitatively on three mechanisms that may result in the spatial (and temporal) scaling of magma plumbing systems.

Firstly, as discussed above, a sufficiently partially molten migmatite terrain may be capable of flowing *en masse* (Fig. 2.7B). If the anatectic region is buoyant and if the overlying crust is weak (as in low viscosity), the top of the migmatite zone may develop into a Rayleigh–Taylor instability (e.g. Berner et al., 1972; Cruden, 1990). Such instabilities have characteristic wavelengths that are determined by density, viscosity and thickness contrasts.

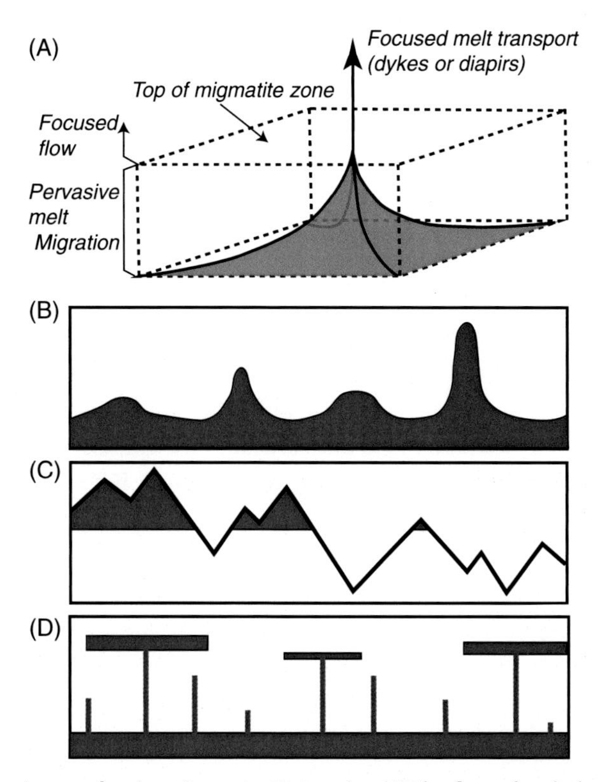

Figure 2.7 ***Melt focusing mechanisms in anatectic terrains.*** (A) The flow of melt that is segregated from pore spaces and then extracted from the source rock must eventually be concentrated by the time it reaches the top of the migmatite zone to feed spaced magma reservoirs higher in the crust (after Weinberg, 1999). (B) Gravity-driven Rayleigh–Taylor instability of the top of the anatectic zone may lead to diapiric structures with a characteristic wavelength or spacing. (C) The isotherms representing the leading edge of the melting front in the source region may show a random variation in depth, like a Brownian walk, leading to periodicity in both the length and time spacing of magmatism (after Pelletier, 1999). (D) Competition between dykes draining melt from the source region may result in a self-organisation of lateral flow within the anatectic zone and eventual spacing of mid- to upper crustal magma reservoirs.

While the instabilities develop, mobile melt can ascend relative to the solid groundmass and accumulate in the apex/cupula of the rising domes. This one-way upward flow of partial melt may become focused via diapirism, which is discussed in more detail in Section 2.3.

Secondly, one can envision that the upward vertical migration of isotherms might control the height to which regions undergoing anatexis might ascend. As the magma rises further within a hot zone and when temperatures are above the magma solidus, it pushes the isotherms further up, allowing further magma rise in a positive feedback loop (Weinberg and Searle, 1998; Leitch and Weinberg, 2002; Section 2.3.3.3). This loop is limited by the faster heat loss from the hot region through diffusion, as the isotherms rise further up in the crust. In the region below the magma front, magma migration can be pervasive as the melt is unimpeded by solidification. If the crust has macroscopic variations of: (a) permeability, (b) thermal conductivity and/or (c) fertility, the isotherm at the top of a melt-rich anatectic zone may evolve with a complex geometry (Fig. 2.7C) (Pelletier, 1999). Such patterns are analogous to those of coffee migrating vertically via capillary action through a paper towel, known as a Brownian walk. Pelletier (1999) applied this concept to a cellular-automaton model that created a distribution of magma supply regions that erupt with equal probability per unit time. The model exhibits statistical self-similarity to observed temporal-spatial data sets of magmatism in the SW USA (Pelletier, 1999).

Finally, melt may attempt to leave the anatectic zone via many dykes that are initially random in space and time (Fig. 2.7D; see also Chapter 3). Many of these dykes are likely to fail due to a lack of sufficient magma supply and solidification. However, some will make it high enough to start feeding a mid-crustal reservoir. Once such a successful connection has been made, and once the segregation network has matured to a point where it can provide sufficient melt flux to sustain the dyke, more melt will start to focus into a single pathway, shutting down adjacent ones. Over time, the system may self-organise in a manner that is similar to the way in which wells are configured to manage water extraction from an aquifer (e.g. Cruden, 2006). Water wells are situated in such a way to minimise aquifer draw-down, given certain pumping and recharge rates (e.g. Domenico and Schwartz, 1998), analogous to the rates of melt extraction and melting, respectively. This will result in a characteristic spacing of plutons observed in many regions and is consistent with pulsed nature of mid- to upper crustal magma systems (see Sections 2.3 and 2.4).

2.3 MAGMA ASCENT MECHANISMS

Once sufficient melt has accumulated in the source, it needs to ascend to higher crustal levels to form a pluton (Fig. 2.8). How this happens has been the subject of considerable debate. Until the 1990s, felsic magma was generally thought to ascend through the deeper crust in bulk via the mechanisms of diapirism, stoping or zone refining (Marsh, 1982, 1984). While it has long been accepted that mafic magmas traverse the crust through dykes, the high viscosities previously assumed for felsic melts were thought to preclude any form of rapid, channelled ascent (Marsh, 1982). However, with the advent of new experimental measurements and theoretical estimates of melt viscosities (e.g. Scaillet et al., 1998; Dingwell, 1999; Caricchi et al., 2007), it is now understood that rapid ascent of felsic magmas at all crustal levels is a viable ascent mechanism that can occur via channelled melt pathways, including dykes (Fig. 2.8B) (Clemens and Mawer, 1992; Petford et al., 1994, 2000). Here, we review the main bulk and channelled magma ascent mechanisms that have been proposed for the transport of both felsic and mafic magmas. A key commonality between all mechanisms is that it is the density difference, or buoyancy, of magma that is the ultimate driver of magma ascent in the crust.

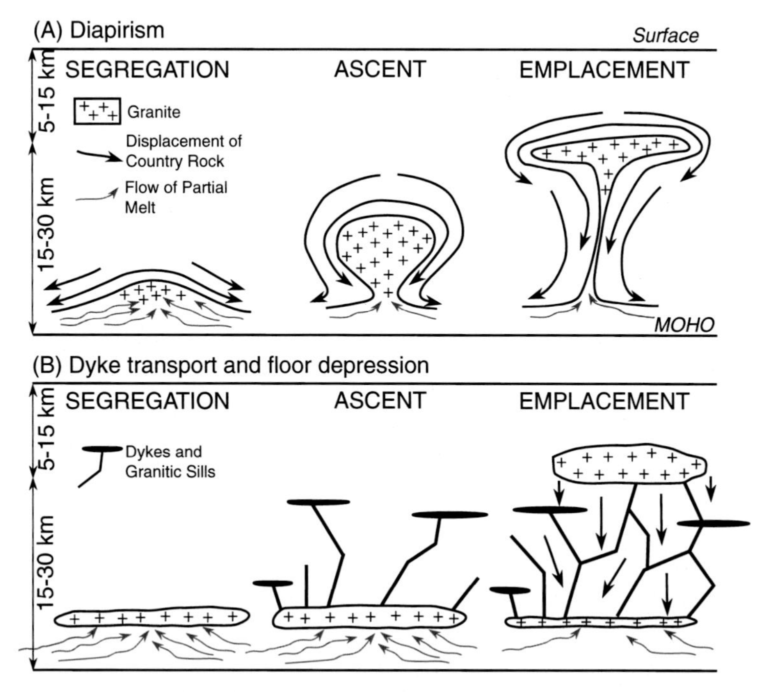

Figure 2.8 ***End-member regimes for magma segregation, ascent and emplacement.*** (A) Diapirism regime. (B) Chanellised ascent and tabular pluton regime. See text for further discussion.

2.3.1 Diapirism

Since the work of Grout (1945) and Ramberg (1970), diapirism has been a popular mechanism used to account for the ascent and emplacement of felsic magma. The concept of diapirism is that a volume of buoyant, mobile and ductilely deformable material forces its way upward through overlying rocks as a blob-like structure (Fig. 2.9A). The term *diapir* is derived from the Greek verb to pierce and, as such, is consistent with the generally stratigraphy-discordant nature of salt diapirs, to which the concept was originally applied. However, the notion of magmatic diapirs has been generally synonymous with *concordant* or *forceful plutons*, associated with deflection of structural markers and significant ductile deformation recorded in the surrounding country rocks (e.g. Brun and Pons, 1981; Cruden, 1990). This contrasts with *discordant* or *passive plutons*, which have been linked to ascent and emplacement via magmatic stoping (e.g. Pitcher, 1979; Marsh, 1982).

As with the formation of salt domes, magmatic diapirs require the accumulation of a sufficient volume of coherent melt in the source for buoyant ascent to commence. Small perturbations in the thickness of a magma source layer lead to periodic Rayleigh–Taylor instabilities (e.g. Berner et al., 1972; Whitehead and Luther, 1975). The fastest growing instabilities, with a preferred or dominant wavelength, amplify at the expense of others to form spaced pillow-like diapirs. The periodic nature of such instabilities has been linked to the spacing

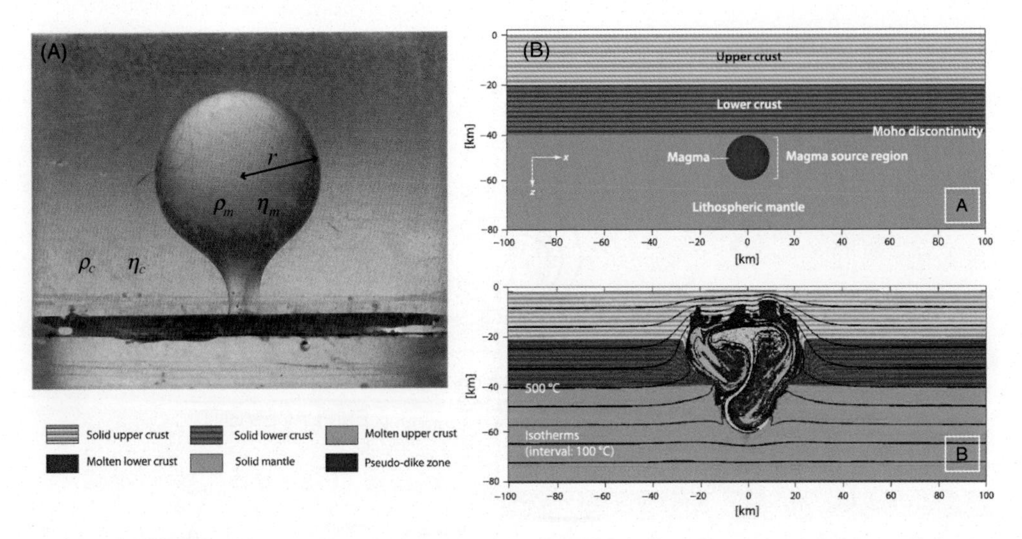

Figure 2.9 (A) Laboratory experiment of low-viscosity diapir with radius r, viscosity η_m and density ρ_m ascending through a higher viscosity and density fluid (η_c, ρ_c). Width of photograph is 10 cm. (B) Elastic–viscous–plastic model of diapirism (from Cao et al., 2016). The legend explaining the different units is located beneath (A).

observed between plutons in orogenic belts and magmatic arcs (e.g. Rickard and Ward, 1981), as well as oceanic islands and hot spots (Whitehead and Luther, 1975; Olson and Singer, 1985).

Laboratory experiments (Whitehead and Luther, 1975; Cruden, 1990) and numerical analyses (Berner et al., 1972; Woidt, 1978) have shown that diapirs less viscous than their surroundings become approximately spherical after rising several diameters in distance. These buoyantly-ascending magma-filled spheres, which may be fully detached from their source or connected via a narrow stalk, are referred to as *Stokes diapirs*.

2.3.1.1 Stokes Diapirs

The ascent velocity of a magmatic diapir can be estimated from Stokes drag formula for creeping flow past a sphere (e.g. Batchelor, 1967). The general expression for the terminal ascent or descent velocity, U, of an isothermal fluid sphere in an infinite, Newtonian viscous medium is given by the Hadamard–Rybczynski equation:

$$U = \frac{1}{3}\frac{\Delta\rho g r^2}{\eta_c}\left(\frac{\eta_c + \eta_m}{\eta_c + \frac{3}{2}\eta_m}\right) \tag{2.1}$$

where r is the diapir radius, g the gravity acceleration, η the kinematic viscosity, and $\Delta\rho$ is the density difference $\rho_c - \rho_m$ between crustal rocks and magma, where subscripts c and m indicate crust and magma, respectively (Fig. 2.9A). Note that the viscosity of the diapir itself has a relatively minor influence on the ascent velocity, such that an inviscid body will only rise two-thirds faster than a rigid body.

For felsic magmatic diapirs in the deep crust ranging in size between $r = 2.5$ and 20 km, with density contrasts $\Delta\rho \sim 700$–200 kg/m^3, crustal viscosities $\eta_c \sim 1 \times 10^{20}$ to 1×10^{18} Pa s and magma viscosities $\eta_m < 1 \times 10^{17}$ Pa s (both felsic and mafic magmas will be many orders of magnitude less viscous than this), we obtain ascent velocities ranging from ~1.3 km/Ma ($r = 2.5$ km, $\Delta\rho = 200$ kg/m^3, $\eta_c = 1 \times 10^{20}$ Pa s) to as high as 287 km/Ma ($r = 20$ km, $\Delta\rho = 700$ kg/m^3, $\eta_c = 1 \times 10^{18}$ Pa s). For each of these end-member diapirs to ascend 15 km from source depths to the mid-crust would take ~12 million to 50 thousand years, respectively.

Because we are concerned with forming a mid-crustal magma reservoir, the diapir must not solidify before it reaches its depth of emplacement. This can be assessed approximately by considering the characteristic solidification time scale, τ, of a sheet-like intrusion (Marsh, 2015), which is approximated by

$$\tau = 0.694\frac{H^2}{\kappa} \tag{2.2}$$

where H is the half-thickness of the intrusion and κ the thermal diffusivity. Taking into account the aspect ratio of the intrusion n, H can be approximated by

$$H = \left(\frac{V}{8n^2}\right)^{1/3} \tag{2.3}$$

where V is the intrusion volume (Marsh, 2015). Taking $\kappa = 1 \times 10^{-6}$ m^2/s, for an equant diapir with $n = 1$, we find that $\tau \sim 0.01$ Ma when $r = 2.5$ km and ~5.7 Ma when $r = 20$ km. For vertically flattened diapirs with $n = 10$, these crystallisation times diminish to ~0.004 and 0.26 Ma, respectively. Comparing these estimates to the ascent rates given by Eq. (2.1) indicates that small diapirs ($r < 5$ km) will freeze after a few hundred metres to kilometres of ascent. However, large diapirs ($r > 10$ km) can theoretically ascend to shallow depths without solidifying.

The above analysis suggests that Stokes diapirism is a thermally viable ascent mechanism for very large magma bodies ascending in weak ductile crust but not for small ones. However, there are several critical factors that have been omitted in the analysis. Firstly, these diapirs must force a very large volume of country rock to flow around them, and experimental studies have shown that a significant strain gradient should extend at least one body radius out from the margin (Cruden, 1988; Schmeling et al., 1988). However, when deformation aureoles are present around plutons, they tend to be very narrow, indicating that ascent-related flow of country rock must be confined to a thin region surrounding the diapir. This may be accounted for by considering the effects of heat transfer from the diapir (i.e. thermal softening) and/or power law creep (i.e. strain rate softening), as discussed below.

Marsh (1982); Ribe (1983); Daly and Raefsky (1985) and Ansari and Morris (1985) have studied the effects of heat transfer from rising, hot spherical diapirs. Marsh (1982) concluded that the viscosity of the lithosphere must be strongly temperature-dependent for diapirically ascending magma to reach within 10 km of the surface and that the process is more efficient if several diapirs rise sequentially along the same path.

Mahon et al. (1988) modelled the ascent of hot diapirs through crust deforming by temperature-dependent power-law creep using experimentally determined flow laws. They also considered various geothermal gradients, the softening effects on country rock due to heat transfer and strain-rate gradients, as well as solidification of the magma itself. In contrast to the simple Stokes diapir approach above, they showed that smaller diapirs rise faster than large ones, but over shorter periods of time. This is because small diapirs must process (soften) a much smaller volume of country rock than large ones, but they cool faster due to their lower heat budget. Mahon et al. (1988) also found that in all diapir scenarios, ascent was limited to ~15 km. This is because of heat loss from the diapir, which diminishes their ability to soften the country rocks over time as well as inducing magma solidification. The progressive decrease in country rock temperature with decreasing crustal depth exacerbates these effects. Miller et al. (1988) applied the results of Mahon et al. (1988) to a quantitative study of a granitic pluton in southeast Australia. These authors concluded that a diapiric ascent mechanism was viable if the viscosity of continental crust were an order of magnitude less than that predicted by laboratory measurements.

Weinberg and Podlachikov (1994) used a fluid-mechanical approach to assess the viability of magmatic diapirism when the country rocks have non-Newtonian and temperature-dependent rheology. By modifying Eq. (2.1), they showed that when the country rocks behave as power-law (strain-rate softening) fluids, diapir ascent rates increase, such that diapirs ascending at 10–100 m/year can reach the middle to upper crust without solidifying. Strain-rate softening, rather than thermal weakening allows diapirs to ascend at these rates, which, except for extreme cases, are several orders of magnitude faster than the Stokes diapirs discussed above. Although Weinberg and Podlachikov (1994) show that the ascent of 'power-law' diapirs is efficient for magma transport in the mantle and lower crust, such models predict narrow strain aureoles around plutons with higher strain intensities than Stokes diapirs (e.g. Cruden, 1988), which are generally not observed.

2.3.1.2 Viscoelastic Diapirs and Dyke–Diapir Hybrids

Following an analysis of dyke propagation through a medium with variable viscous and elastic properties by Rubin (1993), the concept of 'viscoelastic diapirs' was proposed by Paterson and Miller (1998) as an alternative to Stokes, thermal-softening and power-law diapirs. These authors envisage a spectrum from diapiric blobs with narrow ductile strain aureoles, through diapiric ridges with even narrower aureoles to discordant dykes (Fig. 2.1C) (see also Miller and Paterson, 1999; Paterson et al., 2011).

The original formulation by Rubin (1993) was a self-similar solution for a pressurised crack propagating in a linear viscoelastic medium; heat transfer and solidification were not considered. By considering both the elastic response of the country rock and the viscosity contrast between magma and its surroundings, he showed that for host rocks to behave effectively elastically during dyke intrusion they must be 11–14 orders of magnitude more viscous than the magma. For typical dyke thickness/length ratios <0.1, he also found that the viscosity contrast must be >6–8 orders of magnitude. Three important implications arise from this analysis: (1) country rocks will behave elastically during intrusion of basalts and low-viscosity felsic magmas; (2) viscous deformation of the surrounding host rocks will exceed elastic deformation during emplacement of high-viscosity felsic magmas ($\eta_m > 10^8$ Pa s) into hot rocks ($\eta_c < 10^{17}$ Pa s), but these intrusions will have dyke-like shapes; (3) equidimensional, diapir-like felsic plutons emplaced into viscoelastic crust cannot result from viscosity contrasts between liquid granitic magma and hot country rocks alone (Rubin, 1993).

In order to reconcile the above limitations in both hot-Stokes and viscoelastic dyke models, Cao et al. (2016) carried out sophisticated finite-element numerical experiments of pulsed magma ascent in temperature-dependent crust with elastic–viscous–plastic rheology (e.g. Fig. 2.9B). The numerical simulations also permit partial melting of country rocks and use an algorithm to emulate the thermo-mechanical effects of dyking in the crust above a large magma reservoir. A key finding of this work is that both dyking and the injection of multiple magma pulses are required for magma ascent over significant distances and for emplacement into the stiff upper crust. Multiple pulses stop the system from freezing too quickly, and they initiate intrusion of dykes ahead of the main magma body. The latter thermally softens the roof rocks, creating pathways for new pulses of magma.

2.3.2 Magmatic Stoping

Magma ascent and emplacement by magmatic stoping was originally defined by Daly (1903, 1933) as a mechanism that involves continued fracturing of the wall and roof rocks of a magma body, which are incorporated as floating or sinking detached blocks (Burchardt, 2009). Stoping may occur at all crustal levels (Pignotta and Paterson, 2007) as a result of thermally or mechanically induced shear or tensile failure (Roberts, 1970). Since thermal fracturing is the dominant mechanism, magmatic stoping is most efficient in the brittle crust where large temperature gradients between upper crustal rocks and intruding magma are expected (Zak et al., 2006).

Features that are typically attributed to stoping in a pluton include: (1) stoped blocks; (2) mixed populations of xenoliths; (3) stepped intrusive contacts; (4) the lack of penetrative, ductile deformation in the host rock related to pluton emplacement; (5) geochemical evidence for magma contamination (e.g. Paterson and Fowler, 1993; Fowler and Paterson, 1997; Yoshinubo et al., 2003; Dumond et al., 2005; Glazner and Bartley, 2006; Zak et al., 2006). Although these features are often observed in plutons and other intrusions, a number of outstanding questions about stoping are debated (Clarke et al., 1998): (1) By which mechanisms can blocks of host rock become detached from the walls of magma chambers? (2) What is the volumetric significance of stoping for the emplacement of magma bodies? (3) What happens to stoped blocks once detached from the wall or roof of a pluton?

Glazner and Bartley (2006) argued that many of the features commonly attributed to stoping can be explained by other processes and that stoping is therefore not a volumetrically significant pluton-emplacement process (Question 2). Their main argument is based on the unclear fate of stoped blocks (Question 3), since large accumulations of stoped xenoliths are rarely found in plutons. However, Clarke et al. (1998), Clarke and Erdmann (2008) and Yoshinubo and Barnes (2008) have shown that disintegration of xenoliths enclosed within magma by thermal fracturing and explosive exfoliation (spalling), and subsequent dissolution and melting (McLeod and Sparks, 1998), are very effective and do result in contamination of the magma. Stoped blocks can sink rapidly through a magma body at rates dependent on magma viscosity, density contrast and their size (Eq. (2.1)) to accumulate on the pluton floor or other structural levels (e.g. Marsh, 1982; Hawkins and Wiebe, 2004).

In addition to the questions posed above, another critical issue is whether stoping is a viable ascent mechanism. How does the magma remain hot? In evaluating some of the thermal consequences of stoping, Marsh (1982) assessed the cooling effect of spherical blocks taking up different proportions of the intrusion volume. For a low-viscosity mafic magma ($\eta_m = 10^2$ Pa s) half filled with stoped blocks at any time to ascend 15 km in the crust without solidification, the blocks must be >3 m across. For felsic magma with $\eta_m = 10^5$ Pa s, the block size must be >13 m across. For the sizes of xenoliths typically observed in granitic rocks in the field (centimetres to a few metres), plus the suggestion that large blocks will tend to break up into smaller ones (Clarke et al., 1998), these calculations indicate that stoped blocks will have a significant cooling effect. Although theoretical ascent velocities by stoping remain poorly constrained, energetic and thermal considerations suggest that this mechanism is unlikely to allow transport of granite magma more than a few thousand meters before freezing, making it unimportant for crustal-scale magma transport (Marsh, 1982, 1984).

Even though neither the volumetric significance nor the thermal viability of magmatic stoping are well understood, it likely plays an important role as a local vertical material transfer process, resulting in the structural modification of pluton roofs and sides late in their intrusive history. Stoping also contributes to chemical contamination of magma and hence to their compositional evolution (e.g. Dumond et al., 2005; Yoshinubo and Barnes, 2008; see also Chapter 8).

2.3.3 Channelled Magma Ascent

In contrast to the bulk transport mechanisms of diapirism and magmatic stoping, channelled ascent involves the movement of magma through discrete networks of narrow conduits (Fig. 2.8B). Inherent to the establishment of a network of magma channels is that it can be re-used or re-activated multiple times, making it amenable to supplying mid- to upper-crustal reservoirs with multiple pulses of magma, as is frequently observed (e.g. Pitcher, 1979; Deniel et al., 1987; Coleman et al., 2004; Miller, 2008; Brown, 2013). This contrasts to the bulk magma ascent mechanisms, which are unlikely to be associated with pulses, unless the same diapir or stoping pathways are used multiple times (e.g. Marsh, 1982; Paterson et al., 2011; Cao et al., 2016). Here, we describe three possible modes of channelled magma ascent, namely dykes and hydrofractures, ductile fractures and pervasive flow. A more detailed review of dykes and dyke-forming mechanisms is provided in Chapter 3.

2.3.3.1 Dykes and Hydrofractures

A long-standing objection to the dyke transport of granite was that high-viscosity felsic magma would freeze in the dyke before a volume sufficient for filling a pluton could travel through it (Marsh, 1982). However, it is now realised that felsic melts have viscosities in the range 10^3–10^7 Pa s, several orders of magnitude lower than originally expected (Clemens and Petford, 1999; Dingwell, 1999). Consequently, dyke transport of felsic magma may be sufficiently rapid to prevent freezing, and furthermore, dykes can fill pluton-sized upper crustal chambers over geologically rapid times (Petford et al., 1994, 2000; Cruden, 1998, 2006).

This can be assessed using a simple channel flow equation for magma ascent in a tabular dyke:

$$U_{\mathrm{ave}} = \frac{\Delta\rho g w^2}{12\eta_{\mathrm{m}}} \tag{2.4}$$

where U_{ave} is the average velocity within the dyke and w is the dyke width (Petford et al., 1994). It is worth noting that this equation has a similar form to Eq. (2.1) for Stokes diapirs, in that magma ascent is driven by buoyancy contrast $\Delta\rho g$ at a rate governed by the square of the characteristic length scales, w and r, respectively. However, the major difference is that the rate of dyke transport is limited by the magma viscosity η_{m}, whereas Stokes diapirs are controlled by crustal viscosity η_{c}.

Taking a typical density contrast of $\Delta\rho = 300$ kg/m^3 and a felsic magma viscosity of 10^5 Pa s, the average velocity in a 1-m-wide dyke will be 2.45×10^{-3} m/s or 212 m/day. Magma in this dyke will take 78 days to ascend 15 km. Under the same conditions, the average velocity in a 10-m-wide dyke will be 0.245 m/s, taking only 0.7 days to ascend 15 km. These values are in stark contrast to Stokes diapirs, which take on the order of thousands to millions of years to ascend 15 km.

Although dyke ascent is potentially many orders of magnitude faster than even the fastest power-law diapirs, the large surface area and narrow width of dykes makes them susceptible to rapid crystallisation. Petford et al. (1994) assessed this problem by considering the solidification rate of magma flowing vertically in a 30-km-high dyke. Taking reasonable values for temperature contrasts between felsic magma and host rocks, heat capacity, thermal diffusivity and latent heat of crystallisation, they found the critical dyke width required to prevent freezing varies between 2 and 20 m when $\eta_m = 10^4$–10^8 Pa s, respectively. For felsic magma with $\eta_m = 10^6$ Pa s and $\Delta\rho = 200$ kg/m^3 to ascend 15 km without solidifying, the critical dyke width would be ~7 m. From Eq. (2.4), this 15 km of magma ascent would take 21 days.

An alternative to rapid magma ascent in dykes is transport within mobile hydrofractures, which are penny-shaped, magma-filled cracks that ascend via buoyancy by simultaneously propagating their tips and closing their tails (Weertman, 1971; Bons et al., 2001; Lister and Kerr, 2001). Dykes differ from hydrofractures in that they must stay open from top to bottom. Recent experimental work on hydrofractures suggests that they may allow step-wise accumulation of magma pulses into a pluton, leaving very little trace of their ascent in the underlying crust (Bons et al., 2001).

2.3.3.2 Ductile Fractures

According to Rubin (1993), dykes and hydrofractures are associated with purely brittle-elastic deformation of their host rocks. Dyke propagation driven by magma buoyancy occurs by tensile failure (Mode I), which in turn requires stresses at the dyke tip to exceed the fracture toughness of the host rock (Lister and Kerr, 2001; Chapter 3). The previously discussed analysis of viscoelastic dykes by Rubin (1993) was motivated by the fact that the deep crust deforms by ductile creep, although fracture resistance and crack tip processes were neglected in his modelling of ductile host rocks. Rock mechanics and metallurgy studies suggest that with increasing temperature, there is transition from brittle to conjugate ductile (Mode II) shear failure mechanisms (Gandhi and Ashby, 1979). Laboratory and field analyses on quartzites suggest that the transition temperature is as low as ~300 °C (e.g. Hirth and Tullis, 1994). Such ductile or creep fracture mechanisms have been proposed to play a role in the development of magma channels in the deep crust (Weinberg and Regenauer-Lieb, 2010).

Ductile fracturing occurs during rock creep when microscale voids grow and eventually interconnect leading to failure, typically in the form of conjugate shear bands. Microvoids grow as a result of ductile recrystallization. Weinberg and Regenauer-Lieb (2010) suggest that these microvoids may be filled with melt in the deep crust, leading to conjugate ductile fractures that will act as magma channels. They also envisage that once ductile-fracture dykes reach a critical length, magma stresses at dyke tips will exceed fracture toughness, leading to brittle-elastic dyking, as illustrated schematically in Fig. 2.10A. The potential role of conjugate ductile fractures in the early stages of magma migration is supported by both field observations in exhumed deep-crustal sections and numerical modelling (Fig. 2.10B and C; Weinberg and Reigenauer-Lieb, 2010).

2.3.3.3 Pervasive Flow

The concept of pervasive flow of magma arises from field-based structural studies of migmatites (see Section 2; Brown, 1994; Collins and Sawyer, 1996; Weinberg and Searle, 1998; Leitch and Weinberg, 2002). During pervasive flow, magma ascends through

Figure 2.10 ***Magma ascent via ductile fractures.*** (A) Sketch of conjugate ductile fractures converging upwards to form dykes. (B) Example of magma-filled ductile fractures in the field; Pangong injection complex, Himalaya. (C) Numerical model of conjugate ductile fracture formation. Blue colours indicate negligible to low bulk strain, green-yellow are intermediate bulk strains, and red to grey are high bulk strains. Planar intermediate to high strain domains define conjugate sets of ductile fractures comparable to the natural example in (B). After Weinberg and Regenauer-Lieb (2010).

an extensive network of channels, which are formed by active deformation and kept open by a combination of dilational strain and magma pressure (Collins and Sawyer, 1996; Brown and Solar, 1998). This results in an intrusive migmatite terrain or injection complex that may be capped by tabular granitic plutons (e.g. Weinberg and Searle, 1998).

One-dimensional thermal modelling (Leitch and Weinberg, 2002) has shown that such pervasive flow of magma can occur, provided the host rock is above the magma solidus. Although incoming magma has the potential to push isotherms further up in the crust, transport by pervasive flow seems to be limited to only several kilometres in the lower to middle crust. In the field, this will consist of an injection complex made up of granite magma and original crustal rocks that is several kilometres thick.

2.3.3.4 Faults and Shear Zones as Melt Pathways

Although the above sections imply that magma ascent is an inherent consequence of the withdrawal of melt and heat from lower-crustal sources and their transport into the mid- to upper crust (i.e. melt extraction, ascent and emplacement may constitute a closed, internally driven system), numerous structural studies have demonstrated strong connections between regional deformation and associated structures and the ascent and emplacement of both felsic and mafic magmas (e.g. Pitcher, 1979; Castro, 1986; Hutton, 1988; Vigneresse, 1995; Brown and Solar, 1998; de Saint Blanquat et al., 1998; Brown, 2013).

Faults and shear zones are sites of dilation and enhanced porosity, so it is not surprising that they are regarded as favourable pathways for the transport of fluids in the crust, including magmas. Indeed, normal (e.g. Richards and Collins, 2004; Grocott and Taylor, 2002; Grocott et al., 1994, 2009; Hutton et al., 1990), reverse (e.g. Ingram and Hutton, 1994) and strike-slip fault systems (e.g. Guineberteau et al., 1987; Hutton, 1988; Tikoff and Teyssier, 1992) have all been implicated as controlling magma ascent and emplacement in a variety of tectonic settings (see Chapter 7). Transpressional fault systems are strongly associated with magma ascent and emplacement (e.g. McCaffrey, 1992; Brown and Solar, 1998; Benn et al., 1999), which may be a consequence of the excess pressure that occurs in the deep crustal roots of these structures (Robin and Cruden, 1994).

Faults may also be able to capture and arrest magma ascending via dykes and hydrofractures, as demonstrated both theoretically and experimentally (e.g. Le Corvec et al., 2013; Valentine and Krogh, 2006; Gaffney et al., 2007). Similarly, Rubin (1993) proposed that the transition from high-aspect ratio viscoelastic dykes to more equidimensional diapirs may be facilitated by the interaction between ascending magma and crust weakened by active faults.

2.4 GROWTH AND EMPLACEMENT OF MID-CRUSTAL MAGMA RESERVOIRS

The arrest of ascending magma leads to the development of a magma reservoir. This can happen at any depth above the magma source. Here, we focus on plutons emplaced in the middle crust, below ~5 km depth. The formation of upper crustal reservoirs (i.e. sills and laccoliths) is discussed in Chapters 5 and 6.

The concepts of ascent versus emplacement are different for bulk versus channelled magma transport mechanisms. In the case of diapirism and stoping, emplacement occurs when the ascending magma stops moving, most likely by freezing. Here, emplacement processes and related structural signatures are continuous with those of ascent. However, an arresting diapir may flatten out into a tabular body if it encounters a mechanical barrier in the crust (e.g. Brun and Pons, 1981; Cruden, 1990).

When magma ascent is channelled, emplacement starts when magma transport is arrested. At this stage, ascending magma is diverted horizontally into an initial sill-like structure and the pluton subsequently grows by vertical inflation (Fig. 2.11A; see also Chapters 5 and 6).

Field and geochronological observations support a consensus that plutons are constructed incrementally via multiple pulses (e.g. Pitcher, 1979; Deniel et al., 1987; Cruden and McCaffrey, 2001; Coleman et al., 2004; Miller, 2008; Cottam et al., 2010; Brown, 2013; Annen et al., 2015; see also Chapter 12). This is consistent with the episodic nature of magma extraction and transport in channelled networks (cf. Sections 2.2 and 2.3.3). It has also been recognised that plutons thought to have been emplaced by diapirism may

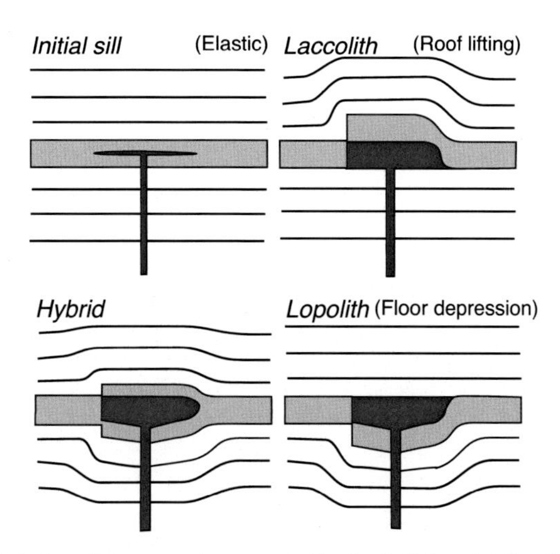

Figure 2.11 ***Modes of tabular pluton emplacement.*** On the left-hand side of each diagram, space for the magma (*red*) is created within the intruded unit (*grey*) by offsets on faults, whereas space is created by ductile deformation on the right-hand side of each diagram.

have assembled incrementally, consistent with the nested geometry of compositional phases observed in many intrusions (e.g. Bouchez and Diot, 1990; Paterson and Vernon, 1995; Paterson et al., 2011).

2.4.1 Shape and Form—Floors, Roofs, Sides and Layering

Areally-extensive granitic plutons were viewed historically as intrusions with steep sides that continued to great depth in the crust (Fig. 2.1C) (e.g. Buddington, 1959; Paterson et al., 1996; Miller and Paterson, 1999). This perspective is hardly surprising, because erosion of several kilometre-high magma bodies in areas of low-to-modest relief will tend to bias preservation of steep marginal contacts. However, field observations of plutons in areas of high relief, combined with the results of geophysical surveys indicate that many granitic plutons are tabular bodies with horizontal dimensions much larger than their vertical extent (Fig. 2.11B) (e.g. Vigneresse, 1995; McCaffrey and Petford, 1997; Cruden, 2006; Cruden et al., 2018 and references therein). Furthermore, in addition to having gently inclined roofs and floors, such tabular intrusions often contain internal layering or sheets that have shallow dips, parallel to pluton roofs and floors (Fig. 2.12).

Two types of pluton floor geometries are observed: funnel or wedge-shaped and flat, tablet-shaped (cf. Vigneresse et al., 1999; Cruden, 2006). Wedge-shaped plutons can be symmetric or asymmetric and typically have one or more root zones, defined by downward-tapering linear deep structures, becoming a narrow cylinder, in gravity models, interpreted to be feeder structures (e.g. Ameglio and Vigneresse, 1999). Their floors dip inward from very shallow angles, defining broad open funnel shapes to steep angles,

Figure 2.12 ***Photographs of pluton roofs, floors and internal layering.*** (A) View of Split Mountain, Sierra Nevada, USA, looking west from Owens Valley. The floor of the Jurassic Tinemaha granodiorite (Jt) is in contact with a septum of Cambrian metasediment (Campito Fm, Cc), which in turn forms the roof of a Jurassic leucogranite (Red Mountain Creek granite, Jrm). See Bartley et al. (2012) for additional information. (B) *A* ~ 1200 m high cliff on the side of Lindenow Fiord, Greenland, exposes a ~ 500 m thick Proterozoic granite sheet emplaced into gneissic country rocks (photograph: courtesy of John Grocott). (C) Cliff exposure (~800 m high) of the Proterozoic Graah Fjeld granite, Greenland. Grey rafts of country rock gneisses define intrusive sheet boundaries (photograph: courtesy of John Grocott; Grocott et al., 1999). (D) Late Cretaceous Chehueque pluton, Coastal Cordillera, Chile. La Pignetta mountain (~2000 m) exposes three separate intrusive units of the Chehueque pluton comprising granite (Pgt), granodiorite (Pgd) and monzonite (Pmz).

defining carrot-like shapes. Tablet-shaped plutons are characterised by almost parallel roofs and floors and steep sides (Fig. 2.12A and B). Some plutons have both wedge- and tablet-shape characteristics.

Field examples of the nature and geometry of pluton floors are relatively uncommon. However, limited observations in Greenland, the North- and South American Cordillera and the Himalaya (Fig. 2.12; e.g. Hamilton and Myers, 1974; Le Fort, 1981; Scaillet et al., 1995; Hogan and Gilbert, 1997; Skarmeta and Castelli, 1997; Grocott et al., 1999; Michel et al., 2008; Bartley et al., 2012) are in general agreement with geophysical data.

Paterson et al. (1996) reviewed the characteristics of mid- to upper-crustal pluton roofs exposed in the Cordillera of North and South America, showing that they consistently have gentle dips to slightly domal morphologies and discordant contact relationships with pre-existing wall-rock structures. Emplacement-related ductile strain in the wall rocks is typically absent to poorly developed, and there is also little evidence that the roofs have been lifted above their pre-emplacement position. Minor amounts of stoped blocks occur beneath the roof, and stoping is a likely candidate for generating the jagged profiles of the roofs, although its role as a major space-making mechanism is debatable (cf. Section 2.3). Other authors report more compelling evidence for upward displacements of pluton roofs (e.g. Morgan et al., 1998; Benn et al., 1999; Grocott et al., 1999), particularly in shallower crustal settings.

Relatively undisturbed roofs, sharp transitions to steeply-dipping walls and the presence of either sharp wall-rock contacts or narrow strain aureoles with evidence for country-rocks-down sense of shear relative to the pluton margin have been used by Paterson et al. (1996), Paterson and Miller (1998) and Miller and Paterson (1999) to argue that most space for emplacement of granites is due to downward transfer of country rock material. Although these authors favour mechanisms such as stoping or return-flow of country rock during diapiric ascent, downward displacement and rotation of wall-rock structural markers and fabrics towards the margins of intrusions in Greenland, Sweden and N. America suggests that floor subsidence may be an important alternative space-making process (Bridgwater et al., 1974; Cruden, 1998; Benn et al., 1999; Grocott et al., 1999; Brown and McClelland, 2000; Culshaw and Bhatnagar, 2001). Pluton-side-down shear sense indicators and roll-over of strata adjacent to some plutons have also been ascribed to late-stage sinking of cooling magma bodies (e.g. Glazner and Miller, 1997). Large-scale tilting of roof pendants and wall rocks in the Sierra Nevada and Boulder batholiths has also been attributed to down-drop of pluton floors during batholith growth and emplacement (Hamilton and Myers, 1974; Hamilton, 1988; Tobisch et al., 2001).

There is increasing evidence that many plutons, including those that are macroscopically homogeneous, are made up of many metre- to kilometre-scale sheets (Fig. 2.12C and D) (e.g. McCaffrey, 1992; Everitt et al., 1998; Cobbing, 1999; Grocott and Taylor, 2002; Coleman et al., 2004; Michel et al., 2008; Grocott et al., 2009; Cottam et al., 2010). Detailed textural observations of intrusions in Maine, SW Australia and South New Zealand suggest that initially sub-horizontal sheets steepen with time during growth of a pluton (Wiebe and Collins, 1998). This is supported by U-Pb studies in the Coast Plutonic Complex, where plutons are interpreted to have grown from the floor upward by stacking of sheets and gradual subsidence and distortion of their floors (Brown and Walker, 1993; Wiebe and Collins, 1998; Brown and McClelland, 2000). Other field and geochronological studies indicate that some tabular plutons are assembled by downward stacking of pulses (Fig. 2.12D) (Michel et al., 2008; Grocott et al., 2009; Leuthold et al., 2012).

2.4.2 Emplacement and Growth of Mid-Crustal Plutons

The degree of ductile wall-rock deformation associated with pluton emplacement appears to be a function of depth and the mechanical properties of the host during intrusion and growth. Plutons emplaced in a ductile environment show evidence for components of lateral and vertical displacement of wall rocks, whereas space for plutons emplaced in brittle environments must be created by vertical translation of country-rock material, due to the absence of strain observed within their lateral margins.

A reasonable generalisation for many, but not all, plutons is that they are emplaced as tabular to wedge-shaped bodies with thicknesses ranging from 1 km to 10 km. This relationship is supported by compilations of available dimensional data from field and geophysical observations, which indicate that there is an empirical scaling relationship between intrusion vertical thickness T and horizontal width L:

$$T = bL^{a} \tag{2.5}$$

where the constants a and b are determined by reduced major axis regression of the data (McCaffrey and Petford, 1997; Cruden et al., 2018). A power law exponent $a > 1$ indicates an emplacement and scaling regime, in which intrusions become thicker vertically faster than they lengthen horizontally. This is the case for laccoliths (Chapter 5), which are characterised by $a \sim 1.5$ (e.g. Rocchi et al., 2002). Intrusions with $a < 1$ tend to lengthen horizontally faster than they grow vertically. This is the case for sills, felsic plutons and layered mafic intrusions (Cruden et al., 2018). Felsic plutons are characterised by $a = 0.81 \pm 0.12$ and $b = 1.08 \pm 1.38$ m, such that the aspect ratio of these intrusions increases with increasing horizontal size L. Such empirical scaling relationships are useful for differentiation between intrusion styles and as constraints for mechanical models of emplacement (e.g. Bunger and Cruden, 2011). They can also be used for the calculation of pluton construction times (Section 2.4.3) and, together with map-based data on intrusion areas, estimating magma fluxes in different tectonic settings (e.g. Cruden, 2006; Paterson and Ducea, 2015).

From the above, the general characteristics of both felsic plutons and layered mafic intrusions are: (1) they are tabular to wedge-shaped with $L \gg T$; (2) they are fed by one or more vertical conduits; (3) they comprise multiple pulses; (4) the bulk of magma flow at the emplacement level is horizontal; and (5) the role of lateral displacement in creating space diminishes with decreasing ductility of the wall rocks. While diapirism may account for some of these characteristics, channelled magma ascent followed by vertical inflation and assembly of sheet-like pulses are the mechanisms favoured by many authors (e.g. Annen et al., 2015; Brown, 2013 and references therein).

Figure 2.11 illustrates three different scenarios for vertical growth of a tabular pluton. Magma ascending in a dyke or hydrofracture is first 'trapped' by an arresting mechanism, such as intersection with a freely slipping horizontal fracture, stopping of the propagating dyke by a ductile horizon or a unit with high fracture toughness, or arrival at a level of neutral buoyancy (e.g. Brisbin, 1986; Corry, 1988; Clemens and Mawer, 1992; Hogan et al., 1998). Once an initial sill has formed (Fig. 2.11A), it can inflate, provided sufficient magma pressure is available (Johnson and Pollard, 1973; Pollard and Johnson, 1973; Corry, 1988). Vertical inflation to plutonic dimensions can occur by roof lifting (Fig. 2.11B; laccolith emplacement), floor depression (Fig. 2.11D; lopolith emplacement) or a hybrid mechanism (Fig. 2.11C) (Cruden, 1998). The dynamics of laccolith emplacement by roof lifting are well established (Chapter 5). Laccoliths are mostly confined to the upper 2–3 km of the crust (Corry, 1988; Cruden et al., 2018), consistent with the requirement that their vertical growth involves interaction with the Earth's surface. The >3 km emplacement levels of most felsic plutons are too deep to lift their roofs. Hybrid intrusions appear to be a transition in emplacement depth, size and mode of vertical growth.

Floor depression, or subsidence, has long been considered as a possible space-making mechanism for felsic plutons (Clough et al., 1909; Cloos, 1923; Hamilton and Myers, 1967; Lipman, 1984). Most early models invoked 'cauldron subsidence' whereby the pluton floor subsided during withdrawal of magma from a deeper chamber or reservoir (see Chapter 10; Branch, 1967; Whitney and Stormer, 1986; Myers, 1975; Bussell et al., 1976; Pitcher, 1979). Downward displacement of pluton floors by ductile flow mechanisms has also been considered (Hamilton and Myers, 1967; Brown and Walker, 1993), possibly aided by isostatic depression of the Moho (Brown and McClelland, 2000). More recent models for floor depression take into account the notions that granitic magmas are likely extracted from partially molten lower crust (e.g. Brown, 1994, 2013; Thompson, 1999), where magma transport is probably channelled and rapid, and that the volumetric flow rates of melt extraction (Q_W), ascent (Q_A) and emplacement (Q_E) must be balanced at the crustal scale (Fig. 2.13; Cruden, 1998, 2006).

The deformation mechanisms that allow the downward transfer of material during pluton growth are not well constrained. However, given that this deformation occurs in the lower to mid-crust (Fig. 2.13) of regions with high heat flow, it is reasonable to assume that the principal mechanism will be by ductile flow, aided by displacements on shear zones in regions of active regional deformation. The viability of this process can be evaluated using two simple end-member wedge and tabular pluton models (Fig. 2.13B and C). The models are circular in the plan view with a radius r, and it is assumed that

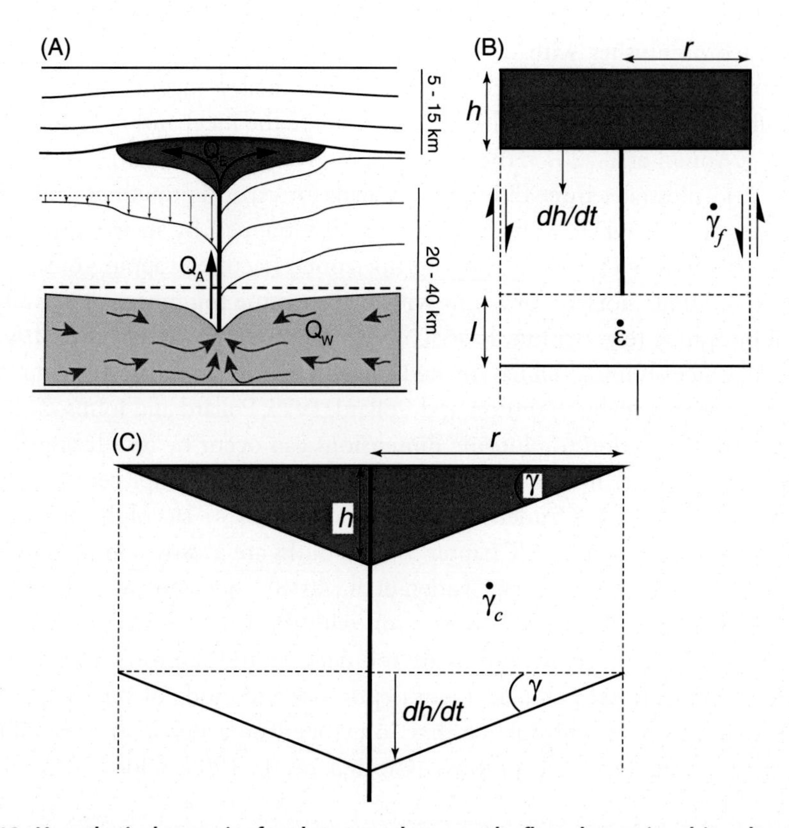

Figure 2.13 ***Hypothetical scenarios for pluton emplacement by floor depression driven by withdrawal of melt from an underlying source region.*** In (A) a symmetric, centrally fed pluton (*red*) is derived from a source region with horizontal dimension far greater than the final width of pluton. *Thin dashed line* on the left below pluton indicates original position of a horizontal marker and *arrows* show its displacement during pluton growth. Lengths of the *red arrows* in source indicate relative flow velocities schematically. Thickness ranges of roof rocks, the pluton and the underlying crust are approximate and not shown to scale. (B) Piston model for pluton growth. (C) Cantilever model for pluton growth. See text for definition of symbols. After Cruden, (2006).

the plutons are fed by a central conduit, and they grow to a maximum thickness h. The models can easily be formulated for asymmetric cases, in which magma is fed from one side (Cruden, 1998) and for plutons that are elliptical in the plan view.

In the cantilever model, pluton growth occurs by tilting (dip angle γ) of the floor about a pivot point located at the edge of the pluton (Fig. 2.13C). The underlying crust deforms by bulk progressive simple shear as it sinks into a region of partial melt in the lower crust. In the piston model, growth occurs by subsidence of a horizontal floor, which is accommodated by a vertical, cylindrical shear zone (or ring dyke) at the perimeter of the pluton (Fig. 2.13B). The piston sinks into a partially molten source region at depth.

The strain rates required for pluton growth by these mechanisms can be evaluated using the volumetric flow rate of magma being delivered via a dyke:

$$Q_A = Q_E = \frac{\Delta\rho}{12\eta_m} g w^3 \beta \tag{2.6}$$

where Q_A is the volumetric flow rate in the feeder dyke (or dykes), Q_E the volumetric filling rate of the pluton and β is the horizontal length of the dyke (Petford et al., 1994; Cruden, 1998). Taking $\Delta\rho$ = 10–400 kg/m^3, η_m = 10^4–10^8 Pa s, w = 1–10 m and β = 1–10 km, gives flow rates ranging from $<10^{-2}$ to $>10^3$ m^3/s. These values are within the range of eruption and magma-chamber filling rates estimated from volcanic systems (e.g. White et al., 2006; Menand et al., 2015).

Using the pluton widths $L \approx$ 5–200 km and thicknesses predicted from Eq. (2.5), both model geometries require strain rates of the compacting source region (piston, $\dot{\varepsilon}$) or intervening crust (cantilever, $\dot{\gamma}$) in the range of 10^{-9} to 10^{-16} s^{-1} (Cruden, 2006), which is consistent with estimated tectonic deformation rates (Pfiffner and Ramsay, 1982; Paterson and Tobisch, 1992). In the piston model, the required strain rates on the bounding faults are considerably faster. For example, a 1-m-wide fault requires a shear strain rate, $\dot{\gamma}_f$ of 10^{-6} s^{-1} for $\dot{\varepsilon}$ = 10^{-10} s^{-1} and the same fault will deform with $\dot{\gamma}_f$ =10^{-10} s^{-1} when $\dot{\varepsilon}$ = 10^{-13} s^{-1}. Such fast rates may be reasonable if the fault is lubricated by magma. However, if they are not, this places a limit on the efficiency of piston sinking for space creation and suggests that the cantilever mechanism may be energetically more favourable.

Space-making mechanisms for granitic plutons have also been linked to dilation within strike slip faults (e.g. Guineberteau et al., 1987; Tikoff and Teyssier, 1992), reverse faults (e.g. Ingram and Hutton, 1994), normal faults (e.g. Hutton et al., 1990; Grocott et al., 1994), transpressional fault systems (e.g. McCaffrey, 1992; Benn et al., 1999) and folds (e.g. Schwerdtner, 1990; Roig et al., 1998; see also Chapter 7). Based on field relationships in the Mesozoic Coastal Cordillera, Chile, Grocott and Taylor (2002) proposed several models that show how displacements on both active normal and reverse faults can accommodate pluton floor depression and roof uplift (Fig. 2.14A and B).

A similar approach can be applied to the case of plutons located in releasing bends of strike fault systems (e.g. Guineberteau et al., 1987; Hutton, 1988). Although published studies have emphasised the role of lateral widening for creating space for magma, the primary space-making mechanism for intrusions in such settings may also be floor depression (Fig. 2.14C), analogous to the formation of sediment accommodation space in pull-apart basins (Sylvester, 1988).

2.4.3 Timescales of Pluton Assembly

Volumetric flow rates associated with magma eruption and chamber filling in volcanic systems from 10^{-2} to $>10^3$ m^3/s (e.g. White et al., 2006; Menand et al., 2015), and the empirical scaling for the size relationships of felsic intrusions (Eq. (2.5)) suggest that

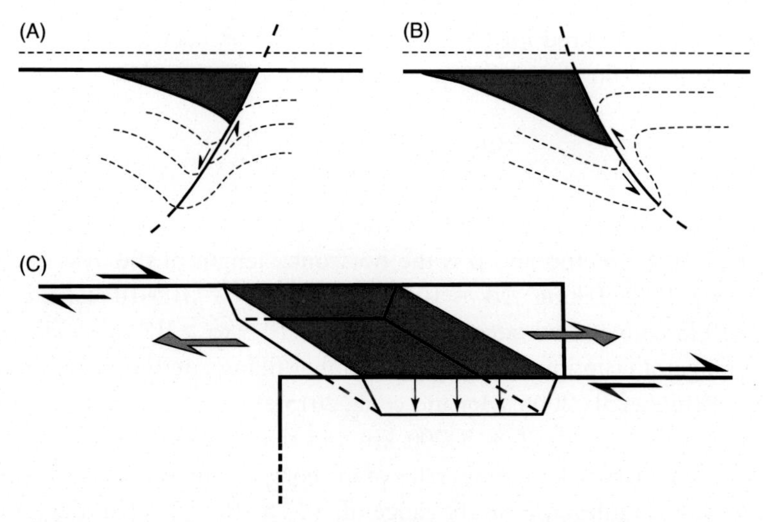

Figure 2.14 ***Links between pluton growth and displacements on active faults.*** Cross-sections (A, B) illustrate how displacements on active normal (A) and reverse (B) dip-slip faults can aid in the growth of a pluton (*red*) by floor depression (after Grocott and Taylor, 2002). (C) Three-dimensional sketch illustrates the structure of a pluton (*red*) emplaced in a transtensional step-over or releasing bend within a dextral strike-slip fault zone (*thick black lines*). Extensional strain (*open grey arrows*) within the step-over results in localisation of magma and facilitates vertical pluton growth (*thin arrows*) due to floor depression.

pluton emplacement is potentially rapid (Petford et al., 2000). For continuous magma emplacement, median pluton-filling times are between 1000 and 10,000 years for smaller plutons ($L \sim 10$ km), 10,000 and 100,000 years for larger plutons ($L \sim 50$ km) and $\gg$1 Myr for batholiths (Cruden, 2006). If a pluton grows by multiple pulses, the filling time can be estimated from

$$t_{\text{fill}} = \sum t_{\text{pulse}} + \sum t_{\text{lag}} \tag{2.7}$$

where t_{pulse} is determined from Q_E (Eq. (2.6)) and $t_{\text{lag}} = \lambda t_{\text{pulse}}$, where λ is a time lag factor. Fig. 2.15A plots filling times for a pluton with $L = 10$ km and $T = 3$ km as a function of λ and a range of magma supply rates. When $\lambda = 1$, the filling time is double compared to the case of constant magma supply. However, for each order of magnitude increase in λ, t_{fill} also increases by 10 (Fig. 2.15A). For the case of a 10-km-wide pluton being emplaced by pulses with $Q_E = 1$ m^3/s (0.032 km^3/year), the pluton-filling time is ~10,000 years when $\lambda = 1$ and this increases to ~1 Myr when $\lambda \sim 100$. The dependence of t_{fill} on pluton size L is shown in Fig. 2.15B.

Hence, although the emplacement times of each magma pulse in a pluton can be rapid, the net filling time can be orders of magnitude greater if pulses are separated by long time gaps. This explains the variability of pluton crystallisation ages determined from U-Pb zircon geochronology, which range from <1 to 8 Ma (e.g. Brown and McClelland, 2000; Glazner et al., 2004; Miller et al., 2007; Leuthold et al., 2012). Such

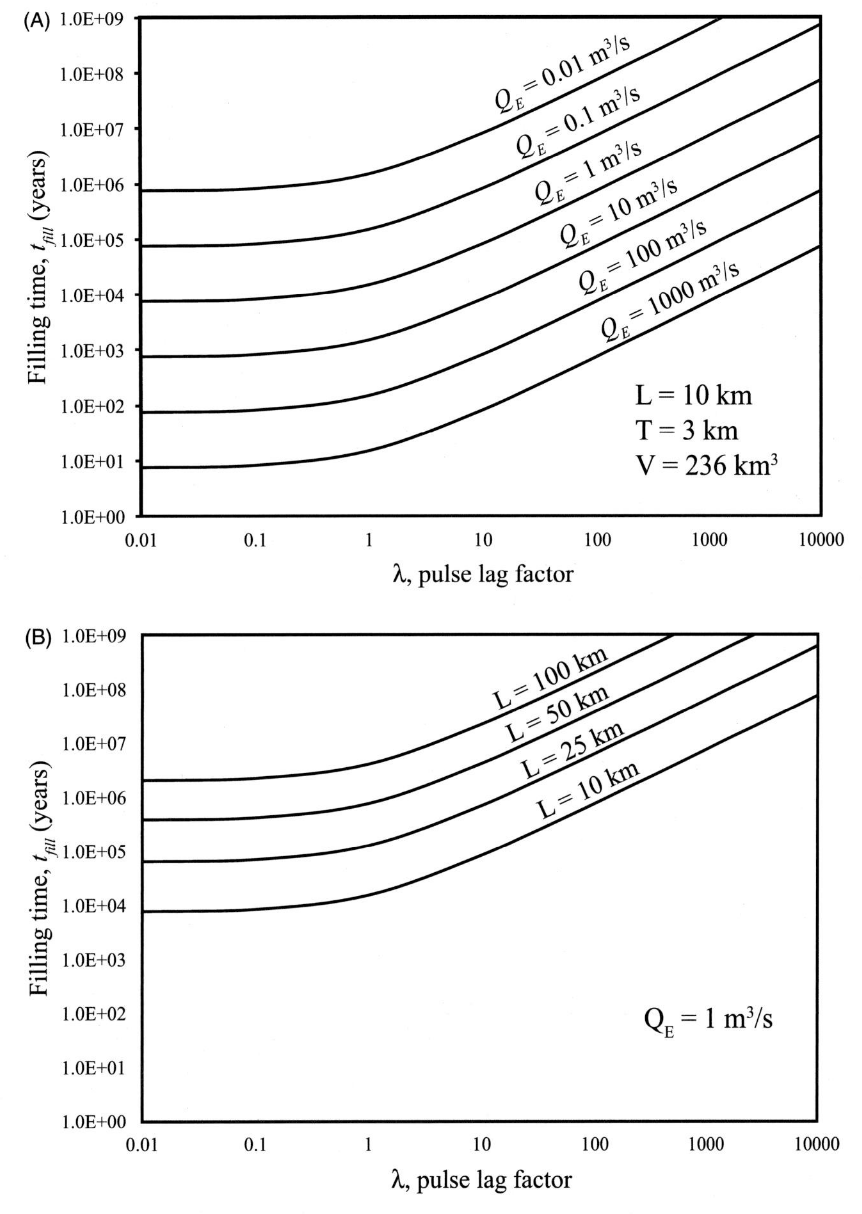

Figure 2.15 ***Calculated pluton filling times as a function of pulse lag factor λ.*** (A) Filling times for a pluton with $L = 10$ km and $T = 3$ km for volumetric emplacement rates $Q_E = 0.01–1000\ m^3/s$ (3.2×10^{-4} to 32 km^3/year). (B) Filling times for constant $Q_E = 1\ m^3/s$ (0.032 km^3/year) and $L = 100–10$ km (and corresponding thicknesses, $T = 8–3$ km) respectively.

potential variability in λ may ultimately be determined by the rate of melting and melt extraction in the source region (see Sections 2.2 and 2.3), and ultimately the heat input or power of the magmatic system. It also explains why some intrusions remain hot for long periods of time, which favours erasure of textural evidence for individual pulses (e.g. Tuolumne pluton, Sierra Nevada batholith; Coleman et al., 2004; Miller et al., 2007; Memeti et al., 2010) versus rapidly emplaced and solidified plutons, in which discrete pulses are well preserved (e.g. plutonic complexes in southern and north central Chile; Michel et al., 2008; Grocott et al., 2009).

2.5 CONCLUDING REMARKS

In this chapter, we have discussed how anatexis of lower crustal source rocks leads to the segregation of melts, and how their extraction leads to the focusing of magmas, which subsequently ascend into the mid-to-upper crust to form magma reservoirs or plutons. We have shown that there is a diversity of views on how magma ascends through the crust (e.g. bulk vs channelled mechanisms) and on the architecture of crustal scale magma plumbing systems (e.g. channelled magma networks, magma mush columns, diapirs with return flow). This diversity of interpretation is due to the range of approaches currently taken to study magma plumbing systems (e.g. structural/physical vs geochemical/petrological) and differences between the geological characteristics of ancient systems exposed in the geological record as well as modern ones observed directly by geophysics. There is broad consensus that the growth of mid-to-upper crustal magma reservoirs occurs by multiple pulses, which implies that magma plumbing systems are repeatedly reactivated or reopened. We suggest that the characteristic time and length scales of these systems are determined in the magma source region by the rates of melting and melt segregation, source rock fertility, H_2O availability, and the dynamics of magma extraction and focusing; magma plumbing systems are therefore self-organised from the bottom up.

REFERENCES

Ameglio, L., Vigneresse, J.L., 1999. Geophysical imaging of the shape of granitic intrusions at depth: a review. Castro, A., Fernandez, C., Vigneresse, J.L. (Eds.), Understanding Granites: Integrating New and Classical Techniques. Special Publications, 168, Geological Society, London, pp. 39–54.

Annen, C., Blundy, J.D., Leuthold, J., Sparks, R.S.J., 2015. Construction and evolution of igneous bodies: towards an integrated perspective of crustal magmatism. Lithos 230, 206–221.

Ansari, A., Morris, S., 1985. The effects of a strongly temperature-dependent viscosity on Stoke's drag law: experiments and theory. J. Fluid Mech. 159, 459–476.

Bachmann, O., Huber, C., 2016. Silicic magma reservoirs in the Earth's crust. Am. Mineral. 101 (11), 2377–2404.

Bartley, J.M., Glazner, A.F., Mahan, K.H., 2012. Formation of pluton roofs, floors, and walls by crack opening at Split Mountain, Sierra Nevada, California. Geosphere 8, 1086–1103.

Batchelor, G.K., 1967. An Introduction to Fluid Dynamics. Cambridge University Press, Cambridge.

Benn, K., Roest, W.R., Rochette, P., Evans, N.G., Pignotta, G.S., 1999. Geophysical and structural signatures of syntectonic batholith construction: the South Mountain Batholith, Meguma Terrane, Nova Scotia. Geophys. J. Int. 136, 144–158.

Berner, H., Ramberg, H., Stephanson, O., 1972. Diapirism in theory and experiment. Tectonophysics 15, 197–218.

Bons, P.D., van Milligen, B.P., 2001. New experiment to model self-organized critical transport and accumulation of melt and hydrocarbons from their source rocks. Geology 29, 919–922.

Bons, P.D., Dougherty-Page, J., Elburg, M.A., 2001. Stepwise accumulation and ascent of magmas. J. Metamorphic Petrol. 19, 627–633.

Bons, P.D., Druguet, E., Castaño, L.-M., Elburg, M.A., 2008. Finding what is now not there anymore: recognizing missing fluid and magma volumes. Geology 36 (11), 851–854.

Branch, C.D., 1967. The source of eruption for pyroclastic flows: caldrons or calderas? Bull. Volcanol. 30, 41–53.

Bouchez, J.L., Diot, H., 1990. Nested granites in question: contrasted emplacement kinematics of independent magmas in the Zaer pluton, Morocco. Geology 18, 966–969.

Bridgwater, D., Sutton, J., Watterson, J., 1974. Crustal downfolding associated with igneous activity. Tectonophysics 21, 57–77.

Brisbin, W.C., 1986. Mechanics of pegmatite intrusion. Am. Mineral. 71, 644–651.

Brown, M., 1994. The generation, segregation, ascent and emplacement of granite magma: the migmatite-to-crustally-derived granite connection in thickened orogens. Earth Sci. Rev. 36, 83–130.

Brown, M., 2013. Granite: from genesis to emplacement. Geol. Soc. Am. Bull. 125 (Aug), 1079–1113.

Brown, E.H., McClelland, W.C., 2000. Pluton emplacement by sheeting and vertical ballooning in part of the southeast Coast Plutonic Complex, British Columbia. Geol. Soc. Am. Bull. 112, 708–719.

Brown, M., Solar, G.S., 1998. Granite ascent and emplacement during contractional deformation in convergent orogens. J. Struct. Geol. 20, 1365–1393.

Brown, E.H., Walker, N.W., 1993. A magma loading model for Barrovian metamorphism in the southeast Coast Plutonic Complex, British Columbia and Washington. Geol. Soc. Am. Bull. 105, 479–500.

Brown, M., Averkin, Y.A., McLellan, E.L., Sawyer, E.W., 1995. Melt segregation in migmatites. J. Geophys. Res. 100 (B8), 15655-11567.

Brun, J.P., Pons, J., 1981. Strain patterns of pluton emplacement in a crust undergoing non-coaxial deformation, Sierra Morena, S. Spain. J. Struct. Geol. 3, 0219–229.

Buddington, A.F., 1959. Granite emplacement with special reference to North America. Geol. Soc. Am. Bull. 70, 671–747.

Bunger, A.P., Cruden, A.R., 2011. Modeling the growth of laccoliths and large mafic sills: the role of magma body forces. J. Geophys. Res. 116. doi: 10.1029/2010JB007648, B02203.

Burchardt, S., 2009. Mechanisms of magma emplacement in the upper crust. Dissertation zur Erlangung des Doktorgrades, der Mathematisch-Naturwissenschaftlichen Fakultäten der Georg-August-Universität zu Göttingen vorgelegt von aus Nordhausen.

Bussell, M.A., Pitcher, W.S., Wilson, P.A., 1976. Ring complexes of the Peruvian Coastal Batholith: a long-standing subvolcanic regime. Can. J. Earth Sci. 13, 1020–1030.

Cao, W., Kaus, B.J.P., Paterson, S., 2016. Intrusion of granitic magma into the continental crust facilitated by magma pulsing and dyke–diapir interactions: numerical simulations. Tectonics 35 (6), 1575–1594.

Caricchi, Burlini, L.L., Ulmer, P., Gerya, T., Vassalli, M., Papale, P., 2007. Non-Newtonian rheology of crystal-bearing magmas and implications for magma ascent dynamics. Earth Planet. Sci. Lett. 264, 402–419.

Cashman, K.V., Sparks, R.S.J., Blundy, J.D., 2017. Vertically extensive and unstable magmatic systems: a unified view of igneous processes. Science 355 (6331). doi: 10.1126/science.aag3055.

Castro, A., 1986. Structural pattern and ascent model in the central Extramadura batholith, Hercynian belt, Spain. J. Struct. Geol. 8, 633–645.

Clarke, D.B., Clarke, G.K.C., 1998. Layered granodiorites at Chebucto Head, South Mountain batholith, Nova Scotia. J. Struct. Geol. 20 (9–10), 1305–1324.

Clarke, D.B., Erdmann, S., 2008. Is stoping a volumetrically significant pluton emplacement process? Comment. Geol. Soc. Am. Bull. 120, 1072–1074.

Clarke, D.B., Henry, A.S., White, M.A., 1998. Exploding xenoliths and the absence of 'elephant's graveyards' in granite batholiths. J. Struct. Geol. 20, 1325–1343.

Clemens, J.D., Mawer, C.K., 1992. Granitic magma transport by fracture propagation. Tectonophysics 204, 339–360.

Clemens, J.D., Petford, N., 1999. Granitic melt viscosity and silicic magma dynamics in contracting tectonic settings. J. Geol. Soc. 156, 1057–1060.
Cloos, H., 1923. Das batholithenproblem. Forschr. Geol. Paleont. 1, 1–80.
Clough, C.T., Maufe, H.B., Bailey, E.B., 1909. The cauldron-subsidence of Glen Coe, and the associated igneous phenomena. J. Geol. Soc. London 65, 611–678.
Cobbing, E.J., 1999. The Coastal Batholith and other aspects of Andean magmatism in Peru. Castro, A., Fernandez, C., Vigneresse, J.L. (Eds.), Understanding Granites: Integrating New and Classical Techniques. Special Publications, 168, Geological Society, London, pp. 111–122.
Coleman, D., Gray, W., Glazner, A.F., 2004. Rethinking the emplacement and evolution of zoned plutons: Geochronologic evidence for incremental assembly of the Tuolumne Intrusive Suite, California. Geology 32, 433–436.
Collins, W.J., Sawyer, E.W., 1996. Pervasive granitoid magma transfer through the lower-middle crust during non-coaxial compressional deformation. J. Math. Phys. 14, 565–579.
Corry, C.E., 1998. Laccoliths: Mechanics of Emplacement and Growth. Special Publication. Geological Society of America, 110 pp. 220.
Cottam, M., Hall, R., Sperber, C., Armstrong, R., 2010. Pulsed emplacement of the mount Kinabalu granite, northern Borneo. J. Geol. Soc. London 167 (1), 49–60.
Cruden, A.R., 1988. Deformation around a rising sphere modeled by creeping flow past a sphere. Tectonics 7, 1091–1101.
Cruden, A.R., 1990. Flow and fabric development during the diapiric rise of magma. J. Geol. 98, 681–698.
Cruden, A.R., 1998. On the emplacement of tabular granites. J. Geol. Soc. London 155, 853–862.
Cruden, A.R., 2006. Emplacement and growth of plutons: implications for rates of melting and mass transfer in continental crust. In: Brown, M., Rushmer, T. (Eds.), Evolution and Differentiation of the Continental Crust. Cambridge University Press, pp. 455–519.
Cruden, A.R., McCaffrey, K.J.W., 2001. Growth of plutons by floor subsidence: implications for rates of emplacement, intrusion spacing and melt-extraction mechanisms. Phys. Chem. Earth A 26, 303–315.
Cruden, A.R., McCaffrey, K., Bunger, A.P., 2018. Geometric scaling of tabular igneous intrusions: implications for emplacement and growth. In: Christoph Breitkreuz, C., Rocchi, S. (Eds.), Physical Geology of Shallow Magmatic Systems. Springer Verlag, pp. 11–38.
Culshaw, N., Bhatnagar, P., 2001. The interplay of regional structure and emplacement mechanisms at the contact of the South Mountain Batholith, Nova Scotia: floor-down or wall-up? Can. J. Earth Sci. 38, 1285–1299.
Daly, R.A., 1903. The mechanics of igneous intrusion. Am. J. Sci. 15, 269–298.
Daly, R.A., 1933. Igneous Rocks and the Depth of the Earth. McGraw-Hill, New York.
Daly, S.F., Raefsky, A., 1985. On the penetration of a hot diapir through a strongly temperature-dependent medium. Geophys. J. R. Astron. Soc. 83, 657–681.
de Saint Blanquat, M., Tikoff, B., Teyssier, C., Vigneresse, J.L., 1998. Transpressional kinematics and magmatic arcs. Holdsworth, R.E., Strachan, R.A., Dewey, J.F. (Eds.), Continental Transpressional and Transtensional Tectonics. Special Publications, 135, Geological Society, London, pp. 327–340.
Deniel, C., Vidal, P., Fernandez, A., Le Fort, P., Peucat, J.-J., 1987. Isotopic study of the Manaslu granite (Himalaya, Nepal): inferences on the age and source of Himalayan leucogranites. Contr. Mineral. Petrol. 96, 78–92.
Dingwell, D.B., 1999. Granitic melt viscosities. Castro, A., Fernandez, C., Vigneresse, J.L. (Eds.), Understanding Granites: Integrating New and Classical Techniques. Special Publications, 168, Geological Society, London, pp. 27–38.
Domenico, P.A., Schwartz, F.W., 1998. Physical and Chemical Hydrogeology. Wiley, New York, 506p.
Dumond, G., Yoshinubo, A.S., Barnes, C.G., 2005. Midcrustal emplacement of the Sjausfjellet pluton, central Norway: ductile flow, stoping, and in situ assimilation. Geol. Soc. Am. Bull. 117, 383–395.
Everitt, R., Brown, A., Ejeckam, R., Sikorsky, R., Woodcock, D., 1998. Litho-structural layering within the Archean Lac du Bonnet Batholith, at AECL's Underground Research Laboratory, Southeastern Manitoba. J. Struct. Geol. 20, 1291–1304.
Finch, M.A., Weinberg, R.F., Hasalová, P., Becchio, R., Fuentes, M.G., Kennedy, A., 2017. Tectono-metamorphic evolution of a convergent back-arc: The Famatinian orogen, Sierra de Quilmes, Sierras Pampeanas, NW Argentina. Geol. Soc. Am. Bull. 129, 1602–1621.

Foster, D., Goscombe, B., 2013. Continental growth and recycling in convergent orogens with large turbidite fans on oceanic crust. Geosciences 3 (3), 354.
Foster, D.A., Gray, D.R., 2000. Evolution and structure of the Lachlan Fold Belt (Orogen) of Eastern Australia. Annu. Rev. Earth Planet. Sci. 28 (1), 47–80.
Fowler, T.K., Paterson, S.R., 1997. Timing and nature of magmatic fabrics from structural relations around stoped blocks. J. Struct. Geol. 19, 209–224.
Gaffney, E.S., Damjanac, B., Valentine, G.A., 2007. Localization of volcanic activity: 2. Effects of pre-existing structure. Earth Planet. Sci. Lett. 263, 323–338.
Gandhi, C., Ashby, M.F., 1979. Fracture-mechanism maps for materials which cleave: F.C.C., B.C.C. and H.C.P. metals and ceramics. Acta Metallurgica 27, 1565–1602.
Gardien, V., Thompson, A.B., Ulmer, P., 2000. Melting of biotite + plagioclase + quartz gneisses; the role of H_2O in the stability of amphibole. J. Petrol. 41 (5), 651–666.
Glazner, A.F., Bartley, J.M., 2006. Is stoping a volumetrically significant pluton emplacement process? Bull. Geol. Soc. Am. 118, 1185–1195.
Glazner, A.F., Miller, D.M., 1997. Late-stage sinking of plutons. Geology 25, 1099–1102.
Glazner, A.F., Bartley, J.M., Coleman, D.S., Gray, W., Taylor, R.Z., 2004. Are plutons assembled over millions of years by amalgamation from small magma chambers? GSA Today 14, 4–11.
Grocott, J., Taylor, G.K., 2002. Magmatic arc fault systems, deformation partitioning and emplacement of granitic complexes in the Coastal Cordillera, north Chilean Andes (25°30′S to 27°00′S). J. Geol. Soc. London 159, 425–442.
Grocott, J., Brown, M., Dallmeyer, R.D., Taylor, G.K., Treloar, P.J., 1994. Mechanisms of continental growth in extensional arcs: an example from the Andean plate-boundary zone. Geology 22, 391–394.
Grocott, J., Garde, A., Chadwick, B., Cruden, A.R., Swager, C., 1999. Emplacement of Rapakivi granite and syenite by floor depression and roof uplift in the Paleoproterozoic Ketilidian orogen, South Greenland. J. Geol. Soc. London 156, 15–24.
Grocott, J., Arevalo, C., Welkner, D., Cruden, A.R., 2009. Fault-assisted vertical pluton growth: coastal Cordillera, northern Chilean Andes. J. Geol. Soc. London 166, 295–301.
Grout, F.F., 1945. Scale models of structures related to batholiths. Am. J. Sci. 243, 260–284.
Guineberteau, B., Bouchez, J.L., Vigneresse, J.L., 1987. The Mortagne granite pluton (France) emplaced by pull-apart along a shear zone: structural and gravimetric arguments and regional implications. Geol. Soc. Am. Bull. 99, 763–770.
Hirth, G., Tullis, J., 1994. The brittle-plastic transition in experimentally deformed quartz aggregates. J. Geophys. Res. 99, 11,731–11,747.
Hamilton, W., 1988. Tectonic setting and variations with depth of some Cretaceous and Cenozoic structural and magmatic systems of the Western United States. Ernst, W.G. (Ed.), Metamorphism and Crustal Evolution of the Western United States: Rubey, 7, Prentice-Hall, New Jersey, pp. 1–40.
Hamilton, W., Myers, W.B., 1967. The nature of batholiths. United States Geological Survey Professional Paper 554(c), 1–30.
Hamilton, W., Myers, W.B., 1974. Nature of the Boulder batholith of Montana. Geol. Soc. Am. Bull. 85, 365–378.
Hawkins, D.P., Wiebe, R.A., 2004. Discrete stoping events in granite plutons: a signature of eruptions from silicic magma chambers? Geology 32, 1021–1024.
Hildreth, W., Moorbath, S., 1988. Crustal contributions to arc magmatism in the Andes of Central Chile. Contrib. Mineral. Petrol. 98 (4), 455–489.
Hogan, J.P., Gilbert, M.C., 1997. Intrusive style of A-type sheet granites in a rift environment: The Southern Oklahoma Aulacogen. Ojakangas, R.W., Dickas, A.B., Green, J.C. (Eds.), Middle Proterozoic to Cambrian Rifting, Central North America. Special Paper, 312, Geological Society of America, pp. 299–311.
Hogan, J.P., Price, J.D., Gilbert, M.C., 1998. Magma traps and driving pressure: consequences for pluton shape and emplacement in an extensional regime. J. Struct. Geol. 20, 1155–1168.
Holness, M.B., Cesare, B., Sawyer, E.W., 2011. Melted rocks under the microscope: microstructures and their interpretation. Elements 7 (4), 247–252.
Huppert, H.E., Sparks, R.S.J., 1988. The generation of granitic magmas by intrusion of basalt into continental crust. J. Petrol. 29, 599–624.

Hutton, D.H.W., 1988. Granite emplacement mechanisms and tectonic controls: inferences from deformation studies. Trans. R. Soc. Edinburgh 79, 245–255.

Hutton, D.H.W., Dempster, T.J., Brown, P.E., Becker, S.M., 1990. A new mechanism of granite emplacement: intrusion into active extensional shear zones. Nature 343, 452–455.

Ingram, G.I., Hutton, D.H.W., 1994. The great Tonalite Sill: emplacement into a contractional shear zone and implications for Late Cretaceous to early Eocene tectonics in southeastern Alaska and British Columbia. Geol. Soc. Am. Bull. 106, 715–728.

Jagoutz, O., Schmidt, M.W., 2012. The formation and bulk composition of modern juvenile continental crust: the Kohistan arc. Chem. Geol. 298–299, 79–96.

Johnson, A.M., Pollard, D.D., 1973. Mechanics of growth of some laccolithic intrusions in the Henry Mountains, Utah, I. Field observations, Gilbert's model, physical properties and flow of the magma. Tectonophysics 18, 261–309.

Le Corvec, N., Menand, T., Lindsay, J., 2013. Interaction of ascending magma with pre-existing crustal fractures in monogenetic basaltic volcanism: an experimental approach. J. Geophys. Res. 118, 1–17.

Le Fort, P., 1981. Manaslu leucogranite: a collision signature of the Himalaya a model for its genesis and emplacement. J. Geophys. Res. 86, 10545–10568.

Leitch, A.M., Weinberg, R.F., 2002. Modelling granite migration by mesoscale pervasive flow. Earth Planet. Sci. Lett. 200, 131–146.

Leuthold, J., Muntener, O., Baumgartner, L.P., Putlitz, B., Ovtcharova, M., Schaltegger, U., 2012. Time resolved construction of a bimodal laccolith (Torres del Paine, Patagonia). Earth Planet. Sci. Lett. 325–326, 85–92.

Lipman, P.W., 1984. The roots of ash flow calderas in western North America: windows into the tops of granitic batholiths. J. Geophys. Res. 89, 8801–8841.

Lipman, P.W., Bachmann, O., 2015. Ignimbrites to batholiths: integrating perspectives from geological, geophysical, and geochronological data. Geosphere 11 (3), 705–743.

Lister, J.R., Kerr, R.C., 2001. Fluid mechanical models of crack propagation and their application to magma transport in dykes. J. Geophys. Res. 96, 10049–10077.

Mahon, K.I., Harrison, T.M., Drew, D.A., 1988. Ascent of a granitoid diapir in a temperature varying medium. J. Geophys. Res. 93, 1174–1188.

Marchildon, N., Brown, M., 2003. Spatial distribution of melt-bearing structures in anatectic rocks from Southern Brittany, France: implications for melt transfer at grain- to orogen-scale. Tectonophysics 364 (3–4), 215–235.

Marsh, B.D., 1982. On the mechanics of igneous diapirism, stoping and zone melting. Am. J. Sci. 282, 808–855.

Marsh, B.D., 1984. Mechanics and energetics of magma formation and ascension. In: Boyd, Jr., F.R. (Ed.), Explosive Volcanism: Inception, Evolution and Hazards. National Academy Press, Washington, DC, pp. 67–83.

Marsh, B.D., 2015. Magmatism, Magma and Magma Chambers, Second ed. Treatise on Geophysics, 6, 273-323.

McCaffrey, K.J.W., 1992. Igneous emplacement in a transpressive shear zone: Ox Mountains igneous complex. J. Geol. Soc. 149, 221–235.

McCaffrey, K.J.W., Petford, N., 1997. Are granitic intrusions scale invariant? J. Geol. Soc. London 154, 1–4.

McKenzie, D., 1984. The generation and compaction of partially molten rock. J. Petrol. 25, 713–765.

McLeod, P., Sparks, R.S.J., 1998. The dynamics of xenolith assimilation. Contrib. Mineral. Petrol. 132, 21–33.

Memeti, V., Paterson, S.R., Matzel, J., Mundil, R., Okaya, D., 2010. Magmatic lobes as "snapshots" of magma chamber growth and evolution in large, composite batholiths: An example from the Tuolumne intrusion, Sierra Nevada, California. Geol. Soc. Am. Bull. 122, 1912–1931.

Menand, T., de Saint Blanquat, M., Annen, C., 2011. Emplacement of magma pulses and growth of magma bodies. Tectonophysics 500, 1–2.

Menand, T., Annen, C., de Saint Blanquat, M., 2015. Rates of magma transfer in the crust: insights into magma reservoir recharge and pluton growth. Geology 43, 199–202.

Michel, J., Baumgartner, L., Putlitz, B., Schaltegger, U., Ovtcharova, M., 2008. Incremental growth of the Patagonian Torres del Paine laccolith over 90 ky. Geology 36, 459–465.

Miller, J.S., 2008. Assembling a pluton…one increment at a time. Geology 36, 511–512.

Miller, R.B., Paterson, S.R., 1999. In defense of magmatic diapirs. J. Struct. Geol. 21 (Sep), 1161–1173.
Miller, C.F., Watson, E.B., Harrison, T.M., 1988. Perspectives on the source, segregation and transport of granitoid magmas. Trans. R. Soc. Edinburgh 79 (Mar), 135–156.
Miller, J.S., Matzel, J.E.P., Miller, C.F., Burgess, S.D., Miller, R.B., 2007. Zircon growth and recycling during the assembly of large, composite arc plutons. J. Volcanol. Geotherm. Res. 167, 282–299.
Morgan, S.S., Law, R.D., Nyman, N.W., 1998. Laccolith-like emplacement model for the Papoose Flat pluton based on porphyroblast-matrix analysis. Geol. Soc. Am. Bull. 110, 96–110.
Myers, J.S., 1975. Cauldron subsidence and fluidization: mechanisms of intrusion of the Coastal Batholith of Peru into its own volcanic ejecta. Geol. Soc. Am. Bull. 86, 1209–1220.
Olson, P., Singer, H., 1985. Creeping plumes. J. Fluid Mech. 158, 511–531.
Paterson, S.R., Ducea, M.N., 2015. Arc magmatic tempos: gathering the evidence. Elements 11 (2), 91–98.
Paterson, S.R., Fowler, Jr., T.K., 1993. Re-examining pluton emplacement processes. J. Struct. Geol. 15, 191–206.
Paterson, S.R., Miller, R.B., 1998. Mid-crustal magmatic sheets in the Cascades Mountains, Washington: implications for magma ascent. J. Struct. Geol. 20, 1345–1363.
Paterson, S.R., Tobisch, O.T., 1992. Rates of processes in magmatic arcs: implications for the timing and nature of pluton emplacement and wall rock deformation. J. Struct. Geol. 14, 291–300.
Paterson, S.R., Fowler, Jr., T.K., Miller, R.B., 1996. Pluton emplacement in arcs: a crustal-scale exchange process. Trans. R. Soc. Edinburgh 87, 115–123.
Paterson, S.R., Okaya, D., Memeti, V., Economos, R., Miller, R.B., 2011. Magma addition and flux calculations of incrementally constructed magma chambers in continental margin arcs: combined field, geochronologic, and thermal modeling studies. Geosphere 7, 1439–1468.
Paterson, S.R., Vernon, R.H., 1995. Bursting the bubble of ballooning plutons: a return to nested diapirs emplaced by multiple processes. Geol. Soc. Am. Bull. 107, 1356–1380.
Patiño Douce, A.E., 1995. Experimental generation of hybrid silicic melts by reaction of high-Al basalt with metamorphic rocks. J. Geophys. Res. 100, 15623–15639.
Pelletier, J.D., 1999. Statistical self-similarity of magmatism and volcanism. J. Geophys. Res. 104, 15425–15438.
Petford, N., Lister, J.R., Kerr, R.C., 1994. The ascent of felsic magmas in dykes. Lithos 32, 161–168.
Petford, N., Cruden, A.R., McCaffrey, K.J.W., Vigneresse, J.-L., 2000. Dynamics of granitic magma formation, transport and emplacement in the Earth's crust. Nature 408, 669–673.
Pfiffner, O.A., Ramsay, J.G., 1982. Constraints on geologic strain rates: arguments from finite strains of naturally deformed rocks. J. Geophys. Res. 87, 311–321.
Pignotta, G.S., Paterson, S.R., 2007. Voluminous stoping in the Mitchell Peak granodiorite, Sierra Nevada Batholith, California. Can. Mineral. 45, 87–106.
Pitcher, W.S., 1979. The nature, ascent and emplacement of granitic magmas. J. Geol. Soc. London 136, 627–662.
Pollard, D.D., Johnson, A.M., 1973. Mechanics of growth of some laccolithic intrusions in the Henry Mountains, Utah, II. Bending and failure of overburden layers and sill formation. Tectonophysics 18, 311–354.
Ramberg, H., 1970. Model studies in relation to plutonic bodies. In: Newall, G., Rast, N. (eds.), Mechanism of Igneous Intrusion, Special Issue of the Geological Journal 2, 261-286.
Reichardt, H., Weinberg, R.F., 2012. Hornblende chemistry in meta- and diatexites and its retention in the source of leucogranites: an example from the Karakoram shear Zone, NW India. J. Petrol. 53 (6), 1287–1318.
Ribe, N.M., 1983. Diapirism in the earth's mantle: experiments on the motion of a hot sphere moving in a fluid with temperature-dependent viscosity. J. Volcanol. Geotherm. Res. 16, 221–245.
Richards, S.W., Collins, W.J., 2004. Growth of wedge-shaped plutons at the base of active half grabens. Trans. R. Soc. Edinburgh 95, 309–317.
Rickard, M.J., Ward, P., 1981. Paleozoic crustal thickness in the southern part of the Lachlan orogen deduced from volcano and pluton-spacing geometry. J. Geol. Soc. Aust. 28, 19–32.
Roberts, J.L., 1970. The intrusion of magma into brittle rocks. In: Newall, G., Rast, N. (Eds.), Mechanism of Igneous Intrusion. Geol. J. Spec. Iss. 2, 287–338.
Robin, P.-Y.F., 1979. Theory of metamorphic segregation and related processes. Geochim. Cosmochim. Acta 43, 1587–1600.

Robin, P.-Y.F., Cruden, A.R., 1994. Strain and vorticity patterns in ideally ductile transpression zones. J. Struct. Geol. 16, 447–466.

Rocchi, S., Westerman, D.S., Dini, A., Innocenti, F., Tonarini, S., 2002. Two-stage growth of laccoliths at Elba Island, Italy. Geology 30, 983–986.

Roig, J.-Y., Faure, M., Truffert, C., 1998. Folding and granite emplacement inferred from structural, strain, TEM and gravimetric analyses: the case of the Tulle antiform, SW French Massif Central. J. Struct. Geol. 20, 1169–1189.

Rosenberg, C.L., Handy, M.R., 2005. Experimental deformation of partially melted granite revisited: implications for the continental crust. J. Metamorphic Geol. 23 (1), 19–28.

Rubin, A.M., 1993. Dykes vs. diapirs in viscoelastic rock. Earth Planet. Sci. Lett. 119, 641–659.

Sawyer, E.W., Cesare, B., Brown, M., 2011. When the continental crust melts. Elements 7 (4), 229–234.

Saleeby, J., Ducea, M., Clemens-Knott, D., 2003. Production and loss of high-density batholithic root, southern Sierra Nevada, Californina. Tectonics 22, 1064.

Scaillet, B., Pêcher, A., Rochette, P., Champenois, M., 1995. The Gangotri granite (Garhwal Himalaya): laccolithic emplacement in an extending collisional belt. J. Geophys. Res. 100, 585–607.

Scaillet, B., Holtz, F., Pichavant, M., 1988. Phase equilibrium constraints on the viscosity of silicic magmas 1. Volcanic-plutonic comparison. J. Geophys. Res. 103 (27), 257–266.

Scaillet, B., Holtz, F., Pichavant, M., 2016. Experimental constraints on the formation of silicic magmas. Elements 12. doi: 10.2113/gselements.12.2.109.

Schmeling, H., Cruden, A.R., Marquart, G., 1988. Finite deformation in and around a fluid sphere moving through a viscous medium: implications for diapiric ascent. Tectonophysics 149, 17–34.

Schwerdtner, W.M., 1990. Structural tests of diapir hypotheses in Archean crust of Ontario. Can. J. Earth Sci. 27, 387–402.

Schwindinger, M., Weinberg, R.F., 2017. A felsic MASH zone of crustal magmas—feedback between granite magma intrusion and in situ crustal anatexis. Lithos 284, 109–121.

Skarmeta, J.J., Castelli, J.C., 1997. Intrusión sintectónica del granito de las Torres del Paine, Andes Patagónicos de Chile. Rev. Geol. Chile 24, 55–74.

Sola, A.M., Hasalová, P., Weinberg, R.F., Suzaño, N.O., Becchio, R.A., Hongn, F.D., Botelho, N., 2017. Low-P melting of metapelitic rocks and the role of H2O: Insights from phase equilibria modelling. J. Metamorph. Geol. 35, 1131–1159.

Sylvester, A.G., 1988. Strike-slip faults. Geol. Soc. Am. Bull. 100, 1666–1703.

Thompson, A.B., 1999. Some time–space relationships for crustal melting and granitic intrusion at various depths. Castro, A., Fernandez, C., Vigneresse, J.L. (Eds.), Understanding Granites: Integrating New and Classical Techniques. Special Publications, 168, Geological Society, London, pp. 7–26.

Tikoff, B., Teyssier, C., 1992. Crustal-scale, en-echelon "P-shear" tensional bridges: A possible solution to the batholithic room problem. Geology 20, 927–930.

Tobisch, O.T., Fiske, R.S., Sorenson, S.S., Saleeby, J.B., Holt, E., 2001. Steep tilting of metavolcanic rocks by multiple mechanisms, central Sierra Nevada, California. Geol. Soc. Am. Bull. 112, 1043–1058.

Valentine, G.A., Krogh, K.E.C., 2006. Emplacement of shallow dikes and sills beneath a small basaltic volcanic center—the role of pre-existing structure (Paiute Ridge, southern Nevada, USA). Earth Planet. Sci. Lett. 246, 217–230.

Veveakis, E., Regenauer-Lieb, K., Weinberg, R.F., 2015. Ductile compaction of partially molten rocks: the effect of non-linear viscous rheology on instability and segregation. Geophys. J. Int. 200 (1), 519–523.

Vigneresse, J.L., 1995. Control of granite emplacement by regional deformation. Tectonophysics 249, 173–186.

Vigneresse, J.-L., Tikoff, B., Ameglio, L., 1999. Modification of the regional stress field by magma intrusion and formation of tabular granitic plutons. Tectonophysics 302, 203–224.

Weertman, J., 1971. Theory of water-filled crevasses in glaciers applied to vertical magma transport beneath ocean ridges. J. Geophys. Res. 76, 1171–1183.

Weinberg, R.F., 1999. Mesoscale pervasive felsic magma migration: alternatives to dyking. Lithos 46 (3), 393–410.

Weinberg, R.F., Hasalová, P., 2015. Water-fluxed melting of the continental crust: a review. Lithos (212-215), 158–188.

Weinberg, R.F., Mark, G., 2008. Magma migration, folding, and disaggregation of migmatites in the Karakoram Shear Zone, Ladakh, NW India. Geol. Soc. Am. Bull. 120 (7-8), 994–1009.

Weinberg, R.F., Podlachikov, Y., 1994. Diapiric ascent of magmas through power law crust and mantle. J. Geophys. Res. 99, 9543–9559.
Weinberg, R.F., Regenauer-Lieb, K., 2010. Ductile fractures and magma migration from source. Geology 38, 363–366.
Weinberg, R.F., Searle, M.P., 1998. The Pangong injection complex, Indian Karakoram: a case of pervasive granite flow through hot viscous crust. J. Geol. Soc. London 155, 883–891.
Weinberg, R.F., Sial, A.N., Pessoa, R.R., 2001. Magma flow within the Tavares pluton, northeastern Brazil: compositional and thermal convection. Geol. Soc. Am. Bull. 113 (4), 508–520.
Weinberg, R.F., Hasalova, P., Ward, L., Fanning, C.M., 2013. Interaction between deformation and magma extraction in migmatites: examples from Kangaroo Island, South Australia. Geol. Soc. Am. Bull. 125 (7-8), 1282–1300.
Weinberg, R.F., Veveakis, E., Regenauer-Lieb, K., 2015. Compaction-driven melt segregation in migmatites. Geology 43 (6), 471–474.
White, A.J.R., Chappell, B.W., 1977. Ultrametamorphism and granitoid genesis. Tectonophysics 43 (1-2), 7–22.
White, S.M., Crisp, J.A., Spera, F.J., 2006. Long-term volumetric eruption rates and magma budgets. Geochem. Geophys. Geosyst. 7, Q03010. doi: 10.1029/2005GC001002.
Whitehead, Jr., J.A., Luther, D.S., 1975. Dynamics of laboratory diapir and plume models. J. Geophys. Res. 80, 705–717.
Whitney, J.A., 1988. The origin of granite: The role and source of water in the evolution of granitic magmas. Geol. Soc. Am. Bull 100, 1886–1897.
Whitney, J.A., Stormer, Jr., J.C., 1986. Model for the intrusion of batholiths associated with the eruption of large-volume ash-flow tuffs. Science 231, 483–485.
Wiebe, R.A., Collins, W.J., 1998. Depositional features and stratigraphic sections in granitic plutons: implications for the emplacement and crystallization of granitic magma. J. Struct. Geol. 20, 1273–1289.
Woidt, W.D., 1978. Finite element calculations applied to salt dome analysis. Tectonophysics 50, 369–386.
Yoshinubo, A.S., Barnes, C.G., 2008. Is stoping a volumetrically significant pluton emplacement process? Discussion. Geol. Soc. Am. Bull. 120, 1080–1081.
Yoshinubo, A.S., Fowler, T.K., Paterson, S.R., Llambias, E., Tickyi, H., Sato, A.M., 2003. A view from the roof: magmatic stoping in the shallow crust, Chita pluton, Argentina. J. Struct. Geol. 25, 1037–1048.
Zak, J., Holub, F.V., Kachlik, V., 2006. Magmatic stoping as an important emplacement mechanism of Variscan plutons: evidence from roof pendants in the Central Bohemian plutonic complex (Bohemian Massif). Int. J. Earth Sci. 95, 771–789.

FURTHER READING

Brown, M., 2012. Open- and closed-system processes in the formation of migmatites and migmatitic granulites. J. Metamorphic Geol. 30 (5), 1–2.
Clemens, J.D., Stevens, G., 2016. Melt segregation and magma interactions during crustal melting: breaking out of the matrix. Earth Sci. Rev. 160, 333–349.
Coleman, D.S., Mills, R.D., Zimmerer, M.J., 2016. The pace of plutonism. Elements 12 (2), 97–102.
de Saint Blanquat, M., Horsman, E., Habert, G., Morgan, S., Vanderhaeghe, O., Law, R., Tikoff, B., 2011. Multiscale magmatic cyclicity, duration of pluton construction, and the paradoxical relationship between tectonism and plutonism in continental arcs. Tectonophysics 500, 20–33.
Sawyer, E.W., 2014. The inception and growth of leucosomes: microstructure at the start of melt segregation in migmatites. J. Metamorphic Geol. 32 (7), 695–712.
Sparks, R.S.J., Cashman, K.V., 2017. Dynamic magma systems: implications for forecasting volcanic activity. Elements 13 (1), 35–40.

CHAPTER 3

Mechanisms of Magma Transport in the Upper Crust—Dyking

Janine L. Kavanagh
University of Liverpool, Liverpool, UK

Contents

3.1 INTRODUCTION

Dykes are magma-filled fractures that discordantly cut across rock layers and bedding surfaces, either creating their own fracture or utilising an existing joint or fault plane. The term 'dyke' is used for both the propagating magma-filled fracture and the solidified intrusion. Dykes are ubiquitous across all tectonic settings and are a regional consequence

Volcanic and Igneous Plumbing Systems. http://dx.doi.org/10.1016/B978-0-12-809749-6.00003-0
Copyright © 2018 Elsevier Inc. All rights reserved.

of plate tectonic stresses (see also Chapter 7). They are key components of volcanic and igneous plumbing systems. Dyke ascent to the surface is facilitated by magma buoyancy, though their interaction with a mechanically heterogeneous crust affects their propagation pathway. Most dykes become arrested at depth during their ascent, either due to magma solidification or the dyke meeting a mechanical barrier such as a rigid or stressed rock layer. Volcanic eruptions are fed by dykes and are a spectacular opportunity to witness what happens when dykes directly breach the surface to form vents.

There are two fundamentally different models of dyke ascent that are dominant in the literature, and these have been derived from field and geophysical observations. Dykes are modelled either as hydraulic fractures, where the dyke is idealised as a fluid-filled fracture that propagates in an infinite elastic material, or as viscous indenters, where the dyke emplacement is accommodated by the intruding magma sheet plastically deforming and shearing the host material. Numerical simulations of dykes are almost exclusively two-dimensional, and laboratory experiments using analogue materials are either two- or three-dimensional.

In this chapter, the geometry of dykes will be described according to their 'length', 'thickness' and 'breadth', where length is the longest dimension of the dyke (typically vertical, the z-plane), thickness is its shortest dimension (the x-plane) and breadth (the y-plane) is its intermediate dimension (typically its lateral extent). The aim of this chapter is to explore the origin of dyke emplacement models and discuss their validity based on the field and geophysical evidence. The methodologies that are used to study dykes are first briefly explained, followed by a description of the key features of dykes, including their geometry, orientation, host deformation (including tip processes), magma flow, and transition into vents that feed volcanic eruptions. In each case, evidence from the field and geophysical surveys are discussed in relation to the insights that have come from laboratory experiments and numerical models. The chapter is aimed at advanced university students, graduates and those with a keen interest in dyke propagation, providing an overview of the latest developments and state of knowledge of the physical processes that control dyke emplacement in the upper crust.

3.2 METHODS TO STUDY DYKE PROPAGATION

3.2.1 Field Geology, Mapping, and Rock Sampling

Field studies of dykes remain a major contributor to our understanding of dyke emplacement mechanisms, informing the development of dyke emplacement models and interpretation of geophysical and geodetic datasets of magma ascent in the crust. Methods that are used in the field include traditional geological mapping, field sampling, and modern imaging techniques. Dykes in the shallow, upper crust are exposed in the field by post-emplacement regional- or local- scale erosion, collapse of a volcanic edifice or mining activities. This provides a direct view and access to the solidified remnants of the magma and the intruded host rock. The depth of dyke exposure can be inferred by relating local stratigraphic constraints with dating of the intrusive event; however, the accuracy of this varies and often it is only possible to approximate emplacement depths to within 1–2 km. A significant challenge to geologists when studying dykes in the field is deciphering pre-, syn- and post-intrusion processes, as outcrops display a restricted view of the dykes and represent a final snapshot of the effects of all processes that occurred during its geological history such that only a partial record of intrusion events are preserved.

Many geological field studies of dykes have focused on their geometry by measuring their length, breadth, thickness, orientation, and relationship to local and regional structures, such as joints and faults. Modern mapping techniques applied to dykes include remote imaging, such as drone photography and airborne or terrestrial LiDAR (light detection and ranging). These methods can be rapidly deployed, provide access to logistically challenging field sites and can map large areas relatively quickly. The 3D high-resolution point cloud data from these techniques enable measurements up to centimetre-scale accuracy to be made. However, despite their excellent potential, airborne surveys need to be supported by ground-based field studies to verify measurement errors and gain additional information that is too small or complex to be assessed remotely. An example would be assessing the compositional variation of the rocks and confirming dykes as 'simple' (one injection) or more complex (e.g., multiple or complex; where 'multiple' refers to a dyke that is composed of successive similar injections, and 'composite' when the dyke is filled with successive injections that differ in chemical composition and/or mineralogy).

Field datasets on the dimensions of dykes need to be precise and large enough to be able to analyse them statistically, and several hundred to thousands of measurements are necessary to achieve high-quality statistical analyses. Another technique used to process field data of dykes is 3D structural modelling, which allows for the reconstruction of the three-dimensional geometry of dykes and their arrangement in the subsurface prior to erosion.

3.2.2 Petrographic, Magnetic, and Geochemical Analyses

A broad range of analytical techniques can be applied to rock samples collected in field studies of dykes. These can reveal insights into the nature of the source region of the magmas, how the physical and chemical properties of the magma have evolved during ascent, the architecture of deeper magma reservoirs from which the dykes have propagated, and how the host rock has responded to the intrusion. Petrographic techniques applied to thin sections of the dyke's host rock, solidified magma, and their contacts, include traditional light microscopy, high-magnification scanning electron microscopy (SEM), and cathodoluminescence (CL). Electron microprobe analysis (EMPA), X-ray fluorescence (XRF), X-ray diffraction (XRD) and inductively coupled plasma mass spectrometry (ICP-MS) provide mineral and bulk-mixture major, trace and isotope compositions; they can be used to support micro-analysis mineral mapping using QEMSCAN (Quantitative Evaluation of Minerals by SCANning electron microscopy) or EBSD (electron backscatter diffraction analysis) to track mineral orientation and crystal lattice deformation (see Chapter 8 for more details). When oriented field samples have been collected, their magnetic properties can be analysed using techniques such as AMS (anisotropy of magnetic susceptibility) or AARM (anisotropy of anhysteretic remanent magnetisation), with the origin of the magnetic carriers identified using the method described by Lowrie (1990) and high-temperature susceptibility experiments.

3.2.3 Geophysical and Geodetic Methods

Geophysical techniques are predominantly used in dyke studies to document dynamic processes that occur during the dyke intrusion event (e.g. Sigmundsson et al., 2014; see also Chapter 11). This is partly because the final, static, vertical to sub-vertical, thin, sheet-like geometry of dykes means that they are difficult to detect with seismic reflection surveys. However, seismic energy is released as the dyke grows and this can be detected by seismometers to determine the depth and lateral position over time. The origin of the seismicity is not always straightforward, but is most often attributed to rock fracturing at the dyke tip. The need for a maintained ground-based network of seismometers, in an area that is likely to be volcanically active and hazardous, means there are relatively few studies where dyking has been recorded in quasi 'real-time'.

As a dyke ascends, it can produce gradual distortion of the Earth's surface and topographic change, as rock is displaced to accommodate the incoming magma. This deformation can be detected geodetically using GPS (global positioning system) and InSAR (interferometric synthetic-aperture radar) (see Chapter 11). Based on the recorded ground deformation pattern, the origin of the deformation is inferred using inverse modelling. In dyke intrusion case studies, the depth and geometry of the dyke are interpreted using model assumptions and a non-unique solution that best represents the observed surface displacement is converged upon (see Chapter 11 for a more detailed description).

3.2.4 Laboratory and Numerical Modelling

Numerical modelling and laboratory experiments are important methodologies in the study of dyke emplacement. Founded on conceptual models, their overarching aim is to reconcile broad patterns that have emerged from numerous field and geophysical observations of dykes rather than an individual case study (although these are often a key inspiration behind the work). Due to the complexities of magma and rock interactions that are evident in the often-limited datasets that measure dyke emplacement in nature, laboratory and numerical dyke ascent models typically explore a subset of observations. Scaling laws and, in the case of laboratory experiments, the use of carefully chosen analogue materials mean it is possible to study the geometrics, kinematics and dynamics of dyke emplacement in nature in a controlled and repeatable manner (Merle, 2015; Galland et al., 2015).

Analogue experiments have been used since the 1970's to study the geometry and propagation dynamics of dykes as magma-filled fractures. Transparent and photo-elastic gelatine solids injected by fluid (such as air, water or a solidifying fluid) have been used extensively to study hydraulic fractures in an elastic medium. Gelatine has controllable viscoelastic properties and is brittle at high strain rates (Kavanagh et al., 2013), with a tensile strength of ~100 Pa based on a bending test (Takada, 1990). Compacted silica flour intruded by a solidifying fluid (vegetable oil) has been used to study dykes as viscous indenters. Silica flour is a cohesive, granular material that can behave in a brittle manner and has the ability to form fractures in shear and in tension. As the silica flour is opaque, the solidified dyke needs to be excavated to observe its final morphology; or thin 2D experimental tanks are used to show a cross-section through the growing intrusion.

Two theoretical frameworks or 'schools of thought' tend to be used to model dyke propagation numerically, based on assumptions about the relevant physics to apply. Firstly, Weertman theory is used to model the dyke as a buoyant magma-filled fracture, and secondly lubrication theory models dyke propagation controlled by magma flow fed by a distance source pressure. However, some numerical models combine both theories. A commonly used numerical approach to represent a pressurised, propagating crack is the boundary element method (BEM), which considers the coupling between pressurised magma and deformation of the surrounding rock deformation.

3.3 DYKE GEOMETRY

Our understanding of the geometry of dykes is largely informed by field studies of ancient intrusions. One of the considerations when modelling dykes is that their geometry is scaled with nature, and in the case of laboratory experiments this influences the combination of analogue magma and crust material that is chosen. The experimental dyke geometries that are created in the laboratory, or in a numerical simulation, need to be compared back to their natural equivalent to assess the success or validity of the model.

3.3.1 Field and Geophysical Data

3.3.1.1 Overall Geometry

Exposures of dykes in the field most frequently comprise two-dimensional planes that cut through the solidified magma and its host rock. Dykes often crop out at the surface as a series of related but offset *en echelon* dyke segments (e.g. Fig. 3.1) that are inferred to have been simultaneously active and are connected at depth. Overall, field observations show that dykes are planar, sheet-like bodies that are very thin compared to their length and breadth. This is supported by geophysical studies that can image to great depths and yet struggle to detect such geometries. The thickness-to-breadth ratio of dykes ranges from 0.01 to 0.0001, meaning a 1-m thick basaltic dyke would typically have a breadth of 100–1000 m and have ascended a few kilometres.

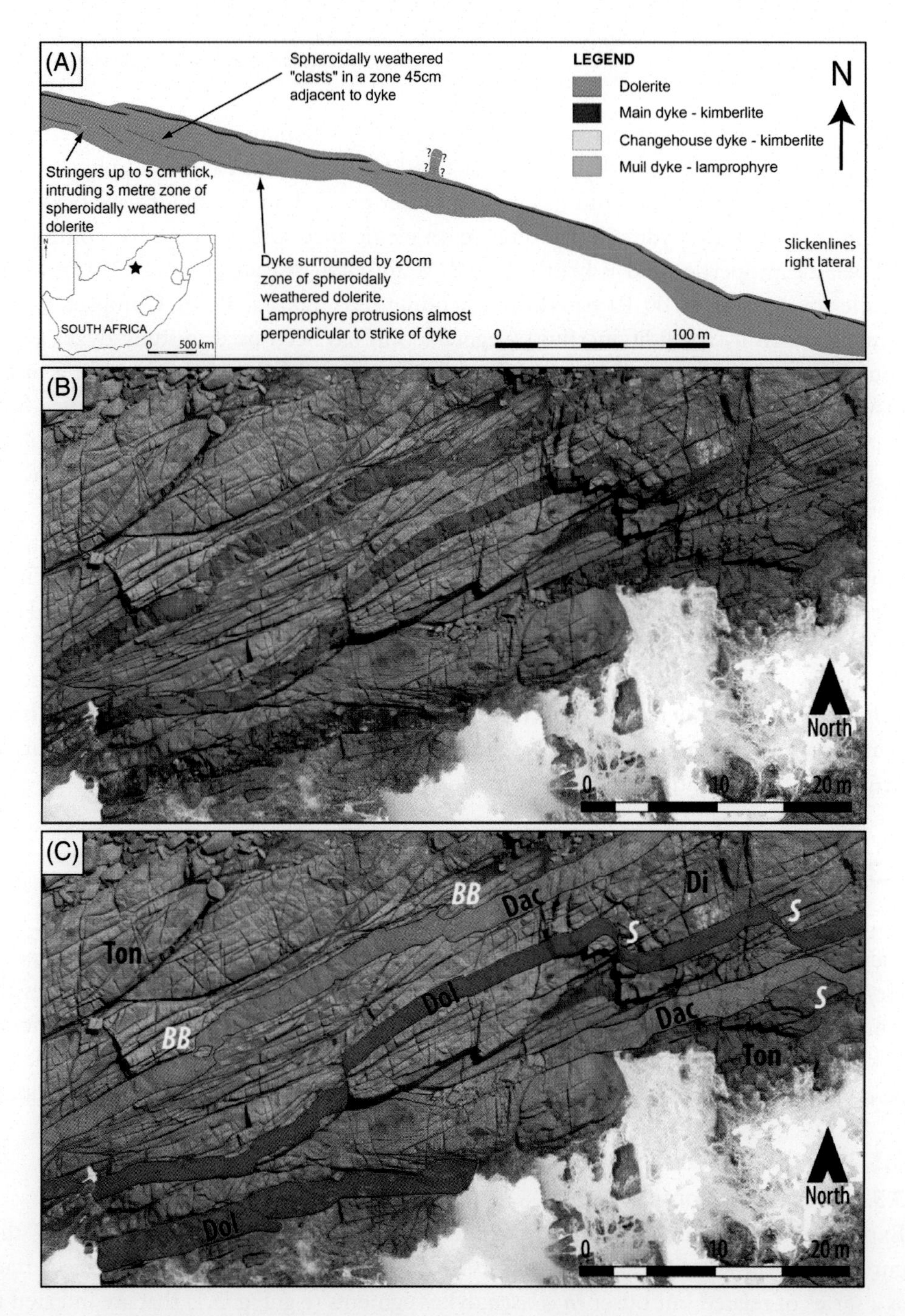

Figure 3.1 ***Well-exposed* en echelon *dyke segments in plan view.*** (A) Geological map of the Swartruggens Kimberlite Dyke Swarm, South Africa, showing kimberlite (*red*) and lamprophyre (*green*) dykes that have intruded a dolerite (purple) host rock (modified after Kavanagh, 2010). The dykes crop out in Edward Shaft, Helam Mine, in South Africa (the star on the inset map indicates the mine location) and they were mapped at ~750 m depth (24th mine level). (B) UAV (unmanned aerial vehicle, also known as 'drone') ortho-images of a dyke-cut beach platform on Bingie Bingie Point, Australia (Cruden et al., 2016). (C) Annotated version of (B), highlighting dykes and structural features: Abbreviations: *Ton*, tonalite; *Di*, diorite; *Dac*, dacite dyke (orange); *Dol*, dolerite dyke (magenta); *BB*, broken bridge structure; *S*, step structure.

Petrographic and geophysical data suggest that magma can ascend rapidly in dykes from their source depth, which could extend into the upper mantle. This suggests that individual dykes, or most likely inter-connected dyke networks, can be tens of kilometres in vertical extent. Kimberlite dykes are particularly deeply sourced as they intrude cratons where the crust is thick, ascending from up to 200 km depth to the surface. However, in outcrop, individual dykes can often only be traced tens of metres vertically, with neither the top nor the bottom contacts seen. In comparison, vertical (x–z-plane) or horizontal (x–y-plane) views spanning the thickness of a dyke are regularly exposed, for example, in a cliff face (Fig. 3.2A and B), or an exposed beach platform (Fig. 3.1B and C). Solidified magma is a relatively hard material, so dykes are often more resistant than their host material and so can produce a 'wall' of rock in the landscape that proffers a glimpse at their three-dimensional morphology. A spectacular example of this is Ship Rock in New Mexico (see Fig. 3.2C), where dykes of minette composition crop out

Figure 3.2 ***Examples of two-dimensional exposures of dykes in the field, providing a cross-section view of the intrusion and showing the contact between solidified magma and host rock.*** (A) Photograph of an exposed cliff face in the Reykjanes Peninsula, Iceland, showing an arrested basaltic dyke (Gudmundsson and Loetveit, 2005). (B) Photograph of a branched kimberlite dyke exposed in a vertical cliff face of Venetia Mine, South Africa. (C) Aerial photograph of the Ship Rock volcanic plug and dyke system (Townsend et al., 2015). (D) Small dyke cross-cutting composite dyke from Adamello, Italy.

through Cretaceous shales and siltstones (e.g. Delaney and Pollard, 1981). In particularly well-exposed areas that have limited cover due to vegetation or scree, individual dyke segments can be traced laterally from tip to tip. Eroded dykes have been recognised in the landscape of other terrestrial planetary bodies, for example, the spectacular dyke swarms of Mars and Venus (Ernst et al., 2001).

Opportunities to appreciate the overall three-dimensional geometry of dykes are scarce. Kimberlite dykes however offer a rare opportunity to study the complex sub-surface three-dimensional geometry of dykes, as being a host to diamonds means the excavation of the solidified magma closely maps the dyke pathway through the rock (see Fig. 3.1A for a geological map of a kimberlite exposed at depth by mining). Exceptional databases of the three-dimensional geometry of these intrusions have provided detailed maps of dyke thickness over several hundred metres of vertical and lateral extent (Kavanagh and Sparks, 2011), detailing how dyke geometry changes through contrasting rock strata (Fig. 3.3).

Thickness measurements of ancient, exposed dykes need to be made with care as complications arise by dyke injection in a region being short-lived or long-lived, producing 'simple' or more complex intrusions (e.g. Fig. 3.2D), and dykes being 'feeder' or 'non-feeder' types. However, dyke thickness is by far the easiest dimension of a dyke to

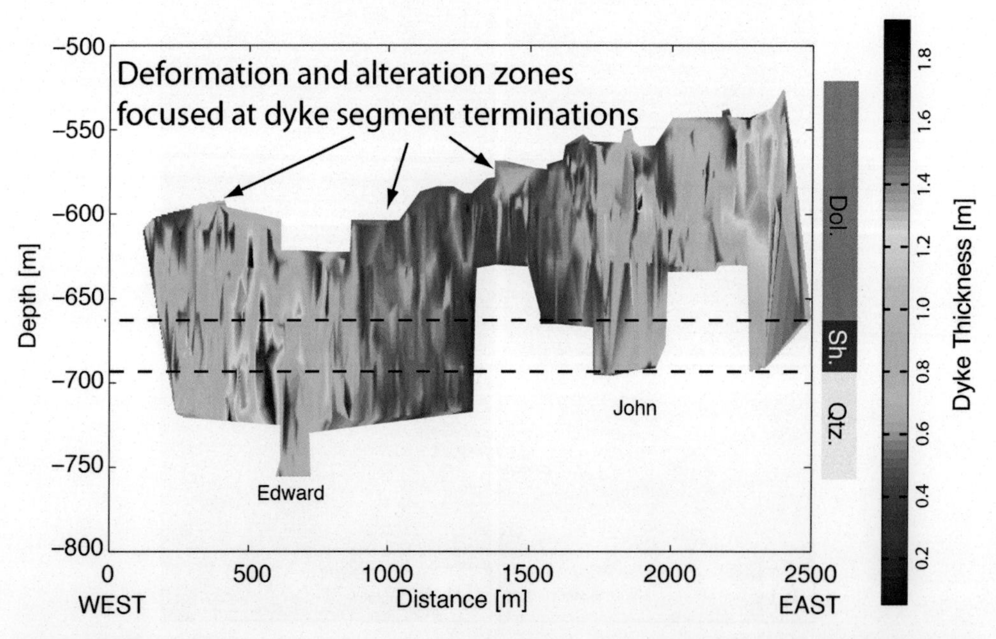

Figure 3.3 ***Three-dimensional geometry of two kimberlite dyke segments (Edward and John) from the Swartruggens kimberlite dyke swarm, South Africa (modified after Kavanagh and Sparks, 2011).*** Dyke thickness was mapped across stratigraphic layers of dolerite, shale and quartz. Measurements are plotted according to their geographic position and depth, and colour represents dyke thickness (dyke thicknesses between measurement points was interpolated). The most intense host-rock deformation was observed in outcrop at dyke segment terminations.

measure, which is why large data sets on dyke thickness exist. The origin of dyke thickness is complex and has been related to: tectonic setting, mechanical properties of the host rock, magma viscosity, and magma overpressure. A survey of approximately 700 individual basaltic dykes from Iceland suggested that their average thickness is 4.1 m (e.g. Gudmundsson, 1983). In comparison, >1500 measurements of two kimberlite dyke segments from South Africa suggest that they have a lower average thickness of 0.64 m (Kavanagh and Sparks, 2011); the difference is thought to reflect the kimberlite magma's relatively low viscosity compared to basalt. A survey and statistical analysis of the thickness of 3676 dykes in different tectonic settings suggested that dyke thickness is primarily controlled by the strength of the host rock, rather than magma viscosity and composition (Krumbholz et al., 2014).

Dyke intrusion events in Kilauea Volcano, Hawaii, have shown that dyke thickness can decrease to approximately half once it has erupted (Wright and Fiske, 1971). This has been explained by the reduction in confining pressure as the magma reaches the surface and suggests that the host rock behaves elastically during dyke emplacement, despite many small-scale heterogeneities. The thickness of an arrested dyke that stalled within the crust may therefore be different to a feeder dyke that breached the surface.

3.3.1.2 Dyke Tip Geometry

As dykes are three-dimensional bodies, the dyke tip includes the entire periphery of the intrusion and includes material at the contact between the host rock and magma. Exposure of dyke tips most frequently occurs at individual dyke segments; but as each dyke segment can be several hundred metres long, exposure of the relatively small dyke tip region is often scarce. This lack of exposure of dyke tips is exacerbated by the fragility of the host-rock tip material that is often highly fractured and weathered, and the tendency of *en echelon* segments to merge together via a 'broken bridge' (handshake) or step structure (Fig. 3.1B and C) and so destroy tips during propagation.

Rare examples of preserved dyke tips are showcased in Fig. 3.4, displaying different features. Fig. 3.4A shows a basaltic dyke from Rum, Scotland, where a cavity between the intruding magma and the crack tip in the host rock has become infilled during post-emplacement processes (Daniels et al., 2012), but it is thought that during dyke propagation it may have been filled with volatiles that exsolved from the magma. Separation of this gas-filled pocket from the intruding magma could be responsible for small phreatic eruptions at the surface that are precursory to explosive events (e.g. the start of the eruption of Mt Unzen, Japan, in 1990; Nakada et al., 1999). It may also be responsible for preconditioning and alteration of the host rock prior to magma injection and play a role in the localisation of volcanic vents (e.g. Brown et al., 2007). A spectacular example of a dyke that came to the surface during an eruption and then stalled is the photograph of a dyke associated with the eruption of Nyiragongo volcano (Democratic Republic of Congo) in 2002 (Komorowski et al., 2003; Fig. 3.4B). The photograph shows a distinctive

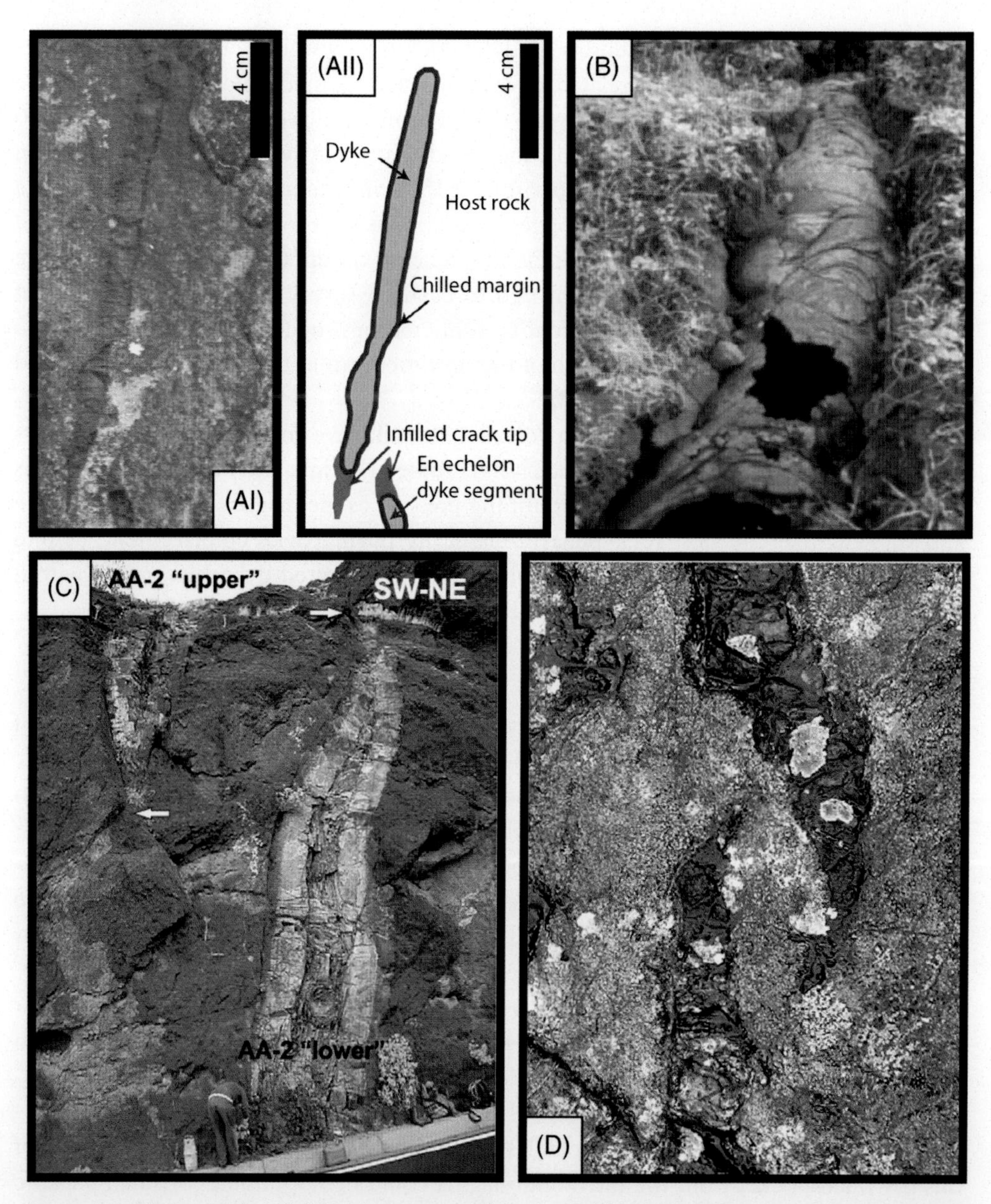

Figure 3.4 ***Photographs of dyke tips preserved in the field.*** (A) The plan view of small basaltic dyke segments on the Isle of Rum, Scotland: (I) photograph and (II) annotated version showing two overlapping dyke segments, the tips of which are filled with sediment (modified after Daniels et al., 2012). (B) The tip of a dyke that stalled at the surface during the 2002 eruption of Nyiragongo, DRC (Komorowski et al., 2003). (C) Photograph of vertical road cutting showing two overlapping mafic dyke segments from the Anaga Massif, Tenerife (Clemente et al., 2007). (D) Photograph of two partially overlapping *en echelon* dyke segments in the Isle of Skye, UK, whose tips bend towards each other (plan view, Tibaldi, 2015). (E) Photograph of splayed basaltic dyke tip from the Rum dyke swarm exposed in Harris Bay, Rum (plan view). The dyke tip transitions into several dykelets that are oriented parallel to but also branch out to cross pre-existing joints in the microgranite host rock (photo credit Steffi Burchardt).

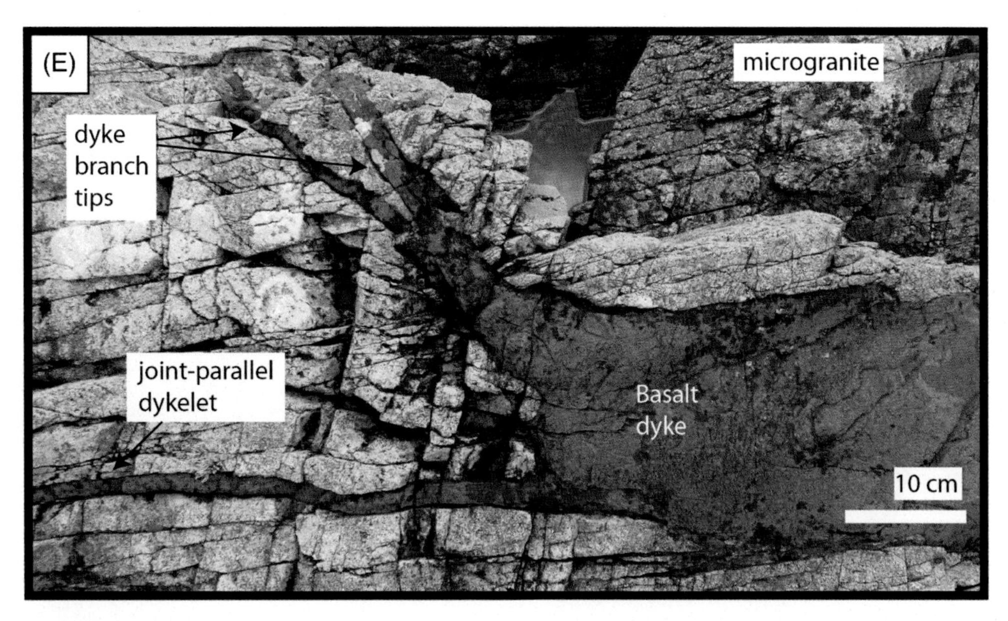

Figure 3.4 (*cont.*)

elongate, rounded pahoehoe surface, where an 'envelope' of quenched magma within the dyke tip has been left behind as a relic as the dyke drained. Fig. 3.4C shows a spectacular example of two *en echelon* dyke segments exposed in a vertical section in Tenerife, and Fig. 3.4D shows two small dyke tips exposed on a beach platform in Scotland having rotated towards one another. The exceptionally well-exposed basaltic dyke tip photographed in Fig. 3.4E from a beach platform in Scotland shows how the tip transitions into several thin dykelets, some of which are aligned with pre-existing fractures in the host rock and others that are at an angle to these and branch into their surroundings.

3.3.2 Dyke Geometry in Numerical Simulations

There are two dominant views that inform mechanical models of dyke emplacement (Rivalta et al., 2015): either that the dyke propagates under the influence of buoyancy as a fixed quantity of fluid ascends disconnected from its source, modelled by Weertman theory (Weertman, 1971); or that the dyke is connected to its source that supplies magma at a specific flux or pressure, modelled by lubrication theory. However, many dykes propagate further laterally than vertically forming horizontal blade-like intrusions, and it can be argued that these should be regarded as an additional type (e.g. Townsend et al., 2017).

The numerical models of dykes consider intrusion into a linear elastic medium, but the dyke geometry that is produced varies depending on the school of thought that has been explored. Fig. 3.5A depicts a fixed volume of magma ascending due to buoyancy that produces a vertical tear-drop geometry in cross-section with a large head region, pinched tail, and circulating fluid within (e.g. Dahm, 2000; Maccaferri et al., 2011). Fig. 3.5B illustrates lubrication theory applied to study dykes, producing upward magma

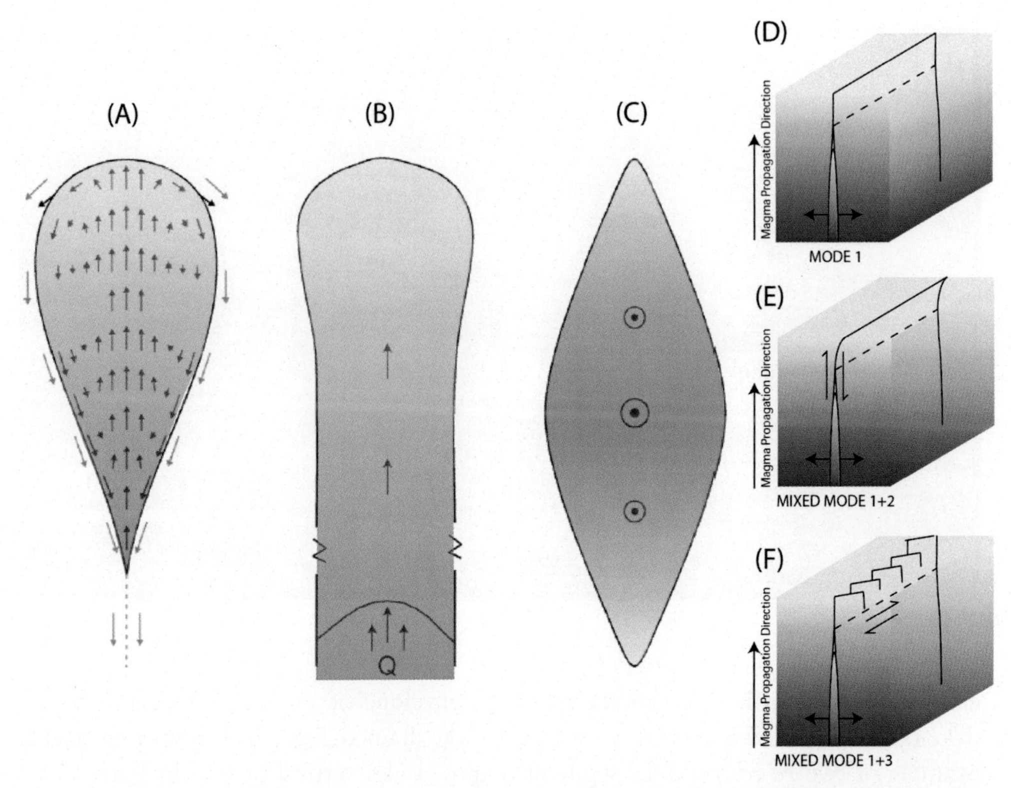

Figure 3.5 ***Schematic illustration of conceptual models of magma flow and crack opening during dyke emplacement.*** (A) A fixed quantity of magma circulates in a vertical fracture, opening at the crack tip and pinching at the tail (modified after Dahm, 2000). (B) The dyke opens at the upper tip and flow is constantly supplied from a distant source at depth (modified after Roper and Lister, 2007). (C) Lateral magma flow in a horizontally propagating vertical fracture that maintains its height and position (modified after Townsend et al., 2017). (D–F) Crack opening and dyke propagation pathways depending on the loading mode (modified after Rubin, 1995): Mode I, tensile loading; Mixed Mode I and II, tensile opening with perpendicular shear; and Mixed Mode I and III, tensile opening with shear parallel to crack front.

flow in a head-and-tail region that are connected by a neck (Roper and Lister, 2007). Fig. 3.5C describes horizontal fluid flow in a vertical fracture modelled as a blade-like dyke, producing stable upper and lower tips whose geometries change depending on a range of factors, such as the mechanical properties of the host rock and density contrasts (Townsend et al., 2017). In all cases, the mode of crack propagation, which relates to the loading conditions (either Mode I tensile, or Mode II or III shear), will be an important additional consideration (Fig. 3.5D–F). Mixed-mode loading, for example, Fig. 3.5F, shows a dyke propagating under Mode I (tensile) and Mode III loading (shear parallel to the crack front), and this could explain the progression of the dyke tip into *en echelon* segments as the intrusion responds to a change in stress field (Rubin, 1995).

Large voids cannot be sustained at depths greater than approximately 1 km, due to the low in situ compressive strength of the rock (e.g. Rubin, 1995). This implies that dyke-shaped cavities do not exist within the crust to be later filled by magma, and so for dykes to form they must prop open existing fractures in the crust or intrude fractures they have created. Strains are high at the dyke tip, but decrease and are small away from the dyke tip (ranging from 10^{-2} to 10^{-4}, and therefore comparable in size to the dyke thickness-to-length ratio). Collectively, this suggests that the deformation of the host rock at the dyke margins is elastic. Models that follow this framework treat dykes as cracks in a linear elastic body, applying linear elastic fracture mechanics (LEFM; see also Chapter 5). Rubin (1995) considers dyke geometry as a static, two-dimensional fluid-filled crack in an elastic medium. The fluid within the crack has an internal pressure $P(z)$, and the dyke is subjected to an ambient compressive stress field $\sigma_{xx}^{0}(z)$ that is perpendicular to the dyke plane. When the loading stresses are uniformly applied, the dyke geometry is elliptical and can be described by the relative magnitudes of the magma overpressure ΔP (sometimes called excess pressure or elastic pressure) and the elastic stiffness M (a mechanical property of the host material and measure of its resistance to deformation):

$$\frac{w(0)}{l} = \frac{\Delta P}{M} \tag{3.1}$$

where

$$\Delta P(z) \equiv P(z) - \sigma_{xx}^{0}(z) \tag{3.2}$$

and

$$M = \frac{G}{(1-\nu)} \tag{3.3}$$

where w is the half-thickness, 0 is the centre point of the intrusion, l is the half-length, G is the elastic shear modulus and ν is Poisson's ratio. For a growing crack, this internal pressure is not uniform and so more complex expressions are required to describe the spatially varying geometry (e.g. Lister and Kerr, 1991). In rock fracturing, crack propagation will occur depending on the fracture criterion. For geological materials, it is commonly stated that crack propagation will occur when $K = K_c$, where

$$K = \Delta P\sqrt{l} \tag{3.4}$$

where K is the crack-tip stress intensity factor due to stress concentration at the crack singularity and K_c is the fracture toughness, which is a material property that describes the ability to resist fracture. This fracture criterion is considered valid if the region of inelastic deformation is small. This relationship between K and ΔP (Eq. (3.4)) means that a longer crack will require less fluid overpressure ΔP for K to exceed K_c, and so longer dykes are 'easier' to grow.

3.3.3 Dyke Geometry in Laboratory Models

The first analogue experiments to study dyke propagation used gelatine solids injected by dyed water to create hydraulic fractures. Fiske and Jackson (1972) created free-standing gelatine blocks moulded into the shape of elongated volcanic rift zones, and injected these with water to study the controls on dyke emplacement in Hawaii. The similarity of their laterally propagating, blade-like experimental dykes compared to the observations of dykes in the field was compelling, and the ability of the laboratory experiments to explain the overall growth of volcanic rifts (see Chapter 4) using physics affirmed the potential of laboratory experiments to model dykes. Gelatine's evident versatility as a transparent elastic medium also paved the way for its use as a crustal analogue material to model dyke propagation in the laboratory ever since.

3.3.3.1 Newtonian Fluids

Takada (1990) used gelatine slabs supported in an experimental tank to study dyke propagation by injecting it with fluids of various density, viscosity and volume. His experiments showed that a buoyant fluid-filled fracture (dyke) propagating in a homogeneous, isotropic, elastic solid with an initially hydrostatic stress field is broadly controlled by the density difference between the injected fluid and the host solid (Fig. 3.6A). The dyke shape changed as it grew due to increased buoyancy; transitioning from penny-shaped in front view (y–z-plane) and elliptical in side view (x–z-plane), to a flat teardrop shape in front view (y–z-plane) and narrow teardrop shape in side view (x–z-plane). This type of behaviour can be attributed to a Peach–Koehler force that acts on dislocations, which Weertman (1971) noted as a 'pseudo-Archimedean force' acting on a fluid-filled crack. The teardrop shape has a distinctive head region where buoyancy forces are important (Fig. 3.6B and C), and its length L_h or h_b depends on the density difference between the injected fluid and host (Taisne and Tait, 2011). Experiments where air and water were injected simultaneously demonstrated how the air collected in the dyke tip; its buoyancy exerted control on the dyke ascent dynamics and geometry, causing an elongated, buoyant head to form (Menand and Tait, 2001).

Mathieu et al. (2008) carried out experiments using sand or fine-grained ignimbrite powder (the crust analogue) and injected honey and golden syrup (the magma analogue). By injecting along the wall of their square-based experimental tank (in the so-called 'half-box' experiments), they directly observed the growth of their intrusions as they propagated through the host material, which was otherwise obscured when injecting into the centre of the experimental tank (in the so-called 'full-box' experiments). Their experimental dykes grew with an irregular but sharp margin, with occasional

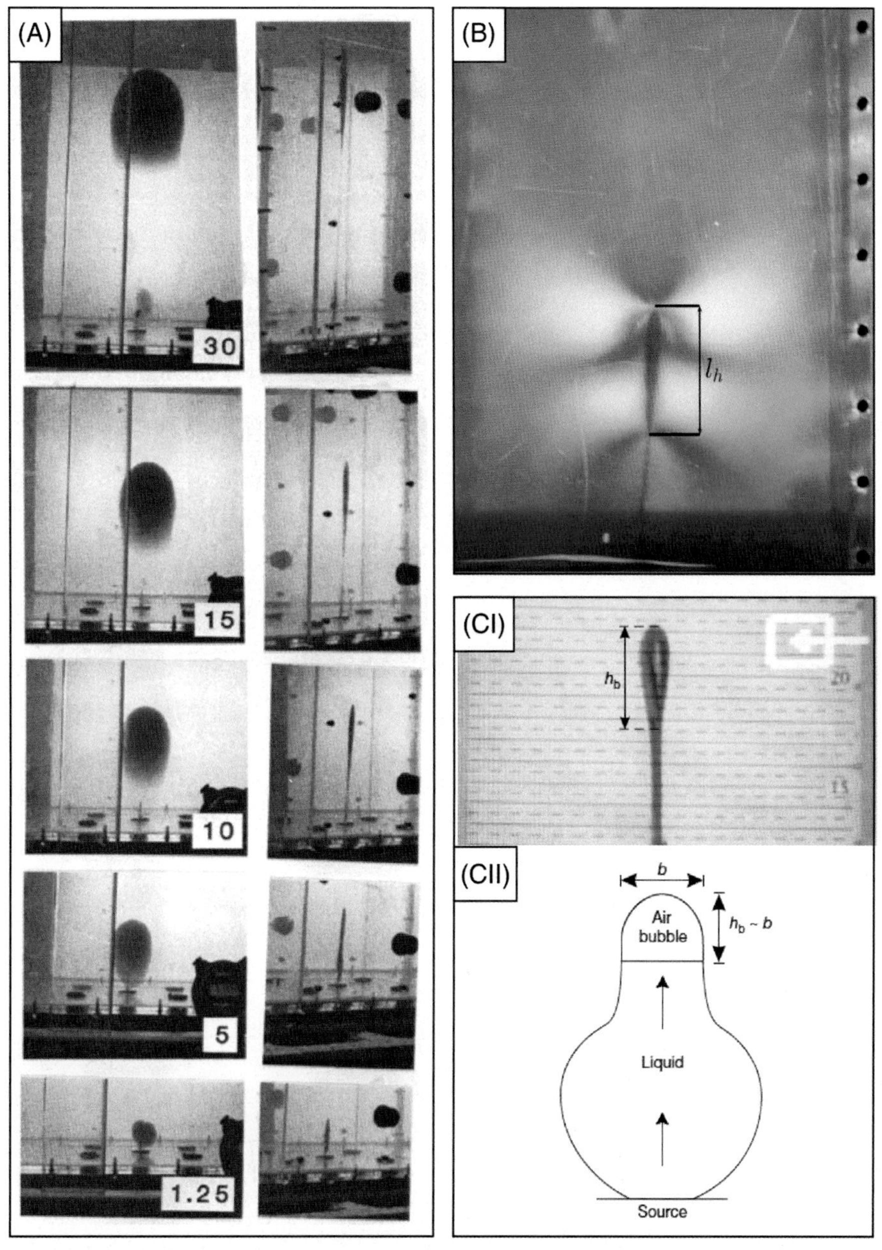

Figure 3.6 ***Laboratory experiments of dyke ascent modelled as a hydrofracture using gelatine as a crust analogue.*** (A) Series of photographs showing injection of buoyant silicon oil into a gelatine solid, viewed in front (*y–z*-plane) and side view (*x–z*-plane). The numbers on the photos are the injected volumes in ml (Takada, 1990). (B) Photograph using polarised light showing injection of fixed volume of buoyant heptane into a gelatine solid viewed from the side (*x–z*-plane). The length of the buoyant dyke head (l_h) is indicated, and interference colours focused at the dyke tip relate to stress concentration in the gelatine (Taisne and Tait, 2011). (C) Photograph (I) and annotated sketch (II) of an air and dilute hydroxyethylcellulose liquid-filled crack propagating in a gelatine solid viewed from the side and front, respectively. Air collects at the dyke tip and, once its buoyancy pressure overcomes the fracture resistance of the gelatine, it controls the crack tip growth (Menand and Tait, 2001).

branching and *en echelon* segmentation during ascent. Abdelmalak et al. (2012) carried out similar experiments but used a form of Hele-Shaw cell (two-dimensional experimental tank) and compacted silica flour as their crust analogue (Fig. 3.7A). The golden syrup injections were described as 'viscous indenters' with a rounded tip when deforming their host by shear failure. Thinner dyke tips and higher cohesion materials produced dykes that caused the host material to fail in tension.

3.3.3.2 Solidification Effects

Experiments where a solidifying fluid is intruded into the crust analogue produce different morphologies to water or air injection, but share similarities independent of the host material. Hot, liquid vegetable oil (often Vegetaline) injected into gelatine has produced morphologies that are thicker than their water counterparts (e.g. Taisne and Tait, 2011;

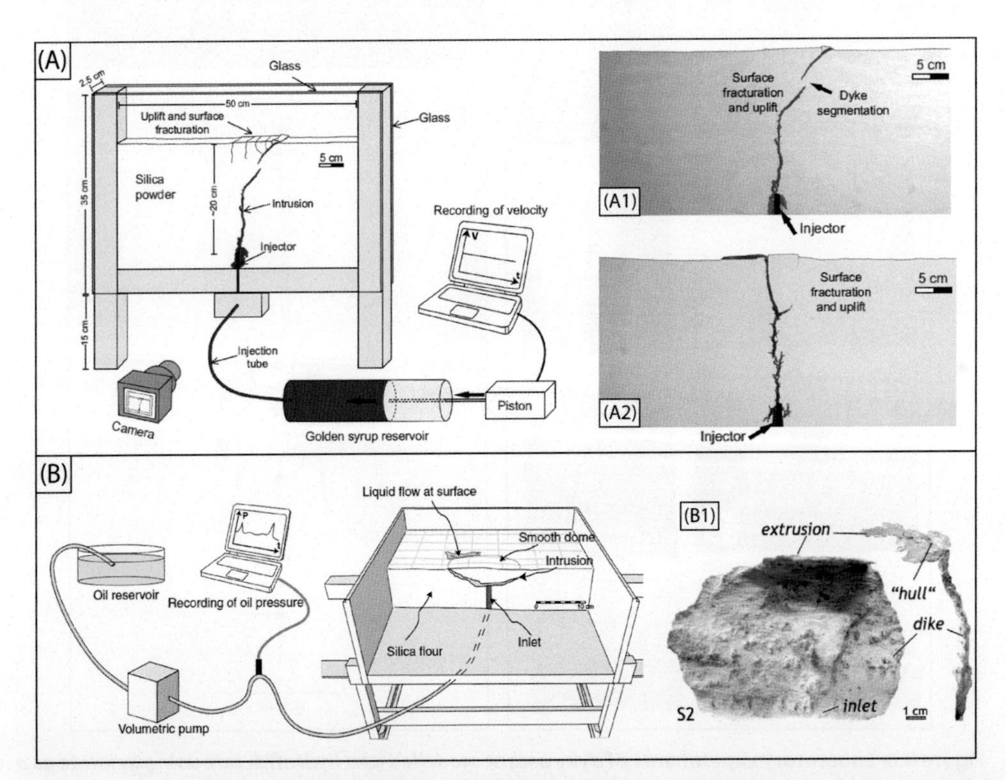

Figure 3.7 ***Dyke emplacement modelled as a viscous indenter in laboratory experiments using compacted silica flour as a crust analogue material.*** (A) Sketch of 2D experiment apparatus where a thin, vertical experimental tank is intruded by golden syrup (Abdelmalak et al., 2012). The growth of the dyke is observed, and photographs show the final stages of emplacement as the dyke fractures and uplifts its host, either propagating to the surface directly (A1) or intruding a surface fracture (A2). (B) Sketch of 3D experimental apparatus where oil is injected into a square-based tank filled with compacted silica flour (Galland et al., 2014). Surface deformation occurs during injection, and post-emplacement solidification enables the dyke to be excavated and its final morphology observed (B1).

Daniels and Menand, 2015). This can be explained by potential melting of the host gelatine during intrusion, or by heating and softening of the surroundings as the intrusions form. These experimental dykes grow by progressive breakout of fluid from the chilled propagation front and often form an irregular *en echelon* tip that is thought to reflect shear stress. Laboratory models studying dykes as viscous indenters have intruded Vegetaline into compacted, crystalline and cohesive silica flour (e.g. Galland et al., 2014). Excavation of the intrusion after the oil has solidified reveals an experimental dyke morphology that is broadly elliptical, with an irregular surface morphology and *en echelon* tips (Fig. 3.7B).

3.4 HOST-ROCK DEFORMATION

There is strong geological and geophysical evidence that suggests dykes damage their host rock as they intrude (e.g. see Chapter 11). This deformation can be manifested as seismicity due to rock fracturing, perturbation of the overlying topography (detected geodetically), the occurrence of joints at the margins of eroded dykes, and contamination of the magma during ascent by the inclusion and assimilation of xenoliths and xenocrysts from the host rock.

3.4.1 Field Exposures and Geophysical Evidence of Rock Deformation Around Dykes

Field exposures of dykes have shown that they can be associated with a range of deformation structures in the host rock. Dyke deformation appears as dyke-parallel joints and/or host-rock brecciation, and it can be localised, regionally present, or absent. Proximity of these structures to the dyke is used to help assess whether this host-rock damage is caused by or pre-exists the dyke emplacement. In the Colorado Plateau, Delaney et al. (1986) described closely-space, dyke-parallel jointing in the sedimentary host rock that was only associated with the dykes they observed. The spacing of the joints was observed to decrease away from the dyke margin. Kavanagh and Sparks (2011) observed localised deformation of the host rock only in the regions between individual dyke segments of kimberlite dykes in South Africa (Kavanagh, 2010; Fig. 3.8A and B). They noted breccia zones (Fig. 3.8C) and joint-bound localised onion-skin style weathering at depth (Fig. 3.8D), which were associated with magmatic gases propagating through the rock ahead of dykes that stalled at depth. Similarly, to Delaney et al. (1986), Kavanagh and Sparks (2011) recorded dyke-parallel joints in the host rock, but they also recorded asymptotic joints that abutted the dyke margin (see Fig. 3.8E). Erosion of the host rock during dyke emplacement was evident by the inclusion of xenoliths within the magma (Fig. 3.8F) originating from a range of stratigraphic depths, being locally sourced or transported from the mantle. Evidence of the assimilation of the xenocrystic and xenolithic cargo into the kimberlite melt is evident geochemically, but also with petrographic textures that show resorption of individual crystals, etched surfaces and rounded margins,

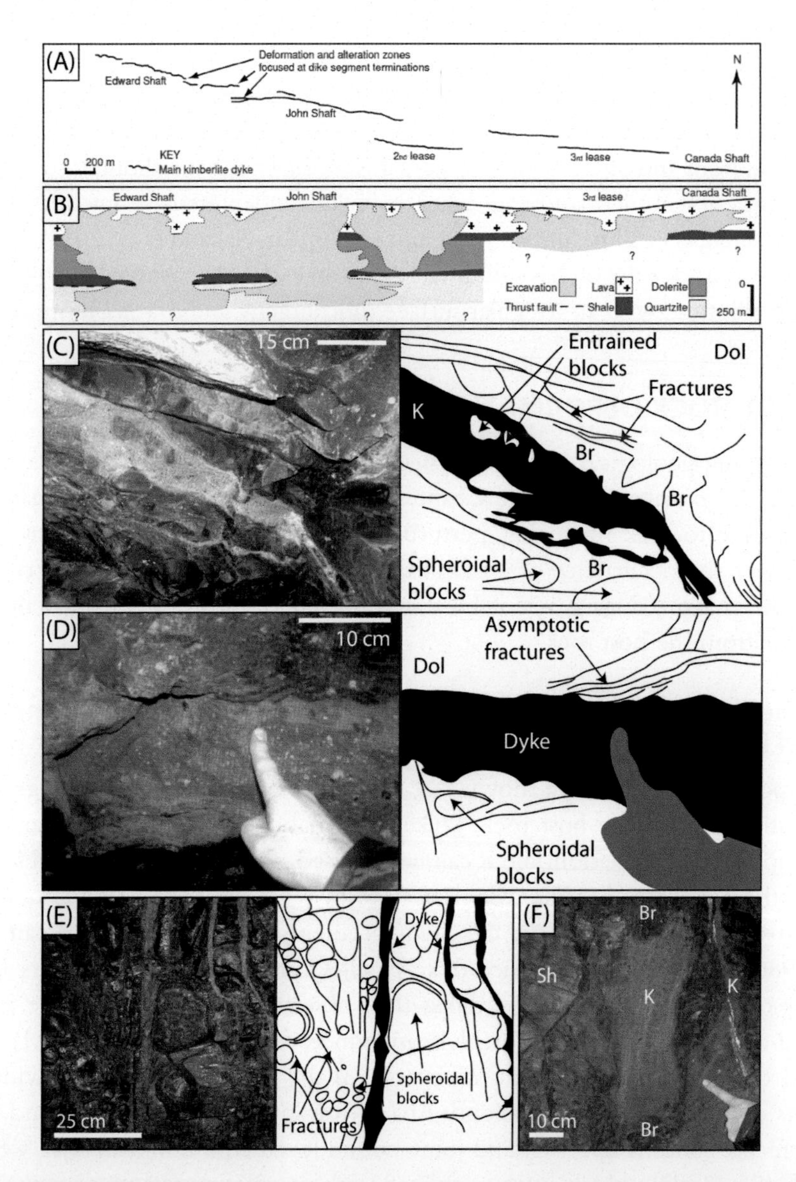

Figure 3.8 ***Variety of host-rock deformation preserved around kimberlite and lamprophyre dykes from the Swartruggens kimberlite dyke swarm exposed at Helam Mine, South Africa.*** (A) Surface expression of *en echelon* kimberlite dyke segments (labelled by their access shaft name), and (B) lateral extent (with depth) of the mined excavation, which matches closely with the kimberlite dyke geometry, and local stratigraphy and structural features. (C–E) Series of outcrop photographs and accompanying sketches from Edward dyke segment: (C) photograph and accompanying sketch of kimberlite dyke with entrained lithic blocks, and fractures, in the dolerite host rock, (D) photograph and accompanying sketch of asymptotic fractures in the dolerite host rock abutting margin of kimberlite dyke, (E) photograph and accompanying sketch of thin kimberlite dykes cutting across and wrapping around fracture-bound spheroidally weathered blocks in dolerite, and (F) photograph of kimberlite magma within an intensely brecciated shale host rock. Abbreviations: *Dol*, dolerite; *Br*, breccia; *Sh*, shale; *K*, kimberlite.

thought to originate from buoyancy-assisted mechanical erosion during dyke ascent (e.g. Russell et al., 2012). Volcanic vent localisation along dyke segments may therefore be preferentially localised by dyke-induced rock deformation and pre-conditioning of the host rock by magmatic gases (Fig. 3.9; see also Chapter 4).

3.4.2 Modelling Host Deformation

3.4.2.1 Elastic Models

Elastic deformation of the host-rock is commonly assumed when modelling dyke emplacement, for example, in nature when interpreting GPS displacement fields and inverse modelling the source of topographic change detected by InSAR (see also Chapter 11). In laboratory experiments, elastic gelatine has been used as a host-rock material for experimental dyke intrusions. In these experiments, the fluid overpressure (ΔP) exceeds the fracture pressure (P_f) of the host rock and causes it to fracture. The experimental results show that the mechanical properties of the host rock are key in controlling dyke ascent and it occurs in a 'toughness-dominated' regime (e.g. Menand and Tait, 2002). Gelatine is photoelastic so stresses within it can be visualised using polarised

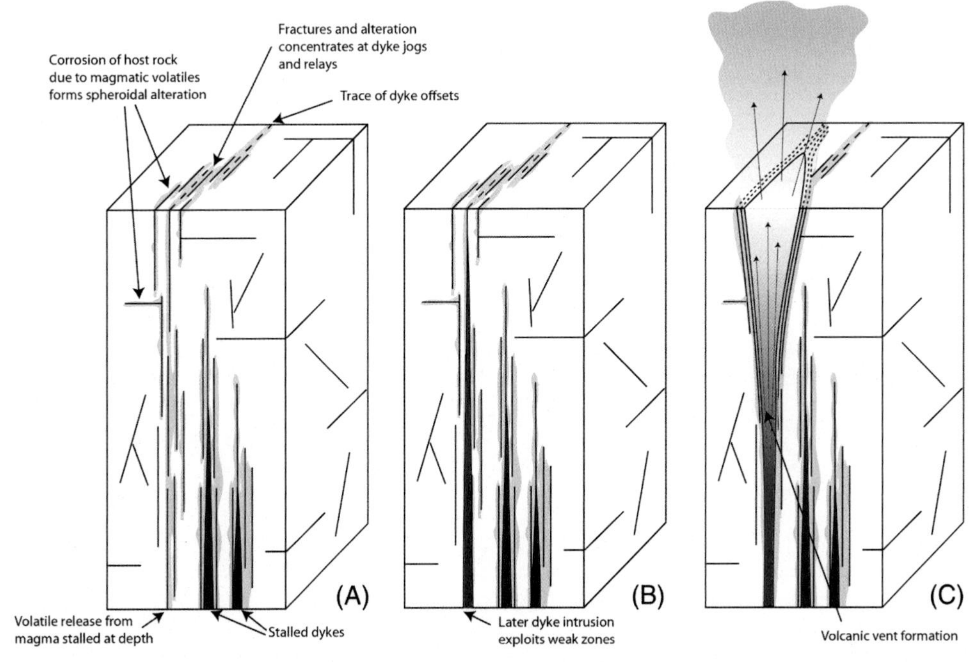

Figure 3.9 ***Schematic illustration of an example of volcanic vent formation from a feeder dyke (credit R.J. Brown).*** (A) Pre-conditioning of the host rock occurs as magma stalls at depth; (B) preferential dyke emplacement along weakened zones that are concentrated at relay zones (regions of the host rock between dyke segments, sometimes known as 'bridges'); and (C) volcanic vent formation from a feeder dyke preferentially at weathered zone between dyke segments.

light. Concentration of stress appears as coloured bands, and hydro-fractures, which form when water or air is injected, have stress concentration at the growing tip that produce a typical 'bow-tie' stress field (see Fig. 3.6B). Early studies attempted to make the qualitative description of stress distribution in gelatine solids quantitative, by assigning stress magnitude to each stress band (Crisp, 1952).

Recent applications of digital image correlation (DIC) have measured incremental and cumulative strain in a gelatine host material as an experimental dyke intrudes. This is done by including passive-tracer particles in the gelatine during its preparation and then illuminating these by a thin vertical laser sheet (e.g. Kavanagh et al., 2015). DIC detects pattern differences between images and uses these to map strain changes over time. These experiments have shown that strain concentration occurs at the tip during dyke propagation, and strain changes can be detected at the surface as the dyke approaches eruption (Fig. 3.10A; Kavanagh et al., 2018). Negative incremental strain gradually appears at depth as the dyke ascends, and this is due to pinching of the dyke

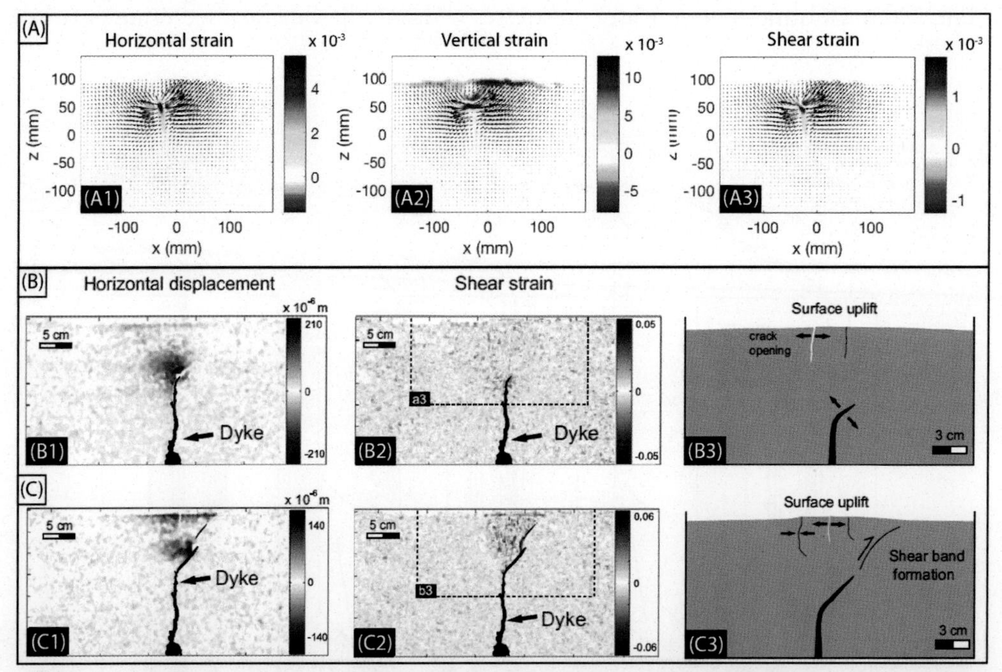

Figure 3.10 ***Experimental near-surface dyke propagation analysed using DIC.*** (A) Side view (*x–z*-plane) of water intruded into gelatine, where horizontal strain (A1), vertical strain (A2) and shear strain (A3) are mapped by colour change, and arrows indicate displacements. Strain is concentrated at the dyke tip, and surface deformation is detected as the dyke approaches eruption (Kavanagh et al., 2018). (B–C) Side view (*x–z*-plane) of dyke intrusion of golden syrup into compacted silica flour, where horizontal displacement (B1, C1) and shear strain (B2, C2)are mapped. Schematic diagrams indicate early propagation by mode I (tensional) opening (B3) of the dyke tip, surface uplift and crack opening; and (C3) subsequent shear band formation (Abdelmalak et al., 2012).

tail as the geometry of the dyke evolves from a penny-shaped crack into a tear-drop shape (Kavanagh et al., 2018). In nature, these stresses and strains could be expressed by seismicity and ground deformation that could be detected by geophysical monitoring networks (see also Chapter 11).

3.4.2.2 Inelastic Models

The relevance of the elastic-deformation models of dyke intrusion has been brought into question based on analysis of dyke geometries preserved in the field. LEFM suggests that the cross-sectional dyke geometry reflects the balance of magma overpressure relative to resistance of the host rock (Eqs. (3.1)–(3.4)). However, when applied to the geometry of dykes in the field, the results have indicated that the elastic model predicts magma overpressures that are much larger than expected to be reasonable. In a study of basaltic dykes from Scotland and kimberlite dykes from South Africa, Daniels et al. (2012) showed that, from centimetre-thick to metre-thick dykes, the overpressures required to explain the dyke aspect ratios was nearly 700 MPa, which is one or two orders of magnitude larger than the experimental values. They surmised that dyke-related deformation of the host rock caused it to be much less competent than the intact rock and that lateral variation in stress (particularly at bridges between dyke segments) could explain the lateral asymmetric geometry of the dykes. Cooling of the magma during intrusion could also be important particularly for small dykes, where quenched magma in the lateral tips helps to prop open the dyke. Localised host-rock erosion, perhaps associated with vent formation at the surface, could also influence the mechanical strength of the host rock around the dykes and so impact the overpressures required to keep the dyke open and propagating.

Inelastic deformation of the host rock during dyke emplacement has also been explored experimentally, where the dyke is modelled as a viscous indenter. Most of these experiments have focused on the progressive surface deformation that is induced as the dyke intrudes (e.g. Galland et al., 2016; Guldstrand et al., 2017). These surface patterns are also evident at depth, as shown by Abdelmalak et al. (2012) who carried out a fracture mode analysis on compacted silica flour experiments injected by golden syrup using DIC (see Figs. 3.7A and 3.10B and C). Shear and tension fractures were formed in the host as the dyke intruded, and a small dome of material was formed at the surface with tensile fractures propagating downwards. Some of the dyke-induced fractures created in the host away from the dyke were later intruded when the growing dyke intersected them.

3.5 IMPACT OF STRESS AND CRUSTAL HETEROGENEITY

Stress and crustal heterogeneities are major factors that influence the orientation and propagation pathway of dykes in the crust. This is true to such an extent that structural geologists often use dykes as 'stress markers', as their opening direction is parallel to the minimum compressive stress σ_3 and the dyke plan occupies the maximum σ_1 and intermediate σ_2 compressive stress plane. Dykes tend to propagate in the direction of maximum compressive stress σ_1, the orientation of which will be different on a regional or local scale. Dykes occur in all tectonic settings, with magma repeatedly injected into the same area either sequentially or simultaneously (see Chapter 4). Local stress loading could arise in a range of settings, including the emplacement of an earlier dyke that is pressurised, or gravitational loading by the accumulation of a volcanic edifice, or presence of an ice sheet. Unloading events could also influence dyke propagation pathways, for example, by the formation of a surface caldera or graben, gravitational subsidence of a volcanic edifice, melting of an ice sheet or glacier, or sector collapse of a volcanic edifice (see Chapter 9). Perturbations in the stress field can therefore occur over short or long timescales, but the relative magnitude of regional and local stress have proved important with the propagating front of the dyke bending to re-align with the new stress field.

3.5.1 Impact of Multiple Dyke Injections

Field exposures of individual dykes and dyke swarms have shown that dyke propagation does not occur in isolation, but with several parallel injections alongside one another. There is also field evidence of multiple injections intruding the same dyke (see Fig. 3.2C) and that offset dyke segments propagate simultaneously. The interaction of co-propagating and successive dykes is therefore of interest and has been studied in the field and laboratory. Watanabe et al. (2002) studied the interaction of co-propagating dykes by injecting silicone oil into gelatine. They found that a pre-existing stalled dyke within the gelatine acted as an attractor to the new injection and that the two may coalesce to form a single dyke.

Exposures of dyke segment tips indicate that they are sometimes curved towards adjacent dyke segments (e.g. Fig. 3.4C and D). This suggests that the presence of one dyke segment influences the propagation pathway of another. A range of field evidence captures this interaction at different stages, with the tip geometry in the 'bridges' between dyke segments (sometimes called 'relay zones') ranging from planar, where segment tips remain in-plane with the dyke segment (e.g. Fig. 3.1A), to rounded, where rotation of the dyke tips has occurred (e.g. Fig. 3.1B and C). Lateral jogs in the dyke segment geometry can be explained by the coalescence of previously separated segments across a 'bridge' to produce a 'broken bridge' (Fig. 3.1B and C).

3.5.2 Extensional Settings

In rift zones, extension of the crust can be accommodated through brittle failure but also by the injection of dykes. Continental exposure of rift zones, such as in Ethiopia or Iceland, have provided important constraints on the influence of extensional stresses on dyke emplacement and the role of dyking in the dynamics of continental rifting (see also Chapter 7). Numerical models have considered the impact of topographically low, elongate rift valleys on dyke ascent in rift zones and have found that the competition between regional tectonic stretching and localised unloading pressures will influence whether dykes intrude the valley or propagate off-rift into the surrounding area (Maccaferri et al., 2014).

Daniels and Menand (2015) carried out laboratory experiments using gelatine as a crustal analogue to study dyke propagation in an extensional setting. To create the extensional environment, prior to injection the gelatine was made to extend by applying a load to the surface. The slab margins migrated outwards to occupy a water-filled margin between the gelatine and the experiment tank as the uniform extensional setting was created. These experiments showed that the dyke orientation was perpendicular to the maximum extensional stress. By creating multiple, sequential dykes into the same experimental tank, Daniels and Menand (2015) also found that a competition between the regional stresses and those from the previous un-erupted and pressurised neighbouring dyke impacted the orientation of subsequent injections, with smaller spacing between injections and a thicker pre-existing dyke more strongly impacting subsequent intrusions. Extensional stresses have also been generated in experiments due to gravitational subsidence, by carrying out the experiment in an unconstrained gelatine solid and using moulds to create different volcanic edifice morphologies (e.g. Fiske and Jackson, 1972).

3.5.3 Lithology, Faults and Joints

Dykes propagate through the crust that is mechanically heterogeneous, due to the presence of different rock layers and weak planes such as bedding planes, faults or joints. Fig. 3.3 shows how the thickness of two dyke segments changes over a depth of 300 m, as the lithology of the host material changed. Kavanagh and Sparks (2011) noted that relatively thick, vertically oriented channels occurred across the breadth of the dyke segments, with the thickest dyke measurement coinciding with the depth of largest mechanical contrast between the host rock layers. The breadth of the dyke segments also was relatively narrow at this depth (Fig. 3.8B) as the dyke progressively ascended through stiff quartzite, then through low stiffness shale into stiff dolerite. Layered gelatine experiments have shown that for a constant fluid pressure within the dyke, the dyke breadth will become pinched when transitioning through layers of different Young's moduli (Kavanagh et al., 2006), and the thickness of the experimental dyke is greater when the Young's modulus of the gelatine layer was reduced.

Crustal stresses will influence the regional distribution of joint planes and faults and affect the overall strength of the crust. The interaction of ascending dykes with pre-existing fractures in the crust was studied in gelatine experiments by Le Corvec et al. (2013a) who propagated air-filled fractures in a pre-cut gelatine. They found that the distance and angle between the dykes and the fractures influence how much they interacted, with the potential to impact dyke trajectory and velocity. Dykes in nature have been documented as parallel to regional faults or joints, and this suggests they may have intruded these rather than creating their own fractures. The alignment of volcanic vents in an intraplate setting, such as a craton (as is the case for kimberlites) or a monogenetic volcanic field, has been inferred based on geological mapping but also statistical analysis of the spatial distribution of vents (e.g. Le Corvec et al., 2013b). This alignment strongly suggests dykes fed these eruptions, and their association with crustal structures detected by magnetic surveys (e.g. Barnett et al., 2013) suggest these played an important role facilitating the magma ascent. Given appropriate stress conditions, dykes can exploit existing crustal weaknesses; but this depends on the orientation of the fracture, and the magmatic overpressure (Delaney et al., 1986).

3.6 MAGMA FLOW

The previous sections have focused on the impact of the host rock and its deformation on dyke propagation. However, coupled to this are the dynamics of magma flow and the ability of magma to be transported within the growing dyke. Magma is a multiphase fluid, comprising melt, crystals and bubbles, and its rheology will be affected by the relative proportions of each phase and their physical properties. The rheology of magma is modelled in different ways; the simplest scenario is a Newtonian fluid (Fig. 3.11A.1), but more complex rheologies, such as pseudo-plastic or shear-thinning (Fig. 3.11A.2) or Bingham fluid (Fig. 3.11A.3) are also relevant.

3.6.1 Magma Viscosity

A detailed and comprehensive review of two-phase magma rheology is provided by Mader et al. (2013), and a three-phase magma rheological model based on experimental work is presented in Truby et al. (2015). Melt viscosity spans several orders of magnitude and is largely affected by composition (for example, low-viscosity carbonatite melts compared to viscous rhyolite melts). Melt composition reflects the degree of melt polymerisation, which is strongly influenced by the relative proportions of network formers, such as silica and aluminium, and network modifiers, such as sodium and potassium. Melt viscosity will change during depressurisation as super-saturation leads to gases being exsolved and crystals being precipitated (see Chapter 8). Magma viscosity is affected by the proportion of solid content that is transported. For example, phenocrysts transported in magma from depth will increase the bulk viscosity. However, the shape of the crystals is also important, as elongate crystals (e.g. feldspar) may be more likely to interact and lock together to make large clots or blockages within the magma. Xenocrysts and xenoliths are additional solid components that will increase magma viscosity. Gases exsolved from the melt will produce bubbles, and these will affect the magma viscosity differently depending on their ability to deform, with small bubbles potentially acting as rigid spheres. Bubbles may coalesce to form larger bubbles, and these are relatively deformable, and so their volume has less impact on the magma viscosity as they cause less resistance during flow.

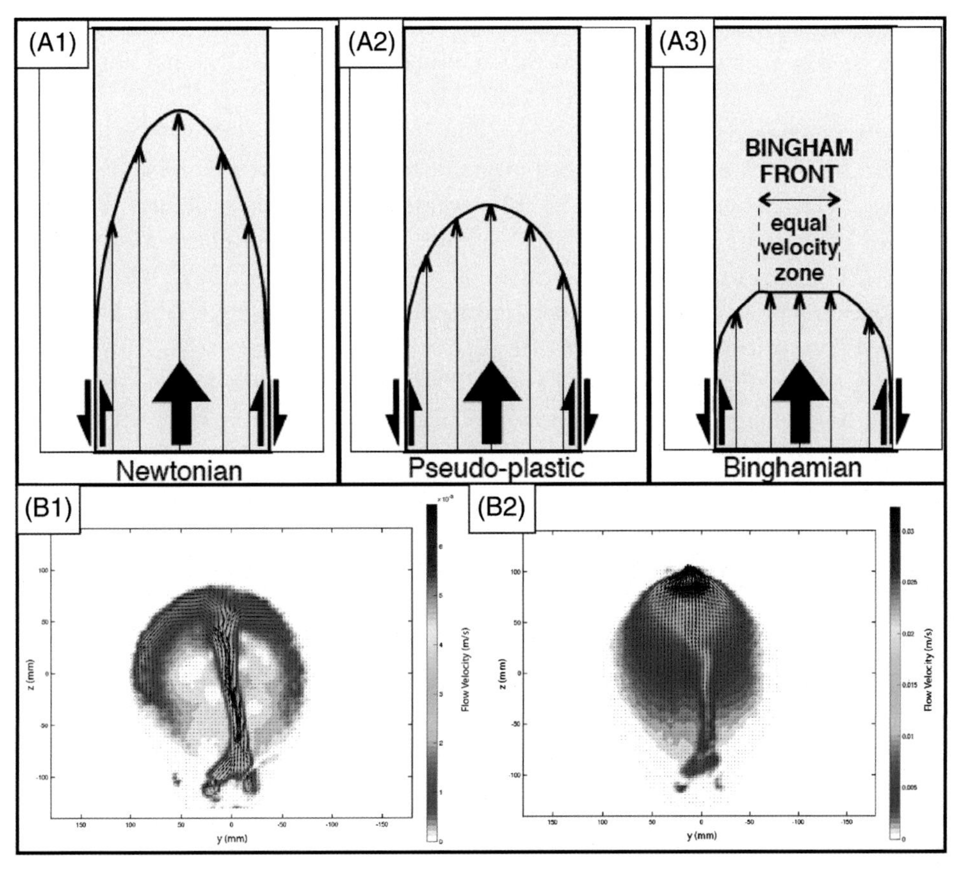

Figure 3.11 ***Magma flow models.*** (A) Velocity profiles of three conceptual model magma rheologies (x–z-plane): Newtonian fluids (A1), pseudo-plastic fluids (A2) and Bingham or viscoplastic fluids (A3) (Correa-Gomes et al., 2001). (B) PIV-mapped velocities within a propagating experimental dyke, formed by injecting water into gelatine (*viewed in the* y–z-plane). As the dyke ascends, it develops a vertical jet with downflow at the dyke margins (B1), and then the fluid is evacuated (B2) by focused vertical flow through a small fissure (Kavanagh et al., 2018).

3.6.2 Magma Transport in Dykes

Magmas of all known erupted composition have been documented in field studies of dykes; and as dykes are the dominant mechanism to feed volcanic eruptions, the erupted magma has almost exclusively moved through a dyke to reach the surface. This leads to a bias in the erupted products, as only 'eruptible' magma will reach the surface, but dyke intrusions will also include the magmas that did not erupt. The crystallised magma within a dyke can be viewed as a record of progressive solidification, with quenched magma at the dyke contact produced when the dyke first formed and when the largest temperature difference between the host rock and magma existed. The magma that has solidified at the dyke margin is therefore interpreted as a record of the first magma to propagate in the dyke. Crystallised magma towards the centre of the dyke is inferred to have formed later, and some magma that transited through the dyke has left no record of its transit.

Macroscopic textures preserved in crystallised magma in dykes displays a range of features that have been inferred to record magma flow; these include flow banding or folding (Fig. 3.12A), scour marks (Fig. 3.12B), development of ropey structures (Fig. 3.12C), stretched vesicles (Fig. 3.12D), shear textures (Fig. 3.12E), phenocryst alignment (Fig. 3.12F) and cataclastic elongation of phenocrysts (Smith, 1987). However, their occurrence and preservation vary widely and are often absent. Magnetic fabrics are inferred to record flow orientations (Herrero-Bervera et al., 2001; Magee et al., 2016), aligning with phenocrysts (Fig. 3.12F; Poland et al., 2004). Variations in flow trajectory preserved in dykes have suggested that dyke flow can be highly variable in space and time as it propagates and solidifies (e.g. Andersson et al., 2016). Evidence of flow focusing by channelisation of magma within the dyke has been inferred by vertically oriented thick regions in the dyke geometry (e.g. Fig. 3.3), and by extrapolating to depth the flow focusing that occurs during vent formation in fissure eruptions at the surface (e.g. Wylie et al., 1999).

3.6.3 Dyke Magma Flow Models

Laboratory experiments have recently been used to map flow velocity within a growing dyke using particle image velocimetry (PIV). Kavanagh et al. (2018) injected water seeded with passive-tracer particles into an elastic gelatine solid to create an experimental dyke. The particles were fluoresced in a thin vertical laser sheet that was oriented in the *y*–*z*-plane of the dyke as it grew. They observed a vertical, rapidly ascending central fluid jet with down-flow at the margins of the growing dyke (Fig. 3.11B). These experiments demonstrate that flow trajectories can vary laterally in a vertically propagating dyke. Numerical models of magma flow in dykes have explored the connection of a dyke to a deep magma reservoir and shallow conduit (Costa et al., 2007) and have found that a dyke can act as a volumetric capacitor that can store and periodically release magma, perhaps controlling pulsatory extrusions of magma in lava dome eruptions. Couplets of bubble-rich and -poor vertical bands in dykes (Kavanagh et al., 2018) could be explained by pressure fluctuations within the dyke caused by pulsatory eruptive behaviour.

3.7 FEEDER AND NON-FEEDER DYKES

Dykes feed volcanic eruptions in two ways, either coming to the surface as a sheet-like fissure eruption or transitioning into a broadly cylindrical or conical vent or conduit (see also Chapter 4). Outcrops showing the transition of a dyke to a volcanic vent (sometime called a 'bud') are rare, but quantifying the proportion of feeder and non-feeder dykes (either locally or regionally) is important for hazard assessments. Most volcanic unrest periods do not result in an eruption and, depending on the tectonic setting, the

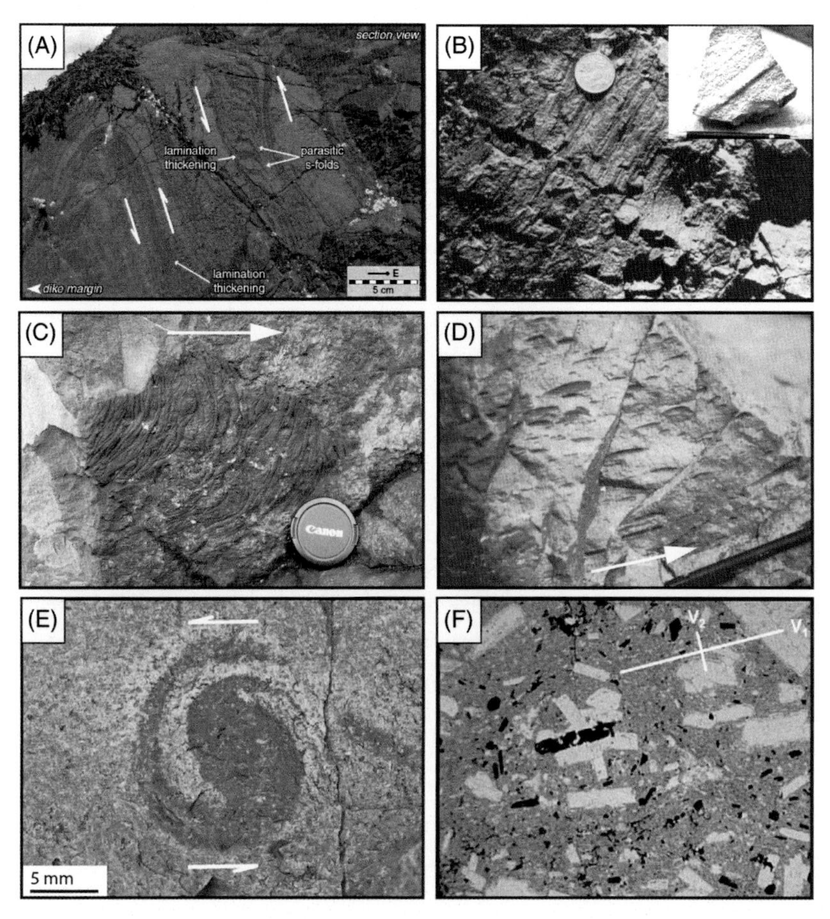

Figure 3.12 ***Photographs of a variety of magmatic fabrics in dykes indicating flow.*** (A) Photograph of the sectional view of the interior of a rhyolite dyke where shear sense indicators are preserved, including asymmetric (left) and parasitic folds (right) (Walker et al., 2017). (B) Photograph of 'hot slickenline' groove and ridge lineations on the surface of a mafic dyke from Troodos Ophiolite, Cyprus (Varga et al., 1998), with inset image showing a close-up view of a similar structure in hand specimen (pen for scale, from Smith, 1987). (C) Photograph of a vertical face in a disused quarry in Skye, UK, showing the exposed chilled margin of a basaltic dyke (*y*–*z*-plane) where the host rock has broken away to reveal ropey structures on the dyke margin (approximate flow direction indicated). (D) Photograph of a mafic dyke from Akaki Canyon, Cyprus, showing sub-horizontal elongated bubbles orthogonal to dyke margin (black pen in the bottom-right of the image for scale). The teardrop shape of the bubble can be used to infer flow direction (image modified after Varga et al., 1998; arrow points in the direction of flow). (E) Rotated lamination within rhyolite dyke indicating direction of simple shear (Walker et al., 2017). (F) Plane-polarised light photomicrograph of magma from a silicic dyke, Summer Coon (USA), where plagioclase petrofabrics approximately align with the sub-horizontal AMS lineation (maximum V_1 and intermediate V_2 susceptibility axes are indicated in white) (Poland et al., 2004).

proportion of feeder to non-feeder dykes is 1:5 to 1:10 (Crisp, 1984), meaning that the majority of dykes never reach the surface. Crustal sections suggest that the frequency of dykes in the crust decreases towards the surface.

Spectacular exposures of feeder and non-feeder dykes in the eroded crater of Miyakejima volcano, Japan (Fig. 3.13A) have shown that the geometry of dykes that are directly connected to erupted products ('feeder-dykes') is different to those that terminate within the rock strata and have become arrested ('non-feeder dykes'). Geshi et al. (2010) used high-resolution photographs to measure the variation in thickness of dykes exposed from 200 m below the surface (Fig. 3.13B). They found that non-feeder dykes had a thickness of 0.5–1 m at depth and this transitioned into a thicker head region that was 1.5–2 m thick towards the surface. Non-feeder dykes could reach within 15–45 m of the surface. In comparison, feeder dykes were a little thicker at depth (1 m), widening to 2–4 m thick at the surface.

Numerical models and laboratory experiments demonstrate that the flow within an erupting fissure is very different to that in an ascending dyke. Kavanagh et al. (2018) used PIV to show that the fluid flow trajectories were most rapid close to the surface as the eruptive fissure formed, and that the flow velocities decreased with increasing depth (e.g. Fig. 3.11B.2). Leading up to eruption, the centralised upward flowing jet of fluid became unstable (Fig. 3.11B.1) and the dyke simultaneously became pinched at its tail. Both fluid dynamics and host deformation were important in the transition into a feeder dyke, and so this emphasises the importance of including coupled host-rock deformation and changing magma flow in future modelling efforts.

3.8 SUMMARY AND FUTURE OPPORTUNITIES

Much can be learned about dyke emplacement by studying solidified dyke intrusions, their host rock, and how magma and host-rock interact. These observations have informed the development of emplacement models, which explore and test the relationships between key physical and chemical parameters and their interdependencies both numerically and in the laboratory. Relating the experimental observations and simplified models back to the natural phenomenon remains the largest challenge to advance our understanding of dyke propagation. Limited information on the nature of the dyke tip is an example of the restricted information available to Earth Scientists who model dyke propagation and the hazards of their eruption. However, if macroscopic flow fabrics preserved in solidified dykes can be linked to eruptive processes, this will enable identification of a feeder dyke without needing direct observations of the vent or tips, thus opening new opportunities for dyke studies and utility of the rock record for regional hazard assessment. To assist with this, more interaction between numerical modellers and experimentalists is needed. A leading aim in the study of dyke emplacement is to aid the interpretation of evidence of dyke propagation happening now, and to assist in real-time hazard assessments and monitoring

Figure 3.13 ***Photographs of the exposed interior of the Miyakejima Volcanic edifice, Japan.*** (A1) A feeder dyke to a lava flow, and later dyke that crosses through the lava; (A2) an annotated version of A1 (Geshi and Oikawa, 2014). (B) Photograph of a feeder dyke with at least 200 m of vertical extent that fed a scoria cone eruption (Geshi et al., 2010). The base of the dyke is not seen, though its position is indicated by the orange arrow.

efforts at volcanoes. Consideration of the coupling between magma flow and host-rock deformation in dyke emplacement models is essential, and experiments exploring the impact of magma rheology on the flow of magma in dykes are needed. An overall multidisciplinary approach will be required to integrate dyke emplacement datasets and combine laboratory, computational and field techniques to lead in the development of the next generation of dyke emplacement models.

FIELD EXAMPLE: THE SWARTRUGGENS KIMBERLITE DYKE SWARM, SOUTH AFRICA

The Swartuggens Kimberlites Dyke Swarm, South Africa (Fig. 3.1A map inset), is a series of Group II kimberlite and lamprophyre dykes that intruded the Kaapvaal craton at the end of the Jurassic. The dyke swarm extends over 7 km and comprises several offset east–west trending *en echelon* segments, each of which is several hundred metres long (Fig. 3.8A; Gurney and Kirkley, 1996; Brown et al., 2007). It is thought that 1–2 km of erosion has occurred since the dykes were emplaced.

The Swartruggens dyke swarm has been mined for diamonds for more than 70 years at Helam Mine, which is located 60 km west of the city of Rustenberg. Diamond mines offer unprecedented exposure of the internal architecture of kimberlite volcanoes and their plumbing systems, due to the progressive extraction of material. At Helam Mine, geologists have collected an exceptional dataset of dyke thickness across several of the kimberlite dyke segments, spanning several hundred metres in lateral extent and down to 750 m below the surface (Kavanagh and Sparks, 2011). The dykes are exposed where they have intruded shallow levels of the upper crust and represent either arrested dykes or the feeding system of kimberlite pipes that have since been eroded.

The economic 'Main' kimberlite is the oldest intrusion of the swarm and is parallel to the younger sub-economic 'Changehouse' kimberlite dyke. A barren alkaline lamprophyre dyke was the last to be intruded, based on cross-cutting relationships. The Main kimberlite dyke has been extensively mined, and the access tunnels provide excellent exposure of the solidified magma and its host rock. The dykes are broadly vertical and discordant to flat-lying strata of quartzite, shale, dolerite and andesite lava from the Proterozoic Pretoria Group, and the dykes cross a large thrust fault in the area (Fig. 3.8B). The three-dimensional geometry of the 'John' and 'Edward' dyke segments of the Main kimberlite is shown in Fig. 3.3 (Kavanagh and Sparks, 2011), showing that the maximum dyke thickness is 1.95 m and average is 0.64 m. Kavanagh and Sparks (2011) showed that the cross-sectional dyke shape was best-fit by a teardrop-shaped geometry and surmised that inelastic processes were responsible for controlling the final dyke morphology. Vertically oriented and thick dyke regions were interpreted as channels of sustained magma flow and localised host-rock erosion.

A wide range of macroscopic host-rock deformation styles associated with the Swartruggens dykes are evident in outcrop at Helam (Kavanagh and Sparks, 2011); ranging from centimetre-spaced dyke-parallel fractures at the dyke margins, to highly fractured and brecciated material in the regions between dyke segments (Fig. 3.8). The frequency of jointing appears to increase in density towards the dyke margin (Kavanagh, 2010). Basson and Viola (2004) carried out a microstructural analysis of fibrous calcite veins associated with the kimberlite intrusions to assess the intrusion mechanisms of the Swartruggens dykes, arguing that the magma passively intruded the upper crust during a period of brittle crustal-dilation.

Joint-bound centimetre- to tens of centimetre-sized spheroidally weathered material occurs locally in the host rock near the dykes, and small dykelets intruded later by partially wrapping around the spheroidal blocks. The occurrence of these spheroidal clasts as xenoliths in the magma suggests the localised host rock fracturing and weathering occurred prior to the magma intrusion. Brown et al. (2007) argued that the spheroidal blocks were formed by preferential in situ chemical erosion in the regions between dyke segments; either by magmatic gases exsolved from magma at depth, or hydrothermal fluids. Petrographic work on the solidified magma has been helpful to constrain the petrogenesis of the Swartruggens dyke swarm (Coe et al., 2008). The Group II kimberlites are rich in mica and comprise xenocrysts (olivine and phlogopite), phenocrysts (phlogopite, olivine and diopside) and a fine-grained crystalline groundmass (phlogopite, calcite, serpentine, diopside, perovskite, opaque oxides and apatite). Coe et al. (2008) interpreted petrographic textures, such as mineral overgrowths and kink-banding in phlogopite crystals, as evidence of MARID (mica-amphibole-rutile-ilmenite-diopside) xenoliths (e.g. Dawson and Smith, 1977); and patterns in major, trace and isotopic compositions support a deep sub-continental lithospheric metasomatised mantle source of the magma.

Independent of magma composition, three-dimensional datasets of dyke geometry are rare. The Swartruggens dykes therefore offers an unusual opportunity to test existing models of dyke emplacement by combining constraints on the three-dimensional geometry of dykes, the relationship with crustal structures and regional strata, macroscopic host-rock deformation, crystal microstructures and magma geochemistry. The evidence suggests that the Swartruggens dykes intruded partly due to passive intrusion into lithospheric-scale structures, but also that magmatic overpressures facilitated magma ascent from great depth. Pre-conditioning of the host rock resulted in brittle to plastic deformation as the magma intruded, and alteration ahead of the intruding magma caused further localised weakening and a potential mechanism to facilitate the transition from a dyke into a volcanic vent.

ACKNOWLEDGEMENTS

The author would like to thank the University of Liverpool for their support. Many thanks also to Catherine Annen, Simon Martin, Suraya Hilmi Hazim and Elliot Wood for their comments on an earlier version of the chapter. Rich Brown is thanked for drafting Fig. 3.9. Special thanks to Steffi Burchardt for the invitation to contribute to this book, for providing a photograph for Fig. 3.4, and for her diligent and considerate editorial support.

REFERENCES

Abdelmalak, M.M., et al., 2012. Fracture mode analysis and related surface deformation during dyke intrusion: results from 2D experimental modelling. Earth Planet. Sci. Lett. 359–360, 93–105.

Andersson, M., Almqvist, B.S.G., Burchardt, S., Troll, V.R., Malehmir, A., Snowball, I., Kübler, L., 2016. Magma transport in sheet intrusions of the Alnö carbonatite complex, central Sweden. Sci. Rep. 6, 27635. doi: 10.1038/srep27635.

Barnett, W., et al., 2013. How structure and stress influence kimberlite emplacement. In: Proceedings of the 10th International Kimberlite Conference. Springer India, New Delhi, pp. 51–65.

Basson, I.J., Viola, G., 2004. Passive kimberlite intrusion into actively dilating dyke–fracture arrays: evidence from fibrous calcite veins and extensional fracture cleavage. Lithos 76 (1–4), 283–297.

Brown, R.J., et al., 2007. Mechanically disrupted and chemically weakened zones in segmented dike systems cause vent localization: evidence from kimberlite volcanic systems. Geology 35 (9), 815–818.
Clemente, C.S., Amorós, E.B., Crespo, M.G., 2007. Dike intrusion under shear stress: effects on magnetic and vesicle fabrics in dikes from rift zones of Tenerife (Canary Islands). J. Struct. Geol. 29, 1931–1942.
Coe, N., et al., 2008. Petrogenesis of the Swartruggens and Star Group II kimberlite dyke swarms, South Africa: constraints from whole rock geochemistry. Contrib. Mineral. Petrol. 156 (5), 627–652.
Correa-Gomes, L.C., Filho, C.S., Martins, C., 2001. Development of symmetrical and asymmetrical fabrics in sheet-like igneous bodies: the role of magma flow and wall-rock displacements in theoretical and natural cases. J. Struct. Geol. 23, 1415–1428.
Costa, A., et al., 2007. Control of magma flow in dykes on cyclic lava dome extrusion. Geophys. Res. Lett. 34 (2), pL02303.
Crisp, J.D., 1952. The use of gelatin models in structural analysis. Proc. Inst. Mech. Eng. B 1 (1–12), 580–604.
Crisp, J.A., 1984. Rates of magma emplacement and volcanic output. J. Volcanol. Geotherm. Res. 20, 177–211.
Cruden, A., Vollgger, S., Dering, G., Micklethwaite, S., 2016. High spatial resolution mapping of dykes using Unmanned Aerial Vehicle (UAV) photogrammetry: new insights on emplacement Processes. Acta Geol. Sin. 90 (Supp. 1), 52–53.
Dahm, T., 2000. On the shape and velocity of fluid-filled fractures in the Earth. Geophys. J. Int. 142 (1), 181–192.
Daniels, K.A., Menand, T., 2015. An experimental investigation of dyke injection under regional extensional stress. J. Geophys Res. Solid. Earth 120, 2014–2035.
Daniels, K.A., et al., 2012. The shapes of dikes: evidence for the influence of cooling and inelastic deformation. GSA Bull. 124 (7–8), 1102–1112.
Dawson, J.B., Smith, J.V., 1977. The MARID (mica-amphibole-rutile-ilmenite-diopside) suite of xenoliths in kimberlite. Geochim. Cosmochim. Acta 41 (2), 309–310.
Delaney, P.T., Pollard, D.D., 1981. Deformation of host rocks and flow of magma during growth of minette dikes and breccia-bearing intrusions near Ship Rock, New Mexico. US Geological Survey Professional Paper 1202.
Delaney, P.T., et al., 1986. Field relations between dikes and joints: emplacement processes and palaeostress analysis. J. Geophys. Res. 91 (B5), 4920–4938.
Ernst, R.E., Grosfils, E.B., Mege, D., 2001. Giant dike swarms: earth, venus, and mars. Annu. Rev. Earth Planet. Sci. 29, 489–534.
Fiske, R.S., Jackson, E.D., 1972. Orientation and growth of Hawaiian volcanic rifts: the effect of regional structure and gravitational stresses. Proc. R. Soc. A 329 (1578), 299–326.
Galland, O., et al., 2014. Dynamics of dikes versus cone sheets in volcanic systems. J. Geophys. Res. doi: 10.1002/2014BJ011059.
Galland, O., et al., 2015. Laboratory modelling of volcano plumbing systems: a review. Advance in Volcanology. Springer, 1-68.
Galland, O., Bertelsen, H.S., Guldstrand, F., 2016. Application of open-source photogrammetric software MicMac for monitoring surface deformation in laboratory models. J. Geophys. Res. 121, 2852–2872.
Geshi, N., Oikawa, T., 2014. The spectrum of basaltic feeder systems from effusive lava eruption to explosive eruption at Miyakejima volcano, Japan. Bull. Volcanol. 76 (3), 95.
Geshi, N., Kusumoto, S., Gudmundsson, A., 2010. Geometric difference between non-feeder and feeder dikes. Geology 38 (3), 195–198.
Gudmundsson, A., 1983. Form and dimensions of dykes in eastern Iceland. Tectonophysics 95 (3–4), 295–307.
Gudmundsson, A., Loetveit, I.F., 2005. Dyke emplacement in a layered and faulted rift zone. J. Volcanol. Geotherm. Res. 144 (1–4), 311–327.
Guldstrand, F., Burchardt, S., Hallot, E., Galland, O., 2017. Dynamics of surface deformation induced by dikes and cone sheets in a cohesive coulomb brittle C rust. J. Geophys. Res. 122 (10), 8511–8524, http://doi.org/10.1002/2017JB014346.
Gurney, J.J., Kirkley, M.B., 1996. Kimberlite dyke mining in South Africa. Afr. Geosci. Rev. 3, 191–201.
Herrero-Bervera, E., et al., 2001. Magnetic fabric and inferred flow direction of dikes, conesheets and sill swarms, Isle of Skye, Scotland. J. Volcanol. Geotherm. Res. 106 (3), 195–210.
Kavanagh, J.L., 2010. Ascent and Emplacement of Kimberlite Magmas. University of Bristol, UK.

Kavanagh, J.L., Sparks, R.S.J., 2011. Insights of dyke emplacement mechanics from detailed 3D dyke thickness datasets. J. Geol. Soc. 168 (4), 965–978.
Kavanagh, J., Menand, T., Daniels, K.A., 2013. Gelatine as a crustal analogue: determining elastic properties for modelling magmatic intrusions. Tectonophysics 582, 101–111.
Kavanagh, J.L., Boutelier, D., Cruden, A.R., 2015. The mechanics of sill inception, propagation and growth: experimental evidence for rapid reduction in magmatic overpressure. Earth Planet. Sci. Lett. 421, 117–128.
Kavanagh, J.L., et al., 2018. Challenging dyke ascent models using novel laboratory experiments: implications for reinterpreting evidence of magma ascent and volcanism. J. Volcanol. Geotherm. Res. doi: 10.1016/j.jvolgeores.2018.01.002.
Kavanagh, J.L., Menand, T., Sparks, R.S.J., 2006. An experimental investigation of sill formation and propagation in layered elastic media. Earth Planet. Sci. Lett. 245 (3–4), 799–813.
Komorowski, J.C., et al., 2003. The January 2002 flank eruption of Nyiragongo volcano (DRC): chronology, evidence for a tectonic rift trigger and impact of lava flows on the city of Goma. Acta Vulcanol. 15 (1–2), 27–62.
Krumbholz, M., Hieronymus, C.F., Burchardt, S., 2014. Weibull-distributed dyke thickness reflects probabilistic character of host-rock strength. Nat. Commun. 5 (3272), 1–7.
Le Corvec, N., Menand, T., Lindsay, J., 2013a. Interaction of ascending magma with pre-existing crustal fractures in monogenetic basaltic volcanism: an experimental approach. J. Geophys. Res. 118 (3), 968–984.
Le Corvec, N., Spörli, K.B., et al., 2013b. Spatial distribution and alignments of volcanic centers: clues to the formation of monogenetic volcanic fields. Earth Sci. Rev. 124 (C), 1–19.
Lister, J.R., Kerr, R.C., 1991. Fluid-mechanical models of crack propagation and their application to magma transport in dykes. J. Geophys. Res. 96 (B6), 10049–10077.
Lowrie, W., 1990. Identification of ferromagnetic minerals in a rock by coercivity and unblocking temperature properties. Geophys. Res. Lett. 17 (2), 159–162.
Maccaferri, F., Bonafede, M., Rivalta, E., 2011. A quantitative study of the mechanisms governing dike propagation, dike arrest and sill formation. J. Volcanol. Geotherm. Res. 208 (3), 39–50.
Maccaferri, F., et al., 2014. Off-rift volcanism in rift zones determined by crustal unloading. Nat. Geosci. 7 (4), 297–300.
Mader, H.M., Llewellin, E.W., Mueller, S.P., 2013. The rheology of two-phase magmas: a review and analysis. J. Volcanol. Geotherm. Res. 257 (C), 135–158.
Magee, C., et al., 2016. Three-dimensional magma flow dynamics within subvolcanic sheet intrusions. Geosphere 12 (3), 842–866.
Mathieu, L., et al., 2008. Dykes, cups, saucers and sills: analogue experiments on magma intrusion into brittle rocks. Earth Planet. Sci. Lett. 271 (1–4), 1–13.
Menand, T., Tait, S.R., 2001. A phenomenological model for precursor volcanic eruptions. Nature 411, 678–680.
Menand, T., Tait, S.R., 2002. The propagation of a buoyant liquid-filled fissure from a source under constant pressure: an experimental approach. J. Geophys. Res. 107 (2306), 177–185.
Merle, O., 2015. The scaling of experiments on volcanic systems. Front. Earth Sci. 3, 1–15, http://doi.org/10.3389/feart.2015.00026.
Nakada, S., Shimizu, H., Ohta, K., 1999. Overview of the 1990–1995 eruption at Unzen Volcano. J. Volcanol. Geotherm. Res. 89, 1–22.
Poland, M.P., Fink, J.H., Tauxe, L., 2004. Patterns of magma flow in segmented silicic dikes at Summer Coon volcano, Colorado: AMS and thin section analysis. Earth Planet. Sci. Lett. 219 (1–2), 155–169.
Rivalta, E., et al., 2015. A review of mechanical models of dike propagation: schools of thought, results and future directions. Tectonophysics 638, 1–42.
Roper, S.M., Lister, J.R., 2007. Buoyancy-driven crack propagation: the limit of large fracture toughness. J. Fluid Mech. 580, 359–380.
Rubin, A.M., 1995. Propagation of magma-filled cracks. Annu. Rev. Earth Planet. Sci. 23 (1), 287–336.
Russell, J.K., et al., 2012. Kimberlite ascent by assimilation-fuelled buoyancy. Nature 481 (7381), 352–356.
Sigmundsson, F., et al., 2014. Segmented lateral dyke growth in a rifting event at BárÐarbunga volcanic system, Iceland. Nature 517 (7533), 191–195.

Smith, R.P., 1987. Dyke emplacement at Spanish Peaks, Colorado, Mafic dyke swarms.
Taisne, B., Tait, S., 2011. Effect of solidification on a propagating dike. J. Geophys. Res. 116 (B1), pB01206.
Takada, A., 1990. Experimental study on propagation of liquid-filled crack in gelatin: shape and velocity in hydrostatic stress condition. J. Geophys. Res. 95, 8471–8481.
Tibaldi, A., 2015. Structure of volcano plumbing systems: a review of multi-parametric effects. J. Volcanol. Geotherm. Res. 298, 85–135.
Townsend, M., et al., 2015. Jointing around magmatic dikes as a precursor to the development of volcanic plugs. Bull. Volcanol. 77 (92).
Townsend, M.R., Pollard, D.D., Smith, R.P., 2017. Mechanical models for dikes: a third school of thought. Tectonophysics (703–704), 98–118.
Truby, J.M., et al., 2015. The rheology of three-phase suspensions at low bubble capillary number. Proc. R. Soc. A 471 (2173), 20140557.
Varga, R.J., et al., 1998. Dike surface lineations as magma flow indicators within the sheeted dike complex of the Troodos Ophiolite, Cyprus. J. Geophys. Res. 103 (B3), 5241–5256.
Walker, R.J., Branney, M.J., Norry, M.J., 2017. Dike propagation and magma flow in a glassy rhyolite dike: a structural and kinematic analysis. Geol. Soc. Am. Bull. 129 (5–6), 594–606. doi: 10.1130/B31378.1.
Watanabe, T., et al., 2002. Analogue experiments on magma-filled cracks: competition between external stresses and internal pressure. Earth Planet. Space 54, 1247–1261.
Weertman, J., 1971. Theory of water-filled crevasses in glaciers applied to vertical magma transport beneath oceanic ridges. J. Geophys. Res. 76 (5), 1171–1183.
Wright, T.L., Fiske, R.S., 1971. Origin of the differentiated and hybrid lavas of Kilauea Volcano, Hawaii 1. J. Petrol. 12 (1), 1–65.
Wylie, J.J., et al., 1999. Flow localization in fissure eruptions. Bull. Volcanol. 60, 432–440.

FURTHER READING

Galland, O., et al., 2015. Laboratory modelling of volcano plumbing systems: a review. Advance in Volcanology. Springer, 1-68.
Gudmundsson, A., 2011. Rock Fractures in Geological Processes. Cambridge University Press.
Rivalta, E., et al., 2015. A review of mechanical models of dike propagation: schools of thought, results and future directions. Tectonophysics 638, 1–42.
Rubin, A.M., 1995. Propagation of magma-filled cracks. Annu. Rev. Earth Planet. Sci. 23 (1), 287–336.
Townsend, M.R., Pollard, D.D., Smith, R.P., 2017. Mechanical models for dikes: a third school of thought. Tectonophysics (703–704), 98–118.

CHAPTER 4

Growth of a Volcanic Edifice Through Plumbing System Processes—Volcanic Rift Zones, Magmatic Sheet-Intrusion Swarms and Long-Lived Conduits

Steffi Burchardt*, Thomas R. Walter, Hugh Tuffen†**
*Uppsala University, Uppsala, Sweden
**Physics of Earthquakes and Volcanoes, GFZ German Research Centre for Geosciences, Telegrafenberg, Potsdam, Germany
†Lancaster Environment Centre, Lancaster University, Lancaster, United Kingdom

Contents

4.1 INTRODUCTION

Hundreds to thousands of magmatic sheet intrusions can be found in eroded volcanoes worldwide. These sheet intrusions are part of the volcanic and igneous plumbing system (VIPS), forming, on the one hand, the internal skeleton of a volcano edifice, and may, on the other hand, reach the surface and feed eruptions. In eroded volcanoes, sheet intrusions are clearly exposed as solidified magma-filled fractures, aligned to form dense swarms. The spacing of these sheets is often so dense that little or no country rock remains visible between individual sheets, but with distance to the main intrusion zone, the spacing gradually increases. The sheer volume of magmatic sheet swarms indicates that volcanoes do not only grow exogenously through repeated eruptions, but also by endogenous injection of sheet intrusions. In fact, the volume of magma intruded may approach scales similar to, or be even larger than, what is erupted. The emplacement of

Volcanic and Igneous Plumbing Systems. http://dx.doi.org/10.1016/B978-0-12-809749-6.00004-2
Copyright © 2018 Elsevier Inc. All rights reserved.

magmatic sheet swarms is therefore a significant process in the evolution of long-lived VIPS that can contribute to localised crustal thickening (Siler and Karson, 2009), surface uplift (Kuenen, 1937) and even gravitational collapse of the volcanic edifice (see Chapter 9).

There are three different types of sheet intrusions that form swarms in central volcanoes: steeply dipping sheets (dykes) that align sub-parallel and define rift zones; concentric sheets that form gently to steeply dipping arrangements surrounding a central source region, a magma chamber or a topographic expression; and sheets that are arranged in a radial pattern relative to a given source or heterogeneity (Fig. 4.1). In addition, there are sills, sheets that intrude roughly concordant to the layering of their host rock. All four types of sheets may occur at different times in the same volcano, or may even occur in chorus, leading to complex patterns of sheet intrusions at volcanoes with long lifespans. In this chapter, we will look at the typical features of the three types of swarm-building sheet intrusions and discuss the specific conditions that control which type of sheet is formed. We also describe some prime field examples of magmatic sheet swarms exposed in the Canary Islands.

In addition to sheet swarms, long-lived central volcanoes often have a central conduit, a pipe-like channel through which summit eruptions occur. We will explore

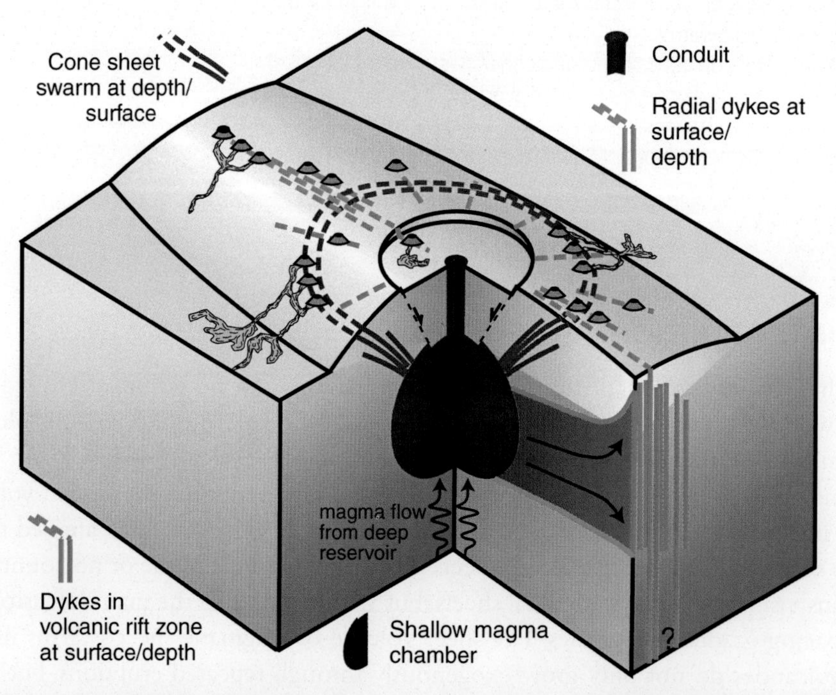

Figure 4.1 ***Schematic sketch of the three different types of swarm-building magmatic sheet intrusions in VIPS, that is, dykes in volcanic rift zones, cone sheets and radial dykes.*** All three types can feed fissure eruptions at the Earth's surface and may occur in the same volcano. Modified from Galland et al. (2014).

how conduits form from sheet intrusions and how they transport magma to the surface of a volcano to feed eruptions. We also highlight examples of conduits exposed in Iceland.

4.2 VOLCANIC RIFT ZONES AND THEIR INTRUSION SWARMS

Many, if not most, of the major basaltic volcanoes develop complexes of parallel and sub-parallel mafic dykes and intrusive sheets (Fornari and Campbell, 1987; Walker, 1992). If a sufficient amount of such aligned sheets is identified, authors commonly refer those as 'volcanic rift zones'. However, no definition has been elaborated yet as to how many sheets and what cumulative dyke thickness is needed to form such a rift zone; as in many more recent volcanic edifices, only the alignment and shape of eruptive vents allows speculating or inferring the presence of a rift zone. Because of the high intensities of dyke intrusions and fissure eruptions, rift zone terrains are often tectonically disrupted and marked by numerous eruptive fissures, scoria cones, open cracks, faults and landslides at the surface (Stearns, 1946).

Rift zones develop on a regional scale linked to plate tectonics (e.g. East-African rift system; see also Chapter 7) and on a local scale on individual volcanic edifices (e.g. Kilauea rift zone; see also Chapter 9). Being located in such different tectonic environments, rift zones linked to divergent plates are bearing many similarities but also many differences to those identified on ocean islands (Rubin, 1995). There are significant differences in the feeding rate, depth, width, length, magma plumbing system and structural development, so both types of rift zones shall not be confused (Wyss, 1980).

4.2.1 How Do Volcanic Rift Zones Develop? Controlling Factors and Geometries

Rift zones at ocean island volcanoes play an important role in the history of growth and collapse of the edifice (e.g. Carracedo, 1996; see also Chapter 9). Rift zones may also host shallow magma chambers or develop as a consequence of existing ones (Lin et al., 2014; Baker and Amelung, 2015). Parameters controlling rift development by dyke intrusions include magma density and viscosity, the pressure in the magma, the tectono-magmatic history and, most importantly, stress magnitudes and orientations in the lithosphere. Therefore, the parameters controlling rift zones are similar to those controlling individual sheet intrusions (see Chapter 3), just with the difference that at rift zones, these allow for sub-parallel intrusion geometries. The geometric arrangements of dykes in rift zones, and the divergence from a linear path, are governed by the stress field. So, the study of such intrusion axes in turn allows inferring the stress field present during the intrusion of the dykes. Understanding the stress field is moreover crucial, because rifting events (dyke intrusions) are known to cause flank deformation and may result in catastrophic collapse of volcano sectors (Elsworth and Voight, 1995; Chapter 9). On the other hand, flank instability and flank collapses affect the stress field within the volcano, thus influencing the geometry of dykes, rift zones and the intrusion frequency and may even lead to migration of the loci of magmatism (Maccaferri et al., 2017). Rift zone development, flank instability and stress field changes in large volcanic edifices go hand in hand (Carracedo, 2011).

As dykes are generally emplaced perpendicular to the direction of the least principal compressive stress (Anderson, 1937; Chapter 3), the orientation of rift zones formed by swarms of parallel dykes also adjusts to this direction (Shaw, 1980). Any change in the stress field may be documented in a re-arrangement of successive intrusions of dykes and formation of rift zones (Zurek et al., 2015). Linear rift zones form preferentially in larger volcanic edifices, indicating that the significance of gravitational stresses controlling the geometry and orientation of linear rift zones increases with the volcano size (height, volume). Based on morphological data sets of seamounts and volcanic islands, Vogt and Smoot (1984) and Mitchell (2001) showed that the length of the rift zones is larger at volcanoes taller than 3 km (Walter and Troll, 2003). This is due to the increasingly asymmetric volcano growth caused by multiple and migrating eruptive centres, flank eruptions and landslides (Mitchell, 2001), so that the growing load of overlapping islands may lead to a stress field favouring rift zone evolution (Walter, 2003). Many ocean islands reach heights above 6 km and are therefore of similar size; however, the lengths, widths and depths of rift zones vary significantly.

Rift zones often form systematic patterns, deviating from a volcanic centre to two or three directions. Different hypotheses have been put forward to explain these divergent rift axes, including (i) the least effort principle during vertical upward loading of a brittle plate (MacFarlane and Ridley, 1968; Luongo et al., 1991; Carracedo, 1994, 1996); (ii) the re-activation of pre-existing crustal structures (Carracedo, 1996); or (iii) the control of the gravitational stress field acting within the volcano edifice (Fiske and Jackson, 1972; Walter and Troll, 2003). Various observations suggest that rift zones evolve and migrate and may start to deviate from a linear feature into two main axes. Such changes are well explained by the evolving gravitational stress field of the growing and eroding edifice that may lead to compressional stress in the sector enclosed by two of the rift axes and to extensional stress in the other sector. Consequently, divergent rift zones may push flanks laterally or lead to the formation of a third rift (Walker, 1992; Walter et al., 2005). Caution has to be exercised when three-armed rifts are identified at ocean islands, as unstable or collapse flanks may obscure the magma plumbing system close to the surface (Becerril et al., 2015).

4.2.2 How Do Volcanic Rift Zones Interact With Volcano Stability?

On large ocean islands, volcano flanks are prone to become unstable and be displaced towards the sea (e.g. Holcomb and Searle, 1991; Oehler et al., 2004 Chapter 9). Flanks of ocean island volcanoes enclosed by two rift zones may experience progressive compression, tilt and an increase in slope steepness, which are considered critical factors that contribute to flank instability (Walker, 1992; Michon et al., 2009). The displacement of unstable flanks may be gradual (slumping, spreading) or sudden and fast (debris avalanches, debris flows; Siebert, 1984). Gradually displaced flanks have been identified at many volcanoes, for instance, at Mount Etna, Italy (e.g. Walter et al., 2005) or at Kilauea Volcano, Hawaii (Delaney et al., 1998), and Piton de la Fournaise (Clarke et al., 2013), Mauna Loa volcano (Miklius et al., 1995), La Palma (González et al., 2010), Arenal (Ebmeier et al., 2010) and Stromboli (Di Traglia et al., 2014). At some of these volcanoes, the lateral movement of a rift-pushed flank is interacting with tectonic faults (Le Corvec and Walter, 2009).

Sudden, rapid landslides in the form of giant debris avalanches have occurred at almost all large volcanoes on the Hawaiian Islands (Moore et al., 1994), the Canary Islands (Krastel et al., 2001) and on Réunion Island (Holcomb and Searle, 1991; Oehler et al., 2004). Also, the geometry of displaced unstable flanks may differ: some being shallow slumps and others being deep-seated block slides. Shallow slumps affect only the upper parts of an ocean island flank, whereas deep-seated block slides affect an entire segment of an edifice and may reach from the surface down to the base of the volcano or even below (Oehler et al., 2005; Chapter 9). At some volcanic islands, such as on Hawaii or El Hierro, interrelations between the geometry and orientation of rift zones and the movement direction and geometry of displaced flank sectors have been discussed for decades (Nakamura, 1980; Swanson et al., 1976; Dieterich, 1988).

The interaction between rift zones and flank instability is proved in two ways. On the one hand, rift zones push a volcano flank off and hence destabilise the edifice, and, on the other hand, creeping, sliding and collapsing volcano flanks change the internal and underlying stress field and affect the volcanic plumbing system throughout the crust and possibly even below. Flank collapses lead to unloading and rebound of volcanic islands. Flank movement causes horizontal expansion and therefore stabilisation of an extensional rift zone, leading to the question whether rift zones are active or passive structures in controlling the depth of intrusions (Cayol et al., 2000). Rift zones circumscribing the expression of a creeping flank have been documented at several places, also leading to recurrent collapses with similar dimensions (Walter and Schmincke, 2002). Upon failure of a volcano flank, a strong morphological change occurs, manifested by a deep depression and a steep amphitheatre. The amphitheatre itself often provides the curved location of dyke intrusions (Delcamp et al., 2012; Tibaldi et al., 2008). Unloading due to flank instability and flank collapse not only depressurises deep magma plumbing systems (Manconi et al., 2009; Longpré et al., 2009), but also causes a change in the dyke paths at depth and therefore migration of volcanic activity into the sector of unloading (Maccaferri et al., 2017). The latter also explains, why at Fogo (Capo Verde), Tenerife (Canaries) and elsewhere the main eruption centre migrated seawards.

4.3 CONE-SHEET SWARMS

4.3.1 How Do Cone Sheets and Cone-Sheet Swarms Look Like?

Cone sheets, also called inclined sheets, represent important components of the shallow plumbing system of many volcanoes (Philipps, 1974). This characteristic shape of cone sheets has been noted since geologists started to map eroded volcanic complexes in the British and Irish Paleogene Igneous Province (BIPIP; e.g. Harker and Clough, 1904; Richey and Thomas, 1930; see also Chapter 1).

Cone sheets usually occur in swarms of hundreds to thousands of sheets in the uppermost kilometres of the crust. Individual intrusions of a cone-sheet swarm are thin (usually <1 m thick) and shaped like inverted cones. In eroded volcanoes, they therefore form elliptical or sub-circular traces at the Earth's surface, while they dip inwards at moderate angles (typically 30 to 60°) towards a common focus at depth (Fig. 4.1). This focus is assumed to be a pressurised magma source (Schirnick et al., 1999), such as the magma chamber that fed the cone sheet, an assumption that is supported by field observations of cone-sheet swarms and by modelling (e.g. Anderson, 1937; Schmincke, 1967; Philipps, 1974; Burchardt and Gudmundsson, 2009; Galland et al., 2014). For instance, in the field, the density of cone sheets relative to their host rocks increases towards solidified magma chambers, next to which cone-sheet density often reaches 100% of the exposed rocks, commonly intruding one next to the other (or even into each other) and uplifting the overburden (Le Bas, 1971).

Cone-sheet swarms can be some tens of kilometres across and are often confined by the size of the volcano. While the density of cone sheets in a swarm is low in the centre, high in a girdle around this centre and then decreases towards its periphery, a swarm can reach considerable volumes and therefore contribute significantly to the growth of volcanic edifices from within (Annen et al., 2001).

So, cone sheets are magma transport channels that feed magma from a shallow magma chamber into the volcanic edifice and in some cases to the volcano surface (Mori and McKee, 1987). When a cone sheet reaches the surface, it can feed a fissure eruption with a curved trace, circumferential to the volcano summit. However, few such circumferential eruptions are observed, which may indicate that most cone sheets remain intrusive. However, the high number of cone sheets observed in eroded swarms indicates that cone-sheet emplacement must be a frequently occurring process during the lifetime of a volcano.

Different varieties of the inverted cone shape have been proposed based on the change in dip with depth (Fig. 4.2): The simplest shape resembles a cocktail glass with a constant dip. If the dip of a cone-sheet decreases with depth, the overall shape resembles a bowl. A trumpet shape results from an increase of the dip with depth. In addition, there are often variations in the density and dip of cone sheets across the same swarm. The reason for the debate about the shape of the cone sheets at depth is partly that individual cone sheets, structures that can be more than 10 km in diameter and several kilometres

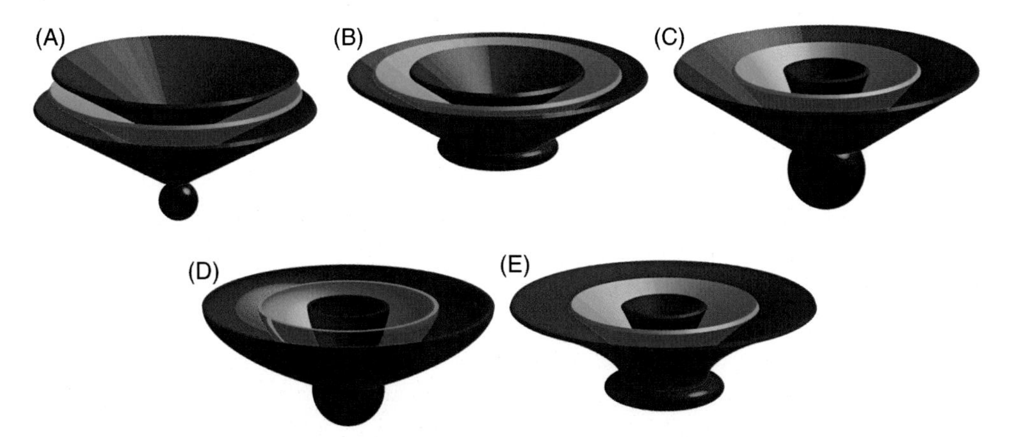

Figure 4.2 ***Schematic illustration of the proposed shapes of cone-sheet swarms.*** Modified from Schirnick et al. (1999) and Samrock (2015). (A–C) Different arrangements of straight cone sheets dipping towards a common source, producing a 'cocktail-glass' shape of the swarm. (D) A 'bowl'-shaped swarm geometry results from cone sheets with dips that decrease down-dip. (E) An increase in dip produces a 'trumpet'-shaped swarm.

high, are often not sufficiently well exposed in nature and too small to be mapped in the subsurface with geophysical methods.

The shape of the cone sheets and cone-sheet swarms are important, because it can be used to reconstruct the location, depth and shape of the underlying, unexposed source magma chamber. Assuming a certain shape of cone sheets results in different estimates for the location of the magma chamber (Burchardt et al., 2011). Such a reconstruction therefore requires careful mapping of the trace and attitudes of cone sheets across the whole volcano. Then 3D structural modelling can be employed to project the cone-sheet traces down-dip, while assuming a realistic overall geometry (Schirnick et al., 1999; Galland, 2012), allowing inference of basic emplacement processes and associated deformation (Guldstrand et al., 2017). The location where the projected cone sheets meet or intersect in the subsurface is assumed to be located within or below the source magma chamber. A cloud of cone-sheet intersections may so approximate the location and shape of the magma chamber (Burchardt et al., 2013).

One of the best exposed cone-sheet swarms worldwide is located on the Ardnamurchan peninsula in western Scotland. There, the traces of cone sheets encircle large parts of the exposed VIPS, which helped the earliest mappers to understand the geometry of cone sheets (Richey and Thomas, 1930). Detailed mapping of the cone-sheet traces and dips inspired Richey and Thomas (1930) to project the cone sheets in two-dimensional cross-sections and to prolong the sheet dips downward. Where the projected sheets met, Richey and Thomas (1930) identified three foci, which they interpreted as three successive magma chambers feeding the cone sheets. Indeed, careful mapping of the cone sheets and their cross-cutting relationships has revealed that volcanoes may have several intersecting cone-sheet swarms (e.g. Burchardt et al., 2011). However, one should bear in mind that magma chambers are not actually focal points at depth. Magma is stored in complex bodies with larger volumes and a variety of shapes (see also Chapters 5 and 6). Likewise, cone sheets are three-dimensional structures, and projecting them in 2D cross-sections represents a strong simplification (cf. Burchardt et al., 2013, appendix). Also, the swarm geometry may change as the pressure source at depth changes. Cone sheets might originate from a larger area or volume that can be assumed to correspond to the source magma body (e.g. Siler and Karson, 2009; Burchardt et al., 2011, 2013).

Studying cone-sheet swarms is therefore useful to unravel the complexity and temporal evolution of the unexposed volcanic plumbing system, including size, shape and location of the source magma chamber, as well as the migration of the location of the magma chamber with time or the existence of several magma chambers.

4.3.2 How are Cone Sheets Emplaced?

Since cone sheets are formed as magma-filled fractures, their overall shape is also a useful indicator of the stress field, in which they form, that is, the stress field around the underlying magma chamber, which controls many of the processes within the volcanic plumbing system. Using the cone sheets as stress-field indicators requires a closer look at the mechanics of cone-sheet emplacement. In the field, the cone sheets are often seen to displace the surrounding host rocks in pure opening mode. Cone sheets are therefore believed to be mode-I (tensile) fractures driven by the fluid overpressure of the magma (see also Chapter 3). By definition, their walls should be parallel to the principal compressive stresses σ_1 and σ_2, and they should open in the direction of σ_3. The overall cone-sheet swarm geometry is thus indicative of the trajectories of the principal stresses in the volcano at the time when the cone sheets were emplaced. Assuming that cone sheets are mode-I fractures is by far the most accepted model and goes back to an analytical model proposed by Anderson based on field observations in the BIPIP (Anderson, 1937). In the Anderson model, the push from a point-shaped magma chamber located at a given depth beneath the free surface creates inwardly inclined mode-I fractures that are steeper directly above the chamber and flattening outwards. Notably, the model predicts that a different type of intrusion, ring dykes, forms when the pressure in the source magma chamber drops. These are bell-shaped, steeply outward dipping and surround a block of host rock that subsides into the underlying magma chamber. This mechanism ultimately leads to caldera collapse (see Chapter 10).

While the Anderson model can explain the shape of the cone sheets and the lower density of sheets in the centre of the swarm, there are a few shortcomings that have inspired alternative models. The most important of these models proposes that the cone sheets are injected as shear-mode or mixed-mode fractures. The geometry of the cone sheets implies that opening of a cone-shaped fracture either compresses the rocks enclosed by the cone horizontally or that these rocks are lifted vertically. The latter requires less energy close to the Earth's surface, that is, at depths typical for cone-sheet

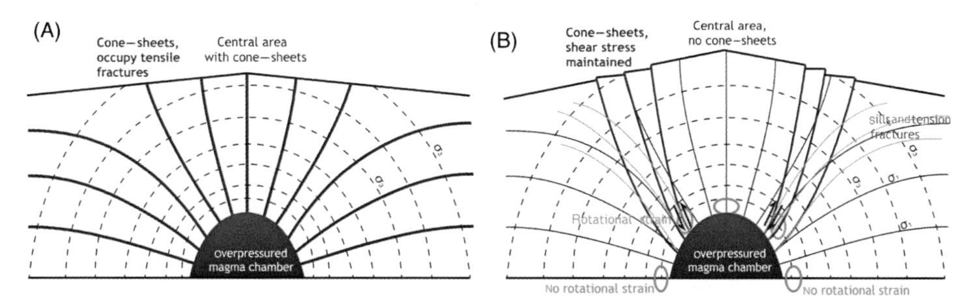

Figure 4.3 Schematic sketches of the proposed stress field around a shallow magma chamber and the resulting path of magmatic sheet intrusions according to (A) Anderson (1936) and (B) Phillips (1974). In both models, overpressure in the magma chamber induces stresses in the host rock. In the classic Anderson (1936) model, cone sheets propagate as tensile (mode I) fractures within the σ_1–σ_2 plane, while in the Phillips (1974) model, cone sheets occupy shear fractures and propagate at an angle to the principal stresses.

emplacement. Sheet opening accommodated by uplift implies displacement parallel to the sheet walls, that is, cone sheets would be shear-mode fractures. A modification of the Anderson model put forward by Phillips (1974) proposes that the cone sheets initiate when the upper lateral ends of the magma chamber fails in shear and that the cone sheets can propagate as either shear or tensile fractures, depending on the stress conditions and the mechanical properties of the host rock (Fig. 4.3). Experimentally, the cone sheets have been produced in materials that allow both tensile and shear failure and produced significant uplift at the model surface (Galland et al., 2014, 2015; Guldstrand et al., 2017).

Notably, the Anderson model and many of its variations imply that the stress field prior to and during cone-sheet emplacement is constant. However, when the cone sheets propagate as magma-driven fractures, they themselves have a very local stress field that is a sum of the interaction of the pressurised fracture walls and the variations in host-rock strength. Hence, cone-sheet propagation is a dynamic process, during which the stress field changes locally (Mathieu et al., 2015).

4.4 RADIAL DYKE SWARMS

In addition to volcanic rift zones and cone sheet swarms, another type of sheet intrusion swarms is arranged in a radiating fashion around shallow magma reservoirs and hence called radial dyke swarms (Fig. 4.1). Radial dykes have been found in many volcanoes on Earth and other terrestrial planets (Ernst et al., 1995). While perfectly radial arrangements in radial dyke swarms do exist (e.g. at Fernandina, Galápagos; Chadwick and Dieterich, 1995; Kliuchevskoi, Kamchatka; Takada, 1997), most volcanoes exhibit a preferred orientation among the radial dykes (e.g. Etna and Stromboli, Italy; Acocella and Neri, 2003; Sakurajima, Japan; Takada, 1997; Erta Ale, Ethiopia; Acocella, 2006a).

Field observations and analyses of flow indicators in radial dykes show that magma is transported laterally or vertically away from the source magma chamber for distances of up to several hundred kilometres (Ernst et al., 1995). Alternatively, radial dykes can be injected from a central conduit (see Section 4.6), from where they may propagate laterally and even downslope. Here, the configuration of the conduit (open or closed) and the degassing of the magma control radial dyke emplacement (Acocella et al., 2006). When reaching the Earth's surface, radial dykes feed fissure eruptions that are radial to the volcano summit (Chadwick and Dieterich, 1995). The resulting radial dyke swarms thus play a significant role in both the growth of the volcanic edifice and in long-distance magma transport.

As radial dykes are generally assumed to propagate as mode-I fractures, they represent useful indicators of the stress field at the time of their emplacement. The radial pattern of the maximum principal compressive stress σ_1 around the source magma chamber of the dykes has been modelled numerically and shown to result from the interplay of the stresses generated by magma pressure in the chamber and the gravitational stresses caused by the load of the volcanic edifice (e.g. Grosfils, 2007). While the magma overpressure in the source chamber would induce dykes that propagate upwards, the load of the edifice causes dyke propagation away from the volcano centre.

Preferred orientations of dykes belonging to a radial swarm can be explained according to Acocella and Neri (2009) (Fig. 4.4) by (i) the influence of regional tectonic stresses (Nakamura, 1977), (ii) the elongated shape of the volcanic edifice, (iii) fault or collapse scars or (iv) instable volcano flanks (e.g. Fiske and Jackson, 1972; Walter et al., 2006).

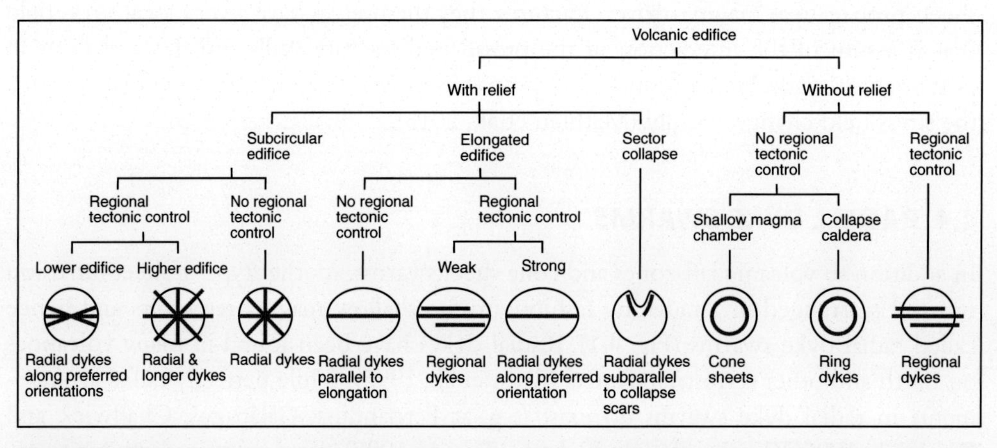

Figure 4.4 ***Main parameters controlling the arrangement of magmatic sheets according to Acocella and Neri (2009).***

4.5 WHY DO DIFFERENT TYPES OF SHEET INTRUSIONS OCCUR IN THE SAME VOLCANO?

While some volcanic plumbing systems predominantly host either rift zones, cone sheets or radial dykes, there are numerous examples where all three types of sheet intrusions occur in the same volcano. Moreover, cross-cutting relationships show that cone sheets and steeply dipping dykes were emplaced roughly contemporaneously and may even originate from the same magma chamber (Walker, 1992). Can we explain why these fundamentally different dyke arrangements form, although all three types suggest a central volcanic source?

Analogue experiments simulating magmatic sheet intrusions have identified two parameters that control the type of sheet intrusion formed (Fig. 4.5) (Galland et al., 2014). The first parameter characterises the geometrical configuration of the plumbing system, specifically the ratio between the depth and the lateral extent of the source, where the source corresponds to the feeder (e.g. the rupture along a magma chamber wall or the tip of a dyke). Here, the cone sheets are preferentially formed when the source is shallow compared to its extent, while dykes initiate from a deeper source.

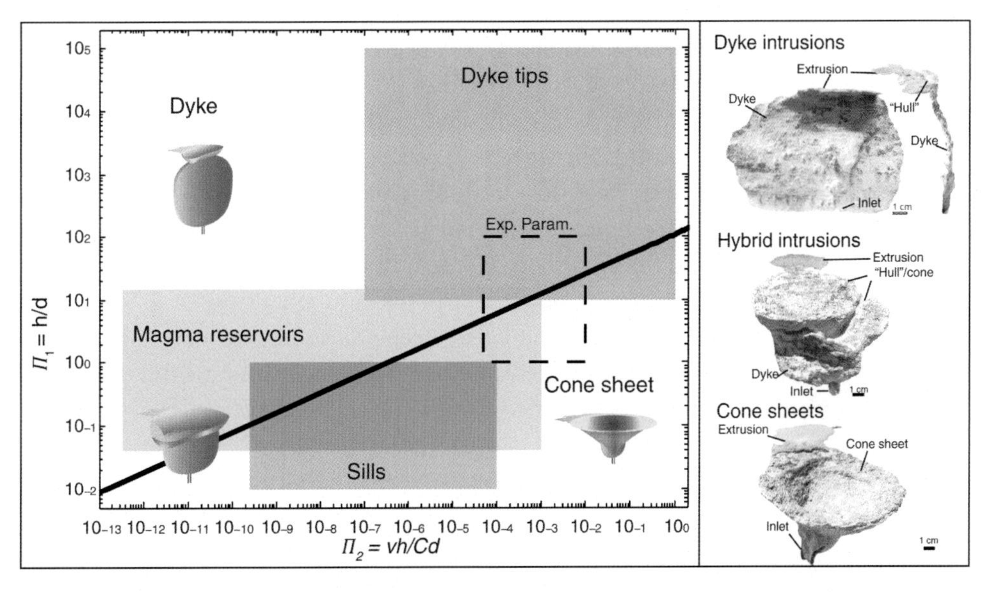

Figure 4.5 ***Experimentally identified parameters Π_1 and Π_2 that control the type of sheet intrusion forming in the same volcano.*** Adapted from Galland et al. (2014). Π_1 describes the geometry of the magmatic feeder, that is, its height to diameter ratio *h*/*d*. The feeder can be, for example, a dyke tip or a fracture in the wall of a magma chamber. Π_2 defines the interaction between the viscous magma stresses (magma viscosity *v*) to the properties of the host rock (host-rock cohesion *C*) in relation to the geometry of the system as *vh*/*Cd*. The column to the right shows photographs of experimentally produced dykes, cone sheets and hybrid intrusions.

The second controlling parameter describes the mechanical properties and their effect on the stresses during sheet initiation, more specifically the ratio between the viscous stresses in the magma and the host-rock cohesion. This second parameter predicts that dykes are preferentially formed at lower viscous stresses relative to the host-rock strength compared to the cone sheets. This implies that the cone sheets should form at higher magma influx rates and higher magma viscosities (i.e. higher viscous stresses).

The predictions that arise from these experiments are generally supported by field observations that show, for example, that the cone sheets preferentially occur in volcanoes with a shallow magma chamber and that dykes tend to become more frequent compared to cone sheets towards the end of the lifetime of the magma chamber when the magma influx is expected to be lower (Geshi, 2005).

Adding the influence of the load of the volcanic edifice, its shape and height, as well as regional tectonic extension, may make the sheets deviate and form elongated rift zones (Nakamura, 1977).

4.6 VOLCANIC CONDUITS

Volcanic conduits are generated by intrusive events that erupt magma to the surface and provide the locus of shallow magma ascent from deeper in the volcanic plumbing system. Conduit-forming eruptions may be either discrete and short-lived or pulsatory, longer-duration events and are typically initiated by the upward propagation of steeply inclined magmatic dykes towards the surface (Costa et al., 2007).

The contrast between deep sheet-like intrusions and well-defined pipe-like conduits and vents close to the surface represented a curiosity already observed by early geologists. So how to form a tube from a sheet? Generally, fluid and thermal flow through a pipe is energetically more efficient compared to flow through a dyke (e.g. Slezin, 2003; Gonnermann and Manga, 2007). Two main concepts have been used to describe the formation of conduits from feeder dykes, one being strictly related to the thermoplastic material flow and another one related to gas-rich magma propagation in a brittle host rock.

When magma is gas-poor, only modest overpressure develops during magma ascent and so only minor fracturing of magma and country rocks occurs. As a result, little damage to the intrusive walls occurs, and so feeder dykes may remain narrow (metres) where they intersect the surface (Delcamp et al., 2014). However, gas-rich silicic magmas develop significant overpressure within hundreds of metres of the surface, driving fracturing of country rock and magmatic fragmentation (Heiken et al., 1988; Stasiuk et al., 1996). This leads to the formation of a dense network of fractures in the feeder dyke walls (Fig. 4.6) and the forceful injection of pyroclastic material within these fractures forming tuffisite veins. The resultant halo of damage facilitates wall erosion, the widening of the magmatic pathway and the generation of a conduit with a more equant,

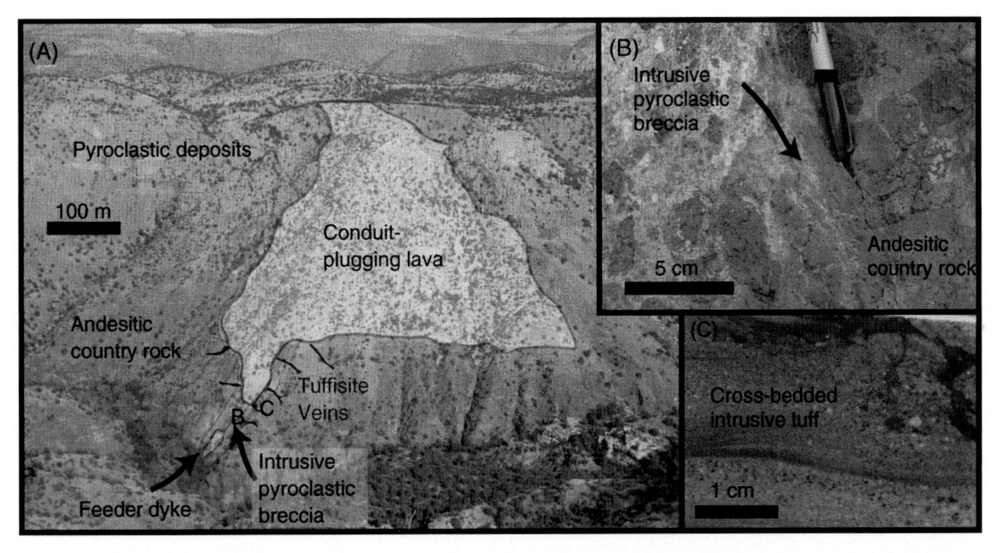

Figure 4.6 ***The dissected rhyolite conduit at Mule Creek, New Mexico (Stasiuk et al., 1996).*** (A) Overview of the conduit and feeder dyke, which are dissected to a depth of 350 m. A screen of intrusive pyroclastic breccia separates the dense conduit-plugging lava from fractured andesitic country rock, which is cut by networks of tuffisite veins traced to >40 m from the conduit wall. (B) Detail of the outer margin of the intrusive pyroclastic breccia, with forceful injection of pyroclastic material creating a dense network of tuffisite veins in the adjacent fractured andesitic country rock. (C) Photomicrograph of sedimentary structures in a tuffisite vein, with delicate cross-bedding that is later intruded by fine-grained dark red material. Such structures record repetitive intrusion of pyroclastic material at the conduit margin, driven by pulses of pressurised magmatic gas.

although not necessarily circular cross-section. The depth of the dyke-conduit transition depends on the magmatic overpressure and flux and the mechanical properties of the country rocks, but is thought to typically range from ~1.5 km to hundreds of metres in silicic eruptions (Fig. 4.6; Stasiuk et al., 1996; Costa et al., 2007; Goto et al., 2008).

Once the magmatic conduit has become established, the main phase of magma discharge then ensues, typically involving a gradual transition from more explosive to more effusive behaviour. The evolving width of the conduit is determined by the competition between country rock erosion and the aggregation of fragmental material onto conduit walls, with the latter becoming predominant in the waning phases of the eruption. Magmatic gas may escape through a combination of connected vesicles and fractures, or through brecciated country rock at conduit margins (Fig. 4.6; Stasiuk et al., 1996; Schauroth et al., 2016). Dense welding of melt-rich material at conduit walls may trigger a drastic reduction in permeability, inhibiting magmatic outgassing and enhancing overpressure.

This mechanism is proposed to have driven the intrusion of a syn-eruptive laccolith at Cordón Caulle during the first month of the 2011–2012 rhyolitic eruption (Castro et al., 2016). The timing of laccolith intrusion coincided with a decrease in conduit size from a linear fissure many hundreds of metres in length to a near-circular feature only ~50 m in diameter. Elastic deformation models indicate that an overpressure of ~10 MPa developed within the shallow conduit, sufficient to drive tensile conduit wall failure and the horizontal injection of approximately 0.8 km^3 of magma at a depth of only ~200 m.

As the intensity of eruptions dwindles, a dense magmatic plug develops, characterised by zones of highly localised magma deformation and compaction of brecciated and/or foamed material (Tuffen and Dingwell, 2005). It is this phase of conduit evolution that is best preserved within dissected conduits (Fig. 4.6), together with the initial conduit-opening phase (Stasiuk et al., 1996), whereas the geological record of the main magmatic phase is largely obliterated by erosion or overprinted by later coherent magma.

Drilling of the conduit that fed the 1990–1995 dacite eruption at Unzen, Japan, has provided a unique opportunity to characterise a conduit associated with an observed eruption. Core samples from ~1500 m depth indicate that coherent conduit-centre dacite is enveloped by a zone of highly brecciated juvenile material. All facies are cut by volcaniclastic dykes attributed to transient opening of fluid-filled hydrofractures (Goto et al., 2008) and similar to those documented around dissected conduits (Stasiuk et al., 1996) and borehole samples from the Inyo rhyolite conduit (Heiken et al., 1988). Such hydrofracture opening is thought to be the source mechanism of the low-frequency volcanic microearthquakes used to monitor subsurface magma movement (Fazio et al., 2017).

4.7 SUMMARY

Magmatic sheet intrusions and conduits are components of the shallow VIPS of many long-lived volcanoes. Pressurised magma source regions, such as crustal magma chambers, may frequently inject sheet intrusions into the volcanic edifice. These sheets can form dense swarms that considerably contribute to the volumetric growth of volcanic edifices.

Sheet intrusions are commonly thought to originate from a volcanic centre, and from there propagate laterally and vertically through the crust, partly shaping systematic intrusion patterns. Intrusions occurring from localised magma sources may lead to either concentrically arranged sheet intrusions (cone sheets) or to radially arranged sheet intrusions diverging away from the source chamber. Upon addition of a remote stress field or a local volcano loading stress, these patterns may become more elongate or even dominate in single or multiple rift zones. In most of the cases, two or three rift axes develop on ocean islands, and their focal point is located above the centre of the underlying magma source. The expression of rift zones and their activity is changing with changing stress fields of the shallow volcano edifice and closely interacts with flank instability and topographic changes. While dyke intrusion into volcanic rift zones contributes to the growth of the rift-zone volcano, the horizontal push from the added magma can destabilise the volcanic edifice and lead to catastrophic flank collapse.

The two other types of magmatic sheet intrusions, radial dykes and cone sheets, may form in very similar environments. The latter are thin magmatic sheets with the shape of an inverted cone and the source magma chamber being located at the tip of the cone. The cone itself may have a variety of geometries, for example, resembling a trumpet, a bowl or a cocktail glass. The geometry of the cone sheets is important, because it may be used to reconstruct the location, shape and size of the underlying, unexposed magma chamber that fed the cone sheets. Changing locations and shapes of magma chambers in theory might change the expression and evolution of sheet intrusions.

Radial dykes form swarms of sub-vertical sheet intrusions with a radial pattern in map view. Their arrangement reflects a volcanic stress field dominated by the load of the volcanic edifice. Radial dykes may occur in the same volcanoes as cone sheets, and which of the two intrusion types forms is controlled by the interplay between the geometry of the magmatic feeder and the viscous stresses that result from the interaction of the magmas viscosity and the host-rock properties.

When a sheet intrusion intersects the volcano surface, it can feed an eruption from a volcanic fissure. Due to fracturing of the wall rocks and thermal erosion within the fissure, magma transport to the surface may focus into a volcanic conduit. Conduits are pipe-like structures that eject magma, volcanic gases and host-rock fragments. This process coincides with shearing, fracturing, heating, cooling, healing and magma and pyroclastic intrusion. Conduits are therefore complex, and their geometry and permeability strongly control the type and magnitude of an eruption.

FIELD EXAMPLE: INTRUSIVE SHEET SWARMS IN THE CANARY ISLANDS

The Canary Islands are a group of ocean island volcanoes located off the coast of western Africa that have been formed by volcanic activity since around 20 Ma ago (Schmincke, 2006). The Canarian archipelago consists of seven major islands and is characterised by long-lived volcanic activity that results from episodically rising blobs of mantle material (Hoernle and Schmincke, 1993; Zaczek et al., 2015). Due to the westward progression of volcanic activity, ocean island volcanoes in different stages of their development can be observed on the different islands. Generally, surface volcanism on ocean island volcanoes is characterised by successive phases of rift-zone and central-type activity (Carracedo, 1994). The Canary Islands comprise examples of both active and eroded volcanoes with a variety of magmatic sheet intrusion swarms.

On Tenerife, there are three volcanic rift zones, extending to the northeast, the northwest, and to the south from the centre of the island, characterised by numerous eruptive vents and fissures fed by rift-zone dykes (Fig. 4.7). While the northwest rift has been

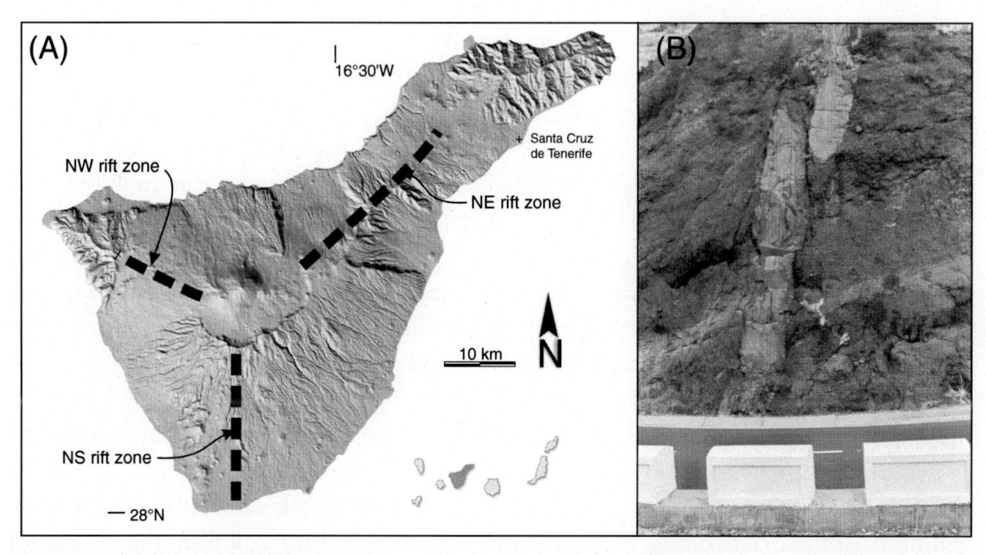

Figure 4.7 (A) Shaded-relief image of the island of Tenerife, Canary Islands, with the orientation of volcanic rift zones marked. Inset shows the location of Tenerife in the archipelago. (B) Photograph of a segmented dyke exposed in a road cut in the NE rift zone. The dyke is about 1.5 m thick where it is cut by the road.

very active during the last 2 Ma (Carracedo et al., 2007), the northeast rift zone is only sparsely active and deeply dissected by several flank collapses (Carracedo et al., 2011). The associated collapse scars and the deep erosion expose the internal structure of the northeast rift zone with numerous sub-vertical dykes that strike mostly sub-parallel to the rift zone (Fig. 4.7; Delcamp et al., 2012). However, locally, dyke orientations deviate from the rift-zone trend, indicating emplacement into a deforming volcano flank in a mature rift prior to catastrophic flank collapse (Delcamp et al., 2012).

On La Palma, the northern part of the island is built up by the remnants of the extinct Taburiente volcano, a large multi-stage shield volcano that experienced at least one major collapse resulting in a landslide and subsequent rebuilding of the edifice (Ancochea et al., 1994; Carracedo et al., 1998; Carracedo, 1999; Guillou et al., 2001). The erosional Caldera de Taburiente (see also Chapter 9) exposes a several hundred-metre-tall section through the volcanic edifice of the Taburiente volcano (Fig. 4.8A). In this section, lava layers and pyroclastic deposits erupted from the volcano alternate. Several hundred dykes cut these layers in what at first glance seems to be a chaotic pattern (Fig. 4.8B and C). However, mapping of the orientation of the dykes reveals a predominantly radial pattern. When a radial dyke intersected the volcano surface, it fed an eruption along a radial fissure, contributing to the growth of the volcanic edifice.

The central part of Gran Canaria consists of the remnants of a large caldera volcano that erupted numerous ignimbrites and lava flows during the Miocene (Fig. 4.9A;

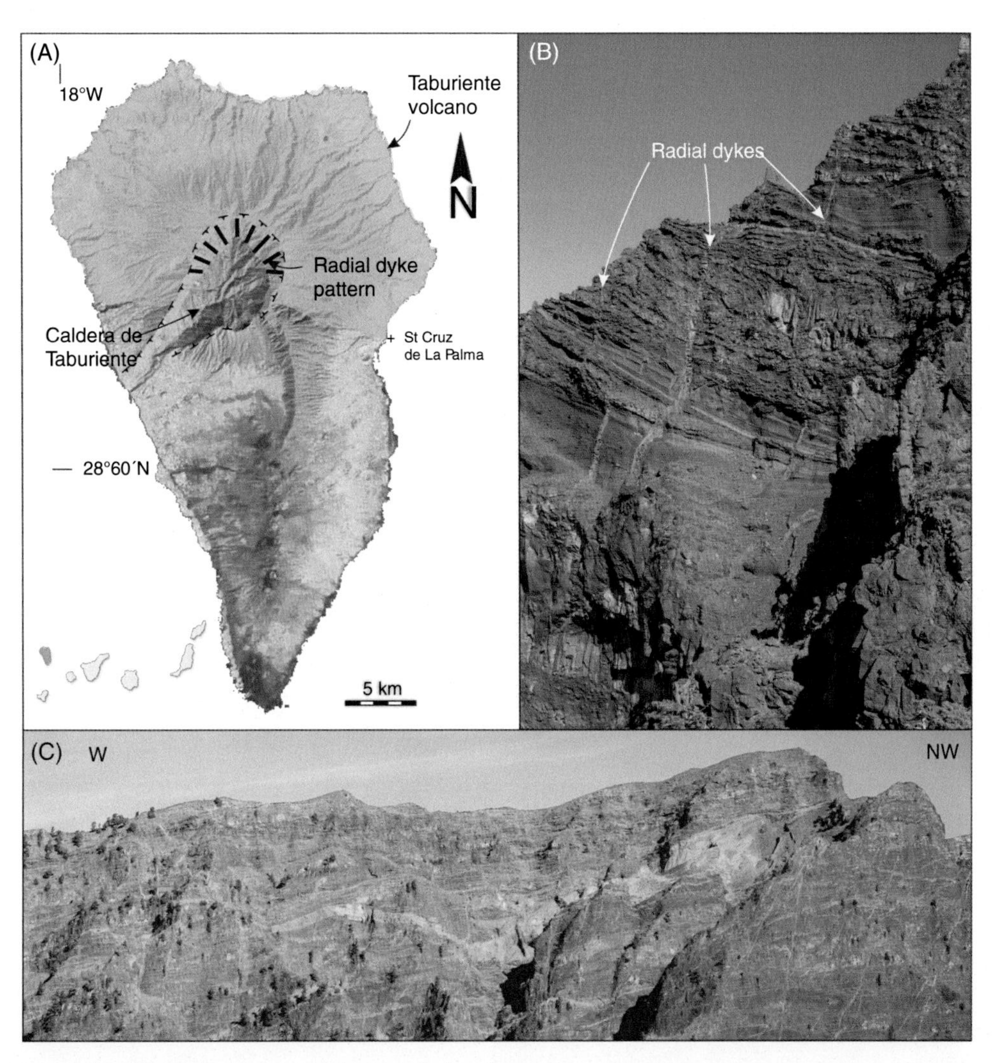

Figure 4.8 (A) Satellite image of the island of La Palma, Canary Islands, with the location of the remnants of the Taburiente shield volcano in orange. Inset shows location of La Palma in the archipelago. (B) Photograph of radial dykes cutting through layers of pyroclastic rocks (*red* and *orange*) and lavas (*grey* and *dark brown*) in the wall of the Caldera de Taburiente. The dykes in the photograph are some dm to about a metre thick. (C) A part of the steep cliff that forms the wall of the Caldera de Taburiente and exposes a section through the Taburiente volcano. Hundreds of dykes that belong to a radial dyke swarm cut the extrusive rocks of the volcano.

Freundt and Schmincke, 1995). During the second stage of the volcano's lifetime (corresponding to the Fataga group), thousands of cone sheets belonging to the so-called Tejeda cone sheet swarm, as well as several syenite plutons, intruded the interior of the volcano and the caldera infill (Fig. 4.9B; Schirnick et al., 1999). The Tejeda cone-sheet swarm is exposed in the deeply eroded central gullies and cliffs of Gran Canaria

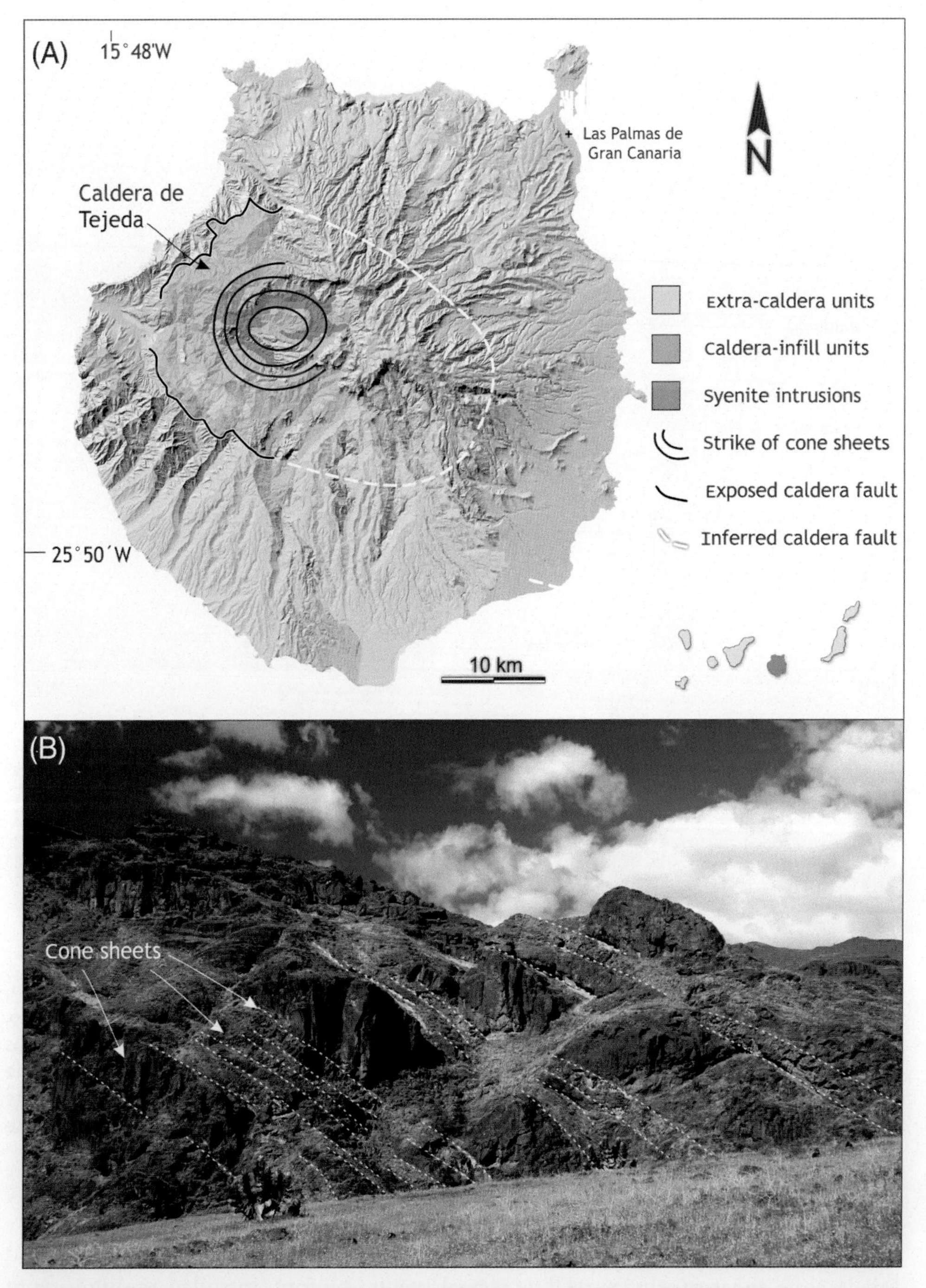

Figure 4.9 (A) Shaded-relief image of the island of Gran Canaria, Canary Islands, with the units of the Miocene Tejeda caldera volcano marked. Inset shows the location of Gran Canaria in the archipelago. (B) Photograph of the cone sheets cutting caldera-infill rocks within the Caldera de Tejeda. Cone sheets in this outcrop are several metres thick.

(Fig. 4.9A). There the swarm extends over 15 km in diameter and almost 1500 m in height, which makes this the probably best-exposed cone-sheet swarm in the world. Mapping of the Tejeda cone sheets (Schirnick et al., 1999; Samrock, 2015) reveals a circular shape of the swarm in the map view. In the swarm centre, the density of the cone sheets is considerably lower (<20% of the exposure), compared to the surrounding girdle, in which 75 to >90% of the exposure is made of the cone sheets. Towards the periphery, cone-sheet density decreases to <60% (Schirnick et al., 1999). Detailed mapping, structural analysis and 3D modelling show that the Tejeda cone sheets exhibit straight conical geometries that give the swarm an overall cocktail-glass shape (see Figs. 4.2 and 4.9C). The source magma chamber that fed the cone sheets was most probably comprised one main ellipsoidal magma chamber, $9 \times 6\ km^2$ in horizontal extend at depths between 800 and 4000 m below the present-day sea level (Fig. 4.9C; Samrock, 2015).

FIELD EXAMPLE: EXPOSED CONDUITS IN CENTRAL ICELAND

Dissected volcanic complexes in Iceland provide superb exposure of the magmatic plumbing system of central volcanoes, which typically comprise a basaltic fissure swarm and a silicic magma chamber (e.g. Gudmundsson, 1995). Dissection to 1–2 km in the East Fjords reveals the deeper plumbing system and relation between gabbroic intrusions and silicic magma chambers (Walker, 1963; Burchardt et al., 2011), whereas Pleistocene rhyolitic conduits at Torfajokull and Krafla are dissected to modest depths of 40–100 m (Tuffen and Dingwell, 2005; Tuffen and Castro, 2009) and preserve key information about the deformation and degassing behaviour of magma during shallow ascent in the conduit. Húsafell central volcano in west-central Iceland is a dissected silicic system with an area of $\sim 120\ km^2$, and eruptive activity that spanned 3–2.5 Ma (Sæmundsson and Noll, 1974). The main phases of silicic activity involved emplacement of extensive, variably-welded rhyolitic ignimbrites, together with dacitic and andesitic lavas. Late-stage rhyolitic intrusions, including sills, dykes and conduits, are well exposed in the Hringsgil and Deildargil stream valleys and highly accessible (Fig. 4.10). Magmatic H_2O contents in glassy intrusion margins indicate dissection to a depth of ~500 m (McGowan, 2016). Detailed characterisation of textural zones within intrusions has allowed reconstruction of the chronology of conduit opening and magma emplacement. This spans initial fracturing of country rock by a pressurised gas–pyroclast mixture, creating tuffisite veins (Fig. 4.10B), followed by intrusion of pyroclastic breccias, which become densely welded towards conduit centres, where they grade into an intact lava plug (Fig. 4.10A). Aspects of interest at Húsafell include the clear influence of country rock on dyke/sill propagation, and the interplay between conduit widening via erosion of the country rock and constriction by the accumulation and welding of pyroclastic debris.

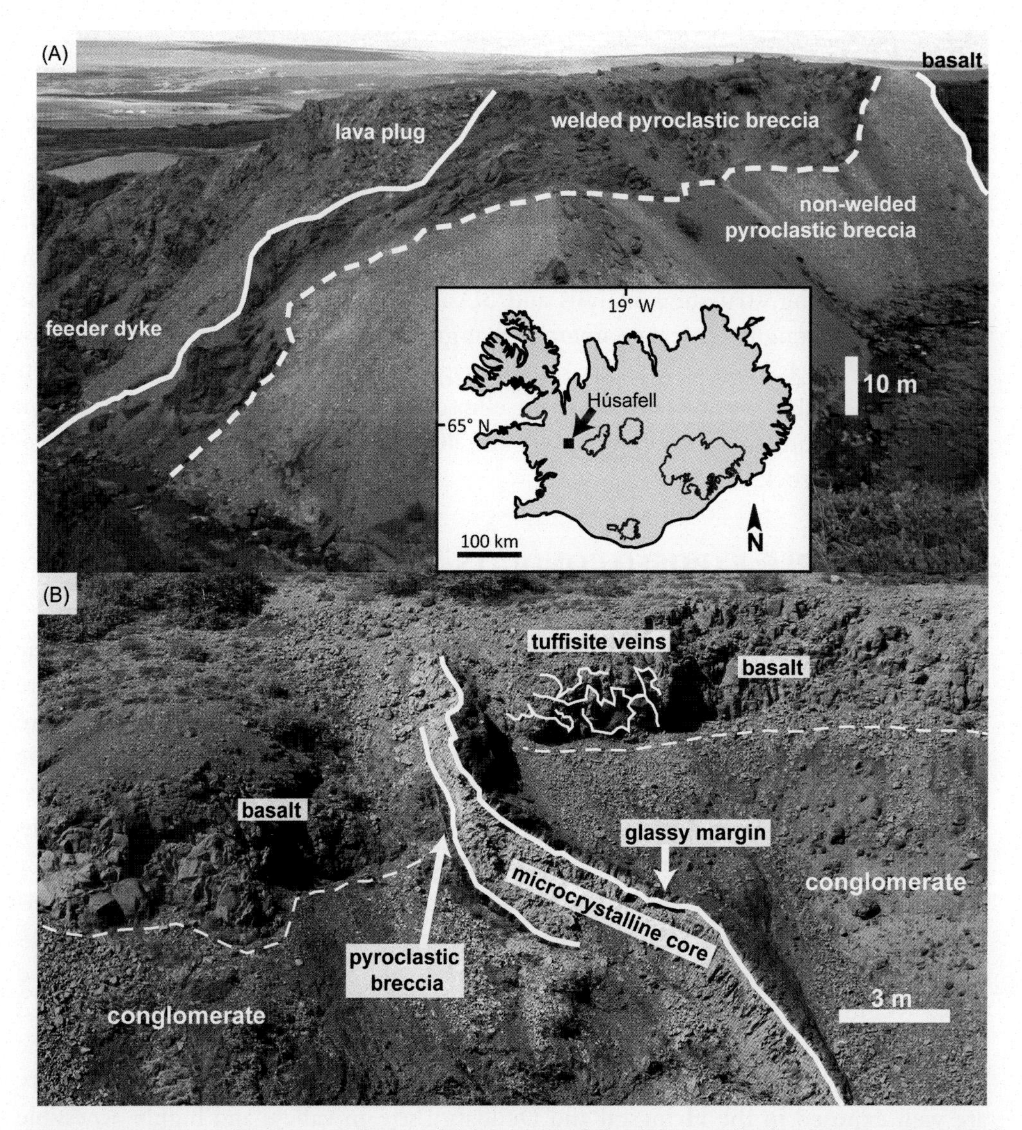

Figure 4.10 Dissected Tertiary conduits and dykes at Húsafell, west-central Iceland (Sæmundsson and Noll, 1974; McGowan, 2016). (A) A shallow upward-flaring rhyolitic conduit at Hringsgil (northwest). Massive pumice-dominated pyroclastic breccias gradually increase in welding intensity towards a coherent lava plug (left-hand side), reflecting focussed magma discharge during the eruption and a transition towards more effusive activity. The steeply inclined contact with basaltic country rock is exposed top right. (B) A rhyolitic dyke in Deildargil (northwest) with well-exposed contacts against basalt and conglomerate country rock. A network of pyroclast-filled fractures (tuffisite veins) records gas-driven dyke propagation, whilst concentric zones of pyroclastic breccia, dense obsidian and microcrystalline rhyolite record the evolution from explosive to effusive activity. Dissection depth ~500 m (McGowan, 2016).

REFERENCES

Acocella, V., 2006a. Regional and local tectonics at Erta Ale caldera, Afar (Ethiopia). J. Struct. Geol. 28 (10), 1808–1820.

Acocella, V., Neri, M., Scarlato, P., 2006. Understanding shallow magma emplacement at volcanoes: orthogonal feeder dikes during the 2002–2003 Stromboli (Italy) eruption. Geophys. Res. Lett. 33 (17).

Acocella, V., Neri, M., 2003. What makes flank eruptions? The 2001 Etna eruption and its possible triggering mechanisms. Bull. Volcanol. 65 (7), 517–529.

Acocella, V., Neri, M., 2009. Dike propagation in volcanic edifices: overview and possible developments. Tectonophysics 471 (1), 67–77.

Ancochea, E., Hernán, F., Cendrero, A., Cantagrel, J.M., Fúster, J., Ibarrola, E., Coello, J., 1994. Constructive and destructive episodes in the building of a young oceanic island, La Palma, Canary Islands, and genesis of the Caldera de Taburiente. J. Volcanol. Geoth. Res. 60 (3–4), 243–262.

Anderson, E.M., 1936. Dynamics of formation of cone-sheets, ring-dykes and cauldron-subsidences. Proc. Roy. Soc. Edinb. 61, 128–157.

Anderson, E.M., 1937. Cone-sheets and ring-dykes: the dynamical explanation. Bull. Volcanol. 1, 35–40.

Annen, C., Lénat, J.-F., Provost, A., 2001. The long-term growth of volcanic edifices: numerical modelling of the role of dyke intrusion and lava-flow emplacement. J. Volcanol. Geotherm. Res. 105, 263–289.

Baker, S., Amelung, F., 2015. Pressurized magma reservoir within the east rift zone of K lauea Volcano, Hawai 'i: Evidence for relaxed stress changes from the 1975 Kalapana earthquake. Geophys. Res. Lett. 42 (6), 1758–1765.

Becerril, L., Galindo, I., Martí, J., Gudmundsson, A., 2015. Three-armed rifts or masked radial pattern of eruptive fissures? The intriguing case of El Hierro volcano (Canary Islands). Tectonophysics 647, 33–47.

Burchardt, S., Gudmundsson, A., 2009. The infrastructure of Geitafell Volcano, Southeast Iceland. In: Thordarson, T., Self, S., Larsen, G., Rowland, S., Hoskuldsson, A. (Eds.), Studies in Volcanology: The Legacy of George Walker. Special Publications of IAVCEI 2. Geological Society, London, pp. 349–370.

Burchardt, S., Tanner, D.C., Troll, V.R., Krumbholz, M., Gustafsson, L.E., 2011. Three-dimensional geometry of concentric intrusive sheet swarms in the Geitafell and the Dyrfjöll volcanoes, eastern Iceland. Geochem. Geophys. Geosyst. 12 (7), Q0AB09. doi: 10.1029/2011GC003527.

Burchardt, S., Troll, V.R., Mathieu, L., Emeleus, H.C., Donaldson, C.H., 2013. Ardnamurchan 3D cone-sheet architecture explained by a single elongate magma chamber. Sci. Rep. 3, 2891.

Carracedo, J.-C., 1994. The Canary Islands: an example of structural control on the growth of large oceanic-island volcanoes. J. Volcanol. Geotherm. Res. 60, 225–241.

Carracedo, J.C., 1996. A simple model for the genesis of large gravitational landslide hazards in the Caray Islands. In: McGuire, W.J., Jones, A.P., Neuberg, J. (Eds.), Volcano Instability on the Earth and other Planets. Geological Society of London Special Publications 110, pp. 125–135.

Carracedo, J.C., 1999. Growth, structure, instability and collapse of Canarian volcanoes and comparisons with Hawaiian volcanoes. J. Volcanol. Geoth. Res. 94, 1–19.

Carracedo, J.C., Day, S., Guillou, H., Rodríguez Badiolas, E., Canas, J.A., Pérez Torrado, F.J., 1998. Hotspot volcanism close to a passive continental margin: the Canary Islands. Geol. Mag. 135, 591–604.

Carracedo, J.C., Guillou, H., Nomade, S., Rodríguez-Badiola, E., Pérez-Torrado, F.J., Rodríguez-González, A., Paris, R., Troll, V.R., Wiesmaier, S., Delcamp, A., Fernández-Turiel, J.L., 2011. Evolution of ocean-island rifts: The northeast rift zone of Tenerife, Canary Islands. Geol. Soc. Am. Bull. 123, 562–584, https://doi.org/10.1130/B30119.1.

Carracedo, J.C., Rodríguez Badiola, E., Guillou, H., Paterne, M., Scaillet, S., Pérez Torrado, F.J., Paris, R., Fra-Paleo, U., Hansen, A., 2007. Eruptive and structural history of Teide Volcano and rift zones of Tenerife, Canary Islands. Geol. Soc. Am. Bull. 119, 1027–1051.

Castro J.M., Cordonnier, B., Schipper, C.I., Tuffen, H., Baumann, T.S., Feisel, Y., 2016. Rapid laccolith intrusion driven by explosive volcanic eruption. Nat. Commun. 7, art. no. 13585.

Cayol, V., Dieterich, J.H., Okamura, A.T., Miklius, A., 2000. High magma storage rates before the 1983 eruption of Kilauea, Hawaii. Science 288 (5475), 2343–2346.

Chadwick, W.W., Dieterich, J.H., 1995. Mechanical modeling of circumferential and radial dyke intrusion on Galapagos volcanoes. J. Volcanol. Geotherm. Res. 66, 37–52.

Clarke, D., Brenguier, F., Froger, J.L., Shapiro, N.M., Peltier, A., Staudacher, T., 2013. Timing of a large volcanic flank movement at Piton de la Fournaise Volcano using noise-based seismic monitoring and ground deformation measurements. Geophys. J. Int. 195 (2), 1132–1140.

Costa, A., Melnik, O., Sparks, R.S.J., Voight, B., 2007. Control of magma flow in dykes on cyclic lava dome extrusion. Geophys. Res. Lett. 34, L02303.

Delaney, P.T., Denlinger, R.P., Lisowski, M., Miklius, A., Okubo, P.G., Okamura, A.T., Sako, M.K., 1998. Volcanic spreading at Kilauea, 1976–1996. J. Geophys. Res. 103 (B8), 18003–18023.

Delcamp, A., Troll, V.R., de Vries, B.V.W., Carracedo, J.C., Petronis, M.S., Pérez-Torrado, F.J., Deegan, F.M., 2012. Dykes and structures of the NE rift of Tenerife, Canary Islands: a record of stabilisation and destabilisation of ocean island rift zones. Bull. Volcanol. 74 (5), 963–980.

Delcamp, A., van Wyk de Vries, B., Stéphane, P., Kervyn, M., 2014. Endogenous and exogenous growth of the monogenetic Lemptégy volcano, Chaîne des Puys, France. Geosphere 10 (5), 998–1019.

Di Traglia, F., Nolesini, T., Intrieri, E., Mugnai, F., Leva, D., Rosi, M., Casagli, N., 2014. Review of ten years of volcano deformations recorded by the ground-based InSAR monitoring system at Stromboli volcano: a tool to mitigate volcano flank dynamics and intense volcanic activity. Earth-Sci. Rev. 139, 317–335.

Dieterich, J.H., 1988. Growth and persistence of Hawaiian volcanic rift zones. J. Geophys. Res. 93 (B5), 4258–4270.

Ebmeier, S.K., Biggs, J., Mather, T.A., Wadge, G., Amelung, F., 2010. Steady downslope movement on the western flank of Arenal volcano, Costa Rica. Geochem. Geophys. Geosyst. 11 (12).

Elsworth, D., Voight, B., 1995. Dike intrusion as a trigger for large earthquakes and the failure of volcano flanks. J. Geophys. Res. 100 (B4), 6005–6024.

Ernst, R.E., Head, J.W., Parfitt, E., Grosfils, E., Wilson, L., 1995. Giant radiating dyke swarms on Earth and Venus. Earth Sci. Rev. 39, 1–58.

Fazio, M., Benson, P.M., Vinciguerra, S., 2017. On the generation mechanisms of fluid-driven seismic signals related to volcano-tectonics. Geophys. Res. Lett. 44, 734–742.

Fiske, R.S., Jackson, E.D., 1972. Orientation and growth of Hawaiian volcanic rifts: the effect of regional structure and gravitational stresses. Proceedings of the Royal Society of London A: Mathematical, Physical and Engineering Sciences, vol. 329, No. 1578. The Royal Society, pp. 299–326.

Fornari, D.J., Campbell, J.F., 1987. Volcanism in Hawaii. US Geological Survey Professional Paper 1 (1350), 109.

Freundt, A., Schmincke, H.-U., 1995. Petrogenesis of rhyolite-trachyte-basalt composite ignimbrite P1, Gran Canaria, Canary Islands. J. Geophys. Res. Solid Earth 100, 455–474. doi: 10.1029/94JB02478.

Galland, O., Burchardt, S., Hallot, E., Mourgues, R., Bulois, C., 2014. Dynamics of dykes versus cone sheets in volcanic systems. J. Geophys. Res. Solid Earth 119, 6178–6192.

Galland, O., Holohan, E., de Vries, B. van W., Burchardt, S., 2015. Laboratory modelling of volcano plumbing systems: a review. In: Advances in Valcanology. Springer, Berlin, pp. 1–65.

Galland, O., 2012. Experimental modelling of ground deformation associated with shallow magma intrusions. Earth Planet. Sci. Lett. 317, 145–156.

Guldstrand, F., Burchardt, S., Hallot, E., Galland, O., 2017. Dynamics of surface deformation induced by dikes and cone sheets in a cohesive Coulomb brittle crust. J. Geophys Res. 122 (10), 8511–8524.

Guillou, H., Carracedo, J.C., Duncan, R.A., 2001. K-Ar, 40Ar-39Ar ages and magnetostratigraphy of Brunhes and Matuyama lava sequences from La Palma. J. Volcanol. Geoth. Res. 106, 175–194.

Geshi, N., 2005. Structural development of dyke swarms controlled by the change of magma supply rate: the cone sheets and parallel dyke swarms of the Miocene Otoge igneous complex, Central Japan. J. Volcanol. Geotherm. Res. 141, 267–281.

González, P.J., Tiampo, K.F., Camacho, A.G., Fernández, J., 2010. Shallow flank deformation at Cumbre Vieja volcano (Canary Islands): implications on the stability of steep-sided volcano flanks at oceanic islands. Earth Planet. Sci. Lett. 297 (3), 545–557.

Gonnermann, H.M., Manga, M., 2007. The fluid mechanics inside a volcano. Annu. Rev. Fluid Mech. 39, 321–356, https://doi.org/10.1146/annurev.fluid.39.050905.110207.

Goto, Y., Nakata, S., Kurokawa, M., Shimano, T., Sugimoto, T., Sakuma, S., Hoshizumi, H., Yoshimoto, M., Uto, K., 2008. Character and origin of lithofacies in the conduit of Unzen volcano, Japan. J. Volcanol. Geotherm. Res. 175, 45–59.

Grosfils, E.B., 2007. Magma reservoir failure on the terrestrial planets: assessing the importance of gravitational loading in simple elastic models. J. Volcanol. Geotherm. Res. 166, 47–75.

GuÐmundsson, A., 1995. Infrastructure and mechanics of volcanic systems in Iceland. J. Volcanol. Geotherm. Res. 64, 1–22.

Harker, A., Clough, C.T., 1904. The Tertiary Igneous Rocks of Skye. HM Stationery Office, Edinburgh.

Heiken, G., Wohletz, K., Eichelberger, J.C., 1988. Fracture filling and intrusive pyroclasts, Inyo domes, California. J. Geophys. Res. 93, 4335–4350.

Hoernle, K., Schmincke, H.-U., 1993. The role of partial melting in the 15-Ma geochemical evolution of Gran-Canaria – A blob model for the Canary Hotspot. J. Petrol. 34, 599–626.

Holcomb, R.T., Searle, R.C., 1991. Large landslides from oceanic volcanoes. Mar. Georesour. Geotec. 10 (1–2), 19–32.

Holcomb, R.T., Searle, R.C., 1991. Large landslides from oceanic volcanoes. Marine Georesources & Geotechnology 10 (1–2), 19–32.

Le Bas, M.J., 1971. Cone-sheets as a mechanism of uplift. Geol. Mag. 108 (5), 373–376.

Le Corvec, N., Walter, T.R., 2009. Volcano spreading and fault interaction influenced by rift zone intrusions: Insights from analogue experiments analyzed with digital image correlation technique. J. Volcanol. Geoth. Res. 183 (3), 170–182.

Kuenen, P.H., 1937. Intrusion of cone-sheets. Geol. Mag. 74 (4), 177–183.

Mathieu, L., Burchardt, S., Troll, V.R., Krumbholz, M., Delcamp, A., 2015. Geological constraints on the dynamic emplacement of cone-sheets: the Ardnamurchan cone-sheet swarm, NW Scotland. J. Struct. Geol. 80, 133–141. doi: 10.1016/j.jsg.2015.08.012.

McGowan, E., 2016. Magma emplacement and deformation in rhyolitic dykes: insight into magmatic outgassing. PhD thesis, Lancaster University, UK, 357 pp.

Krastel, S., Schmincke, H.U., Jacobs, C.L., Rihm, R., Le Bas, T.P., Alibes, B., 2001. Submarine landslides around the Canary Islands. J. Geophys Res. 106 (B3), 3977–3997.

Maccaferri, F., Richter, N., Walter, T.R., 2017. The effect of giant lateral collapses on magma pathways and the location of volcanism. Nat. Commun. 8 (1), 1097.

MacFarlane, D.J., Ridley, W.I., 1968. An interpretation of gravity data for Tenerife, Canary Islands. Earth Planet. Sci. Lett. 4 (6), 481–486.

Manconi, A., Longpré, M.A., Walter, T.R., Troll, V.R., Hansteen, T.H., 2009. The effects of flank collapses on volcano plumbing systems. Geology 37, 1099–1102.

Miklius, A., Lisowski, M., Delaney, P.T., Denlinger, R.P., Dvorak, J.J., Okamura, A.T., Sakol, M.K., 1995. Recent inflation and flank movement of Mauna Loa volcano. Mauna Loa Revealed, 199–205.

Michon, L., Cayol, V., Letourneur, L., Peltier, A., Villeneuve, N., Staudacher, T., 2009. Edifice growth, deformation and rift zone development in basaltic setting: Insights from Piton de la Fournaise shield volcano (Réunion Island). J. Volcanol. Geoth. Res. 184 (1), 14–30.

Mitchell, N.C., 2001. Transition from circular to stellate forms of submarine volcanoes. J. Geophys. Res. 106, 1987–2003.

Moore, J.G., Normark, W.R., Holcomb, R.T., 1994. Giant hawaiian landslides. Annu. Rev. Earth Planet Sci. 22 (1), 119–144.

Mori, J., McKee, C., 1987. Outward-dipping ring-fault structure at Rabaul caldera as shown by earthquake locations. Science 235, 193–196.

Lin, G., Amelung, F., Lavallée, Y., Okubo, P.G., 2014. Seismic evidence for a crustal magma reservoir beneath the upper east rift zone of Kilauea volcano, Hawaii. Geology 42 (3), 187–190.

Longpré, M.A., Troll, V.R., Walter, T.R., Hansteen, T.H., 2009. Volcanic and geochemical evolution of the Teno massif, Tenerife, Canary Islands: Some repercussions of giant landslides on ocean island magmatism. Geochem. Geophys. Geosyst. 10 (12).

Luongo, G., Cubellis, E., Obrizzo, F., Petrazzuoli, S.M., 1991. A physical model for the origin of volcanism of the Tyrrhenian margin: the case of Neapolitan area. J. Volcanol. Geotherm. Res. 48 (1–2), 173–185.

Nakamura, K., 1977. Volcanoes as possible indicators of tectonic stress orientation—principle and proposal. J. Volcanol. Geotherm. Res. 2, 1–16.

Nakamura, K., 1980. Why do long rift zones develop in Hawaiian volcanoes: a possible role of thick oceanic sediments. Bull Volcanol Soc Japan 25, 255–269.

Oehler, J.F., Labazuy, P., Lénat, J.F., 2004. Recurrence of major flank landslides during the last 2-Ma-history of Reunion Island. Bull. Volcanol. 66 (7), 585–598.

Phillips, W.J., 1974. The dynamic emplacement of cone sheets. Tectonophysics 24, 69–84.

Richey, J.E., Thomas, H.H., 1930. The geology of Ardnamurchan, North-west Mull and Coll: Geol. Survey Scotland Mem.

Rubin, A.M., 1995. Propagation of magma-filled cracks. Ann. Rev. Earth Planet. Sci. 23 (1), 287–336.

Sæmundsson, K., Noll, H., 1974. K/Ar Ages of Rocks from Húsafell, Western Iceland, and the Development of the Húsafell Central Volcano. Jökull 24, 40–59.

Samrock, L.K., 2015. 3D Modelling of the Tejeda Cone-Sheet Swarm, Gran Canaria, Canary Islands, Spain. Degree Project at the Department of Earth Sciences. ISSN 1650-6553 No. 343.

Schauroth, J., Wadsworth, F.B., Kennedy, B., et al., 2016. Conduit margin heating and deformation during the AD 1886 basaltic Plinian eruption at Tarawera volcano, New Zealand. Bull. Volcanol. 78, 12. doi: 10.1007/s00445-016-1006-7.

Schirnick, C., van den Bogaard, P., Schmincke, H.-U., 1999. Cone sheet formation and intrusive growth of an oceanic island—the Miocene Tejeda complex on Gran Canaria (Canary Islands). Geology 27, 207–210.

Schmincke, H.U., 1967. Cone sheet swarm, resurgence of Tejeda Caldera, and the early geologic history of Gran Canaria. Bull. Volcanol. 31 (1), 153–162.

Schmincke, H.-U., 2006. Volcanism. Springer, Berlin, 324 pp.

Shaw, H.R., 1980. The fracture mechanisms of magma transport from the mantle to the surface. Phys. Magma. Processes 64, 201–264.

Siebert, L., 1984. Large volcanic debris avalanches: characteristics of source areas, deposits, and associated eruptions. J. Volcanol. Geoth. Res. 22 (3–4), 163–197.

Siler, D.L., Karson, J.A., 2009. Three-dimensional structure of inclined sheet swarms: implications for crustal thickening and subsidence in the volcanic rift zones of Iceland. J. Volcanol. Geotherm. Res. 188, 333–346.

Slezin, Y.B., 2003. The mechanism of volcanic eruptions (a steady state approach). J. Volcanol. Geotherm. Res. 122, 7–50.

Stasiuk, M.V., Barclay, J., Carroll, M.R., Jaupart, C., Ratté, J.C., Sparks, R.S.J., et al., 1996. Degassing during magma ascent in the Mule Creek vent (USA). Bull. Volcanol. 58, 117–130.

Stearns, H.T., 1946. Geology of the Hawaiian islands. Bulletin of the USGS 8. http://pubs.er.usgs.gov/publication/70160866.

Swanson, D.A., Duffield, W.A., Fiske, R.S., 1976. Displacement of the south flank of Kilauea Volcano; the result of forceful intrusion of magma into the rift zones (No. 963). US Govt. Print. Off.

Takada, A., 1997. Cyclic flank-vent and central-vent eruption patterns. Bull. Volcanol. 58 (7), 539–556.

Tibaldi, A., Corazzato, C., Kozhurin, A., Lagmay, A.F., Pasquarè, F.A., Ponomareva, V.V., Vezzoli, L., 2008. Influence of substrate tectonic heritage on the evolution of composite volcanoes: predicting sites of flank eruption, lateral collapse, and erosion. Glob. Planet. Change 61 (3), 151–174.

Tuffen, H., Castro, J.M., 2009. The emplacement of an obsidian dyke through thin ice: Hrafntinnuhryggur, Krafla Iceland. J. Volcanol. Geotherm. Res. 185, 352–366.

Tuffen, H., Dingwell, D., 2005. Fault textures in volcanic conduits: evidence for seismic trigger mechanisms during silicic eruptions. Bull. Volcanol. 67, 370–387.

Vogt, P.R., Smoot, N.C., 1984. The Geisha Guyots: Multibeam bathymetry and morphometric interpretation. J. Geophys Res. 89 (B13), 11085–11107.

Walker, G.P.L., 1963. The BreiÐdalur central volcano, eastern Iceland. Q. J. Geol. Soc. London 119, 29–63.

Walter, T.R., 2003. Buttressing and fractional spreading of Tenerife, an experimental approach on the formation of rift zones. Geophys. Res. Lett. 30 (6).

Walter, T., Schmincke, H.U., 2002. Rifting, recurrent landsliding and Miocene structural reorganization on NW-Tenerife (Canary Islands). Int. J. Earth Sci. 91 (4), 615–628.

Walter, T.R., Troll, V.R., 2003. Experiments on rift zone evolution in unstable volcanic edifices. J. Volcanol. Geotherm. Res. 127 (1), 107–120.

Walker, G.P.L., 1992. Coherent intrusion complexes in large basaltic volcanoes—a new structural model. J. Volcanol. Geotherm. Res. 50, 41–54.

Walter, T.R., Acocella, V., Neri, M., Amelung, F., 2005. Feedback processes between magmatic events and flank movement at Mount Etna (Italy) during the 2002-2003 eruption. J. Geophys Res. 110 (B10).

Walter, T.R., Klügel, A., Münn, S., 2006. Gravitational spreading and formation of new rift zones on overlapping volcanoes. Terra Nova 18 (1), 26–33.

Wyss, M., 1980. Hawaiian rifts and recent Icelandic volcanism: expressions of plume generated radial stress fields. J. Geophys. 47, 19–22.

Zaczek, K., Troll, V.R., Cachao, M., Ferreira, J., Deegan, F.M., Carracedo, J.C., Soler, V., Meade, F.C., Burchardt, S., 2015. Nannofossils in 2011 El Hierro eruptive products reinstate plume model for Canary Islands. Sci. Rep. 5, 7945. doi: 10.1038/srep07945.

Zurek, J., Williams-Jones, G., Trusdell, F., Martin, S., 2015. The origin of Mauna Loa's N nole Hills: Evidence of rift zone reorganization. Geophysical Research Letters 42 (20), 8358–8366.

CHAPTER 5

Storage and Transport of Magma in the Layered Crust—Formation of Sills and Related Flat-Lying Intrusions

O. Galland*, H.S. Bertelsen*, C.H. Eide, F. Guldstrand*, Ø.T. Haug*, Héctor A. Leanza†, K. Mair*, O. Palma‡, S. Planke*, O. Rabbel*, B. Rogers*, T. Schmiedel*, A. Souche*, J.B. Spacapan§**

*Physics of Geological Processes, University of Oslo, Oslo, Norway
**University of Bergen, Bergen, Norway
†CONICET, Argentinian Museum of Natural Sciences, Buenos Aires, Argentina
‡CONICET Y-TEC and Universidad Nacional de La Plata, La Plata, Argentina
§YPF, Buenos Aires, Argentina

Contents

5.1 INTRODUCTION

Volcanic and igneous plumbing systems exhibit numerous components of various shapes, including sheet intrusions (dykes, cone sheets and sills) and more massive intrusions (laccoliths, plugs, plutons, etc.). Even though dykes are the main magma pathways from the lower crust to the Earth's surface, recent research has highlighted that igneous sills represent significant

Volcanic and Igneous Plumbing Systems. http://dx.doi.org/10.1016/B978-0-12-809749-6.00005-4
Copyright © 2018 Elsevier Inc. All rights reserved.

parts of volcanic plumbing systems (Planke et al., 2005) and substantially contribute to lateral magma transport and storage at various levels in the Earth's crust (Magee et al., 2016).

Sills have been known and described for more than a century (Gilbert, 1877; Tweto, 1951). However, it is only the last two decades of research that have highlighted the abundance of sills worldwide and thus their global scientific significance. Indeed, large-scale seismic surveys dedicated to hydrocarbon exploration along continental passive margins revealed the presence of voluminous sill complexes in numerous sedimentary basins (Planke et al., 2005; Svensen et al., 2012; Magee et al., 2016). This chapter describes sills as essential components of volcanic plumbing systems, how they form and how they contribute to magma transport and storage in the Earth's crust.

5.2 WHAT ARE SILLS? DEFINITIONS AND MORPHOLOGIES

The original definition of a sill is a tabular sheet intrusion that is concordant to the layering of the host rock. This layering includes strata in sedimentary basins, layers of lava and other erupted volcanic products in volcanic areas as well as foliation in metamorphic rocks. Since sills are often found in sedimentary strata, they are commonly subhorizontal (Fig. 5.1). However, in cases of deformed sedimentary sequences, sills can be inclined, and even vertical.

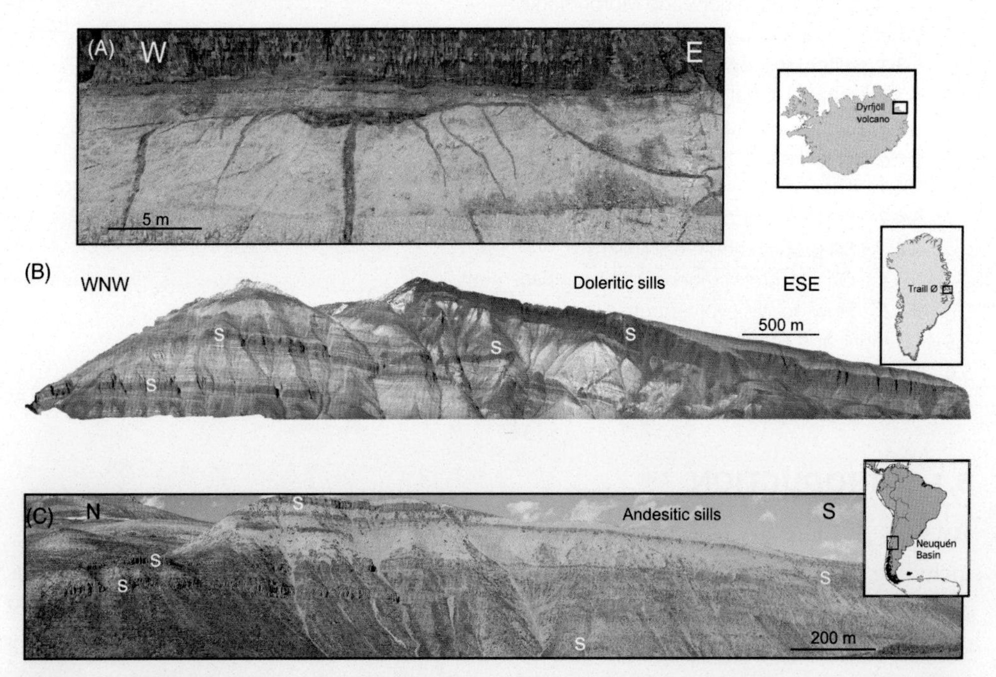

Figure 5.1 (A) Field photograph of basaltic dyke-fed sill emplaced in lake sediments (yellow formation) in the Breidavik volcano, NE Iceland (photograph: Elodie Saubin). (B) Orthomozaic of geological section exposing doleritic sills (white 's'), Traill Ø, East Greenland (photographs: Sverre Planke). (C) Panoramic image of Miocene andesitic sills (white 's') emplaced in organic-rich shale, northern Neuquén Basin, Mendoza province, Argentina (photographs: Olivier Galland) (Senger et al., 2017).

This rigorous definition is based on outcrop-scale observations, which are usually limited in extent (tens of meters to hundreds of meters). However, sills can extend over kilometres, and even tens of kilometres (Fig. 5.1), and it is common to observe that sills are locally discordant (Polteau et al., 2008b). Therefore, a more general definition entails sills as intrusions that are dominantly concordant with the layering of the host rock.

Depending on their degree of concordance to the host-rock layering, observations of sills worldwide lead to a classification of distinct types of intrusions (Planke et al., 2005; Jackson et al., 2013) (Fig. 5.2), mostly based on interpretation of seismic data of sill complexes.

- Strata-concordant sills fulfil the original definition of sills, that is, they correspond to continuous sheets that remain concordant with the layering of the host rock. These intrusions commonly form deeper than the other types of sills described below.
- Transgressive sills are sheets that step to stratigraphically higher levels with oblique angle with respect to host-rock layering. The transgressive parts of the intrusions can be straight,

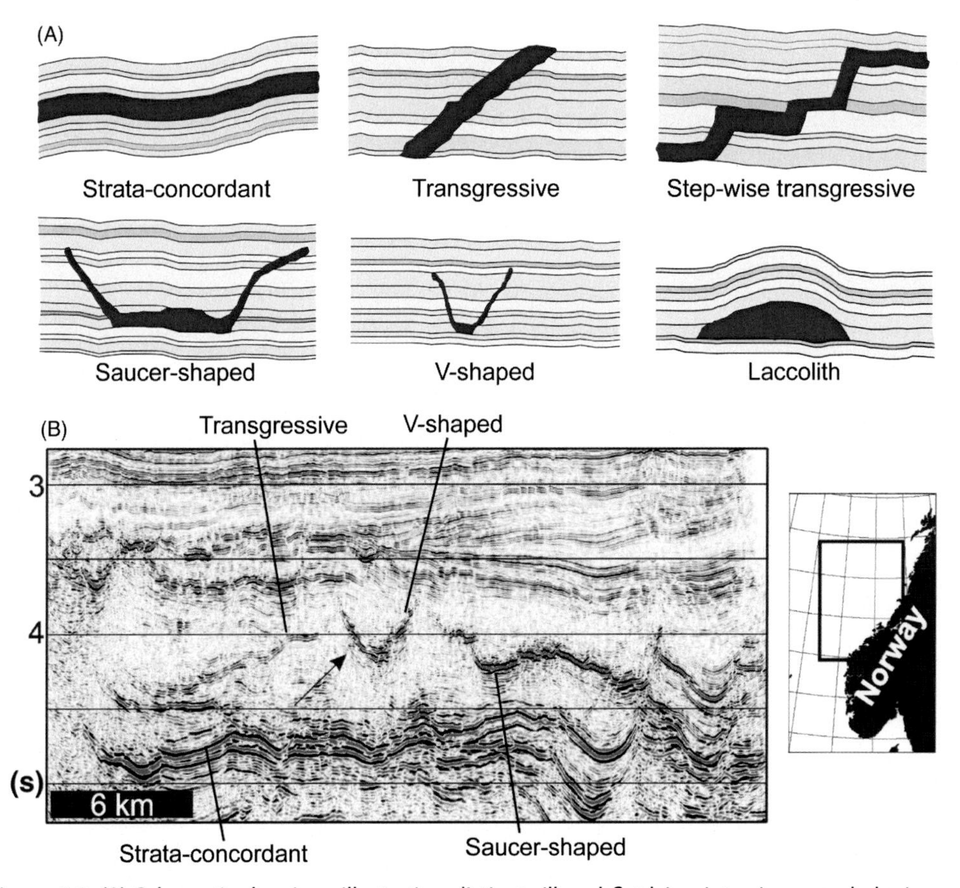

Figure 5.2 (A) Schematic drawings illustrating distinct sill and flat-lying intrusion morphologies as defined from seismic data. See text for morphological description. (B) Characteristic seismic profile of Gleipne Sill Complex, Vøring Basin, mid-Norwegian margin, exhibiting several sill and flat-lying intrusion (strong reflectors) morphologies (Planke et al., 2005).

indicating an overall discordant sheet at an angle to the host-rock layering, or step-wise, the latter indicating alternating concordant and steeply dipping discordant segments.

- Saucer-shaped sills are roughly symmetric intrusions with (1) a concordant, lower inner sill, (2) transgressive, climbing outer inclined sheets, which (3) flatten out to outer sills. Saucer-shaped sills are usually thicker at their centre and taper toward their tips.
- V-shaped intrusions correspond to saucer-shaped sills, the inner sills of which are limited in extent, such that mostly the inclined sheets are present. These intrusions commonly form at shallow levels of the crust.
- Hybrid sills exhibit mixed characteristics of transgressive, saucer-shaped and V-shaped intrusions.

Other types of flat-lying igneous intrusions include laccoliths. Similar to sills, their basal contact is strata-concordant and usually flat (Gilbert, 1877; Corry, 1988). The pressure and volume of magma are such that the overlying rocks are forced upward, giving the roof of the laccolith a dome-like shape (see Chapter 6). It is assumed that laccoliths initiate from the emplacement of one or several sills that subsequently grow and inflate as magma influx continues (Jackson, 1997). However, a robust definition of the transition between a sill and laccolith does not exist, which is why many igneous bodies can be called both sills and laccoliths. The characteristics of laccolith intrusions are addressed in Chapter 6.

5.3 WHERE DO SILLS OCCUR?

Sills are found in various geological settings where magmatism has occurred: central volcanoes, sedimentary basins and the layered lower crust (Fig. 5.3). Note that the distinction between sills emplaced in these settings is done for classification purposes only and does not represent exclusive relations. For example, central volcanoes can be located in sedimentary basins (Indonesia, East African rift, Kamchatka) and it is likely that lower crust sills are part of the deep plumbing systems of central volcanoes (see Chapter 2).

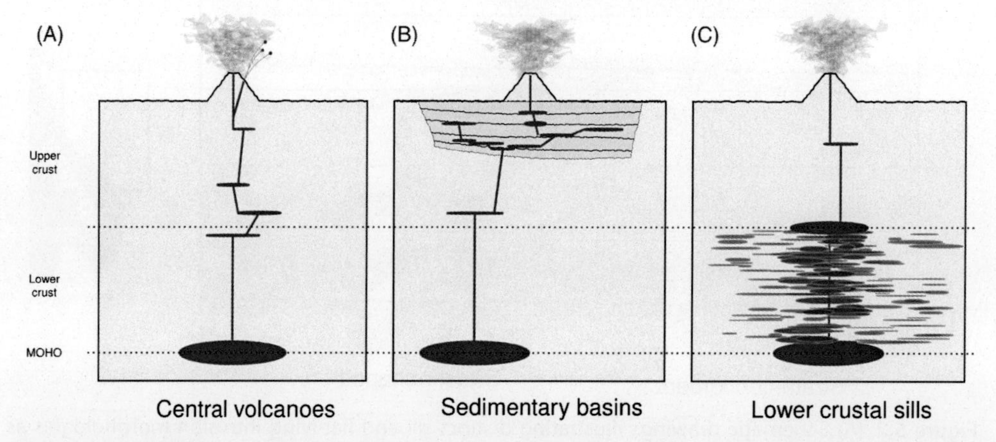

Figure 5.3 Schematic drawings showing the main geological settings where sills are encountered: (A) central volcanoes, such as in volcanic rifts and oceanic islands, (B) sedimentary basins and (C) the layered lower crust.

5.3.1 Sills in the Lower (Layered) Crust

Crustal-scale seismic profiles often show strong sub-horizontal reflectors in the lower crust (White et al., 2008). Such reflectors are usually not present in the lower crust of cratonic regions, however they are imaged in geological settings that involve thermal activation of the lithosphere, that is, magmatic input (Allmendinger et al., 1987). This seismic layering is therefore interpreted to indicate the occurrence of mafic sills emplaced at the base of the crust. The emplacement of these lower crustal mafic sills, e.g. offshore Scotland and the Faroe Islands, is probably the main process of magma underplating as a result of massive input of magma related to the North Atlantic large igneous province, prior to continental break-up. Offshore Scotland and the Faroe Islands, where continental break-up was preceded by massive input of magma related to the North Atlantic large igneous province, the supposed emplacement of large lower crustal sills is a plausible mechanism for magmatic underplating (White et al., 2008).

Obviously, outcrops of the lower layered crust are not common. It is therefore challenging to collect direct geological observations of the structures responsible for the seismic layering. An exceptional locality is the Oman ophiolite, where an oceanic lithosphere has been obducted on top of the margin of the Arabian plate. There, geological studies highlight the presence of numerous mafic sills close to the former Moho. These sills have greatly contributed to the migration of magma from the partially molten mantle to the crust (see also Chapter 2).

5.3.2 Sills in Central Volcanoes

Classic views of volcanic and igneous plumbing systems have considered dykes and central conduits as the main feeders of fissure eruptions and summit eruptions at central volcanoes (see also Chapter 4). However, plumbing systems in active central volcanoes appear much more complex, including numerous, co-existing intrusions of contrasting shapes, such as sills.

In eroded volcanic systems, the main difficulty in observing sills is that they commonly have the same lithology as their host rock. It can therefore be challenging to distinguish a thick lava flow from a sill. Detailed field observations, however, show the presence of abundant sills in central volcanoes, for example, in Iceland (Pasquare and Tibaldi, 2007; Burchardt, 2008).

In active central volcanoes, the obvious challenge is that the magmatic plumbing system is inaccessible in the sub-surface and thus not directly observable. However, geophysical and geodetic monitoring techniques provide considerable insights into the dynamics and shapes of magmatic intrusions. In particular, geodetic surface deformation patterns monitored at numerous active volcanoes highlight that the measured deformation can often be interpreted in terms of a growing sill in the sub-surface (Amelung et al., 2000; Pagli et al., 2012). In the case of Eyjafjallajökull volcano in Iceland, the successive emplacement of sills is thought to have triggered the infamous 2010 eruption, which disrupted the air traffic over Europe and North America (Sigmundsson et al., 2010). In numerous examples, the emplacement of sills at depth preceded an eruption, which suggests that sills act both as shallow-depth magma storage reservoirs and as feeders of successive eruptions (Kavanagh et al., 2015). Hence, they are important components of volcanic plumbing systems and need to be considered for volcanic hazard assessments. While the examples above relate to individual magma injections, geodetic surveys at several large volcanic systems (e.g. Yellowstone and Sorrocco magma body, USA; Campi Flegrei, Italy) suggest that magma reservoirs likely exhibit sill-like shapes.

As sills represent significant components of volcanic plumbing systems in central volcanoes, substantial lateral transport of magma happens at the root of these volcanoes (Galland et al., 2007), in contrast to the classic idea that magma migrates mainly vertically up from depth.

5.3.3 Sills in Sedimentary Basins

Petroleum exploration in numerous sedimentary basins worldwide unravelled the presence of voluminous sill complexes (Planke et al., 2005; Magee et al., 2016), which appear as strong reflectors on seismic data (Fig. 5.2). The most voluminous sill complexes emplaced in sedimentary basins are those resulting from the rapid emplacement of large igneous provinces (LIPs). Good examples are the Karoo-Ferrar igneous province (Fig. 5.4), the North Atlantic igneous province (Rockall basin offshore Ireland, Møre and Vøring basins offshore Norway, Greenland margin) and the Siberian igneous province (Tunguska basin) (Magee et al., 2016, and references therein). In these settings, sills usually occur as columnar jointed doleritic intrusions of up to several hundred metres thickness. These sills are likely the main feeders of flood basalts provinces associated with LIPs (Muirhead et al., 2014), such as those of the Deccan Traps and Lesotho Plateau (Fig. 5.4).

Sills also represent essential constituents of volcanic plumbing systems of arc and back-arc volcanoes within sedimentary basins. In contrast to LIP-type volcanism, arc- and back-arc-related sills usually exhibit andesitic compositions (Fig. 5.1B). Classic examples are the sills associated with the type locality of laccoliths in the Henry Mountains (Gilbert, 1877) and in the San Rafael subvolcanic field (Walker et al., 2017), Utah. There, the sills were dominantly emplaced in sandstone formations. Another spectacular example is the Neuquén Basin, Argentina, where voluminous Miocene back-arc volcanism led to the emplacement of numerous sills in the Mesozoic sedimentary sequences of the basin, dominantly in organic-rich shale formations (Rossello et al., 2002; Spacapan et al., 2017; Witte et al., 2012) (Fig. 5.1C).

5.4 HOW DO SILLS FORM?

Because of the scale and shape of sills, conclusive field studies of sill emplacement mechanisms require favourable outcrop conditions, which are hard to find. Conversely, most recent studies of sill complexes are based on seismic interpretation, which offers insights into the three-dimensional geometry of sills, but suffers from critical resolution problems

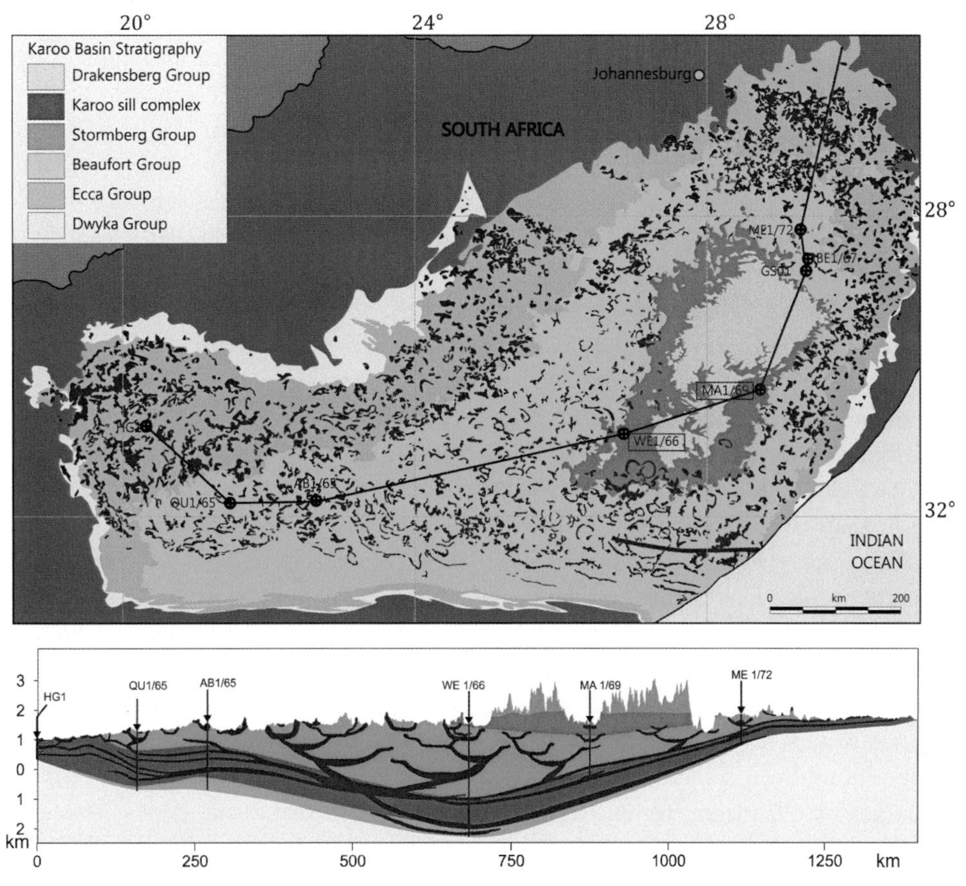

Figure 5.4 Simplified geological map (*top*) and cross-section (*bottom*) of a characteristic example of sills resulting from LIP emplaced in sedimentary basins, the Karoo Basin, South Africa (modified from Svensen et al., 2012). The numerous Karoo sills (red) have invaded the whole sedimentary sequence of the Karoo Basin.

and imaging challenges (Eide et al., 2017). Therefore, large parts of the geology and physics of sill emplacement mechanisms and evolution remain poorly understood. In the following paragraphs, we describe and discuss the processes governing sill emplacement in a chronological order, that is, from sill initiation, sill propagation, to subsequent sill evolution to inclined sheets or laccoliths (Galland et al., 2018).

5.4.1 The Feeders of Sills

A key aspect for understanding the initiation of sills is the nature of their feeders and the feeder-to-sill relationships. However, there are very few direct outcrop observations of feeder-to-sill relationships. A common tool to study sill complexes is seismic imaging, but because this method has been designed for imaging sub-horizontal structures, it is not suitable for imaging feeders, especially if these are sub-vertical dykes.

In most laboratory and numerical models of sill emplacement, dykes are assumed to be the main feeders for sills (Kavanagh et al., 2006, 2017). However, clear examples of dykes feeding sills are rare. The clearest direct field observations of dykes feeding sills have been recently provided by Eide et al. (2016) (see also Fig. 5.1B). Spectacular images of a 25-km-long section unambiguously display sub-vertical dykes feeding sills, the dyke-to-sill transition being very sharp (Fig. 5.5A). Indirect structural and geochemical evidence of a dyke feeding saucer-shaped sills have been collected in the Karoo basin by Galerne et al. (2011), who showed that the elliptical saucer-shaped sills of the Golden Valley Sill Complex have been likely fed by dykes located below the long axis of the ellipse (see also Fig. 5.12).

Conversely, field and seismic studies suggest that sills, in particular, inclined sheets, can feed other sills through a sill complex. Several examples in Greenland (Eide et al., 2016) and Antarctica (Muirhead et al., 2012; Airoldi et al., 2016) show how sills are connected to underlying sills via gently dipping inclined sheets (Fig. 5.5B). Such a structural relationship has been widely described on seismic data of sill complexes that display features that resemble junctions between sills (Hansen et al., 2004). The wide occurrence of such junction features may even suggest that all sills of a sill complex were interconnected through sill-to-sill feeding relationships at the time of their emplacement (Cartwright and Hansen, 2006). However, systematic interpretations of sill junctions and associated feeding relationships require some caution. Galerne et al. (2011) provided field and geochemical evidence that show that two sills in contact in the Golden Valley Sill Complex, Karoo Basin, South Africa, were emplaced at distinct times and did not exhibit the same geochemical compositions. Even if such a contact would appear as a sill junction on seismic data, it is not indicative of a magma feeding relationship. This implies that the junctions between sills visible on seismic data should not be interpreted as feeding junctions by default. To summarize, sill complexes can be fed by both dykes and underlying sills and inclined sheets.

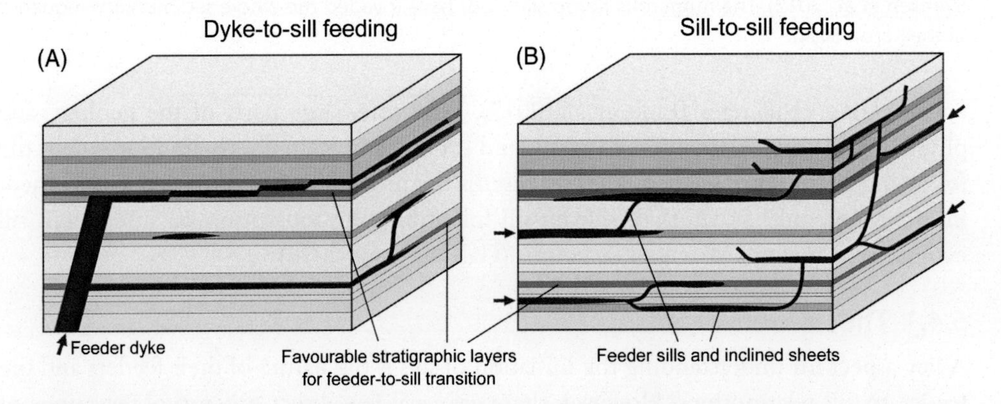

Figure 5.5 (A) Schematic drawing of a dyke feeding sills. (B) Schematic drawing of sills feeding sills via inclined sheets (modified from Airoldi et al., 2016). Note that in both cases, feeder-to-sill transitions occur abruptly at favourable stratigraphic levels.

Depending on the geometric relation between a sill and its feeder, several models for sill-feeding mechanisms have been proposed. A common model assumes that sills, and in particular saucer-shaped sills, are fed centrally such that the magma flows radially outward and upward (Polteau et al., 2008b). This model is supported by morphological flow indicators highlighted by 3D seismic data (Magee et al., 2016) and by anisotropy of magnetic susceptibility (AMS) data (Polteau et al., 2008a). The central feeding model explains, for instance, the symmetrical shapes of saucer-shaped sills. In contrast, geological observations show that sills can also be fed from the sides (Eide et al., 2016), leading to asymmetrical sill morphologies (Fig. 5.5).

5.4.2 Factors Controlling Sill Initiation

A classic model for sill emplacement states that the magma spreads horizontally along its level of neutral buoyancy (LNB) (Francis, 1982). Even if this model has been popular, numerous arguments show that it does not apply for the emplacement of sills, for the following reasons (see detailed discussion by Thomson and Schofield, 2008):

- When present, sills are emplaced at many stratigraphic levels, implying unrealistically many different LNBs.
- If the magma reaches the LNB, it should have no driving pressure, that is, the magma should stop propagating (Hogan et al., 1998). However, field observations and seismic data show that sills commonly trigger doming of their overburden, showing that the magma was over-pressured and not neutrally buoyant.
- Inclined sheets fed by sills show that even if magma spreads horizontally, it is sufficiently 'buoyant' to keep ascending subsequently.

Another classic model states that sills open parallel to the least principal stress σ_3: the horizontal orientation of sills implies that σ_3 is vertical, which is typical for compressional tectonic stresses (Hubbert and Willis, 1957). However, most sills associated with LIPs formed in un-stressed basins, often prior to rifting, where the stress is dominantly lithostatic and σ_3 is horizontal. Sills are even emplaced in active rifts, where σ_3 is horizontal parallel to extension and is expected to control the formation of vertical dykes. Even if a compressional tectonic stress regime might favour sill emplacement, it is likely not the primary controlling factor.

Alternative models that explain the emplacement of sills are based on the key observations that most sill complexes are found in layered host rock, either sedimentary strata or layered volcanic deposits, and that sills are overall strata-concordant. These first-order observations strongly suggest that sills form when a magma feeder hits a layer that is favourable for the magma to flow along. There are several possible mechanisms that explain how mechanical layering may lead to sill formation:

- Static mechanical models calculate the elastic stresses induced by a pressurized feeder dyke, the tip of which reaches a boundary between two layers, the upper one being more rigid, that is, higher Young's modulus, than the lower one (Barnett and Gudmundsson, 2014). These models calculate complex stress distributions, with rotation of the principal stresses that become favourable to sill growth along the boundary between the two layers.

- In addition to the rigidity contrast of the host-rock layers above and below an interface, the strength of the interface itself may play a significant role during sill initiation (Burchardt, 2008). In particular, weak interfaces between elastic rock layers strongly favour the rotation of a feeder dyke to a sill when the dyke reaches the interface (Kavanagh et al., 2017).
- The models above only consider the elastic properties of the layered host rock on sill initiation. However, sills are common within shale formations, the deformation of which can be strongly inelastic. It is therefore likely that inelastic deformation, that is, brittle shear deformation and ductile deformation, control the initiation of sills to a large degree (Schofield et al., 2012; Spacapan et al., 2017). The observation that sills are abundant in organic-rich shale formations also suggests that pore fluid pressure resulting from the maturation of organic matter might control sill initiation (Gressier et al., 2010). In this model, the local fluid overpressures lead to significant vertical fluid pressure gradients associated with fluid migration. Such pressure gradients generate vertical seepage forces that significantly reduce the vertical effective stress, such that the vertical effective stress becomes smaller than the horizontal effective stress, which is favourable for horizontal fracturing.

5.4.3 Sill Propagation Mechanisms

Various models of sill propagation have been proposed. Each model is associated with contrasting shapes of intrusion tips and associated structures in the host rock (Fig. 5.6). The following paragraphs describe the characteristics of each model and the expected intrusion shapes and structures in the host rock.

The most established propagation model, also called the splitting model, assumes that sills propagate through tensile linear elastic fracture mechanics (LEFM; Fig. 5.6A). Since sills are sheet intrusions, they are assumed to be hydraulic fractures propagating through a purely elastic host rock (Kavanagh et al., 2006; Bunger and Cruden, 2011; Michaut, 2011; Galland and Scheibert, 2013). In the splitting model, the tip propagates by tensile opening of the host rock, such that the opening vector is dominantly perpendicular to the contacts. The displaced host rock layers should therefore mimic the intrusion shape, and in the case of concordant sills, the strata should have the same thickness along the sill as ahead of the sill tip (Fig. 5.6G). This also implies that the displaced layers are expected to have a constant thickness along strike. The LEFM splitting model assumes that sill tips are sharp and thin. Hence, viscous magma should be unable to flow into the narrow, sharp tips of sills, which should result in a tip cavity to form between the magma front and the intrusion tip. This cavity is expected to be filled with volatiles either exsolved from the magma or from the host rock (Lister, 1990).

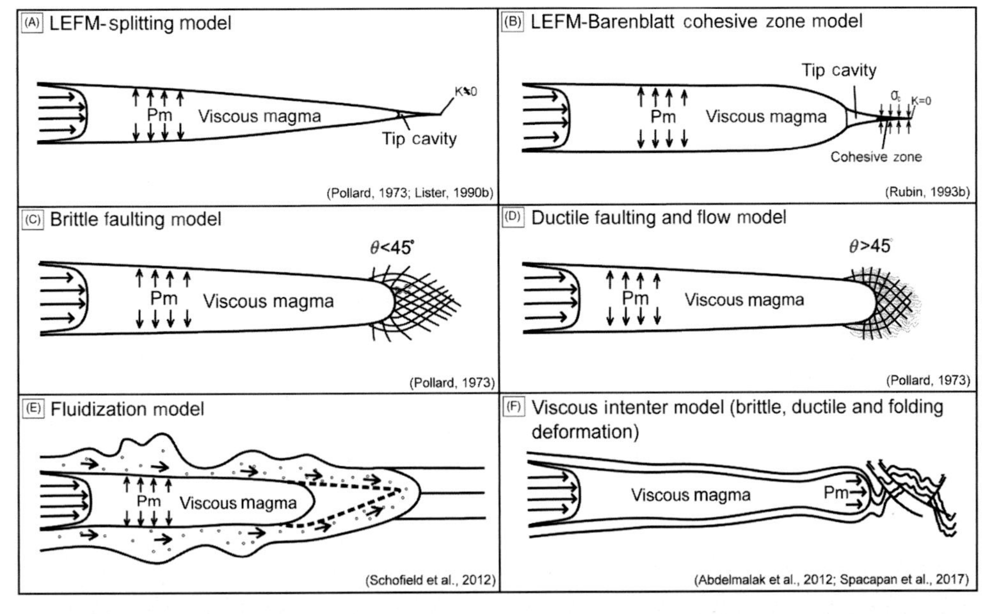

Figure 5.6 Schematic drawings of existing sill propagation mechanisms. (A) Linear elastic fracture-splitting model (e.g. Lister, 1990), which considers sill propagation by tensile failure, purely elastic deformation of the host rock, sharp tip of the propagating sill and the presence of a tip cavity between the magma front and the sill tip. (B) Elastic tensile fracture with a Barenblatt cohesive zone (Rubin, 1993). (C) Brittle faulting model and (D) ductile faulting model (Pollard, 1973). In both models, the magma pushes its host rock, which fails in shear manner. (E) Fluidization model triggered by rapid boiling of pore fluids as magma intrudes the host rock (Schofield et al., 2012). (F) Viscous indenter model (Merle and Donnadieu, 2000; Spacapan et al., 2017). In this model, the shear stresses due to highly viscous magma flow lead to shear (brittle and ductile) failure of the host rock. In this mode, the magma appears as rigid as, or even more rigid than, the host rock. (G) Helicopter field photograph of a sill tip in Traill Ø, Eastern Greenland. The outcrop shows that the tip of the mafic sill (dark) is sharp, and its propagation is dominated by tensile opening and elastic bending of the layered sandstone host rock.

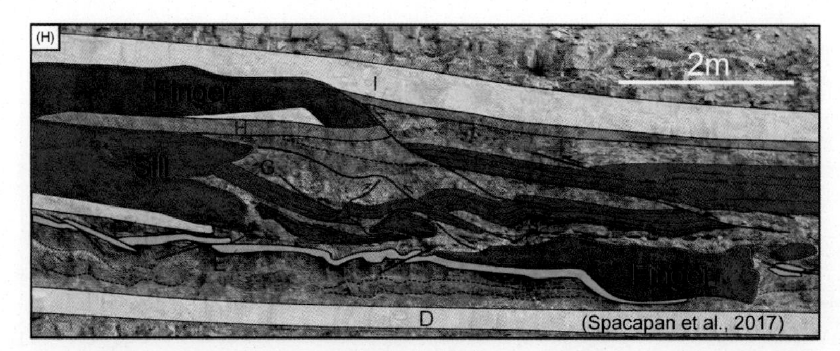

Figure 5.6 (*Cont.*) (H) Interpreted field image of outcrop exposing a sheet-like sill, magmatic fingers and the associated structures in the shale-carbonate host rock, Cuesta del Chihuido, Mendoza Province, Argentina (Spacapan et al., 2017). The outcrop shows that the sill tip is round or blunt and that substantial compressional inelastic deformation (brittle faulting of the carbonate layers, ductile flow of the shale layers) accommodates the emplacement and propagation of the sill.

Common rocks of the brittle crust that host sills, such as volcanic tuffs and sedimentary shale formations, do not deform elastically, and field observations show that significant inelastic deformation can accommodate sill propagation and emplacement. Thus, several propagation mechanisms accounting for inelastic deformation have been proposed. (1) the LEFM-Barenblatt cohesive zone model (Fig. 5.6B) is an extension of the LEFM splitting elastic model (Rubin, 1993). In this model, intrusion propagation occurs by tensile failure with simultaneous plastic deformation in a cohesive zone in the host rock beyond the tip of the propagating sill (Scheibert et al., 2017). (2) The brittle and ductile faulting–viscous indenter models (Fig. 5.6C and D) suggest that the host rock ahead of the intrusion tips fails in a shear manner, that is, is faulted, by the propagation of the magma. The brittle and ductile faulting models (Pollard, 1973) represent two end-member models of host-rock deformation, that is, brittle and ductile, respectively. However, they are phenomenologically similar in that they account for the push of the magma, which 'bulldozers' the host rock at the intrusion tip, so that the structures that form at the tip of the intrusions accommodate compression (Spacapan et al., 2017) (Fig. 5.6H). The viscous indenter model assumes both brittle (Fig. 5.6C) and ductile (Fig. 5.6D) shear failure of the host rock and implies that the viscous shear stresses near the tip of a propagating intrusion are high enough to overcome the strength of the host rock. Hence, the propagating magma pushes its host rock ahead like an indenter with a blunt or rectangular tip. In this model, the deformation associated with tip propagation consists of conjugate shear faults that accommodate shortening of the host ahead of the tip. The viscous-indenter model has been mainly applied to viscous magmas, for example, of rhyolitic compositions (Merle and Donnadieu, 2000). However, similar features have been observed associated with dykes of lower viscosity magma, that is, of basaltic and andesitic compositions. (3) Finally, the fluidization model (Fig. 5.6E) assumes that the heat brought by magma into the host rock can generate pore fluid pressure build-up in the host at the vicinity of the intrusion (Schofield et al., 2012). The pressure build-up is such that it can trigger fluidization of the host, the transport of which accommodates the emplacement of the magma. Rock fluidization produces incoherent disruption of the fluidized rocks, easily recognizable in the field.

Sill propagation does not usually occur as one coherent magmatic sheet, but instead as multiple, slightly offset segments connected closer to the feeder. These segments are equivalent to *en-échelon* dyke segments commonly observed in the field (see Chapter 3). Sill segments are assumed to be elongated in the magma propagation direction and can therefore be used as magma flow indicators, in particular, on 3D seismic data (Fig. 5.7C) (Thomson, 2007). The morphology of sill segments and their connections strongly depends on the lithology and rheology of the host rock. If the host is competent, the sections of the magma segments are usually rectangular, the transition between them being very sharp and accommodated by complex brittle bridge structures (Fig. 5.7A) (Hutton, 2009). Conversely, if the host is much weaker, significant ductile deformation is expected, and the segment sections are elliptical and called magma fingers (Figs. 5.6H and 5.7B) (Pollard et al., 1975; Schofield et al., 2012; Spacapan et al., 2017).

5.4.4 Other Factors Influencing Sill Emplacement

Sill emplacement is largely controlled by the layering of the host rock. It is therefore likely that other types of mechanical heterogeneities, such as faults, also influence the emplacement of sills. Indeed, sills emplaced in rifted basins show clear evidence that normal faults affect the morphology of sills, as concordant sill segments are connected by intrusive segments emplaced along faults (Magee et al., 2013). In foreland basin settings, sill emplacement may also be controlled by thrust faults and ramps (Galland et al., 2007; Ferré et al., 2012).

Magma viscosity is another important parameter that controls sill emplacement. Most sills are of mafic to intermediate compositions (basaltic to andesitic), that is, composed of low-viscosity magma. In contrast, felsic intrusions (rhyolitic) dominantly exhibit laccolithic shape due to the high viscosity of the magma (Bunger and Cruden, 2011). Moreover, cooling of the magma at the contact with the cold host rock leads to significant viscosity increase near the propagating tips of sills, which considerably affects emplacement and propagation dynamics. This phenomenon results in thicker sills (Thorey and Michaut, 2016), which can also develop lobate morphology (Chanceaux and Menand, 2014).

5.4.5 How Do Sills Evolve? From Sills to Saucer-Shaped Sills or Laccoliths

Numerous sills in sedimentary basins resemble saucers in shape, that is, they consist of a flat-lying lower, inner sill connected to outer inclined sheets and outer sills. Typical examples are found in the Karoo basin, South Africa, offshore mid-Norway in the Vøring and Møre basins, and offshore Scotland in the Rockall Basin (see review by Polteau et al., 2008b). The most characteristic feature of saucer-shaped sills is the sharp

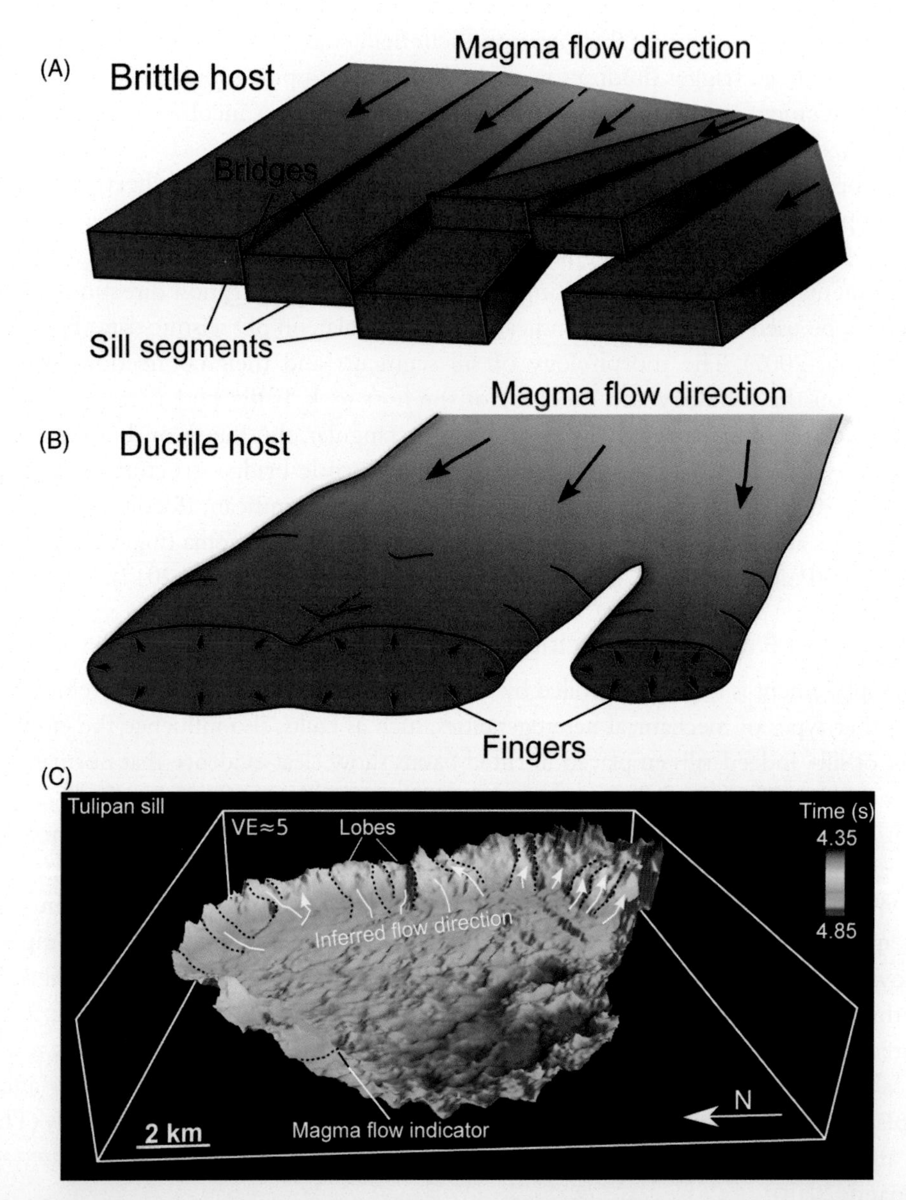

Figure 5.7 (A) Schematic drawing of the relations between sill segments, bridges between the segments and magma flow direction. The straight and geometric shapes of the segments highlight the brittle deformation of the host rock. (B) Schematic drawing of the relations between sill fingers and magma flow direction. The rounded sections of the fingers highlight the ductile deformation of the host rock. (C) 3D visualization of the saucer-shaped geometry of the top Tulipan sill horizon, Møre Basin, Offshore Norway (Schmiedel et al., 2017). Radial magma flow indicators mark edges of reflection segments representing upward, outward transgressing, igneous inclined sheets.

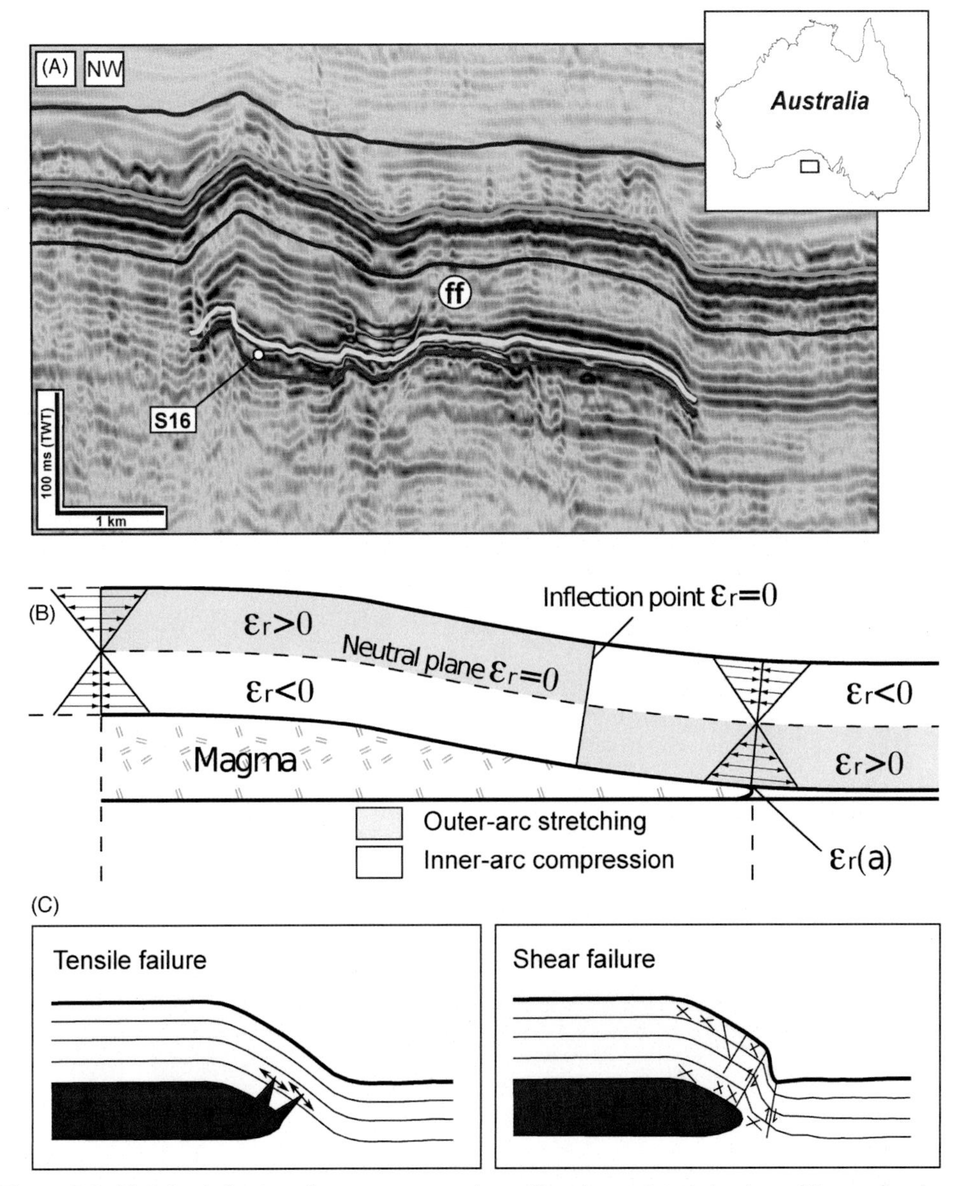

Figure 5.8 (A) Seismic image of a strata-concordant sill and associated doming of its overburden, a so-called forced fold, offshore southern Australia modified from (Jackson et al., 2013). (B) Schematic drawing of the stress distribution in forced folds induced by a sill to a thin elastic overburden (modified from Galland and Scheibert, 2013). This drawing shows that tensile stress occurs at the vicinity of the sill tip. (C) Schematic drawing of tensile (left) and shear (right) failure modes of sill overburden at the vicinity of sill tip modified from Haug et al., 2017.

transition from the flat inner sill to the inclined sheets (Fig. 5.2A). Systematic observations of saucer-shaped sills worldwide show that (1) they are almost systematically associated with dome structures in their overburden (Fig. 5.8) and (2) the diameter of the inner sill increases with increasing emplacement depth, the deepest sills being strata-concordant (Polteau et al., 2008b). These observations led to the hypothesis that saucer-shaped sills result from the mechanical interaction between the growing sill and the near free surface.

Saucer-shaped intrusions are proposed to form in the following way (Fig. 5.9): (1) a shallow sill grows and spreads along a layer, parallel to the free surface, (2) the sill deforms its overburden by doming, (3) differential uplift at the edges of the dome results in asymmetric stresses at the sill tip and (4) when the inner sill reaches a critical diameter, an inclined sheet

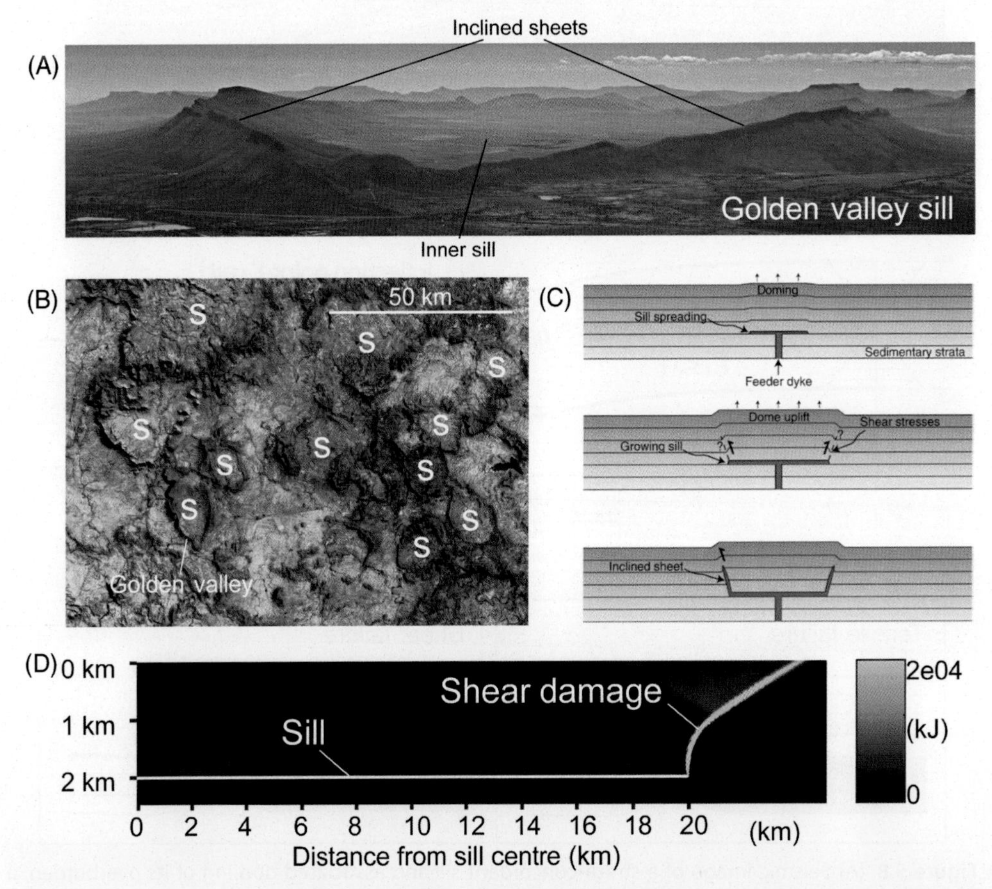

Figure 5.9 (A) Aerial photograph of the Golden Valley Sill, Karoo Basin, South Africa (photograph: Stéphane Polteau). (B) Satellite image of the part of Karoo Basin in the area of Golden Valley sill. The image displays numerous rings, which are exposed inclined sheets of saucer-shaped sills. (C) Schematic diagram illustrating emplacement mechanism of saucer-shaped sills (modified from Galland et al., 2009). (D) Example of rigid perfect plastic modelling showing the expected distribution of shear damage in the overburden of a 20-km-radius over-pressurized sill at 2 km depth (Haug et al., 2017). The shear damage zone mimics inclined sheets, suggesting that it is an important mechanical precursor for their emplacement.

initiates. The most common mechanical model of the transition from inner sill to inclined sheet considers tensile elastic stresses in the host rock at the vicinity of the propagating sill tip: when the stress reaches the tensile strength of the host rock, tensile failure occurs and guides the upward propagation of an inclined sheet (Fig. 5.8B and C) (Goulty and Schofield, 2008; Galland and Scheibert, 2013). Nevertheless, seismic and field observations suggest that shear failure, at least partly, controls the initiation of inclined sheets from sills (Figs. 5.8C and 5.9). In addition, rigid perfect plastic models show that over-pressurized shallow sills can trigger inelastic damage zone that mimics the shape of inclined sheets (Fig. 5.9D) (Haug et al., 2017); these models strongly suggest that shear damage can be a first-order controlling factor for the emplacement of inclined sheets (Fig. 5.9).

The formation of saucer-shaped sills likely influences the distribution of volcanism at the Earth's surface. In the Central Andes of northern Chile, for example, an elliptical distribution of stratovolcanoes may suggest that they are fed by inclined sheets connected to a sill-like reservoir at depth (Mathieu et al., 2008). In the Danakil depression, Afar rift, Ethiopia, radial lava flows are fed by volcanic vents concentrated at the edge of the Alu dome, suggesting that the volcanic feeders are inclined sheets fed by multiple sills that jacked up the dome (Magee et al., 2017).

It is interesting to note that igneous saucer-shaped sills exhibit the same shapes as those of saucer-shaped sand injectites that result from the injection of fluidized sand due to large pore fluid pressures (Szarawarska et al., 2010). The occurrence of both igneous and sedimentary intrusions of saucer shapes show that such structures are natural and fundamental results of growing flat-lying intrusions at shallow depth.

Large flat-lying intrusions do not systematically produce saucer shapes, especially when the magma has relatively felsic compositions (andesitic to rhyolitic), that is, of significantly higher viscosity. High-viscosity magma cannot flow over large distances, but it accumulates and inflates by substantial doming of the overburden, leading to the formation of laccoliths. The mechanisms associated to laccolith emplacement will be addressed in Chapter 6.

5.5 SCIENTIFIC AND ECONOMIC RELEVANCE OF SILLS

The above sections highlight that sills are essential components of volcanic plumbing systems. Due to their sub-horizontal shapes, they greatly contribute to lateral magma transport and magma stalling and accumulation at depth, strongly impacting volcanic activity at the surface. Accounting for sills in models of volcanic plumbing systems is a major improvement with respect to classic textbook views of vertical magma pathways.

The last two decades of research have shed light on the tremendous implications of sills emplaced in sedimentary basins. When sills are emplaced in prospective sedimentary basins, that is, those that contain all the components of petroleum systems, they may have strong effects on exploration and production of hydrocarbons (Fig. 5.10) (Rateau

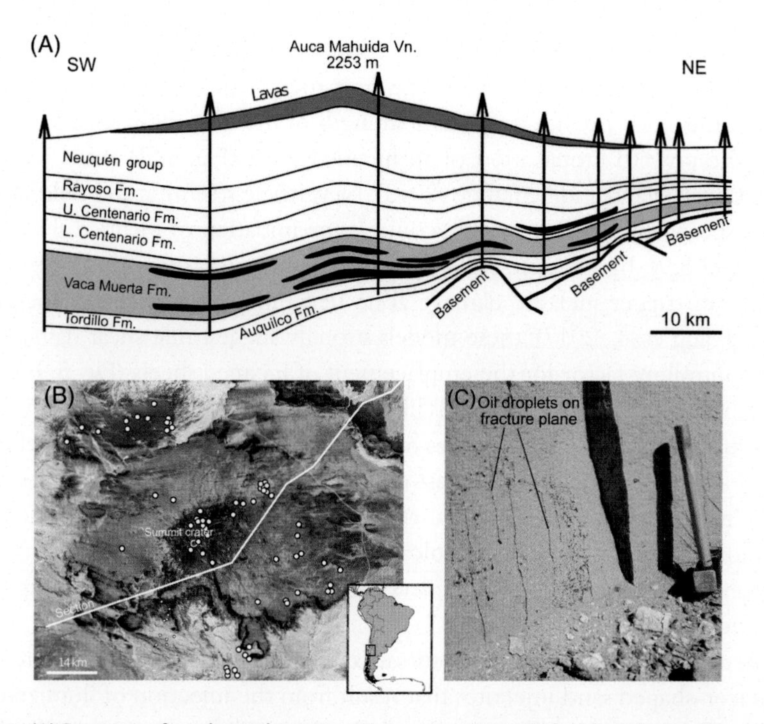

Figure 5.10 (A) Drawing of geological cross-section across Auca Mahuida Volcano, Neuquén Province, Argentina (see location on map B) (Rossello et al., 2002). Quaternary basaltic sills are dominantly emplaced in the Lower Cretaceous organic-rich shale Vaca Muerta Formation. Notice that sills in the area are gas reservoirs. (B) Satellite Landsat image of the Auca Mahuida Volcano, locating the cross-section in (A) and the numerous producing hydrocarbon wells white dots. (C) Field photograph of oil droplets on fracture planes within andesite sills emplaced in the organic-rich shale Vaca Muerta Formation, northern Neuquén Basin, Mendoza Province, Argentina. This photograph illustrates that sills can be fractured hydrocarbon reservoirs.

et al., 2013; Senger et al., 2017). These effects concern all elements of the petroleum system, including source rock maturation, fluid migration, reservoir, trap and seal. (1) Sills are preferentially emplaced along stratigraphic levels such as organic-rich shale, that is, hydrocarbon source rocks, where the heat provided by the magma locally maturates the organic matter in the surrounding sediments. (2) The injection of high-pressure magma may cause uplift and deformation of the host rock, forming broad domes, or 'forced folds', in their overlaying strata (Fig. 5.8), some of which represent hydrocarbon traps. (3) The damage induced by the emplacement of magma in the host rock produces abundant fractures that enhance permeability and fluid flow (Fig. 5.8). (4) Solidified igneous intrusions are heavily fractured due to cooling, for example, the characteristic columnar jointing of sills. Therefore, the intrusions can be highly productive hydrocarbon reservoirs and migration pathways for hydrocarbons. Since sills have good reservoir properties, they also act as aquifers in arid parts of the world, such as the Karoo basin, South Africa (Chevallier et al., 2004).

The voluminous sill complexes resulting from LIPs within sedimentary basins were emplaced during short geological time intervals (<1 Ma). Several times during the Earth's history, the enormous amount of heat brought by the sills triggered rapid degassing of the sedimentary host-rock formations, which resulted in the voluminous release of methane, CO_2 and/or poisonous gases (Fig. 5.11). These gases are released into the atmosphere through numerous hydrothermal vent complexes rooted in the metamorphic aureoles

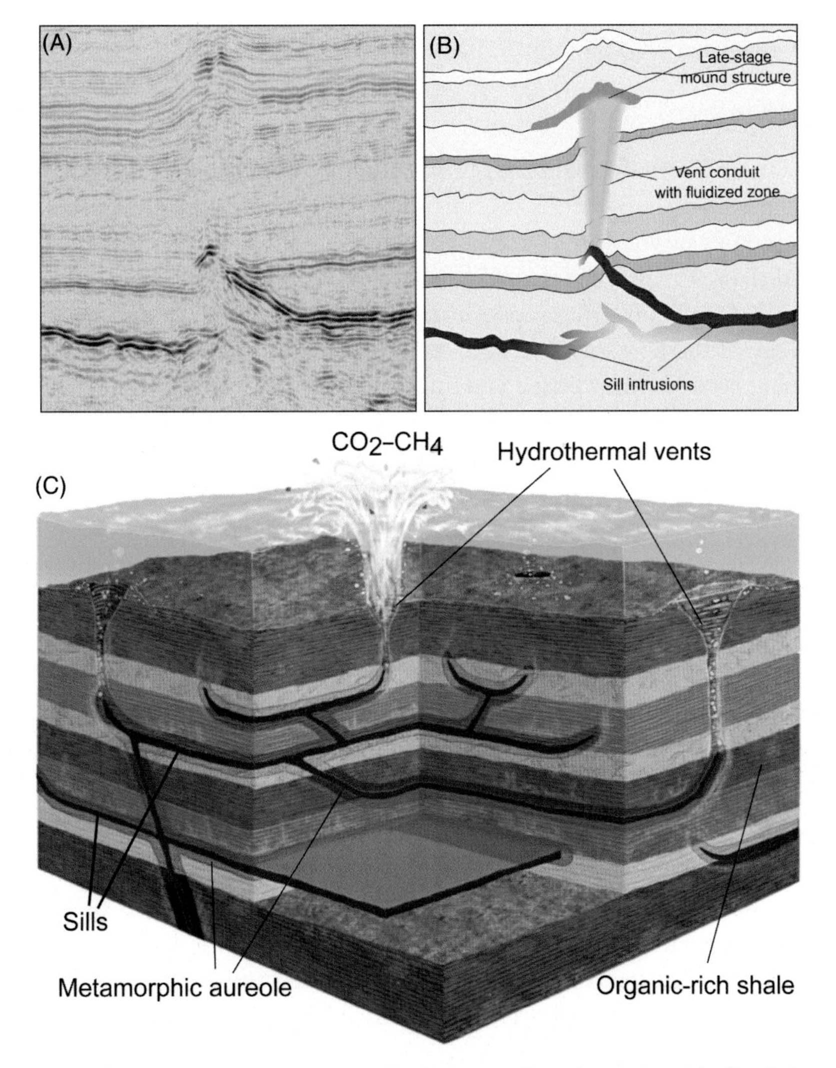

Figure 5.11 (A) Characteristic seismic image displaying a sill and associated inclined sheet, and a hydrothermal vent rooted at the tip of the inclined sheet. (B) Interpretative drawing of (A) modified from (Kjoberg et al., 2017). (C) Artistic drawing illustrating the implications of sill emplacement within organic-rich sediments, with the formation of hydrothermal vents and release of CO_2 and CH_4 in the atmosphere.

around the sills (Fig. 5.11) (Jamtveit et al., 2004). This mechanism had catastrophic consequences in the Earth's climate and life conditions (Svensen et al., 2004), probably triggering major mass extinctions in Earth's history (Courtillot and Renne, 2003). The most emblematic example is the Permo-Triassic mass extinction coeval with the Siberian LIP emplaced in the Tunguska Basin, Siberia, which triggered the extinction of 95% of the marine and 70% of the continental species. Other examples are the end-Triassic extinction associated with the Central Atlantic Magmatic Province, the early Toarcian climatic excursion associated with the Karoo-Ferrar Magmatic Province and the late Paleocene climatic excursion associated with the North Atlantic Igneous Province.

5.6 SUMMARY

Igneous sills are common magmatic intrusions and mainly occur in the layered crust. Sills have been studied in various volcanic settings, principally in sedimentary basins and central volcanoes in volcanic rifts and oceanic islands. Given that sills mainly form in the layered crust, it is assumed that the mechanical layering of the host rock dominantly controls their formation. The propagation mechanisms highly depend on the lithology and rheology of the host rock and span from tensile elastic fracturing for sills emplaced in competent rocks to shear brittle and ductile failure in weaker rocks. At shallow depth, growing sills may develop typical saucer shapes because of the mechanical interaction between the growing sill and the doming of the overburden. Better understanding of sill emplacement and evolution proves essential for assessing volcanic hazards in active volcanoes and exploring hydrocarbons in sedimentary basins. The massive and fast emplacement of sills associated with LIPs in sedimentary basins triggered catastrophic climate changes and mass extinctions in Earth's history.

FIELD EXAMPLE: THE GOLDEN VALLEY SILL COMPLEX, KAROO BASIN, SOUTH AFRICA

The Karoo Basin in South Africa is likely the best natural laboratory to investigate sill emplacement mechanisms in a layered sedimentary basin because erosion under arid conditions has exposed hundreds of saucer-shaped sills (Fig. 5.4). Among these, the Golden Valley Sill Complex has excellent exposures of four main sills and some dykes of basaltic composition (Fig. 5.12). The Golden Valley Sill Complex has been the subject of ambitious research efforts integrating structural mapping, measurements of AMS and geochemical composition through extensive sampling campaigns, such that it is likely the best-studied sill worldwide. The multi-disciplinary research implemented at the Golden Valley Sill Complex intended to (1) constrain magma flow directions in the sills to understand their emplacement dynamics, (2) test whether nearby sills are connected to constrain sill-feeding mechanisms, (3) study post-emplacement magma differentiation within sills and (4) better understand the permeability of sill intrusions.

The main method to constrain magma flow directions is AMS. The basis of the AMS method is to measure the preferred orientation of magnetic minerals (i.e. the magnetic fabric) in a rapid and accurate manner. The magnetic fabric corresponds to the petrofabric and can be interpreted in terms of viscous shear induced by magma flow. Therefore, the AMS method applied to the Golden Valley Sill is used to estimate the magma flow direction during emplacement. A total of 665 samples were collected in situ at 113 localities that cover homogeneously the Golden Valley Sill and the connections with adjacent sills (Polteau et al., 2008a). The AMS results indicate that the emplacement of magma in saucer-shaped sills is an extremely dynamic process with inflation and deflation cycles. The AMS measurements at opposite margins of the finger-like structures show an outward magma flow direction. The fingers may have represented long-term magma channels with active magma flow at the time when the remaining sill was crystallizing.

To test whether the different sills of the complex were connected at the time of their emplacement, that is, whether they were feeding each other, statistical analysis of geochemical compositions from 327 rock samples was performed (Galerne et al., 2008). Geochemical variation diagrams for the different sills of the Golden Valley Sill Complex showed similar compositional ranges for the sills, but some differences in ratios between incompatible elements. These differences are statistically relevant and imply distinct, characteristic geochemical signatures among the sill populations. Each geochemical signature most likely represents a magma batch of distinct chemical characteristics. Notably, most sills of the Golden Valley Sill Complex exhibit significantly distinct chemical signatures, showing that each sill was fed by a distinct magma batch, with the exception of two sills that were likely connected.

The post-emplacement magma differentiation within the sills was studied through chemical analyses of compositional variations along profiles from the Golden Valley Sill (Galerne et al., 2010). In total, 18 whole-rock compositional profiles were sampled along the inclined sheets of the saucer-shaped sill (Fig. 5.12). The profiles show that different compositional patterns, previously described in distinct mafic–ultramafic sills elsewhere, may be found in different parts of a single saucer-shaped sill. The detailed examination of the mineral grain assemblage and compositions suggests that processes within 100-m-thick sills relate to early and late fractional crystallization. The observations in the Golden Valley Sill suggest that a significant part of the fractionation takes place at a late stage of cooling when a crystalline skeleton or mush zone forms. These results suggest that the process of post-emplacement melt flow regionally overprinted compositional patterns produced by earlier crystal segregation from the cooling magma at fluid-like stages during the emplacement.

To improve the understanding of fractures associated with intrusive bodies emplaced in sedimentary basins, quantitative analysis of fracturing in the sills and in the nearby host rock was performed through integration of field data, high-resolution Lidar virtual

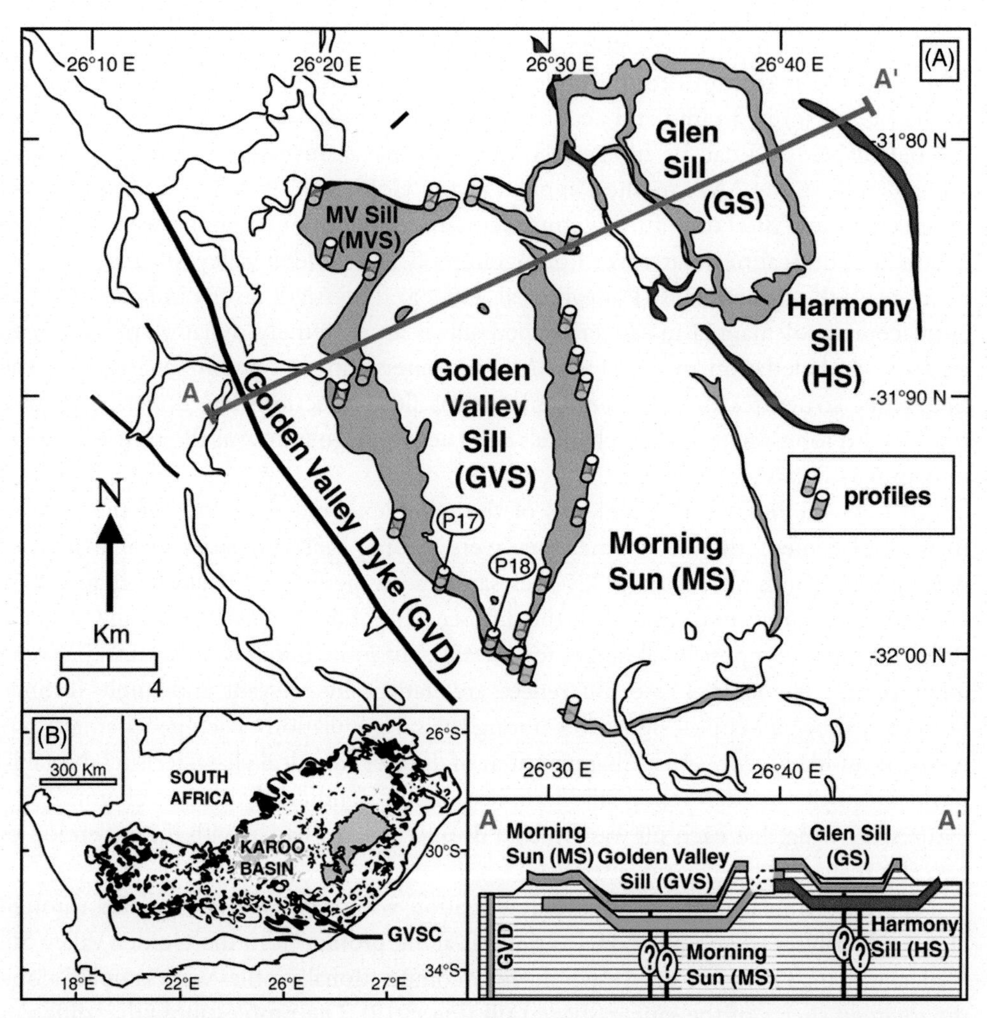

Figure 5.12 Summary of the geochemical architecture of the Golden Valley Sill Complex, after Galerne et al. (2010). (A) Simplified geological map showing the four different magma batches (different colours) identified in the complex. Line A–A′ locates geological cross-section (bottom right). (B) Map showing the location of the Golden Valley Sill Complex in the Karoo basin among dolerite intrusives (black) and the Lesotho Flood basalt remnant (*grey*).

outcrop models and image processing applied to key outcrops of the Golden Valley Sill (Senger et al., 2015). Fracture mapping highlights two main fracture sets oriented parallel and perpendicular to the contact, respectively. Finally, the fracture frequency in the host rock increases towards the sill contacts, showing permeability enhancement and a high potential for fluid flow channelling along the intrusion–host rock interfaces. The study of Senger et al. (2015) highlights how sills can considerably affect the permeability architecture of sedimentary basins.

REFERENCES

Airoldi, G.M., Muirhead, J.D., Long, S.M., Zanella, E., White, J.D.L., 2016. Flow dynamics in mid-Jurassic dikes and sills of the Ferrar large igneous province and implications for long-distance magma transport. Tectonophysics 683, 182–199.

Allmendinger, R.W., Oliver, J., Hauge, T.A., Hauser, E.C., Potter, C.J., 1987. Tectonic heredity and the layered lower crust in the Basin and Range Province, western United States. Geol. Soc. London. Spec. Pub. 28, 223–246.

Amelung, F., Jonsson, S., Zebker, H., Segall, P., 2000. Widespread uplift and "trapdoor" faulting on Galápagos volcanoes observed with radar interferometry. Nature 407, 993–996.

Barnett, Z.A., Gudmundsson, A., 2014. Numerical modelling of dykes deflected into sills to form a magma chamber. J. Volcanol. Geotherm. Res. 281, 1–11.

Bunger, A.P., Cruden, A.R., 2011. Modeling the growth of laccoliths and large mafic sills: role of magma body forces. J. Geophys. Res. 116, B02203.

Burchardt, S., 2008. New insights into the mechanics of sill emplacement provided by field observations of the Njardvik Sill, Northeast Iceland. J. Volcanol. Geotherm. Res. 173, 280–288.

Cartwright, J., Hansen, D.M., 2006. Magma transport through the crust via interconnected sill complexes. Geology 34, 929–932.

Chanceaux, L., Menand, T., 2014. Solidification effects on sill formation: an experimental approach. Earth Planet. Sci. Lett. 403, 79–88.

Chevallier, L., Gibson, L.A., Nhleko, L.O., Woodford, A.C., Nomquphu, W., Kippie, I., 2004. Hydrogeology of fractured-rock aquifers and related ecosystems within the Qoqodala dolerite ring and sill complex, Great Kei catchment, Eastern Cape, Water Res. Com. S. Afr. p. 127.

Corry, C.E., 1988. Laccoliths: mechanisms of emplacement and growth, Geol. Soc. Am. Spec. Pap. 220, 110.

Courtillot, V.E., Renne, P.R., 2003. On the ages of flood basalt events. Comptes Rendus Geosci. 335, 113–140.

Eide, C.H., Schofield, N., Jerram, D.A., Howell, J.A., 2016. Basin-scale architecture of deeply emplaced sill complexes: Jameson Land. East Greenland. J. Geol. Soc. London, 174, 23–40. doi: 10.1144/jgs2016-018.

Eide, C.H., Schofield, N., Lecomte, I., Buckley, S.J., Howell, J.A., 2017. Seismic interpretation of sill-complexes in sedimentary basins: the 'sub-sill imaging' problem. J. Geol. Soc. London.

Ferré, E., Galland, O., Montanari, D., Kalakay, T., 2012. Granite magma migration and emplacement along thrusts. Int. J. Earth Sci., 1–16. doi: 10.1007/s00531-012-0747-6.

Francis, E.H., 1982. Magma and sediment. I. Emplacement mechanism of late carboniferous tholeiite sills in northern Britain. J. Geol. Soc. London 139, 1–20.

Galerne, C.Y., Neumann, E.R., Planke, S., 2008. Emplacement mechanisms of sill complexes: information from the geochemical architecture of the Golden Valley Sill Complex, South Africa. J. Volcanol. Geotherm. Res. 177, 425–440.

Galerne, C.Y., Neumann, E.R., Aarnes, I., Planke, S., 2010. Magmatic differentiation processes in saucer-shaped sills: evidence from the Golden Valley Sill in the Karoo Basin, South Africa. Geosphere 6, 163–188.

Galerne, C.Y., Galland, O., Neumann, E.R., Planke, S., 2011. 3D relationships between sills and their feeders: evidence from the Golden Valley Sill Complex (Karoo Basin) and experimental modelling. J. Volcanol. Geotherm. Res. 202, 189–199.

Galland, O., Scheibert, J., 2013. Analytical model of surface uplift above axisymmetric flat-lying magma intrusions: implications for sill emplacement and geodesy. J. Volcanol. Geotherm. Res. 253, 114–130.

Galland, O., Cobbold, P.R., de Bremond d'Ars, J., Hallot, E., 2007. Rise and emplacement of magma during horizontal shortening of the brittle crust: insights from experimental modeling. J. Geophys. Res. 112 (B6). doi: 10.1029/2006JB004604.

Galland, O., Planke, S., Neumann, E.R., Malthe-Sørenssen, A., 2009. Experimental modelling of shallow magma emplacement: application to saucer-shaped intrusions. Earth Planet. Sci. Lett. 277, 373–383.

Galland, O., Holohan, E.P., van Wyk de Vries, B., Burchardt, S., 2018. Laboratory modelling of volcano plumbing systems: a review. In: Breitkreuz, C., Rocchi, S. (Eds.), Physical Geology of Shallow Magmatic Systems—Dykes, Sills and Laccoliths. Springer, Berlin, Heidelberg, pp. 1–68.

Gilbert, G.K., 1877. Report on the geology of the Henry Mountains. U.S. Geogr. Geol. Surv. Rocky Mountain Region (Powel), p. 160.

Goulty, N.R., Schofield, N., 2008. Implications of simple flexure theory for the formation of saucer-shaped sills. J. Struct. Geol. 30, 812–817.

Gressier, J.B., Mourgues, R., Bodet, L., Matthieu, J.Y., Galland, O., Cobbold, P.R., 2010. Control of pore fluid pressure on depth of emplacement of magmatic sills: an experimental approach. Tectonophysics 489, 1–13.

Hansen, D.M., Cartwright, J.A., Thomas, D., 2004. 3D seismic analysis of the geometry of igneous sills and sill junctions relationships. In: Davies, R.J., Cartwright, J.A., Stewart, S.A., Lappin, M., Underhill, J.R. (Eds.), 3D Seismic Technology: Application to the Exploration of Sedimentary Basins. Geological Society, London, pp. 199–208, Memoirs.

Haug, Ø.T., Galland, O., Souloumiac, P., Souche, A., Guldstrand, F., Schmiedel, T., 2017. Inelastic damage as a mechanical precursor for the emplacement of saucer-shaped intrusions. Geology 45, 1099–1102.

Hogan, J.P., Price, J.D., Gilbert, M.C., 1998. Magma traps and driving pressure: consequences for pluton shape and emplacement in an extensional regime. J. Struct. Geol. 20, 1155–1168.

Hubbert, M.K., Willis, D.G., 1957. Mechanics of hydraulic fracturing. AIME Trans. 210, 153–168.

Hutton, D.H.W., 2009. Insights into magmatism in volcanic margins: bridge structures and a new mechanism of basic sill emplacement, Theron Mountains, Antarctica. Petroleum Geosci. 15, 269–278.

Jackson, M., 1997. Processes of laccolithic emplacement in the Southern Henry Mountains, Southeastern Utah. In: Friedman, J.D., Huffman, A.C. (Eds.), Laccolith Complexes of Southeastern Utah: Time of Emplacement and Tectonic Setting—Workshop Proceedings. U.S. Geological Survey Bulletin 2158, pp. 51–59.

Jackson, C.A.L., Schofield, N., Golenkov, B., 2013. Geometry and controls on the development of igneous sill-related forced-folds: a 2D seismic reflection case study from offshore southern Australia. Geol. Soc. Am. Bull. 125, 1874–1890.

Jamtveit, B., Svensen, H., Podladchikov, Y.Y., Planke, S., 2004. Hydrothermal vent complexes associated with sill intrusions in sedimentary basins. In: Breitkreuz, C., Petford, N. (Eds.), Physical Geology of High-level Magmatic Systems. Special Publication. Geological Society, London, pp. 233–241.

Kavanagh, J.L., Menand, T., Sparks, R.S.J., 2006. An experimental investigation of sill formation and propagation in layered elastic media. Earth Planet. Sci. Lett. 245, 799–813.

Kavanagh, J.L., Boutelier, D., Cruden, A.R., 2015. The mechanics of sill inception, propagation and growth: experimental evidence for rapid reduction in magmatic overpressure. Earth Planet. Sci. Lett. 421, 117–128.

Kavanagh, J.L., Rogers, B.D., Boutelier, D., Cruden, A.R., 2017. Controls on sill and dyke–sill hybrid geometry and propagation in the crust: the role of fracture toughness. Tectonophysics 698, 109–120.

Kjoberg, S., Schmiedel, T., Planke, S., Millett, J., Jerram, D.A., Galland, O., Lecomte, I., Schofield, N., Haug, Ø.T., Helsem, A., 2017. 3D structure and formation of hydrothermal vent complexes at the Paleocene-Eocene transition, the Møre basin, min-Norwegian margin. Interpretation 5, SK65–SK81.

Lister, J.R., 1990. Buoyancy-driven fluid fracture. The effects of material toughness and of low-viscosity precursors. J. Fluid Mech. 210, 263–280.

Magee, C., Jackson, C.A.-L., Schofield, N., 2013. The influence of normal fault geometry on igneous sill emplacement and morphology. Geology 41, 407–410.

Magee, C., Muirhead, J.D., Karvelas, A., Holford, S.P., Jackson, C.A.L., Bastow, I.D., Schofield, N., Stevenson, C.T.E., McLean, C., McCarthy, W., Shtukert, O., 2016. Lateral magma flow in mafic sill complexes. Geosphere 12, 809–841.

Magee, C., Bastow, I.D., van Wyk de Vries, B., Jackson, C.A.-L., Hetherington, R., Hagos, M., Hoggett, M., 2017. Structure and dynamics of surface uplift induced by incremental sill emplacement. Geology 45, 431–434.

Mathieu, L., van Wyk de Vries, B., Holohan, E.P., Troll, V.R., 2008. Dykes, cups, saucers and sills: analogue experiments on magma intrusion into brittle rocks. Earth Planet. Sci. Lett. 271, 1–13.

Merle, O., Donnadieu, F., 2000. Indentation of volcanic edifices by the ascending magma. In: Vendeville, B., Mart, Y., Vigneresse, J.-L. (Eds.), Salt, Shale and Igneous Diapirs in and around Europe. Geological Society, 174. Special Publications, London, pp. 43–53.

Michaut, C., 2011. Dynamics of magmatic intrusions in the upper crust: theory and applications to laccoliths on Earth and the Moon. J. Geophys. Res. 116, B05205.

Muirhead, J.D., Airoldi, G., Rowland, J.V., White, J.D.L., 2012. Interconnected sills and inclined sheet intrusions control shallow magma transport in the Ferrar large igneous province, Antarctica. Geol. Soc. Am. Bull. 124, 162–180.

Muirhead, J.D., Airoldi, G., White, J.D.L., Rowland, J.V., 2014. Cracking the lid: sill-fed dikes are the likely feeders of flood basalt eruptions. Earth Planet. Sci. Lett. 406, 187–197.

Pagli, C., Wright, T.J., Ebinger, C.J., Yun, S.-H., Cann, J.R., Barnie, T., Ayele, A., 2012. Shallow axial magma chamber at the slow-spreading Erta Ale Ridge. Nat. Geosci. 5, 284–288.

Pasquare, F., Tibaldi, A., 2007. Structure of a sheet-laccolith system revealing the interplay between tectonic and magma stresses at Stardalur Volcano, Iceland. J. Volcanol. Geotherm. Res. 161, 131–150.

Planke, S., Rasmussen, T., Rey, S.S., Myklebust, R., 2005. Seismic characteristics and distribution of volcanic intrusions and hydrothermal vent complexes in the Vøring and Møre basins. In: Doré, A.G., Vining, B.A. (Eds.), Proc. 6th Petrol. Geol. Conf. Geological Society, London.

Pollard, D.D., 1973. Derivation and evaluation of a mechanical model for sheet intrusions. Tectonophysics 19, 233–269.

Pollard, D.D., Muller, O.H., Dockstader, D.R., 1975. The form and growth of fingered sheet intrusions. Geol. Soc. Am. Bull. 86, 351–363.

Polteau, S., Ferré, E.C., Planke, S., Neumann, E.-R., Chevallier, L., 2008a. How are saucer-shaped sills emplaced? Constraints from the Golden Valley Sill, South Africa. J. Geophys. Res. 113, 113.

Polteau, S., Mazzini, A., Galland, O., Planke, S., Malthe-Sørenssen, A., 2008b. Saucer-shaped intrusions: occurrences, emplacement and implications. Earth Planet. Sci. Lett. 266, 195–204.

Rateau, R., Schofield, N., Smith, M., 2013. The potential role of igneous intrusions on hydrocarbon migration, West of Shetland. Petroleum Geosci. 19, 259–272.

Rossello, E.A., Cobbold, P.R., Diraison, M., Arnaud, N., 2002. Auca Mahuida (Neuquén basin, Argentina): a quaternary shield volcano on a hydrocarbon-producing substrate, 5th ISAG, Extended Abstracts, Toulouse, pp. 549–552.

Rubin, A.M., 1993. Tensile fracture of rock at high confining pressure: implications for dike propagation. J. Geophys. Res. 98, 15919–15935.

Scheibert, J., Galland, O., Hafver, A., 2017. Inelastic deformation during sill and laccolith emplacement: insights from an analytic elasto-plastic model. J. Geophys. Res.: Solid Earth 122, 923–945. doi: 10.1002/2016JB013754.

Schmiedel, T., Kjoberg, S., Planke, S., Magee, C., Galland, O., Schofield, N., Jackson, C.A.-L., Jerram, D.A., 2017. Mechanisms of overburden deformation associated with the emplacement of the Tulipan sill, mid-Norwegian margin. Interpretation 5, SK23–SK38.

Schofield, N., Brown, D.J., Magee, C., Stevenson, C.T., 2012. Sill morphology and comparison of brittle and non-brittle emplacement mechanisms. J. Geol. Soc. London 169, 127–141.

Senger, K., Buckley, S.J., Chevallier, L., Fagreng, Å., Galland, O., Kurz, T.K.O., Planke, S., Tveranger, J., 2015. Fracturing of doleritic intrusions and associated contact zones: insights from the Eastern Cape, South Africa. J. Afr. Earth Sci. 102, 70–85.

Senger, K., Millett, J., Planke, S., Ogata, K., Eide, C.H., Festøy, M., Galland, O., Jerram, D.A., 2017. Effects of igneous intrusions on the petroleum system: a review. First Break 35, 47–56.

Sigmundsson, F., Hreinsdottir, S., Hooper, A., Arnadottir, T., Pedersen, R., Roberts, M.J., Oskarsson, N., Auriac, A., Decriem, J., Einarsson, P., Geirsson, H., Hensch, M., Ofeigsson, B.G., Sturkell, E., Sveinbjornsson, H., Feigl, K.L., 2010. Intrusion triggering of the 2010 Eyjafjallajökull explosive eruption. Nature 468, 426–430.

Spacapan, J.B., Galland, O., Leanza, H.A., Planke, S., 2017. Igneous sill and finger emplacement mechanism in shale-dominated formations: a field study at Cuesta del Chihuido, Neuquén Basin. Argentina. J. Geol. Soc. London 174, 422–433. doi: 10.1144/jgs2016-056.

Svensen, H., Planke, S., Malthe-Sorenssen, A., Jamtvelt, B., Myklebust, R., Eldem, T.R., Rey, S.S., 2004. Release of methane from a volcanic basin as a mechanism for initial Eocene global warming. Nature 429, 542–545.

Svensen, H., Corfu, F., Polteau, S., Hammer, Ø., Planke, S., 2012. Rapid magma emplacement in the Karoo Large Igneous Province. Earth Planet. Sci. Lett. 325–326, 1–9.

Szarawarska, E., Huuse, M., Hurst, A., De Boer, W., Lu, L., Molyneux, S., Rawlinson, P., 2010. Three-dimensional seismic characterisation of large-scale sandstone intrusions in the lower Palaeogene of the North Sea: completely injected vs. in situ remobilised sandbodies. Basin Res. 22, 517–532.

Thomson, K., 2007. Determining magma flow in sills, dykes and laccoliths and their implications for sill emplacement mechanisms. Bull. Volcanol. 70, 183–201.

Thomson, K., Schofield, N., 2008. Lithological and structural controls on the emplacement and morphology of sills in sedimentary basins. In: Thomson, K., Petford, N. (Eds.). Geological Society, London, Special Publications, pp. 31–44.

Thorey, C., Michaut, C., 2016. Elastic-plated gravity currents with a temperature-dependent viscosity. J. Fluid Mech. 805, 88–117.

Tweto, O., 1951. Form and structure of sills near Pando, Colorado. Geol. Soc. Am. Bull. 62, 507–532.

Walker, R.J., Healy, D., Kawanzaruwa, T.M., Wright, K.A., England, R.W., McCaffrey, K.J.W., Bubeck, A.A., Stephens, T.L., Farrell, N.J.C., Blenkinsop, T.G., 2017. Igneous sills as a record of horizontal shortening: The San Rafael subvolcanic field. Utah. Geol. Soc. Am. Bull 129, 1052–1070. doi: 10.1130/b31671.1.

White, R.S., Smith, L.K., Roberts, A.W., Christie, P.A.F., Kusznir, N.J., 2008. The rest of the iSIMM team, Lower-crustal intrusion on the North Atlantic continental margin. Nature 452, 460–464.

Witte, J., Bonora, M., Carbone, C., Oncken, O., 2012. Fracture evolution in oil-producing sills of the Rio Grande Valley, northern Neuquén Basin, Argentina. AAPG Bull. 96, 1253–1277.

FURTHER READING

Menand, T., 2008. The mechanics and dynamics of sills in layered elastic rocks and their implications for the growth of laccoliths and other igneous complexes. Earth Planet. Sci. Lett. 267, 93–99.

CHAPTER 6

Pascal's Principle, a Simple Model to Explain the Emplacement of Laccoliths and Some Mid-crustal Plutons

Sven Morgan
Iowa State University, Ames, IA, United States

Contents

6.1 INTRODUCTION

Field, petrological/geochemical, and magnetic data, supported by thermal modelling indicate that many plutons, laccoliths, and sills are often emplaced into the middle and upper crust incrementally (Wiebe, 1993; Miller and Paterson, 2001; Coleman et al., 2004; Michel et al., 2008; Morgan et al., 2008; Miller et al., 2011; Stearns and Bartley, 2014; Menand et al., 2015, Chen and Nabelek, 2017). This model, based on evidence collected on ancient intrusive systems exposed at the Earth's surface, is supported by the data collected from the subsurface which points to pulsed magma chamber growth beneath active volcanoes (Saunders et al., 2012) and also through examination of chemical and textural analysis on volcanic assemblages (Cashman and Blundy, 2013; see also Chapter 12). What is also becoming clear is that it is increasingly difficult to find evidence of these increments the bigger the intrusion, the deeper the level of emplacement, and

Volcanic and Igneous Plumbing Systems. http://dx.doi.org/10.1016/B978-0-12-809749-6.00006-6
Copyright © 2018 Elsevier Inc. All rights reserved.

the more homogeneous the magma. This is because the only way to preserve contacts where the magma pulsing is homogeneous in composition is to cool rapidly. The rate of cooling has to be more rapid than the rate of sheet emplacement (de Saint Blanquat et al., 2006; see Chapter 2). Even if contacts do form, the textural evidence can be destroyed or modified by continued crystallisation, intrusion of later pulses and subsequent annealing, gravity settling (fractional crystallisation) or compaction driven melt segregation (Zeig and Marsh, 2012). Thermal cycling, which is expected during multiple episodes of magma injection (Mills et al., 2011; Glazner and Johnson, 2013) can also modify textures and destroy contacts. If internal contacts do form, they may only be observed when schlieren are crosscut by a contact and this contact may not be detectable away from the schlieren layering (e.g., Mahan et al., 2003). Internal contacts are also known to exist when lenses of host rock ('ghost stratigraphy', Pitcher et al., 1958) or known sedimentary rock (Horsman et al., 2005) or screens of host rock terminate (Miller and Paterson, 2001; Anderson et al., 2013) and leave one sheet against another. Obviously it can be very difficult to identify internal contacts when the composition does not change between pulses (Morgan et al., 2017). Therefore, large plutons of similar composition, or where compositional changes are gradational over tens to hundreds of meters may also be composed of many discrete pulses (Coint et al., 2013), or it may be that magma fluxes are sufficient to sustain temperatures above the solidus during the growth of the chamber (cf. Chapter 2).

Magma rises through the crust most efficiently as dikes (see Chapter 3), and rising magma will cease to rise for a variety of reasons (Gudmundsson, 2002). However, assuming magma pressure is large enough to produce a magma chamber or a sill, decreasing the driving pressure (Lister and Kerr, 1991) and/or density differences (Ryan, 1987) are probably not the reasons for discontinuing the magma's vertical path. Some of the more plausible reasons for magma moving outward and intruding as sills is changing the rheological boundaries (Kavanagh et al., 2006); either by changing viscosities of higher level country rocks (Menand, 2008) or through crystallisation of early pulses of diking (Annen et al., 2006; see Chapter 5).

This chapter will deal primarily with how the construction of a magma/crystal mush body occurs after the magma has been forced to move laterally. The focus of the first part of this chapter will be on the intrusions in the Henry Mountains for three reasons: 1) the Henry Mountains intrusions have been emplaced into the largely undeformed Colorado Plateau and therefore the shapes of intrusion, textures, and fabrics can all be interpreted in terms of forceful emplacement of magma, as tectonic deformation is not a variable. 2) There is a range of volumes of intrusions exposed, and the larger intrusions preserve parts of their initial growth history, so that the larger intrusions preserve parts of their younger and smaller beginnings (Horsman et al., 2009). For example, sills grow through the amalgamation of magma fingers and sheets (see Chapter 3), sheets stack to form laccoliths, and laccoliths inflate to produce bysmaliths/plutons. 3) Because the intrusions have been emplaced into flat-lying sedimentary rocks, the variables dictating their shapes and growth histories are reduced, and the importance of magma pressure is revealed.

The similar ratios between the length and thickness for many plutons and laccoliths have been discussed as indicating that there is a scale-independent mechanism controlling their shape, and therefore the mechanism of emplacement (McCaffrey and Petford, 1997; Cruden and McCaffrey, 2001; Rocchi et al., 2010). A simple model is introduced that helps explain these ratios and the emplacement of laccoliths based on Pascal's principle, whereby small magma pressures, consistent with magma pressures determined for modern volcanic systems, are able to produce large vertical forces when surface areas are increased. Two stages of magma emplacement are necessary: 1) the sill/sheet like initial pulse of magma, which must expand in area so that the vertical force can be multiplied, and 2) the inflation stage, whereby sheet-like additions are abandoned and upward expansion occurs. This model is extended to mid-crustal plutons, and specifically the Birch Creek pluton, where detailed mapping reveals the transition between sheeting and inflation.

6.2 HOW DO LACCOLITHS AND OTHER LARGE MAGMA BODIES GROW FROM SILLS? EXAMPLES FROM THE HENRY MOUNTAINS

The Henry Mountains (Fig.6.1) are a classic locality to study the shallow emplacement of magma. The term "laccolite" originated here with the pioneering work of G.K. Gilbert (1877) who on his horseback and mule expedition to southern Utah, realised that the igneous rock of the Henry Mountains originated as intrusive magma and was visionary enough to conclude that the magma had deformed its wall rocks (see also Chapter 1). The emplacement of these intrusions, whether they were stocks (Hunt, 1953) or laccoliths (Jackson and Pollard, 1988, 1990) and the details of incremental emplacement have been worked on since that time (Horsman et al., 2005, 2009; Saint Blanquat et al., 2006; Morgan et al., 2008, 2017).

The friable and poorly-cemented Mesozoic sandstones that the late Oligocene magma intruded into and the desert climate often results in three-dimensional exposures of contacts and preserved original emplacement shapes. The igneous rock is andesite porphyry with large (up to 7 mm) plagioclase and hornblende phenocrysts (see Chapter 8). Hornblende is almost always completely replaced by a mixture of calcite, clay, and Fe-oxides. Most of the detailed work has focused on three intrusions varying by size and therefore degree of inflation: the Maiden Creek Sill (initial finger-sheet phase), the Trachyte Mesa sill/laccolith (stacking sheets stage), and the Black Mesa Bysmalith (inflation stage). All three of these intrusions intrude Upper Jurassic sandstones and shales. Moreover, all of them are termed "satellites", because they are found outside of, and separated from, the five main intrusive centres that comprise the bulk of the Henry Mountains.

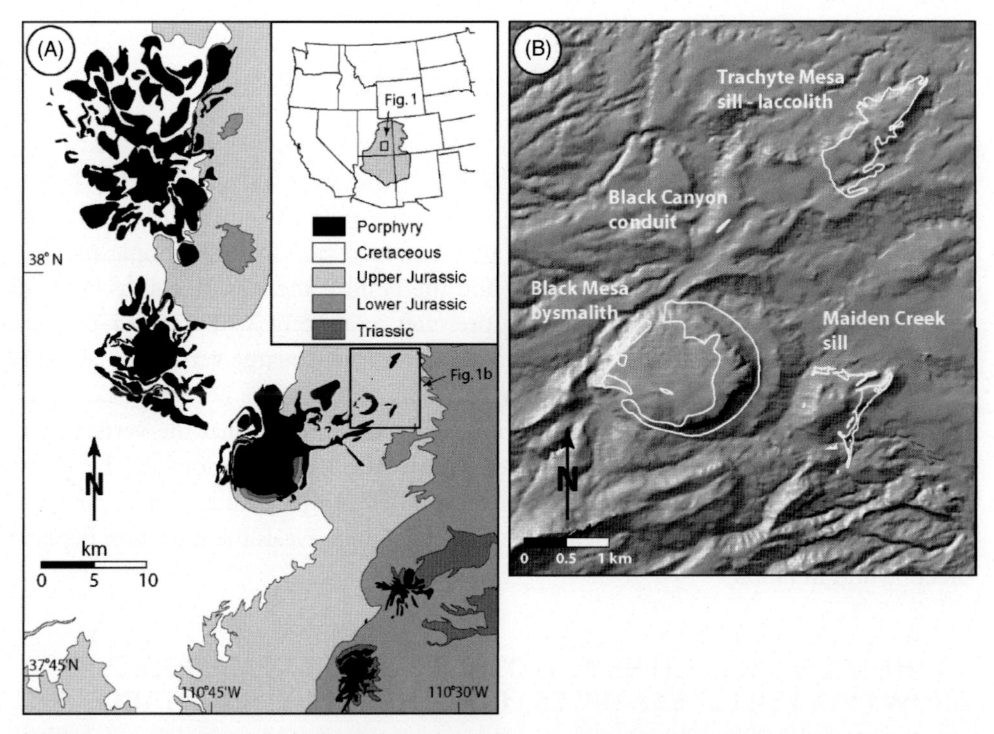

Figure 6.1 ***Location map of the Henry Mountains in southern Utah, USA.*** (A) Henry Mountains intrusions are in black, mostly flat lying sedimentary rocks of the Colorado Plateau in shades of grey and white. Inset shows area of the Colorado Plateau. (B) Shaded relief map of the area outlined in A, illustrating the locations of the intrusions discussed in this paper. *After Horsman et al., 2009.*

6.2.1 The Maiden Creek Sill

The Maiden Creek sill (MCS) intruded the Jurassic Entrada sandstone and was initially studied by Horsman et al. (2005). The MCS is a small intrusion (~1 km in diameter), only partly exposed along a hill side, and is sheet-like in shape with four elongate fingers protruding from a central base. Horsman et al. (2005, 2009) proposed a conceptual model for the emplacement of the MCS whereby the fingers could be the beginning stages to the growth of a larger sill (see Chapter 5), similar to the model developed by Pollard et al. (1975) for the magma fingers observed at Shonkin Sag. Pollard et al. (1975) documented the intense deformation within the shales on the lateral margins of fingers extending from the Shonkin Sag laccolith and determined that the lateral expansion of these fingers resulted in the growth of a sill.

The hornblende mineral lineation on the top of the MCS, as well as the magnetic lineation are consistent and subparallel to the long axis of the intrusive fingers, which supports transport of magma along the long/finger axis. Lateral deformation is also locally observed at the margins of the MCS; most intensely in a zone between where two fingers that emanate from the main body (Horsman et al., 2009). There is upward deflection of the surrounding sandstone layers as the contact with the sill is approached, but it is not enough to account for the height of the sill. This observation indicates that lateral expansion of the initial sill must be accommodating a significant amount of the volume needed, at least near the contacts, similar to that observed adjacent to the fingers at Shonkin Sag. (Pollard et al., 1975) state that permanent deformation above and below the fingers at Shonkin Sag is restricted to compaction. Significant strain and compaction has also been observed in the sandstone layers immediately above the Trachyte Mesa sill/laccolith.

6.2.2 The Trachyte Mesa Sill/Laccolith

The Trachyte Mesa (TM) sill/laccolith is 2 km long and 0.5 km wide. In the SW, the intrusion is 50 m thick and resembles a laccolith with multiple sheets stacked upon one another. In the NE, the intrusion is much more sill-like with evidence for 1-3 stacked sheet-like sills and is only 5-10 m thick. Morgan et al. (2008) and Wilson et al. (2016) studied the emplacement of the intrusion, the fabrics, and the deformation of the overlying Entrada Sandstone and concluded that there were multiple steps in the emplacement. The magnetic lineation pattern of the long axis of the AMS ellipsoid (Morgan et al., 2008; see Chapter 3 for more details on the Anisotropy of Magnetic Susceptibility method) documents a channel of magma flowing parallel to the long axis of the intrusion. Off to the sides, the lineation fans outward, becoming sub-perpendicular to the channel axis. Multiple small fingers emanate from the margins. The magnetic lineation pattern is consistent with the growth of sheets by outward expansion of magma fingers. In this case, one long finger grew along the long axis, and then multiple fingers grew perpendicular to the long axis. Then "filling in" occurred in between these fingers (cf. Thomson and Schofield, 2008). The remnants of some of these fingers remain on the margins, where infilling appears to have been incomplete. Pollard et al. (1975) observed "cusp-shaped grooves" on the top surface of the TM, consistent with these textures resulting from finger growth before "filling-in" occurred. Similar cusp-shaped grooves are visible in cross-sectional views between fingers at Shonkin Sag. This pattern of perpendicular, infilling flow is also observed on the very top of the Black Mesa Bysmalith (BM) which has been interpreted to be the top sheet and the oldest sheet in a much more inflated intrusion. On the top of the BM, mineral lineation trends NW, away from a possible central feeder oriented NE (Saint Blanquat et al., 2006).

In the SW corner, the TM intrusion is at its thickest (~50 m) and is unique in that the contact with the overlying Entrada Sandstone is preserved as it is deflected upward by the intrusion. Here, the intrusion exhibits multiple sub-horizontal, 1-2 cm-thick shear zones. Beginning at the base of the outcrop and moving upward to the top contact, these shear zones are found between 50 cm and 2 m apart. The shear zones are defined by intense brittle fragmentation and stretching of the plagioclase crystals and have been interpreted as contacts between separate sheets, intruding at slightly different times

(Morgan et al., 2008). The deformation of the sedimentary section immediately above the intrusion is partitioned into shearing and flattening, both types of strain occurring simultaneously and accommodating the emplacement. This deformation is discussed in detail below, because it supports the model of sheet-stacking for the initial emplacement.

A well-developed foliation and lineation is only found within the top 1 cm of the intrusion. The lineation is within the foliation, which is parallel to the contact and defined by stretched and broken plagioclase crystals and also by aligned hornblende crystals. The lineation is oriented NW, perpendicular to the outer margin of the laccolith, which trends NE. The foliation below this brittle-ductile shear zone is less well developed, completely magmatic, and oriented sub-perpendicular to the top intrusion margin. Within 5 cm of the contact, the foliation rotates into parallelism with the top contact. The rotation of the foliation has been interpreted as being dragged into parallelism with the top contact as the sheet moves outward, parallel to the lineation, resulting in a shear-sense indicator (Morgan et al., 2008).

The immediately overlying sandstone layer is intensely fractured and best defined as a mega-breccia (Fig. 6.2). The shales and sandstones within one meter above the contact exhibit structures associated with intense simple shear strain (Fig. 6.2). Low-angle faults that cross the beds dip down towards the outer edge of the laccolith and exhibit thrust-sense offsets, consistent with drag of the underlying magma sheets. Contacts between shale and sandstone beds are also sheared and exhibit grooves and slicken lines for several meters above the contact. One bed of massive red Entrada Sandstone, which is ~6 m thick and found 2 m above the contact, was examined in detail for its changes in thickness and finite strain. This sandstone bed is lifted from its regional position and translated and rotated up over the margin to the top of the laccolith (Fig. 6.3). The bed changes in thickness from 6 m, to ~7m as it begins to be uplifted, to 4 and 5m as the top of the laccolith is approached. Strain ratios, determined from normalised Fry analyses on sandstone grains, begin at 1 and increase to 1.6 near the top of the intrusion. The long axes of the strain ellipses are parallel to bedding. Porosity, determined from image analysis from the same images as the strain, decreases from 40% to 20% along the same interval. Microscopic images from the sandstone document increases in grain fracturing and grain-boundary sliding as pore spaces collapsed and filled with clasts and accommodated the flattening (Morgan et al., 2008). This bed also exhibits R-shears, which are only found along the part of this bed that has been rotated upward and seem to be accommodating extension of the bed in the opposite sense to the thrust-like faults found 3 m below, near the contact.

In the lowest part of this SW outcrop, which is assumed to be near the base of the intrusion, thin (20-40 cm thick) sandstone beds are rotated to vertical and individual magmatic sheets at the edge of the intrusion are preserved breaking through the vertical beds. These sheets have bulbous, rounded margins at the front, features also observed at other locations in the Henry Mountains, and intense fracturing is observed on all

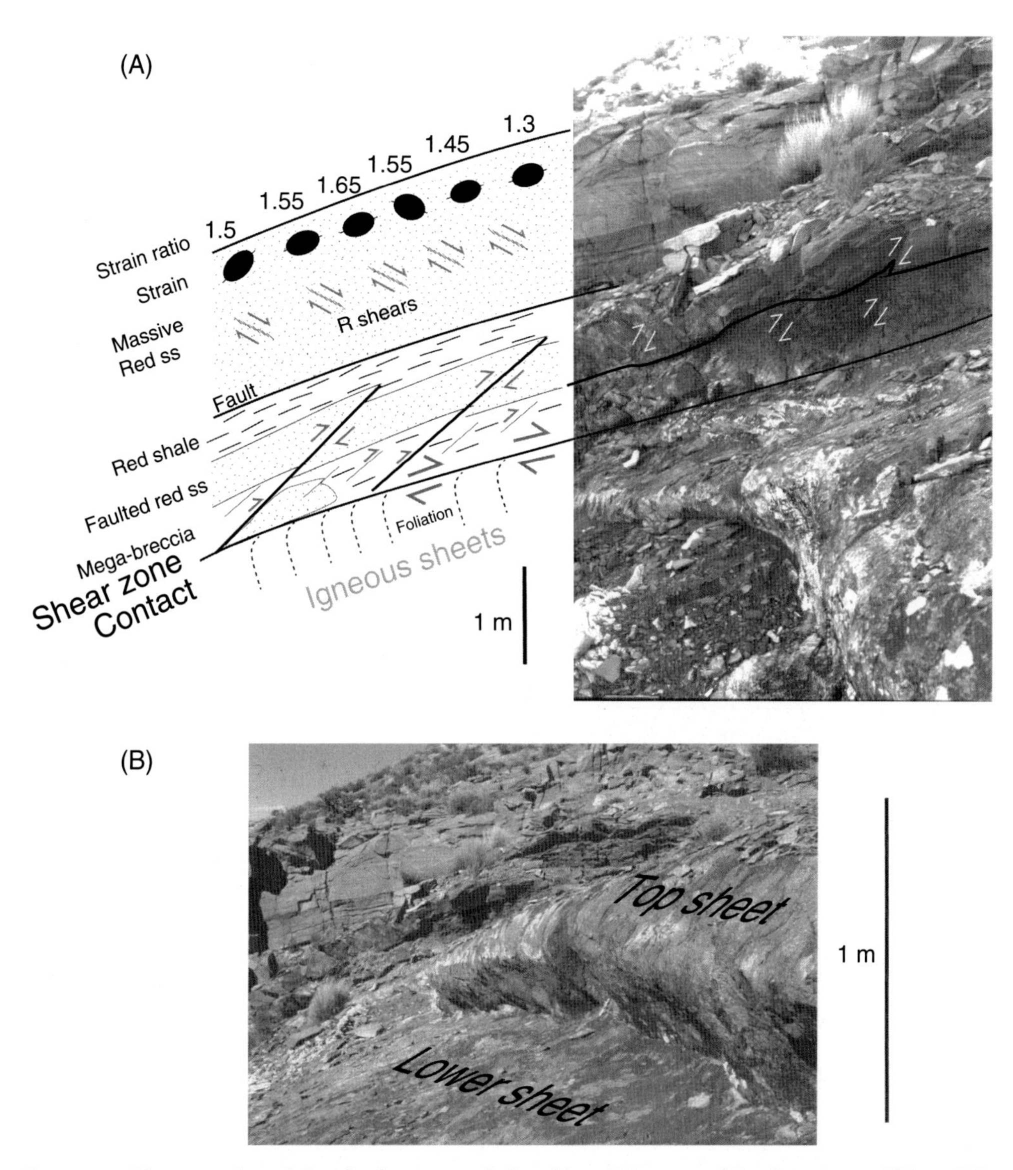

Figure 6.2 ***Photograph and sketch of contact relationships, SW corner of Trachyte Mesa sill/laccolith.*** (A) Details of deformation and strain in the sandstones (ss) and shales for the first six meters above the contact. (B) Same location but different angle illustrating the top two sheets immediately below the contact. *Adapted from Morgan et al., 2008.*

margins within the sandstones. The upturned sandstones have been interpreted to have been rotated to steep angles in front of sheets that were incrementally stacked upon one another and which all terminated at the same lateral extent, an observation also observed at Maiden Creek sill. These thinner beds could deform more readily and conform to the

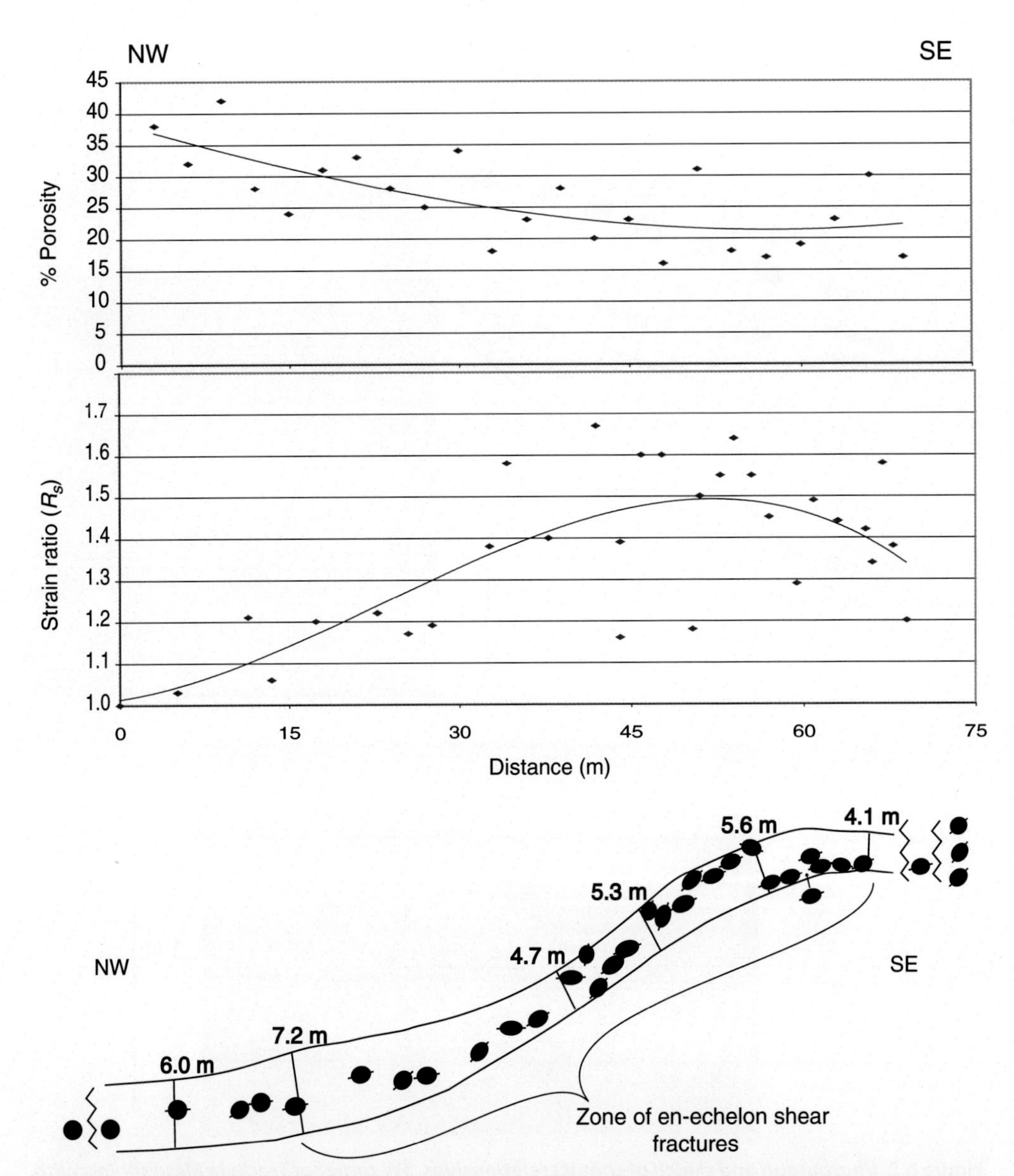

Figure 6.3 ***Strain ratios and changes in thickness and porosity in the massive red sandstone above the contact at the SW margin of the Trachyte Mesa sill/laccolith.*** *After Morgan et al., 2008.*

sub-vertical margin of the stacked sheets. The overlying thicker and more rigid red sandstone layer could not bend as readily, and therefore a triangular zone of lower pressure was created immediately in front of the intruding magma sheets as the stacking continued. This low-pressure zone was filled by these later magma sheets, probably emanating from the original margin (Morgan et al., 2008; Fig. 6.4).

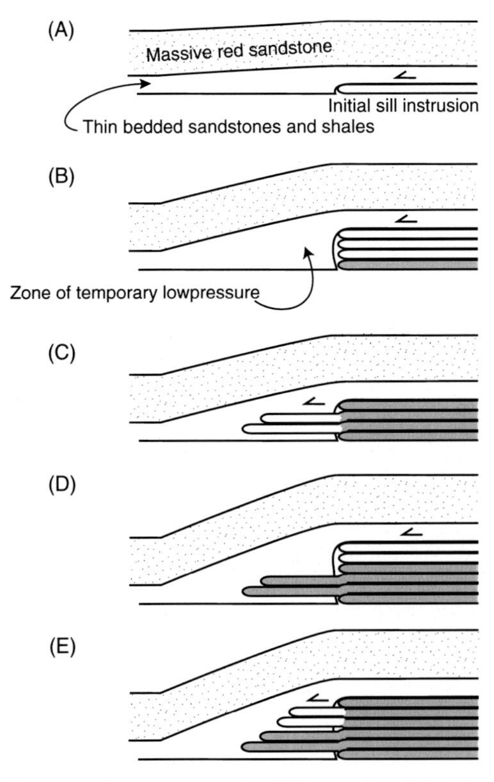

Figure 6.4 ***Sequence of sheet emplacement at the SW margin of the Trachyte Mesa sill/laccolith.*** Vertical stacking of sheets (newly added sheets in white) produces a temporary zone of low pressure (path of least resistance) which is filled by later pulses emanating from the front of the intrusion. Same area as in Figs 6.2 & 6.3. *After Morgan et al., 2008.*

This last observation is important, because it illustrates the importance of how the magma behaved as a fluid that obeyed Pascal's principle, at these crustal levels, and specifically that the fluid pressure is transmitted throughout the fluid, or magma body. It is also clear that the magma followed the path of least resistance, and that path changes as the intrusion grows. Therefore, even though the intrusion is composed of individual sheets, which are defined by thin crystallised margins, the intrusion behaved as one connected fluid mass late in the construction history. This argument will be constructed further with more data from the Black Mesa Bysmalith.

6.2.3 The Black Mesa Bysmalith

The Black Mesa bysmalith (laccolith), studied by Saint Blanquat et al. (2006), exhibits the largest degree of magma inflation and is the largest of the three intrusions described here. It is 1.7 km in diameter and between 150 m and 250 m thick. It has a distinctly round shape in map view and a cylinder-like shape in three dimensions, as it rises out of the desert (Fig. 6.5).

Figure 6.5 ***Photograph of the Black Mesa bysmalith. Looking to the SW.***

The sedimentary wall rocks are horizontal to within ~100 m of the intrusion. As the intrusion is approached, the wall rocks gradually rotate upward, and even become sub-vertical on the SE margin. Small-scale faults increase in number as the contact is approached. The outer margins of the Mesa are very close to the actual outer margin of the intrusion and cliff-like on three sides. A thin veneer of flat-lying sedimentary cover is preserved over much of the top. These sedimentary roof rocks are relatively undeformed except for the regional jointing. The intrusion itself is largely homogeneous in major and in trace-element composition and exhibits a weak fabric that changes from sub-horizontal near the top and bottom to steep in the middle (Saint Blanquat et al., 2006). The foliation is rarely well defined. Magmatic layering and subtle internal contacts are occasionally observed near the top and are sub-horizontal. Sharp shear-zone contacts defining individual sheets observed at Maiden Creek sill and Trachyte Mesa sill/laccolith are not observed. However, we do observe the transition in fabrics between initial horizontal sheet-like emplacement and the inflation stage. The very top surface of BM is characterized by a solid-state stretching lineation trending WNW (Fig. 6.6).

Commonly, fractured plagioclase crystals are pulled apart, defining the lineation and characteristics of many of the top contacts in the Henry Mountains, including the MC and the TM. Just below this solid-state shear zone, the lineation is magmatic and parallel to the top contact. Centimetres to metres below this top zone, the lineation changes orientation to trend NNW and even NE. Saint Blanquat et al. (2006) attributed this change in orientation to the rapid changes in temperature beneath the contact, which changed the viscosity of the magma and therefore preserved different parts of the deformation

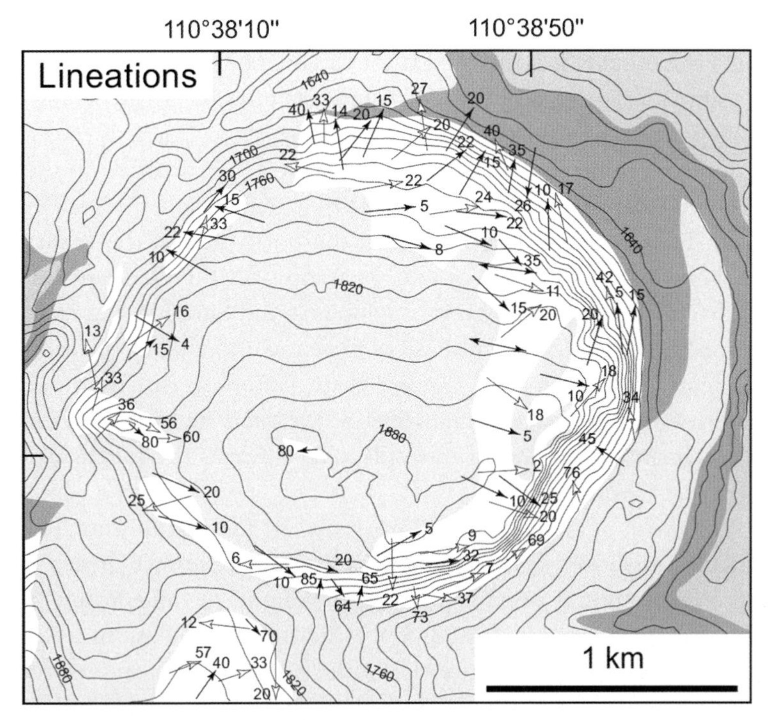

Figure 6.6 ***Mineral (black arrow heads) and Anisotropy of Magnetic Susceptibility (white arrow heads) lineations from the Black Mesa bysmalith.*** Unshaded area is the exposed andesite, shaded rocks are sedimentary layers. Notice how the lineations switch from WNW trending at the top to NE trending below that. *After Morgan et al., 2005.*

(displacement at the top contact versus stretching below). An alternative model is that the top shearing is a record of the initial-sheet like emplacement but that at deeper levels, the change in lineation is recording the inflation stage, which may or may not be sheet-like. The interior portions of the BM are not exposed and therefore it is impossible to examine those fabrics. The outer margin with the wall rocks is exposed and examination of these walls does not reveal shear zones or other features attributable to internal sheet-like contacts.

6.3 CONCEPTUAL MODEL FOR THE GROWTH OF LACCOLITHS FROM SILLS

A conceptual model has been developed to attempt to explain the dimensional similarities between many laccoliths using Pascal's principle and the common theme of sheeting. Pascal's principle states that changes in fluid pressure are transmitted through the fluid everywhere equally so that pressure is always equal, as long as the fluid is connected ($P_1 = P_2$). The force applied by the fluid is a function of the area exposed to the fluid pressure, and this is the method by which the force can be multiplied, even though the pressure remains constant and is the basis for the hydraulic press (Fig. 6.7). The magma pressure is assumed to originate from tectonic overpressure in the lower crust and/or upper mantle where melting occurs.

The reasoning behind applying Pascal's principle is based partly on the observation that many inflated intrusions are circular or partly circular in map view. The Black Mesa bysmalith, the Table Mountain bysmalith on the north side of the Henry Mountains, and the southern half of the large central intrusion of Mt. Hillers, are almost perfectly circular. In the Highwood Mountains of Montana, the Shonkin Sag laccolith is 70 m thick and also almost perfectly circular. All of these intrusions are emplaced between flat-lying sedimentary rocks with little regional deformation, so that the fluid pressure has acted on media where the major planar zones of weakness are horizontal bedding planes. Therefore, mechanical weakness of the wall rocks governs the initial sill-like phase of emplacement (see Chapter 5).

The second phase of emplacement is governed by magma pressure. The connection between circular intrusions and fluid/magma pressure is surface area: a circular sheet maintains the least amount of surface area per volume of fluid, whereby the highest fluid/magma pressures are maintained. Since the inflated intrusions, such as BM and Table Mountain, have faults as their vertical walls/outer contacts (at least on half of their outer margins), the only direction the intrusions grew during their inflation stage was upward. Therefore, the intrusions must have first expanded to their horizontal terminations before inflating upward. The obvious reason being is that the magma did not have enough surface area to increase the upward force to overcome the lithostatic load until it grew to the current size as a simple sheet or small stack of sheets. To maintain the same magma pressure as the surface area is increased, the force on the overlying sedimentary layer must be increased at the same rate as the area is increased. Once a large enough surface area is obtained so that the force equals the lithostatic load (or the minimum

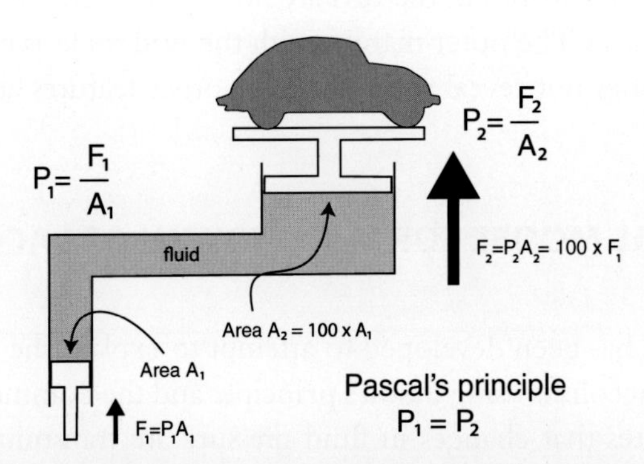

Figure 6.7 ***Pascal's principle illustrating how constant pressure is transmitted resulting in the force being multiplied with increasing area to produce a hydraulic press.***

horizontal stress for mid-crustal intrusions), the intrusion can grow vertically (or laterally for dykes and mid-crustal intrusions).

In the conceptual model presented here, it is assumed that the magma will always follow the path of least resistance, and this path is only planar and horizontal at the initial stage. After the initial sill is emplaced, sheets are still favoured, because the initial sheets cool more rapidly at their margins (including the top and bottom) and result in shear zones, similar to those observed at TM. This layering ensures that the zones of weakness continue to be horizontal and promotes further sheeting. After a certain amount of sheeting occurs, the cooling rate is reduced and contact shear zones cannot form at sheet margins because crystallisation does not occur before the next magma pulse arrives. The wall rocks are also rotating to higher angles at the margin due to the stacking and eventually layer-parallel extension leads to extensional faulting at the margin. At this point, sheeting is not favoured due to the lack of planar zones of weakness, the cohesion is lost at the sedimentary margins due to the deformation and full inflation is allowed to occur. If the surface area is large enough and therefore the fluid/magma force on the overlying rocks is equal to or greater than the lithostatic load, then the intrusion can grow vertically. Inflation of the magma chamber occurs, but the dynamics of this infilling is unknown (i.e. whether sheeting continues).

Using Pascal's principle the forces and magma pressures associated with the emplacement of the Black Mesa bysmalith are calculated below. Pollard et al. (1975) assumed that the depth of emplacement to the BM bysmalith was between 2.5 and 2.7 km based on the preserved thickness of the overlying sedimentary units that are pre-Miocene in age. But this assumption assumes no differential erosion (and consistent thickness) to the section directly over the top of the BM at the time of emplacement, which may or may not hold. The other variables are as follows:

Black Mesa

Assumed depth of emplacement = 2600 m

Area of intrusion = 2.16 km^2

Volume uplifted = 5.62 x 10^9 m^3

Density of high porosity sandstones overlying intrusion = 2380 kg/m^3

Mass = 1.34 x 10^{13} kg

Force needed to lift mass = 1.31 x 10^{14} N

Magma Pressure (F/A) determined given the area of intrusion = 61 MPa

Magma pressures determined for the growth of the lava dome at Mt. St. Helens between 2004 and 2006 were determined to be 13 MPa (Iverson et al., 2006) and 32 MPa (Mastin et al., 2008). The Mt St. Helens magma has a very similar composition to the intrusions in the Henry Mountains. The MC sill has an approximate area of 0.72 km^2 and the TM has an area of 0.65 km^2, both much less than the area of the BM, and apparently not enough to induce enough vertical force for upward inflation.

If we apply the same analysis to the Shonkin Sag laccolith, which has an area of 6.15 km^2 and was assumed to be emplaced at a depth of 1.4 km (Barksdale, 1937), the magma pressure is determined to be 32 MPa, exactly similar to that determined by Mastin et al. (2008) for the dome growth at Mt. St. Helens. If the same analysis is applied to the Roundstone pluton in western Ireland, also a circular intrusion modelled to be a punched laccolith (McCarthy et al., 2015), and emplaced between 4 and 8 km (using 6 km as the emplacement depth), the estimated magma pressure is 45 MPa.

One of the critical variables in these calculations is the depth to emplacement, which dictates the mass that needs to be uplifted, and therefore as the depth increases, the area uplifted must also increase. One of the limiting conditions is that as the initial area of the sill increases with depth, cooling and crystallisation affects the furthest-travelled magma and will lock up the ends of the sill before the equilibrium area is reached. Another problem is that the rate of uplift also decreases rapidly the bigger the area, and for Pascal's Principal to work, the entire system must retain its fluid-like connectivity. This has apparently been achieved for the BM and for the Shonkin Sag laccolith. Neither intrusion has evidence indicating there was suficient crystallisation to form internal contacts during the emplacement, and detailed petrologic arguments indicate that the Shonkin Sag laccolith underwent intense differentiation and crystal settling (Congdon, 1991) post-emplacement. Thermal modelling suggests that the BM bysmalith was fully emplaced in less than 100 years (Saint Blanquat et al., 2006).

Another observation to support, and help to explain, the conceptual model is that it has been noticed that the upward displacement of the roof layer overlying the initial finger and sheet-like intrusions of the Henry Mountains and the Highwood Mountains intrusions is small to negligible. Much of the volume required for these initial intrusions comes from internal strain with in the surrounding sedimentary rocks (cf. Chapter 5). This is well documented by Pollard et al. (1975), Morgan et al. (2008), and also noted by Horsman et al. (2005), indicating that these small initial intrusions did not have enough force to overcome the lithostatic load. The TM laccolith/sill does not fit this model well, because it is hard to explain the rotated roof rocks and uplift on the SW margin unless the entire load over the laccolith is uplifting. However, the other margins of the TM are more sill-like and have less volume to displace and are therefore less problematic. One possible solution is that the TM intruded into the side of a structural basin (Wetmore et al., 2009), and volume was created through volume loss/compaction that is far-field but horizontally-directed, consistent with the asymmetric uplift on the margins of the TM.

Pollard and Johnson (1973), Koch et al. (1981), Jackson and Pollard (1988), Kerr and Pollard (1998), Pollard and Fletcher (2005), and Bunger and Cruden (2011) model the emplacement of laccoliths assuming that elastic properties of the overlying sedimentary layers dictate the shape of the laccolith (see also Chapter 5). This beam-bending approach may be problematic for several reasons. As addressed previously (Morgan et al., 2008;

Bunger and Cruden, 2011), the flat-topped roofs of many laccoliths are not consistent with growth model of the bending of a beam, where the roof rocks are modelled to be pinned at the margins of the laccolith and uplifting at the centre. In addition, as discussed in detail, the margins of the laccoliths studied in the Henry Mountains exhibit intense finite strain, and this strain only occurs within a very narrow region at the margin. This shear sense at the contact is consistent with emplacement and stacking of sheets. The beam bending model must rely on elastic strain acting on the layer over the length of the layer and the concentrated deformation we observe along a very narrow region at the margin may not be consistent with this approach.

6.4 PASCAL'S PRINCIPLE APPLIED TO THE EMPLACEMENT OF MID-CRUSTAL INTRUSIONS

With mid-crustal and deeper intrusions, Pascal's principles might still be applicable, but the fundamental difference is that the direction of the least principle stress is not vertical, especially in regions undergoing transpression. Much of the evidence for horizontal and outward growth of mid-crustal plutons comes from a combination of vertical contacts, vertical fabrics, internal sheeting with vertical geometries, vertical slices of country rocks within plutons, and many of these examples exhibiting outwardly-bowed geometries (see also Chapter 2; Pitcher et al., 1958; McNulty et al., 1996; Cruden et al., 1999; Tikoff et al., 1999; Miller and Paterson, 2001; Anderson et al., 2013; Morgan et al., 2013; Stearns and Bartley, 2014).

6.4.1 The Birch Creek Pluton

One example of outward inflation from an initial dyke within the mid-crust will be discussed in detail because of mapping that allows for the transition between sheeting and inflation to be recorded. The Birch Creek pluton (Fig. 6.8) was originally studied by Nelson and Sylvester (1971) who mapped the deflection of faults and regional-scale folds around the north and west side of the pluton and developed the idea of forceful emplacement based on these deflections (Fig. 6.9). Barton (2000a,b) used highly detailed mapping (locally on the meter-scale) to document multiple magma pulses within the pluton and tied these pulses to periods of deformation and fluid fluxes in the immediate aureole (<200 m from the contact). This work is highlighted because Barton's mapping (2000a,b) of subtle internal contacts reveals an initial series of dyke-like intrusions, parallel and along a regional fault, which transition in later pulses to semi-circular internal contacts parallel to foliation and interpreted as documenting outward inflation, consistent with the deflection of the wall rocks.

The Birch Creek pluton intruded into the southeastern part of the southerly-plunging White Mountain anticline (Nelson, 1962, Nelson, 1966, Fig. 6.8). The Birch Creek pluton is a two-mica granite, 82 Ma (Barton, 2000a), and has intruded a thick

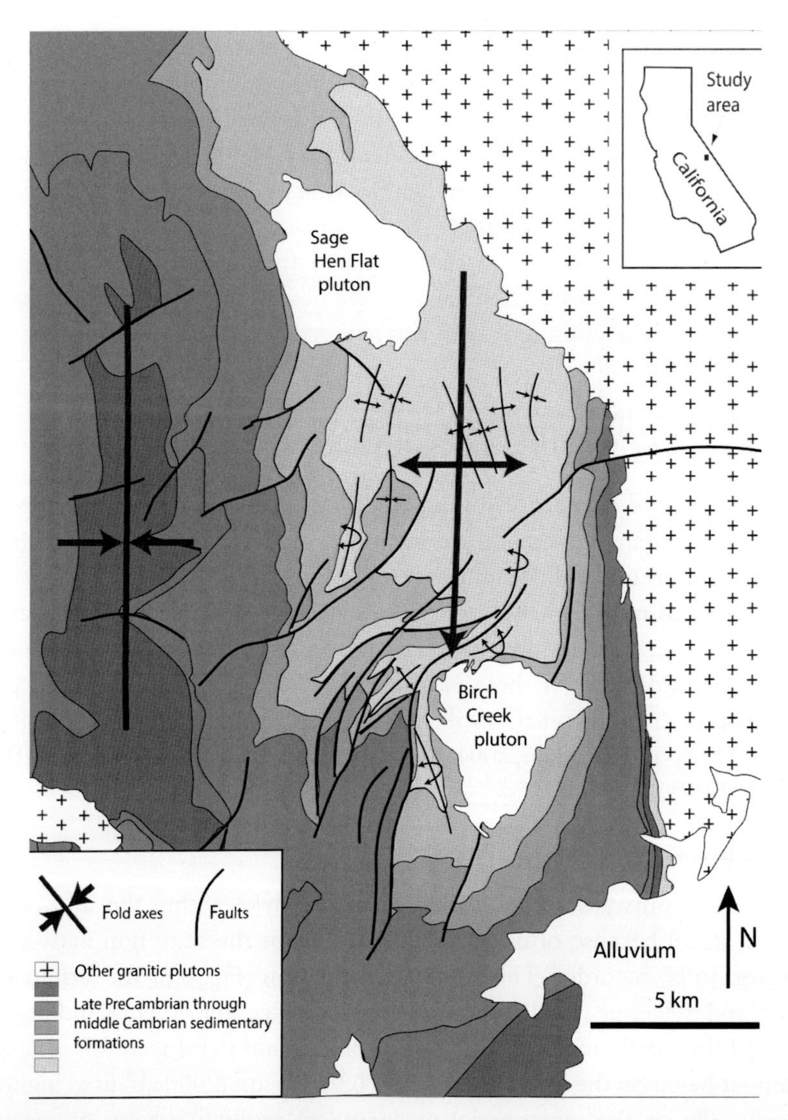

Figure 6.8 ***Location map of the Birch Creek and Sage Hen Flat plutons in the White Mountains of eastern California.*** *After Nelson and Sylvester, 1971.*

sedimentary package spanning the PreCambrian-Cambrian boundary (Nelson, 1962). The pluton is exposed closer to its top-half, as evident from bedding in the surrounding wall rocks and the foliation within the margins of the pluton that all dip away from the pluton (Fig. 6.9). Bedding planes to the north of the pluton have northerly strikes and dip steeply to the east or are overturned and dip steeply to the west. Within one km north of the pluton, on the northeastern margin, the beds rotate 90° to become east-west striking and parallel to the northern margin of the pluton (Fig. 6.9). As the

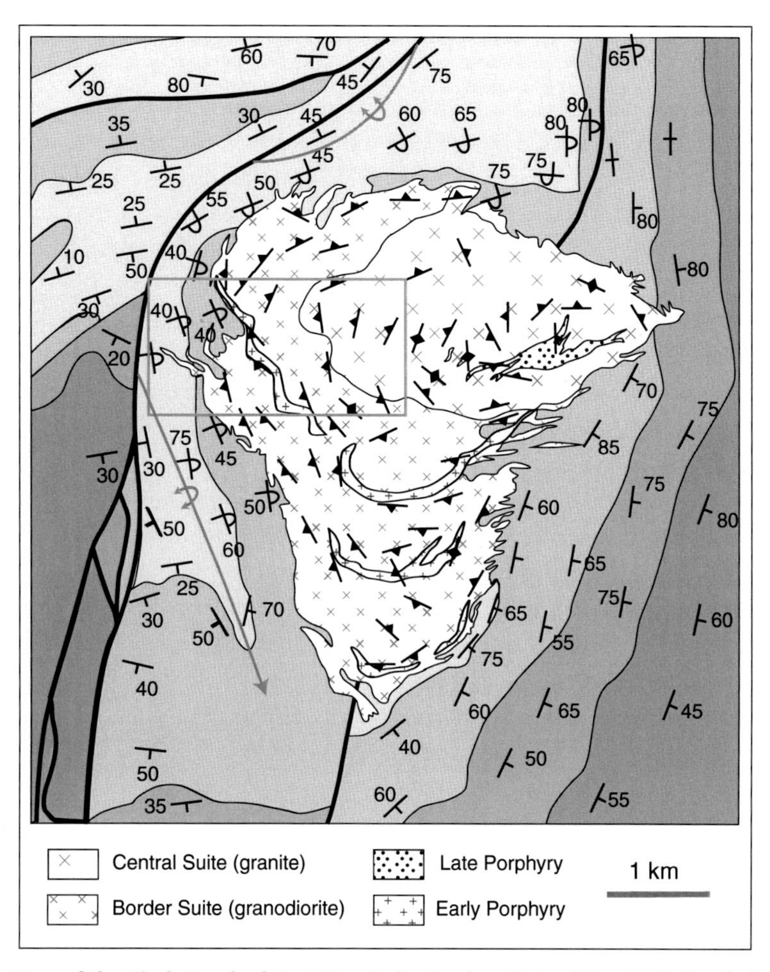

Figure 6.9 ***Map of the Birch Creek pluton (Box indicates location of Fig. 6.10).*** Symbols with numbers show orientation of foliation in host rocks. Symbols with triangles show orientation of foliation in the pluton. *After Nelson and Sylvester, 1971 and Barton, 2000.*

beds rotate clockwise, they become more shallowly dipping to the north and more overturned, dipping as low as 30° to the north (Nelson and Sylvester, 1971), as they are probably exposed near the top of the pluton contact. This overturned sequence of beds wraps concordantly around the NW corner of the pluton to become NE striking, parallel to the contact with the pluton and still overturned but dipping to the SW (Nelson and Sylvester, 1971). Overall, bedding is deflected around the pluton as far away as 3 km and grossly parallel to the contact with the pluton but can be highly discordant on the outcrop scale (Nelson and Sylvester, 1971; Barton, 2000b). Faults and smaller scale folds associated to the White Mountain anticline are also rotated around this NW corner of the pluton, and folds become tighter and closer spaced (Nelson and Sylvester, 1971).

A solid-state (i.e. non-magmatic) foliation, defined by quartz and feldspar and aligned micas, is strongest within 30 m of the NW margin and is parallel to the margin. Foliation is well developed in the marbles surrounding the NW margin and parallel to the pluton-wall rock contact. The marbles exhibit a NW trending mineral elongation lineation, which is best developed on the NW margin near the contact (Nelson and Sylvester, 1971). Early quartz veins in the aureole (associated with dykes rom the Border Suite) are boudinaged and some applites also exhibit chocolate-tables boudinage (Barton, 2000b).

Barton (2000a,b) focused his work on the NW margin of the pluton where the wall rock deflections are greatest. His detailed mapping (1:6,000 and in critical areas, 1:300 or 1:60) reveals four main suites of magma emplacement and associated fluid fluxes and that all, or almost all, of the deformation in the immediate wall rocks and within the pluton occurred synchronous with consecutive pulses of magma. These four main suites, which are defined by minor changes in composition, texture, and position within the pluton, can each be divided into many more individual magma pulses (Barton, 2000a). The bulk of the pluton is divided into a Border Suite which wraps around the western and southern part and the Central Suite on the eastern and central portions of the pluton (Fig. 6.9). The Border Suite is dominantly composed of a series of granodiorite dykes, 20 to 100 m wide, which can be further divided into four main texturally distinct rock types. Later Border Suite dykes are more granitic, similar in composition to the Central Suite. The internal contact between the Border and Central Suites is generally concordant to the exterior contact and is marked by a strong contrast in foliation development from more well developed in the Border Suite to weakly developed in the Central Suite and a marked increase in k-feldspar megacrysts in the Central Suite (Barton, 2000b; Fig. 6.10).

The contact between the Border and the Central Suites is a zone 50–100 m wide marked by a series of textural "fronts" that are defined by dendritic k-feldspars aligned perpendicular to the internal contact. These comb-layers, or "unidirectional solidification fronts" (USTs; Barton, 2000b) have been interpreted as indicating the direction of more liquid-rich magma (Central Suite) versus more solidified crystal mush (Border Suite, Barton, 2000b). The USTs have also been interpreted to form during decompression (McCarthy and Müntener, 2016), consistent with fracturing of a more solid carapace (Border Suite) and release of fluids and formation of cross-cutting aplite dykes (Barton, 2000a). The observation that there are multiple USTs defining the contact between the two suites indicates multiple fluid pressure build-ups at the contact with eventual decompression, fracturing of the Border Suite, and fluid expulsion events resulting in abundant aplite dyke swarms originating from these USTs.

K-feldspar megacrysts are common in the Central Suite near the contact with the Border Suite (Barton, 2000a; Fig. 6.10). In one case, abundant k-feldspar megacrysts are observed filling the entrance to a dyke originating from a UST zone and which can be

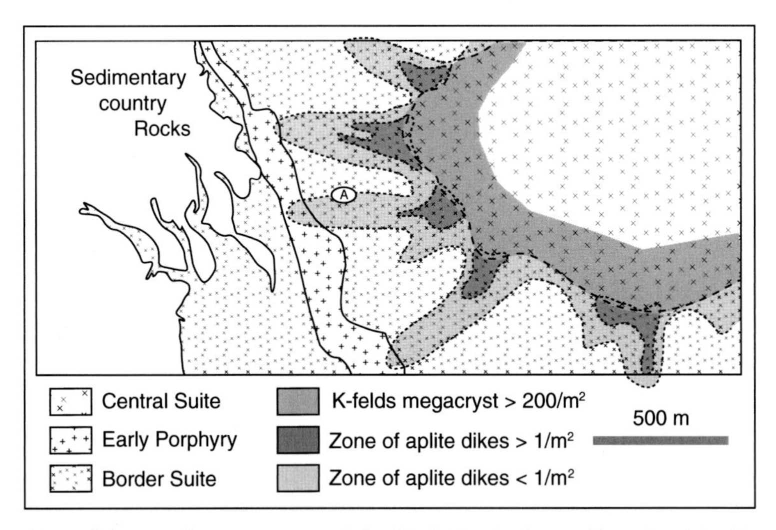

Figure 6.10 ***Map of the northwestern part of the Birch Creek pluton.*** Note the dyke-like contacts of the Border Suite with the sedimentary country rocks as well as the contact between the Early Porphyry and the Border Suite in contrast to the concentric internal contact between the Border Suite and the Central Suite. Circle with A is location of photograph in Fig. 6.11. Location of the map area is found in Figure 6.9. *Adapted after Barton, 2000.*

mapped crossing the Border Suite (Barton, 2000b). In general, dykes and quartz veins are relatively rare crossing the Central Suite. The Central Suite does not have internal contacts that are dyke-like, in contrast to the Border Suite, and foliations are weak to absent (Barton, 2000b).

There are also two suites of porphyry dykes, an early one that intruded into the Border Suite but prior to the emplacement of the Central Suite and a late set that intrudes into the Central Suite and marks the termination of magma emplacement. These are isotopically and chemically distinct from the Border and Central Suites (Barton, 2000b). The early ones are foliated and cut by dikes from the Central Suite (Barton, 2000b). The late set is not associated with ductile deformation, but it did generate abundant aplite dykes and veins.

The key to deciphering the complex timing relationships between various magma pulses and deformational events are the cross-cutting nature of aplite dyke swarms (Fig. 6.10) and associated veins, which intruded at almost all stages (Barton, 2000b). In several areas, the timing of the intense deformation in the outer part of the pluton (solid state foliation and folding of early aplites/veins) is constrained to have occurred prior to the solidification of the adjoining interior pulses of magma because aplite dykes that cross-cut the foliation (Fig. 6.11) can be traced to the interior to these later pulses. For example, the early quartz veins in the outer deformed margin of the pluton have been isoclinally folded and their axial planes are parallel to the foliation. The early quartz veins and the solid-state foliation are cut by unfolded and little-foliated aplite dykes that can

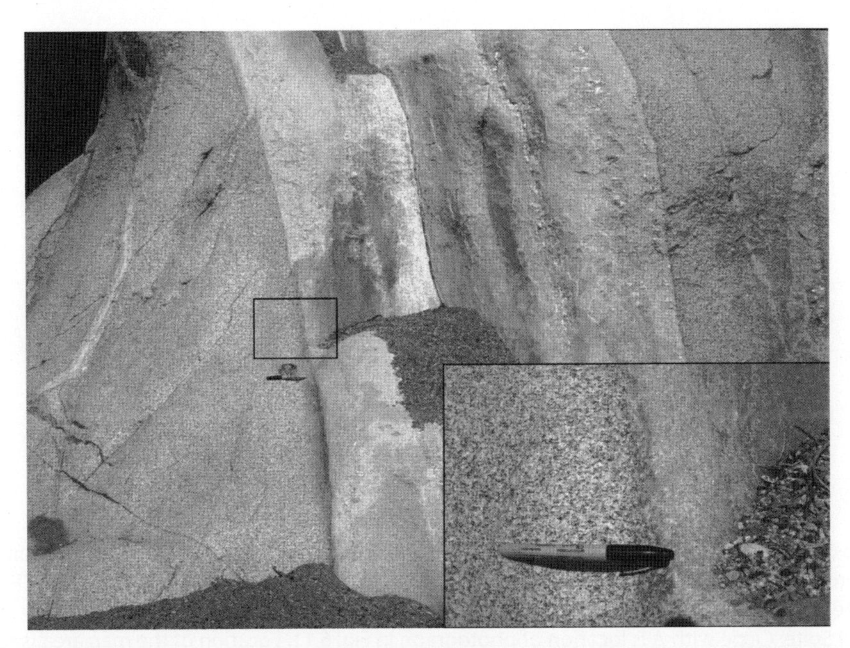

Figure 6.11 ***Photograph of unfoliated aplite dike cutting well developed magmatic foliation in the Border Suite, Birch Creek pluton.*** Foliation is parallel to pen. Dike can be traced into the interior of the pluton where it formed in the poorly foliated Central Suite.

be traced for over 100 m to the interior where the foliation is much less intense and magmatic (Barton, 2000a). This kind of detailed mapping documents the switch from dyke-like emplacement to inflation and deformation.

In summary, the work of Nelson and Sylvester (1971) and of Barton (2000a,b) reveal several important aspects on the emplacement of the Birch Creek pluton:

1) The transition from dyke-like emplacement of magma to radial expansion is recorded in the pluton from the earliest pulses of the Border Suite to the radial expansion/inflation of the Central Suite. It is assumed that the magma originally rose up through this part of the crust along a fault because the pluton is found along, and cuts across, a fault (Nelson and Sylvester, 1971). There is evidence that three out of the four main pulses of magma initially were emplaced as dykes (Border Suite, early and late porphyries). There is no record of the initial geometry of the Central Suite pulses, only the latest inflation geometry is preserved. The Border Suite contact with the sedimentary wall rocks is defined by dykes and the internal contacts within the Border Suite are also dyke-like (Barton, 2000b). In contrast, the contacts between the Border Suite and the Central Suite are concentric, subparallel to the outer margins. The incremental emplacement of the Central Suite magmas deformed the Border Suite, resulting in a concordant foliation and the outward translation of the wall rocks. The Border Suite has a stronger foliation than the Central Suite at the contact with Central Suite, indicating that the deformation took place late in the crystallisation of the Border Suite (crystal mush) but early in the crystallisation of the Central Suite (liquid-rich magma). The crystallisation in the Border Suite had come to a completion near the wall rock contact because the foliation grades into a solid stated foliation within 30 m of the contact.

2) The outward translation of wall rocks cannot explain all of the volume of the pluton (Nelson and Sylvester, 1971). Palinspastic restoration of the deflection of the wall rocks reveals that some percentage of the pluton volume cannot be explained by translation of the wall rocks. Given the assumption that the magma initially rose along a fault, and given that there are dyke-like contacts to the early phases of intrusion, it is interesting that there is no evidence of the magma prying apart the fault it supposedly intruded, based on examining the fault as it protrudes from the northern and southern margins of the pluton. In fact, it seems as if the pluton cuts the fault. If all of the volume of the pluton was to be explained by translation of the wall rocks, the western, outer contact with the wall rocks would have to be the original fault/dyke contact, if the magma rose up along the fault. Obviously, other processes occurred at the level of exposure in addition to radial, in situ, expansion (e.g., Marko and Yoshinobu, 2011).

A possible consequence of this analysis is that the difference in dimensional ratios between plutons and laccoliths (Rocchi et al., 2010) might be due to an incorrect method of measuring the length and thickness for plutons. If Pascal's principle holds, and mid-crustal intrusions begin as dykes and only inflate when their surface area has expanded to the point where the force is equal to the minimal horizontal stress, then their thickness needs to be measured in the direction of expansion, which can be horizontal, and not vertical (e.g. Dinkey Creek pluton, Cruden et al., 1999).

6.4.2 Metamorphic Evidence for Two-Phases of Emplacement

Other evidence for initial dyke and/or sill-like emplacement followed by inflation within the mid-crust comes from examination of the metamorphic assemblages and mineral textures in the surrounding wall rocks from the mid-crust. Several well-known plutons have aureole assemblages with contact metamorphic porphyroblasts that exhibit an initial passive growth, where the core overgrows a pre-existing regional fabric, and then a dynamic rim, where the late stage growth is syn-deformational with respect to the strain associated with emplacement. Inclusion trails document this transition by exhibiting planar trails within the core regions and then wrapping around or rotating from the regional fabric to the deformational fabric associated with the inflation of the pluton (Fig. 6.12). This is best documented for plutons that intrude greenschist facies terrains with pelitic sedimentary wall rocks.

For example, the inclusion trails within the cores of andalusite porphyroblasts in the aureole surrounding the Papoose Flat pluton have been geometrically matched to the regional slaty cleavage fabric that pre-dates the emplacement of the pluton (Morgan et al., 1998). The regional angle between bedding and cleavage has the same angle and

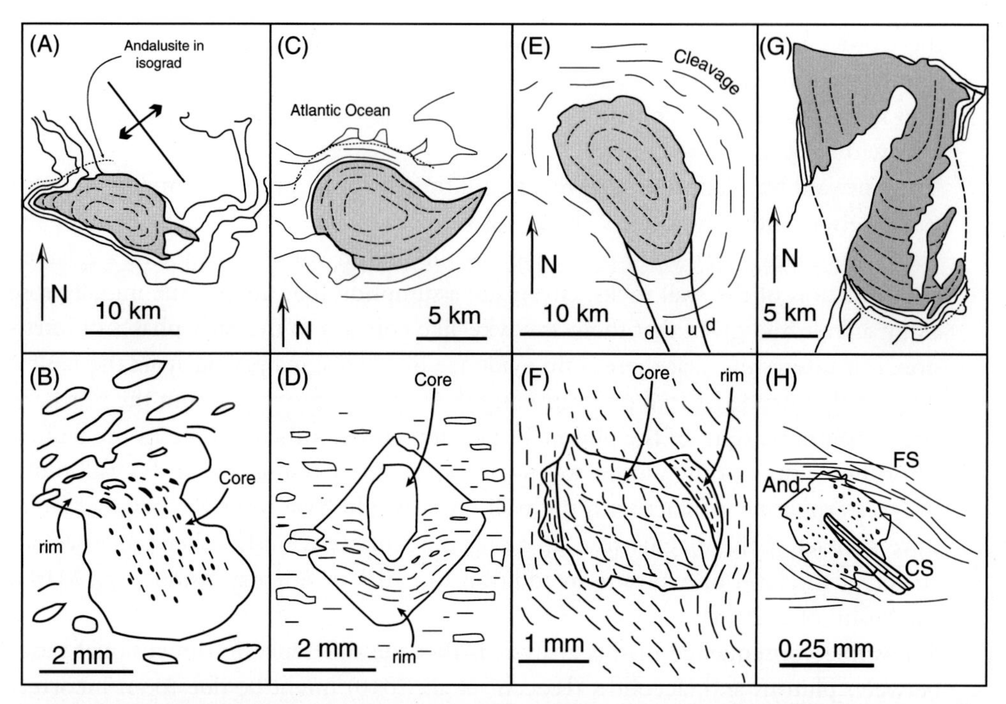

Figure 6.12 ***Sketches of maps of plutons and porphyroblast-matrix relationships from their contact metamorphic aureoles.*** (A & B), Papoose Flat pluton, USA. Cores of andalusite porphyroblasts record the regional slatey cleavage fabric. Rims record the transition from passive growth to dynamic growth associated with inflation. Adapted after Morgan et al., 1998. (C & D), Ardara pluton, Ireland. Some andalusite porphyroblast cores contain fine-grained inclusions oriented at an angle to the exterior foliation that are surrounded by a rim containing inclusion trails that are deflected around the core. The rim inclusion trails are continuous into the matrix where the foliation is coarser grained than in the porphyroblast. Adapted after Akaad, 1956. (E & F), Cannibal Creek pluton, Australia. Within the country rocks external to the contact aureole, an earlier crenulation cleavage is developed which is only preserved in the structural aureole as inclusion trails within cores of andalusite porphyroblasts and in low strain zones (Davis, 1993). The cores of these andalusite porphyroblasts are surrounded by rims which overgrew the margin-parallel foliation. Adapted after Davis, 1993. (G & H), Eureka Valley, Joshua Flat, Beer Creek (EJB) pluton, USA. Early andalusite (And) is replaced by later crystalline sillimante (CS) and fibrolite sillimanite (FS) found in shear bands.

orientation as between inclusion trails and compositional layering within cores of andalusite porphyroblasts. The rim inclusion trails rotate from the cores and are associated with the inflation deformation and intense attenuation of the aureole.

For plutons with concordant aureoles, this suggests that the first pulse of magma was parallel to the surrounding bedding, suggesting that sheeting or sill-like emplacement resulted in the first thermal pulse, prior to deformation. Core porphyroblasts grow in the absence of deformation and overgrow the regional fabric. During the inflation stage, rims grow synchronous with inflation-related aureole deformation, often associated with attenuation of the thermally-softened metamorphic aureole, resulting in dynamic rims.

6.4.3 Other Mechanisms of Mid-Crustal Pluton Emplacement

It is also very clear that not all emplacement of magma involves inflation, and there are convincing examples of where other processes like magmatic stoping (Yoshinobu et al., 2003; Zak and Paterson, 2006; Burchardt et al., 2012) and/or downward rotation of wall rocks occurs (See Chapter 2; Cruden and McCaffrey, 2001). One of these examples is only 10 km to the north of the Birch Creek pluton (Fig. 6.8). The Sage Hen Flat pluton is another small granitic pluton, which intruded the same sedimentary section as the Birch Creek pluton, but is 100 million years older (175 Ma, Coleman et al., 2003). Bilodeau and Nelson (1993) mapped and studied the pluton and determined that emplacement was accommodated by magmatic stoping due to the abrupt termination of the sedimentary bedding against the pluton, with no deflection of the bedding as the contact is approached, combined with little evidence for a concordant foliation at the margin of the pluton. In addition, irregular blocks of the adjacent sedimentary section are found floating within the pluton and the contact is marked by abundant small intrusive veins and dykes (Fig. 6.13). Althouth, as Glazner and Bartley (2006), point out, except for a narrow strip adjacent to the contact, the pluton contains very few interior xenoliths. The only other possible model for emplacement is that the Sage Hen Flat pluton was emplaced in a manner similar to the Black Mesa bysmalith, and the margins are faults. It may be that in the hotter environment of the mid-crust, the magma stayed hotter longer and interacted with the contact zone, destroying any evidence of faulting and/or marginal deflection at the contact.

6.5 CONCLUSIONS

It is clear that magmas are emplaced into the crust as pulses, and many times these pulses are sheet-like in shape. Intrusions in the Henry Mountains preserve these initial sheet-like geometries as they grow from sill to laccolith to bysmalith/pluton. These initial

Figure 6.13 ***Contact relationship with the sedimentary wall rocks (dark grey) surrounding the Sage Hen Flat pluton (light coloured), USA.***

sheets form as fingers, fingers coalesce into sheets, and sheets spread laterally. A simple, two-stage emplacement model is proposed using Pascal's principle for the multiplication of force whereby an initial sill grows in lateral extent until the upward force is equal to the lithostatic load. At this point, uplift and inflation occurs. It is unclear if magma pulsing continues in the inflation stage. This conceptual model is extended to mid-crustal plutons, and the Birch Creek pluton of eastern California is used as an example because detailed mapping has revealed the transition from initial dyke to inflation stages. Inflation of magmatic sheets is one model that seems to work for many shallow and mid-crustal plutons, although there are also plutons where contact relationships dictate that other processes are also operating.

ACKNOWLEDGEMENTS

These ideas are built on discussions, mostly in the field, with great collaborators. I am in dept to Eric Horsman, Rick Law, Peter Nabelek, Michel de Saint Blanquat, Jim Student and Basil Tikoff, although they might not agree with all the ideas presented here. I also want to thank Steffi Burchardt for allowing me to contribute and for her editorial assistance and patience. This research was supported by NSF grants EAR-0003574 and EAR-1220340.

REFERENCES

Akaad, M.K., 1956. The northern aureole of the Ardara pluton of County Donegal. Geological Magazine 93, 377–392.

Anderson, H.S., Yoshinobu, A.S., Nordgulen, Ø., Chamberlain, K., 2013. Batholith tectonics: Formation and deformation of ghost stratigraphy during assembly of the mid-crustal Andalshatten batholith, central Norway. Geosphere: June 2013 9 (3), 1–24. doi:10.1130/GES00824.1.

Annen, C., Scaillet, B., Sparks, R.S.J., 2006c. Thermal constraints on the emplacement rate of a large intrusive complex: the Manaslu leucogranite, Nepal Himalaya. Journal of Petrology 47 (1), 71–95.

Barksdale, J.D., 1937. The Shonkin Sag Laccolith. The American Journal of Science 33, 321–359.

Barton, M.D., 2000a, Guide for Field Trip Day 1 Birch Creek, White Mountains, California: in Dilles, J.H., Barton, M.D., Johnson, D.A., Proffett, J.M., and Einaudi, M.T., editors, 2000, Contrasting Styles of Intrusion Associated Hydrothermal Systems: Society of Economic Geologists Guide Book Series, v. 32, 27-43.

Barton, M.D., 2000b, Overview of the lithophile-element-bearing magmatic-hydrothermal system at Birch Creek, White Mountains, California: in Dilles, J.H., Barton, M.D., Johnson, D.A., Proffett, J.M., and Einaudi, M.T., editors, 2000, Contrasting Styles of Intrusion Associated Hydrothermal Systems: Society of Economic Geologists Guide Book Series, v. 32, p. 9-26.

Bilodeau, B.J., and Nelson, C.A., 1993, Geology of the Sage Hen Flat pluton, White Mountains, California: Geological Society of America Map and Chart Series, Map MCH077, 18 p.

Bunger, A.P., Cruden, A.R., 2011. Modeling the growth of laccoliths and large mafic sills: Role of magma body forces. Journal of Geophysical Research 116, B02203. doi:10.1029/2010JB007648.

Burchardt, S., Tanner, D., Krumbholz, M., 2012. The Slaufrudalur pluton, southeast Iceland- An example of shallow magma emplacement by coupled cauldron subsidence and magmatic stoping. Geological Society of America Bulletin 124, 213–227.

Cashman, K., Blundy, J., 2013. Petrological cannibalism: the chemical and textural consequences of incremental magma body growth. Contributions to Mineralogy and Petrology 166, 703–729, 10.1007/s00410-013-0895-0.

Chen, Y., Nabelek, P.I., 2017. The influences of incremental growth on magma crystallinity and aureole rheology: numerical modeling of growth of the Papoose Flat pluton, California. Contributions to Mineralogy and Petrology 172, 89. https://doi.org/10.1007/s00410-017-1405-6.

Coint, N., Barnes, C.G., Yoshinobu, A.S., Chamberlain, K.R., Barnes, M.A., 2013. Batch-wise assembly and zoning of a tilted calc-alkaline batholith: Field relations, timing, and compositional variation. Geosphere 9 (6), 1729–1746, 2013.

Coleman, D.S., Briggs, S., Glazner, A.F., Northrup, C.J., 2003. Timing of plutonism and deformation in the White Mountains of eastern California. Geological Society of America Bulletin. 115, 48–57.

Coleman, D.S., Gray, W., Glazner, A.F., 2004. Rethinking the emplacement and evolution of zoned plutons: Geochronologic evidence for incremental assembly of the Tuolumne Intrusive Suite, California. Geology 32, 433–436.

Congdon, R.D., 1991, The solidification of the Shonkin Sag laccolith: Mineralogy, petrology, and experimental phase equilibria. Unpublished PhD thesis, The Johns Hopkins University.

Cruden, A.R., McCaffrey, K.J.W., 2001. Growth of plutons by floor subsidence: Implications for rates of emplacement, intrusion spacing, and melt-extraction mechanisms. Physics and Chemistry of the Earth (A) 26, 303–315.

Cruden, A.R., Tobisch, O.T., Launeau, P., 1999. Magnetic fabric evidence for conduit-fed emplacement of a tabular intrusion: Dinkey Creek pluton, central Sierra Nevada batholith, California. Journal of Geophysical Research 104 (10), 511–610, 530.

Davis, B.K., 1993, Mechanism of emplacement of the Cannibal Creek Granite with special reference to timing and deformation history of the aureole: Tectonophysics, v. 224. p. 337-362.

Gilbert, G.K. 1877. Report on the Geology of the Henry Mountains. U.S. Geological Survey. 170 pp.

Glazner, A.F., Bartley, J.M., 2006. Is stoping a volumetrically significant pluton emplacement process? Geological Society of America Bulletin 118, 1185–1195.

Glazner, A.F., Johnson, B.R., 2013. Late crystallization of K-feldspar and the paradox of megacrystic granites. Contributions to Mineralogy and Petrology 166 (3), 777–799.

Gudmundsson, A., 2002. Emplacement and arrest of sheets and dykes in central volcanoes. Journal of Volcanology and Geothermal Research 116, 279–298.

Horsman, E., Tikoff, B., Morgan, S.S., 2005. Emplacement-related fabric in a sill and multiple sheets in the Maiden Creek sill, Henry Mountains, Utah. Journal of Structural Geology 27, 1426–1444.

Horsman, E., Morgan, S., Saint-Blanquat (de), M., Habert, G., Hunter, R.S., Nugent, R., Tikoff, B., 2009. Emplacement and assembly of shallow plutons through multiple magma pulses, Henry Mountains, Utah. Earth and Environmental Science Transactions of the Royal Society of Edinburgh 100, 117–132.

Hunt, C. 1953. Geology and geography of the Henry Mountains region, Utah. US Geological Survey Professional Paper 228.

Iverson, R.M., Dzurisin, D., Gardner, C.A., Gerlach, T.M., LaHusen, R.G., Lisowski, M., Major, J.J., Malone, S.D., Messerich, J.A., Moran, S.C., Qamar, A.I., Schilling, S.P., Pallister, J.S., Vallance, J.W., 2006. Dynamics of seismogenic volcanic extrusion at Mount St Helens in 2004-05. Nature 444. doi:10.1038/nature05322.

Jackson, M.D., Pollard, D.D., 1988. The laccolith-stock controversy: new results from the southern Henry Mountains, Utah. Geological Society of America Bulletin 100, 117–139.

Jackson, M.D., Pollard, D.D., 1990. Flexure and faulting of sedimentary host rocks during growth of igneous domes, Henry Mountains, Utah. Journal of Structural Geology 12, 185–206.

Kavanagh, J.L., Menand, T., Sparks, R.S.J., 2006. An experimental investigation of sill formation and propagation in layered elastic media. Earth Planet. Sci. Lett. 245, 799–813.

Kerr, A.D., Pollard, D.D., 1998. Toward more realistic formulations for the analysis of laccoliths. Journal of Structural Geology 20. 1783-1793.

Koch, F., Johnson, A., Pollard, D., 1981. Monoclinal bending of strata over laccolithic intrusions. Tectonophysics 74, T21–T31. doi:10.1016/0040-1951(81)90189-X.

Lister, J.R., Kerr, R.C., 1991. Fluid-mechanical models of crack propagation and their application to magma transport in dikes. Journal of Geophysical Research 96, 10049–10077.

Mahan, K.H., Bartley, J.M., Coleman, D.S., Glazner, A.F., Carl, B.S., 2003. Sheeted intrusion of the synkinematic McDoogle pluton, Sierra Nevada, California. Geological Society of America Bulletin 115, 1570–1582.

Marko, W.T., & Yoshinobu, A.S., 2011, Using restored cross sections to evaluate magma emplacement, White Horse Mountains, eastern Nevada, U.S.A., Tectonophysics, 500, p. 98-111.

Mastin, L.G., Roeloffs, E., Beeler, N.M., and Quick, J.E., 2008, Constraints on the size, overpressure, and volatile content of the Mount St. Helens magma system from geodetic and dome-growth measurements during the 2004-2006+ eruption, in A Volcano Rekindled: The Renewed Eruption of Mount

St. Helens, 2004-2006 (Sherrod, D.R., Scott, W.E., & Stauffer, P.H., eds). U.S. Geological Survey Professional Paper 1750, 2008.

McCaffrey, K.J.W., Petford, N., 1997. Are granitic intrusions scale invariant? Journal of the Geological Society, London 154, 1–4.

McCarthy, A., Müntener, O., 2016. Comb layering monitors decompressing and fractionating hydrous mafic magmas in subvolcanic plumbing systems (Fisher Lake, Sierra Nevada, USA). J. Geophys. Res. Solid Earth 121, 8595–8621. doi:10.1002/2016JB013489.

McCarthy, M.S., Petronis, M.S., Reavy, R.J., & Stevenson, C.T., 2015, Distinguishing diapirs from inflated plutons: an integrated rock magnetic fabric and structural study on the Roundstone pluton, western Ireland, Journal of the Geological Society. doi:10.1144/jgs2014-067.

McNulty, B.A., Tong, W., Tobisch, O.T., 1996. Assembly of a dike-fed magma chamber: The Jackass Lakes pluton, central Sierra Nevada, California. Geological Society of America Bulletin 108, 926–940.

Menand, T., 2008. The mechanics and dynamics of sills in elastic layered media and their implications for the growth of laccoliths. Earth Planet Sci. Lett. 267, 93–99.

Menand, T., Annen, C., and de Saint Blanquat, M., 2015, Rates of magma transfer.

Michel, J., Baumgartner, L., Putlitz, B., Schaltegger, U., Ovtcharova, M., 2008. Incremental growth of the Patagonian Torres del Paine laccolith over 90 ky. Geology 36, 459–462.

Miller, R.B., Paterson, S.R., 2001. Construction of mid-crustal sheeted plutons: Examples from the North Cascades, Washington. Geological Society of America Bulletin 113, 1423–1442.

Miller, C.F., Furbish, D.J., Walker, B.A., Claiborne, L.L., Koteas, C., Bleick, H.A., Miller, J.S., 2011. Growth of plutons by incremental emplacement of sheets in crystal-rich host: evidence from Miocene intrusions of the Colorado River region, Nevada, USA. Tectonophysics 500, 65–77.

Mills, R.D., Ratner, J.J., Glazner, A.F., 2011. Experimental evidence for crystal coarsening and fabric development during temperature cycling. Geology 39, 1139–1142.

Morgan, S.S., Law, R.D., Nyman, M.W., 1998. Laccolith-like emplacement model for the Papoose Flat pluton based on porphyroblast-matrix analysis. Geological Society of America Bulletin 110, 96–110.

Morgan, S., Horsman, E., Tikoff, B., Saint-Blanquat, M.de, and Habert, G., 2005, Sheet-like emplacement of satellite laccoliths, sills, and bysmaliths of the Henry Mountains, southern Utah, in Pederson, J., Dehler, C.M., eds., Interior Western United States: Geological Society of America Field Guide 6. doi:10.1130/2005.fl d006.(14).

Morgan, S.S., Stanik, A., Horsman, E., Tikoff, B., Saint Blanquat, Habert, G., 2008. Emplacement of multiple magma sheets and wall rock deformation: Trachyte Mesa intrusion, Henry Mountains, Utah. Journal of Structural Geology 30, 491–512.

Morgan, S.S., Law, R.D., Saint Blanquat, M., 2013. Forceful Emplacement of the EJB composite pluton into a structural basin in eastern California: internal structure and wall rock deformation. Tectonophysics 608, 753–773.

Morgan, S.S., Jones, R., Conner, J., Student, S., Schaner, M., Horsman, E., de Saint Blanquat, M., 2017. Magma sheets defined with magnetic susceptibility in the Maiden Creek sill, Henry Mountains, Utah, USA. *Geology* 45, 599–602.

Nelson, C.A., 1962. Lower Cambrian-Precambrian succession, White-Inyo Mountains, California. Geological Society of America Bulletin 73, 139–144.

Nelson, C.A., 1966, Geologic map of the Blanco Mountain Quadrangle, Inyo and Mono counties, California: U.S. Geological Survey, Geologic Quadrangle Map GQ-0529.

Nelson, C.A., Sylvester, A.G., 1971. Wall rock decarbonation and forcible emplacement of Birch Creek pluton, southern White Mountains, California. Geological Society of America Bulletin 82, 2891–2903.

Pitcher, W.S., Read, H.H., Cheesman, R.L., Pande, I.C., Tozer, C.F., 1958. The Main Donegal Granite. Quarterly Journal of the Geological Society 114, 259–300. https://doi.org/10.1144/gsjgs.114.1.0259.

Pollard, D.D., Fletcher, 2005. Fundamentals of Structural Geology. Cambridge University Press, New York.

Pollard, D.D., Johnson, A.M., 1973. Mechanics of growth of some laccolithic intrusions in the Henry Mountains, Utah II. Bending and failure of overburden layers and sill formation, Tectonophysics 18, 311-354.

Pollard, D.D., Muller, O.H., Dockstader, D.R., 1975. The form and growth of fingered sheet intrusions. Geological Society of America Bulletin 3, 351–363.

Rocchi, S., Westerman, D.S., Dini, A., Farina, F., 2010. Intrusive sheets and sheeted intrusions at Elba Island. Geosphere 6 (3), 225–236. DOI:10.1130/GES00551.1.

Ryan, M.P., 1987. Neutral buoyancy and the mechanical evolution of magmatic systems, in Magmatic Processes: Physicochemical Principles (Mysen, B.O., Ed), The, Geochem, Soc., Spec., Publ., 1, 259-287.

Saunders, K., Blundy, J., Dohmen, R., Cashman, K., 2012. Linking Petrology and Seismology at an Active Volcano. Science 336, 1023. DOI: 10.1126/science.1220066.

Saint Blanquat, M., Habert, G., Horsman, E., Morgan, S.S., Tikoff, B., Launau, P., Gleizes, G., 2006. Mechanisms and duration of non-tectonically assisted magma emplacement in the upper crust: The Black Mesa pluton, Henry Mountains, Utah. Tectonophysics 428, 1–31.

Stearns, M.A., Bartley, J.M., 2014. Multistage emplacement of the McDoogle pluton, an early phase of the John Muir intrusive suite, Sierra Nevada, California, by magmatic crack-seal growth. Geological Society of America Bulletin 126, 1569–1579.

Thomson, K. and Schofield, N., 2008. Lithological and structural controls on the emplacement and morphology of sills in sedimentary basins. In : Thomson, K., Petford, N., (Eds.) Structure and Emplacement of High-Level Magmatic Systems. Geological Society London, Special Publication 302, 31-44.

Tikoff, B., de Saint Blanquat, M., Teyssier, C., 1999. Translation and resolution of the pluton space problem. Journal of Structural Geology 21, 1109–1117.

Wetmore, P.H., Connor, C.B., Kruse, S.E., Callihan, S., Pignotta, G., Stremtan, C., Burke, A., 2009. Geometry of the Trachyte Mesa intrusion, Henry Mountains, Utah: Implications for the emplacement of small melt volumes into the upper crust: Geochemistry. Geophysics, Geosystems 10, 1525–2027. doi:10.1029/2009GC002469 ISSN.

Wiebe, R.A., 1993. The Pleasant Bay layered gabbro-diorite, coastal Maine: Ponding and crystallization of basaltic injections into a silicic magma chamber. Journal of Petrology 34, 461–489.

Wilson, P.I.R., McCaffrey, K.J.W., Wilson, R.W., Jarvis, I., Holdsworth, R.E., 2016. Deformation structures associated with the Trachyte Mesa intrusion, Henry Mountains, Utah: Implications for sill and laccolith emplacement mechanisms. Journal of Structural Geology 87, 30–46. doi: 10.1016/j.jsg.2016.04.001.

Yoshinobu, A.S., Fowler, K.T., Paterson, S.R., Llambias, E., Tickyj, H., Sato, M.A., 2003. A view from the roof: magmatic stoping in the shallow crust, Chita pluton, Argentina. Journal of Structural Geology 25, 1037–1048.

Zak, J., Paterson, S.R., 2006. Roof and walls of the Red Mountain Creek pluton, eastern Sierra Nevada, California (USA): implications for process zones during pluton emplacement. Journal of Structural Geology 28, 575–587.

Zeig, M.J., Marsh, B.D., 2012. Multiple reinjections and crystal-mush compaction in the Beacon sill, McMurdo Dry Valleys, Anartica. Journal of Petrology 53, 2567–2591.

CHAPTER 7

Tectonics and Volcanic and Igneous Plumbing Systems

Benjamin van Wyk de Vries, Maximillian van Wyk de Vries
Laboratoire Magmas et Volcans, Université Blaise Pascal, Clermont-Ferrand, France

Contents

7.1 INTRODUCTION

7.1.1 What is the Role of Plate Tectonics in Volcanic and Igneous Plumbing Systems?

Plate tectonics and mantle convection set the scene for magmatism, and magma is an essential product of plate tectonics. Magma ultimately produces the lithospheric portion of plates. The different tectonic environments creating magmatism are illustrated in Fig. 7.1.

Magma is produced principally at divergent plate boundaries by spreading at rifts and mid-ocean ridges (about 90% of all magmatic activity is at mid-ocean ridges). New oceanic lithosphere is generated by magma accumulation and eruption in a continually growing and expanding plumbing system.

Continental rifts can also produce magma and eventually evolve into full oceanic spreading systems. Many of these rifts start as passive rifts, with no initial asthenospheric convection. These produce no magma until stretching thins the lithosphere sufficiently for the

Volcanic and Igneous Plumbing Systems. http://dx.doi.org/10.1016/B978-0-12-809749-6.00007-8
Copyright © 2018 Elsevier Inc. All rights reserved.

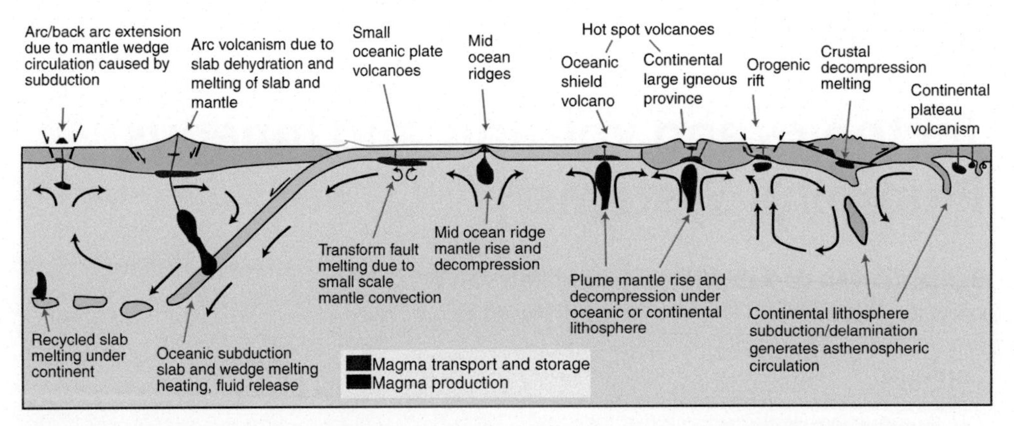

Figure 7.1 ***Plate tectonics and magma production.*** **Conceptual figure showing the Earth's plate tectonic system, with the principal magma generation, transport and storage systems.**

asthenosphere to melt through decompression. Active rifts, which are associated with early asthenospheric convection, begin with large-scale volcanism and uplift followed by rifting.

Hotspots are zones of rising asthenospheric convection or circulation and/or zones where the mantle is especially prone to melting (Foulger, 2010). At hotspots asthenospheric or lithospheric melts are produced through excess heat and create a variety of volcanic fields ranging from the continental flood basalt provinces (Colobia River, USA), huge oceanic islands (Hawai'i) and large volcanic fields (El Pinacate, Mexico) to small, disperse monogenetic fields (southeast Australia).

At convergent plate boundaries, subduction of oceanic lithosphere causes slab dehydration, volatile release and melting of the overlying mantle. This process has worked to create continental crust since at least 2.4 billion years ago.

In orogenic belts, melting of the crust and mantle lithosphere occurs by heating (from basaltic magma underplating), decompression (by removal of load through erosion or rifting/crustal stretching) or from the addition of volatiles that lower the melting point (solidus).

Subduction of continental material does not cause the same scale of melting as oceanic subduction; instead, the subducted continental material is assimilated and may be recycled at a later date. The mantle circulation perturbed by the subduction may also lead to the asthenosphere rising and decompressing, thus generating additional melt.

Once magma is produced, it collects and moves through the lithosphere in bodies that make up an igneous plumbing system (see, e.g. Chapter 2). Depending on the relevant tectonic environment, the configuration of these plumbing systems varies widely. We will consider magma plumbing at mid-ocean ridges, continental rifts, hotspots, oceanic subduction zones (both oceanic and ocean–continental), continental convergence and transform zones.

We will then take a closer look at very shallow plumbing systems, where topographic stresses and sedimentary/volcanic lithology and rheology are important. This includes superficial structures, which only become significant at very shallow levels. Finally, we will consider the possibility of a unified system for tectonics and igneous plumbing systems.

7.2 HOW IS MAGMA TRANSPORTED AND STORED IN DIVERGENT GEODYNAMIC SETTINGS?

7.2.1 Volcanic and Igneous Plumbing Systems at Mid-Ocean Ridges

Mid-ocean ridges are the longest, largest and most voluminous magmatic environment on Earth. Ridges are the site of new lithospheric and crustal production that may be subsequently subducted into the mantle and recycled, or involved in magma-producing dehydration reactions that slowly build up continental crust (Fig. 7.2).

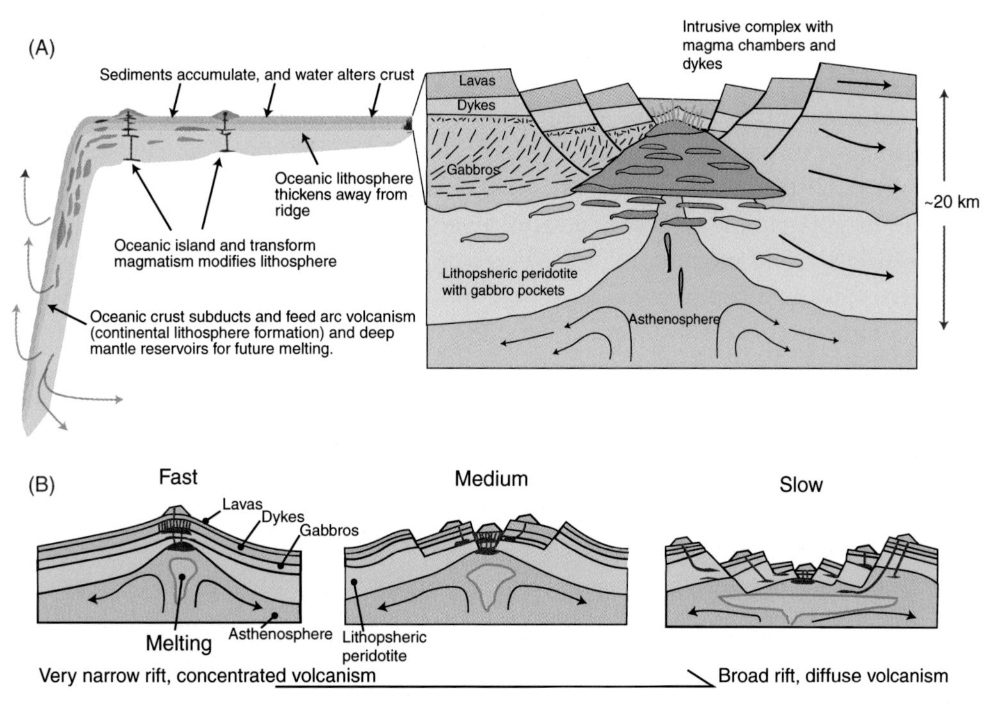

Figure 7.2 ***The mid-ocean ridge plumbing system.*** **(A)** The structure of the oceanic crust resulting from the development of magma chambers, dykes and lavas, as well as their subsequent deformation and transport away from the ridge centre. Effectively, the whole oceanic crust is made up of plumbing system components. The new crust travels away from the rift zone to eventually subduct and later form arc magmas. Oceanic sediment collects on the crust, and hydrothermal alteration (creating serpentinites), oceanic island and transform volcanism modify the crust on its journey. (B) Magma-dominant (fast) ridges construct crust made only of plumbing system components (dykes, sills and their eruptive products), magma-poor (slow) ridges stretch by faulting and exhume lithosphere and asthenosphere. The crust in such situations is a mix of magmatic addition and stretched mantle. Volcanism at fast ridges is concentrated at the axis, while at slow ridges it can be broadly distributed and intrusions may follow other pathways to erupt at the rift shoulders.

The mid-ocean ridge system is made up of diverging lithosphere over rising asthenosphere (Fig. 7.2A). The latter decompresses and melts with volumes related to its ascension rate. At fast spreading ridges, such as the East Pacific Rise, the rapid spreading rate (up to 20 cm/year) produces large amounts of magma as decompression is fast in relation to cooling, whereas at slow spreading ridges, magma production is very low as cooling dominates over decompression.

In ocean-ridge plumbing systems, this means that at fast spreading ridges, magma is the dominant material added to make new crust (typical ophiolite sequence), whereas at slow spreading ridges the asthenosphere may rise to the surface without any magmatic crust forming (serpentinitic crust; Fig. 7.2C). One consequence of this is that at fast spreading ridges, the crustal structure is controlled by magma and volcanics, whereas at very slow spreading ridges, faulting and stretching dominate, allowing exhumation of the asthenosphere. The oceanic crust at slow spreading ridges is thus thinner than fast ones and includes exhumed mantle. The thickness relation between the fast and slow ridges can be seen on Moho depth maps. Departures from this pattern happen where other influences, such as mantle plumes, add additional magmatic input. For example, in Iceland, the magmatic production is high compared to the spreading rate, leading to unusually thick crust (see maps in Artemieva et al., 2017).

The magma plumbing system at a mid-ocean ridge starts with an area of partial melt generation and migration within the rising asthenosphere. When enough magma is collected, it can move upwards as a body, through buoyancy (see Chapter 2). Magma may continue to rise as a dyke and either erupt or cool in the crust, or it may feed into a magma chamber. A conceptual magmatic system is show in Fig. 7.2, which shows that often the whole oceanic crust is made up of magma plumbing elements, topped by lavas and sediments. The thickness and the extent of the oceanic crust relate to the melt flux and the spreading rate, such that at slow spreading ridges some crust will be composed of serpentinised mantle, whereas at fast ridges (or ones with excess melting, such as Iceland), the crust would be exclusively built out of plumbing elements.

Along a mid-ocean ridge, magma collects into discrete centres, each of which concentrates magma and distributes it outwards via a plumbing system of dykes and sills (Fig. 7.2B). The Axial seamount on the Juan de Fuca Ridge is a good example of this (Sigmundsson, 2016) (see also Chapter 11).

Consequently, the intrusive complex can be considered to extend from the axial magma chamber throughout the entire plate, making up the largest intrusive systems on Earth (Fig. 7.2B).

Two types of field sites provide evidence of mid-ocean ridge plumbing. The older parts of Iceland represent exposed sections of magmatic systems, where analogues to mid-ocean ridges can be seen. These have central intrusive systems with cone-sheet intrusions and dyke swarms (see Chapter 4). There are generally many more dykes below 2 km depth than above (80% are below this level). In rifting episodes, such as at Krafla volcano in the 1970s and 1980s (Einarsson and Brandsdóttir, 1980), the majority of dykes identified in seismic data did not reach the surface. While exposures such as those in the east and west of Iceland provide a view into the upper few kilometres of the oceanic crust, deeper levels remain unexposed. However, in ophiolites (obducted oceanic crust), entire sections from the mantle upwards can be seen. In these ophiolites, a complex plutonic sequence of coarse gabbros that transition up into sills and dykes, and finally into lavas (Moores, 2003) is exposed, representing typical fast spreading mid-ocean ridge segments.

The mid-ocean ridge system is driven by several forces, in particular, from asthenospheric flow as the subducting oceanic lithosphere sinks. Known as slab pull, this force is thought to be the main driving force for ocean spreading. The expansion from intruded dykes, and the gravitational force from the topography and magmatism and the slope of the crust–mantle boundary also contribute. All the forces acting from around a mid-ocean ridge can be grouped together and termed 'ridge push', with restrictive forces being the rheology of the lithosphere/asthenosphere, friction and inertia.

7.2.2 How Do Continental Rift Volcanic and Igneous Plumbing Systems Work?

Mid-ocean ridges are born either directly within the oceanic crust as plate motions shift or by the splitting apart of continental crust. In both cases, initial stretching creates a rift that develops into a full oceanic basin with new crust formed by extension. In the case of a continental rift, the process is a long drawn-out evolution that thins the continental lithosphere and slowly leads to the development of oceanic crust.

The continental rift plumbing system differs from the mid-ocean ridge, as continental crust is present as a medium into which the magmas are intruded, rather than magmas forming their own crust (Fig. 7.3).

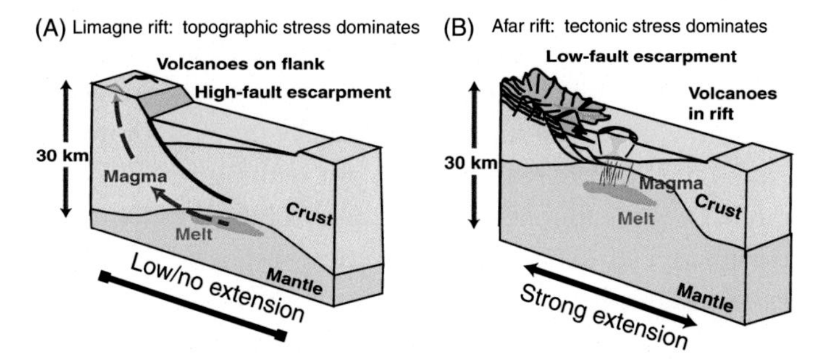

Figure 7.3 ***Two examples of continental rift plumbing systems.*** (A) The example of a deep rift with low tectonic stress (equivalent to a slow spreading ridge). Here volcanism starts only after large degrees of extension and is concentrated on the rift margin as the rift deepens. (B) The example of a mature rift with higher tectonic stress, where extension and magmatism are concentrated in the centre of the rift. Here the stretching is mostly accommodated by dyking and superficial faulting, although sills may also have a major role in weakening and stretching the lithosphere. A sedimentary basin within the rift also provides local lithological control on the type of plumbing.

Nevertheless, the main driving forces and conditions are similar. The continental crust varies more in terms of density and rheology, and its greater thickness means that magma has more volume into which to intrude. The corresponding intrusive systems thus also exhibit more variety. Magma can reside in the crust and assimilate it and also tends to form large central magmatic complexes in which magmas can collect and evolve. These are characteristic of well-developed rifts such as in East Africa (e.g. Nongorogoro) or the Rio Grande (e.g. Valles Caldera, the Socorro magma body).

Rifts have distinct contexts and evolutionary sequences that correlate with different magmatic plumbing styles. Two main evolutionary pathways are possible for rifts:

1.1.1 Passive Rifts

One such pathway is passive, where divergence occurs without initial asthenospheric convection or circulation. Such rifts exhibit no magmatism until the lithosphere is thinned enough to allow decompression melting. Once this occurs, then small-degree melts feed small-scale systems, which are typically monogenetic fields.

Such passive rifts may have small-scale volcanic and igneous plumbing systems, with isolated dykes and sills, the location of which are strongly controlled by local lithology and structure. The strong linkage between tectonics and volcanic grouping and alignment has been explored by Le Corvec et al. (2013), who studied many monogenetic fields and found that most have alignments controlled by local structures. They also noted that as fields became more concentrated (with closer volcanoes), then this association became more evident. The most concentrated volcanic fields were also the most compositionally varied, relating to the progressive development of crustal magma storage, and magma intruded closer in time and space (Fig. 7.3).

As such, passive rifts are not associated with large magmatic systems, and if divergence continues, they can become non-magmatic continental margins and slower oceanic spreading ridges, where the asthenosphere has risen to the surface.

Passive rifts begin with no volcanism and gradually develop limited intra-rift plumbing that may extend out over a broad area, not just within the rift. Passive rifts are often continuously infilled with sediments. This results in subdued topography, however thinning of the lithosphere and externally driven asthenospheric flow can cause uplift; in this case, increased magmatism may accompany regional uplift. Uplift causes the rift sediments to be eroded, allowing topographic stresses to concentrate magmatic activity towards the rift margins (Fig. 7.3A).

1.1.2 Active Rifts

The other path is that of active rifting, where additional asthenospheric uplift provides greater initial melt volume. Uplift tends to occur early in active rifts, and is accompanied by voluminous magmatic activity. Large igneous provinces such as flood basalts can also be produced at this stage. Active rifting can be associated with large-scale magmatic under-plating and in later stages, crustal melting and silicic magmatism, producing large caldera systems such as Yellowstone Caldera. Where this occurs, it is progressively concentrated on the developing rift margin.

Shallow sedimentary basins in active rifts contain large sills interpreted as shallow plumbing systems, such as in the Ethiopian or Antarctic rifts, or in East Greenland (see Chapter 5, Fig. 7.1). Deeper plumbing is not well represented in the surface record, so can only be indirectly imaged. Overall, large lower crustal–mantle lithospheric sill reservoirs are probable, as suggested by gravity anomalies and isostatic uplift.

Central volcanoes and magmatic systems can be associated with the later evolved stage, such as the Messum Centre in Namibia (Korn and Martin, 1955; Jerram et al., 1999), or the North Atlantic Palaeocene igneous province. In East Africa, the later stage of flood basalt eruption is exposed in centres, such as Ras-Dashan in Ethiopia (Williams, 2016).

As the rift further develops into a basin, rift shoulder magmatic systems develop (such as Mt Kenya, Kenya; and Kilimanjaro, Tanzania). The setting of these on the rift shoulder may be linked to magma migration up faults, direct ascent from the edge of rifts or even by the control of magma pathways by a balance of topographic and tectonic stresses (Maccaferri et al., 2014).

As the rift develops, strain and stress are concentrated in the thinning lithosphere, and magmatism becomes more and more concentrated within the rift with first a generation of central volcanoes (often with large calderas and silicic volcanism), then progressively more fissural activity on basaltic systems, with smaller magma storage centres. This can trend towards mid-ocean ridge style elongated magma systems, as seen in northern Ethiopia (Erta Ale) and Iceland. Finally, this type of rift can develop into full mid-ocean ridge conditions as the continental crust is fully replaced. Corti (2013) provides a good description of this process for the East African Rift.

Active rifts form when abundant melting occurs either independently, or caused by the divergence over an area of especially fertile mantle (Foulger, 2010). This classically occurs associated with continental hotspots, or mantle plumes, where rising asthenosphere can melt even before rifting begins. The Ethiopian trap series and the Afar traps are examples of this.

Back arc, or arc rifts, and mountain belt-related rifts can also be associated with abundant magmatic activity. In a continental collision foreland setting, the Auvergne and Eiffel provinces in the European Cenozoic rift system are examples. The Nicaraguan and El Salvador Depression in Central America are examples of arc extension, and the Payun province in Argentina is an example where abundant magmatism is associated with back-arc rifting.

7.3 IN MANTLE PLUMES HOW IS MAGMA STORED AND TRANSPORTED?

Mantle plumes are rising areas of buoyant asthenosphere. There is a certain amount of controversy surrounding their origin and real nature. For example, Foulger (2010) argued that the model of plumes originating at the core–mantle boundary does not fit with much geological data and rather proposed a more diverse range of mantle convection, circulation and differential melting capacity can better explain what we see. However, for the purposes of this chapter, this argument can be bypassed to focus on the consequences of mantle melting caused by asthenosphere circulation of any type. Whatever their origin and exact nature, mantle plumes produce large volumes of magma over a point or area, which may or may not remain stable with respect to overlying plate motions (Fig. 7.4).

In the case of intraplate oceanic settings, classic sites are found in Hawaii, Galapagos, La Reunion, the Canary Islands and the Azores. In each of these cases, there are major differences in the style of volcanism and structure of the islands. All of these relate to underlying plumbing systems and in part to tectonic settings, although the main features of all types are regrouped in Fig. 7.4A.

In the case of the Azores, they coincide with strike-slip faulting at the European-African plate boundary, which leads to highly elongated edifices and rift zones. In the case of the Canary Islands, the setting is a continental margin without any distinct structural influence. Hawaii and La Reunion are located on fast moving oceanic plates, which has led to a periodic shift in magmatism from one edifice to the next, forming long island chains. In these sites, there may be a periodic magmatic evolution from voluminous basaltic shield building to subsequent more evolved alkaline magmatism.

In all cases, the volcanoes are very large (up to 12 km tall and 100 km wide), so large plumbing systems build up within them. Thus, the tectonics of the volcanic edifice has a major control on its internal plumbing system. Conversely, the plumbing will also have an effect on the volcano's tectonics and a very active role in shaping the volcano (e.g. Borgia, 1994; Klügel et al., 2015; Delcamp et al., 2012).

Sills can be favoured when the sagging of an edifice creates a horizontal maximum compression and layered crust is present. The sills can lead to relative uplift that reverses the sagging process. Uplifted submarine rocks or even emerged oceanic crust (e.g. Klügel et al., 2015) shows that at many sites, intrusions have caused uplift. Uplift can lead to higher slopes and to higher rates of sliding on the laterally spreading upper edifice.

For example, rifts such as those found in Hawaii, the Canaries and La Reunion may be correlated with gravity spreading, where a basal unit, like pelagic clay, allows outwards sliding that negates the effect of flexure (e.g. Nakamura, 1977; Borgia et al, 2000; see also Chapter 4). Flexure or 'sagging' tends to squeeze the volcano, creating compression, while spreading leads to extension—the balance is vital in determining the type of plumbing system within the volcano.

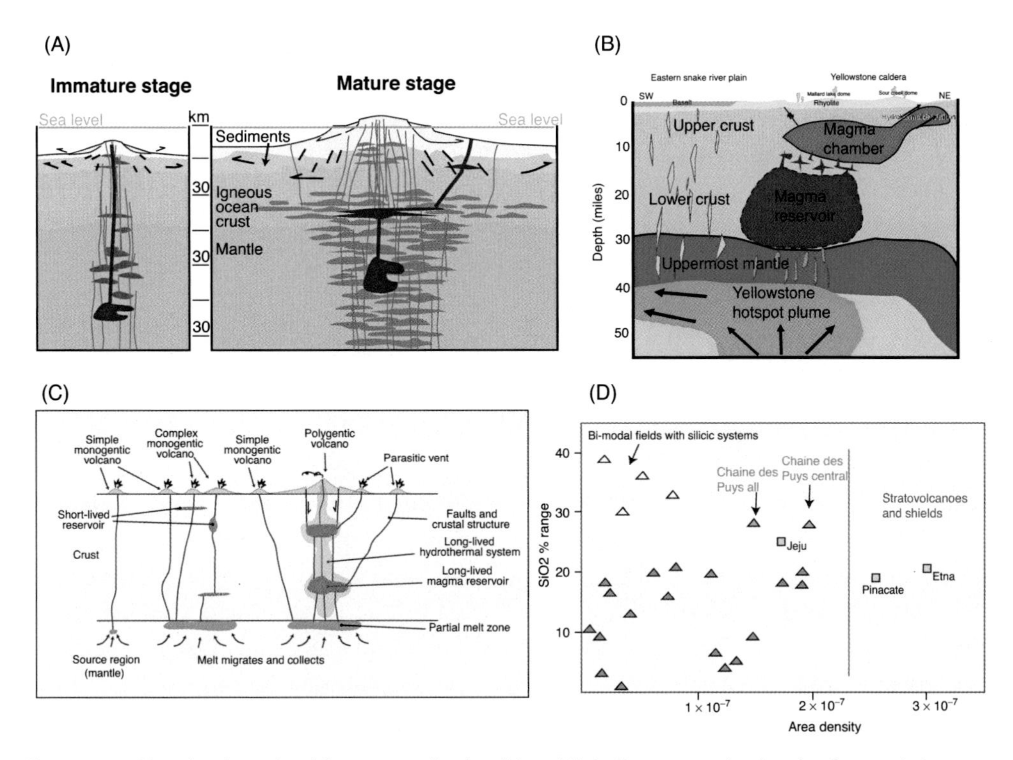

Figure 7.4 ***Mantle plume/melting anomaly plumbing.*** (A) A diagrammatic sketch of oceanic hotspot plumbing from Klügel et al. (2015) showing the evolution of a young to old oceanic hotspot-related plumbing system. (B) A diagrammatic sketch of a continental hotspot plumbing system: the Yellowstone Caldera system (Smith et al., 2009). (C) Diagram of a smaller scale intraplate monogenetic volcanic system with the evolution of plumbing from monogenetic to polygenetic. (D) Graph of volcano spacing in monogenetic volcanic systems vs silica range, showing that widely spaced systems are generally basaltic (no chance to evolve), while more concentrated fields have a wider range of silica concentrations (magmas can collect and evolve in proto-polygenetic systems). Outlying fields are also highly disperse fields with high silica ranges (related to major silicic caldera systems, Los Humeros, Yellowstone), and highly concentrated fields with low silica range (related to polygenetic volcanic systems with multiple vents such as El Pinecate and Jeju)

The Galapagos Islands have a silicic basal sediment layer, which does not deform as easily as the clay layer below Hawai'i. Because of this, McGovern et al. (2014) suggested that the large calderas and radial dyke swarms of these islands were caused by a lack of spreading and preponderance of sagging. The Galapagos islands may also have larger intra-volcano magma plumbing systems and fewer deep sills, such that sagging is predominantly linked to uplift from underplating.

In continental areas, hotspots can lead either to large-scale basaltic magmatism (e.g. Columbia River, Siberian Traps) or large-scale under-plating of magma and crustal melting, such as at Yellowstone (Fig. 7.4B). In these areas, large basaltic ocean shield-type volcanoes can exist (Jeju, South Korea is an example, El Pinacate, Mexico another).

Many monogenetic fields around the world have been tentatively associated with weaker plumes, however more and more of these are also being interpreted as resulting from other processes, such as lithospheric delamination or asthenospheric flow under the edges of the continental lithosphere.

There is a clear progression from small volumes of mantle being associated with small and sparse volcanism, to more concentrated volcanism and then polygenetic volcanoes, flood basalts and large caldera systems (Fig. 7.4C and D).

7.4 HOW IS MAGMA TRANSPORTED AND STORED IN CONVERGENT GEODYNAMIC SETTINGS?

7.4.1 Oceanic Subduction Zones and Volcanic and Igneous Plumbing Systems

Magma production at oceanic or oceanic/continental convergent plate boundaries is linked to dehydration of the subducting slab and melting of the overlying asthenosphere. The asthenospheric flow caused by slab descent and may also cause decompression melting especially in the back-arc (Sternai et al., 2014). The location and volume of melt depends on the tectonic context. Where the slab is steeply dipping, magma is produced in a narrow zone and leads to a narrow arc of concentrated volcanoes (e.g. Central America). In shallower subduction, melting occurs over a much broader area and leads to a broad arc (e.g. Mexican volcanic belt). When subduction changes from shallow to steep, the 'slab rollback' can create major extension in the arc and back-arc, and the inflowing mantle will generate additional volcanism. The opposite, when a slab changes from steep to shallow, occurs when a hotter or thicker oceanic crust is subducted. An example of this is southern Central America, where the Cocos Ridge is subducted below Costa Rica and Panama. Here, the arc volcanoes are very large, and broad and associated with contraction, while in neighbouring Nicaragua, the steep slab is related to a very narrow arc, of smaller volcanoes.

Under the arc, magma tends to gather near the Moho, where in oceanic environments it can create gabbroic chambers that then feed upper plumbing systems (Fig 7.5). In continental zones, such underplating may be accompanied with crustal melting and the creation of partial melt zones, such as in the Andes (Schilling et al., 1997). The underplating can then generate uplift and produce high topography, such as in the Puna and Altiplano. This uplift is caused either by the support of the underplated magma or by its high-grade metamorphism. This process causes a density increase and then the sinking of the underplated material by delamination, which allows new asthenosphere to flow in. The uplift of the Sierra Nevada has previously been explained in this way (Saleeby et al., 2004).

Magma in the lithosphere can provide low resistance layers that can concentrate deformation (Fig. 7.5). Such intrusion-related strain localisation can also happen at the

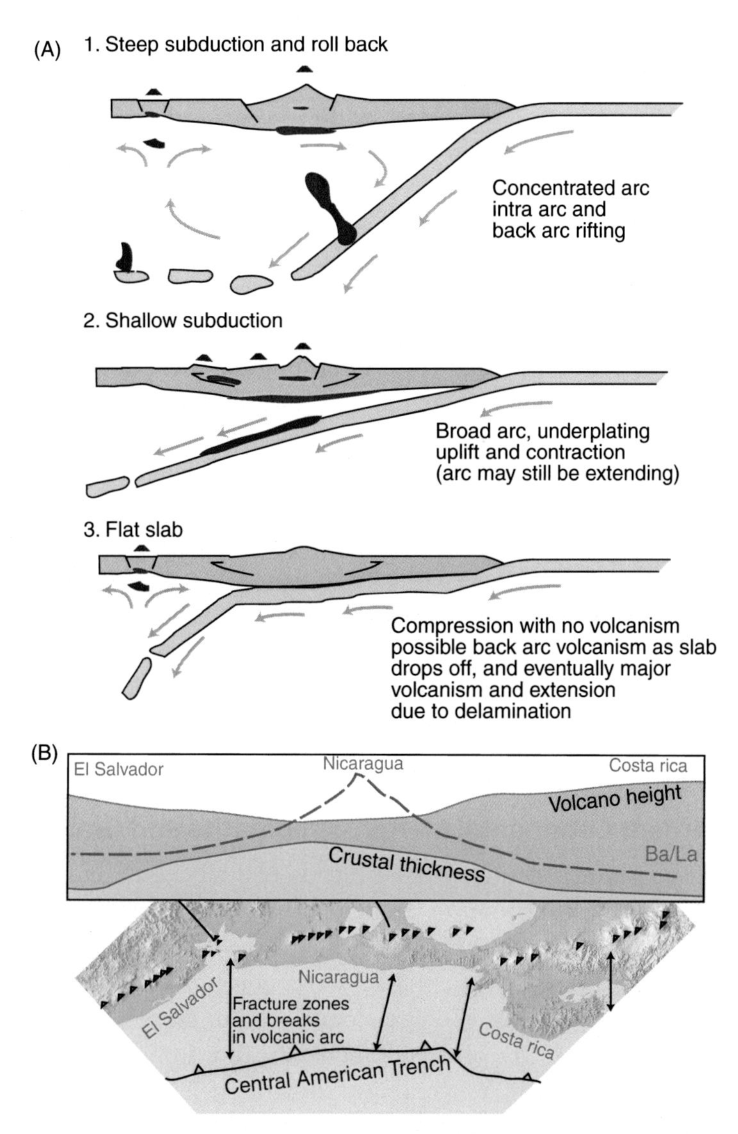

Figure 7.5 ***Convergent oceanic plumbing: the oceanic subduction plumbing system.*** (A) Cartoon showing the different plumbing systems that develop in steep subduction, shallow subduction and flat subduction. Note that a change from shallow to flat subduction can create major extension in the arc, and high rates of magmatism as the mantle flows in behind the retreating slab. Flat subduction has no associated volcanism, except potentially in the far back arc. A very shallow slab can eventually founder and delaminate, as may have occurred under the Colorado plateau. (B) Example of Central America (El Salvador to Costa Rica) with the segmented subduction of the Cocos plate, leading to changes in plumbing style and along arc volcanism. Crustal depth, volcano height and Ba/La values (a proxy for the amount of subducted sediment) are shown, which indicate how crustal structure and the amount of subducted sediment also correlates with changes in plumbing (Carr et al., 2003).

scale of individual volcanoes, for example, at Tromen, Argentina, where thrusting is concentrated at the base of the edifice (Galland et al., 2007).

Subduction zones are often segmented and the associated magmatic systems follow this trend. This is particularly clear in Central America (Fig. 7.5), where subduction segment boundaries are either zones of no volcanism (Nicaragua/Costa Rica boundary) or voluminous activity, such as between Nicaragua and El Salvador, or in central Nicaragua at Masaya (van Wyk de Vries et al., 2001; Carr et al., 1982). Transverse fault zones that cut arcs provide pathways for magma plumbing. In such zones, the magmas tend to have shallower fractionation trends and the volcanoes tend to be monogenetic. Where polygenetic they are formed of multiple vents, indicating that the plumbing is built out of many separate pathways, which may merge (van Wyk de Vries, 1993).

In oblique subduction, the transverse component of movement is taken up along the volcanic arc. This is most clearly illustrated in Sumatra along the great Sumatra fault, but is also seen in northern Central America with the Montagua fault or in Celebes (van Wyk de Vries and Merle, 1996). In these areas, there is a chicken and egg situation where magma in the lithosphere may reduce its effective strength and thus concentrate faulting along the arc, but also the fault concentrates magma into the zone of deformation. Pull-apart basins can be produced around intrusive centres, and the resulting extension may further concentrate magma (Girard and van Wyk de Vries, 2004). The largest calderas, such as Toba, Raung, Tondano, Amatitlan and Masaya, all lie within such strike-slip pull apart basins (see also Chapter 10).

7.4.2 Convergent Continental Margins: Migmatite and Granite Plumbing

Magmatism in convergent continental settings is typically less voluminous than any other setting, especially in relation to erupted volumes. Magmas can be produced in such settings via heating and decompression of crustal rocks during burial, respectively and exhumation. However, they can also be a consequence of continental subduction and delamination. This causes asthenosphere circulation to bring hot mantle material to lower depths, where it can melt or cause delamination away from the main orogenic belt. It can also generate lithospheric stretching and rifting (Fig. 7.6). In the latter case, within mountain plateaus such as Tibet, this forms monogenetic fields generally associated with strike-slip fault zones relating to tectonic escape.

Crustal rock that is heated undergoes metamorphism, dehydration and melting of more mobile minerals. This forms migmatites that separate melt from residue. The melt can be transported out of the zone, collect and form granitic magmas (see also Chapter 2). In the Andes, underplating from subduction-derived magmas has led to a partially melted lower crust (Schilling et al., 1997), and in the Himalayas, around 21–24 Ma years ago, a partially melted layer (migmatite) formed due to decompression melting. The collected melt from this event generated a chain of granites throughout the Himalayas including the Mansalu granite (Fig. 7.6). The accumulation large volumes of melt in the crust allows enhanced strain and lateral flow of the lower crust (Hall and Kisters, 2016).

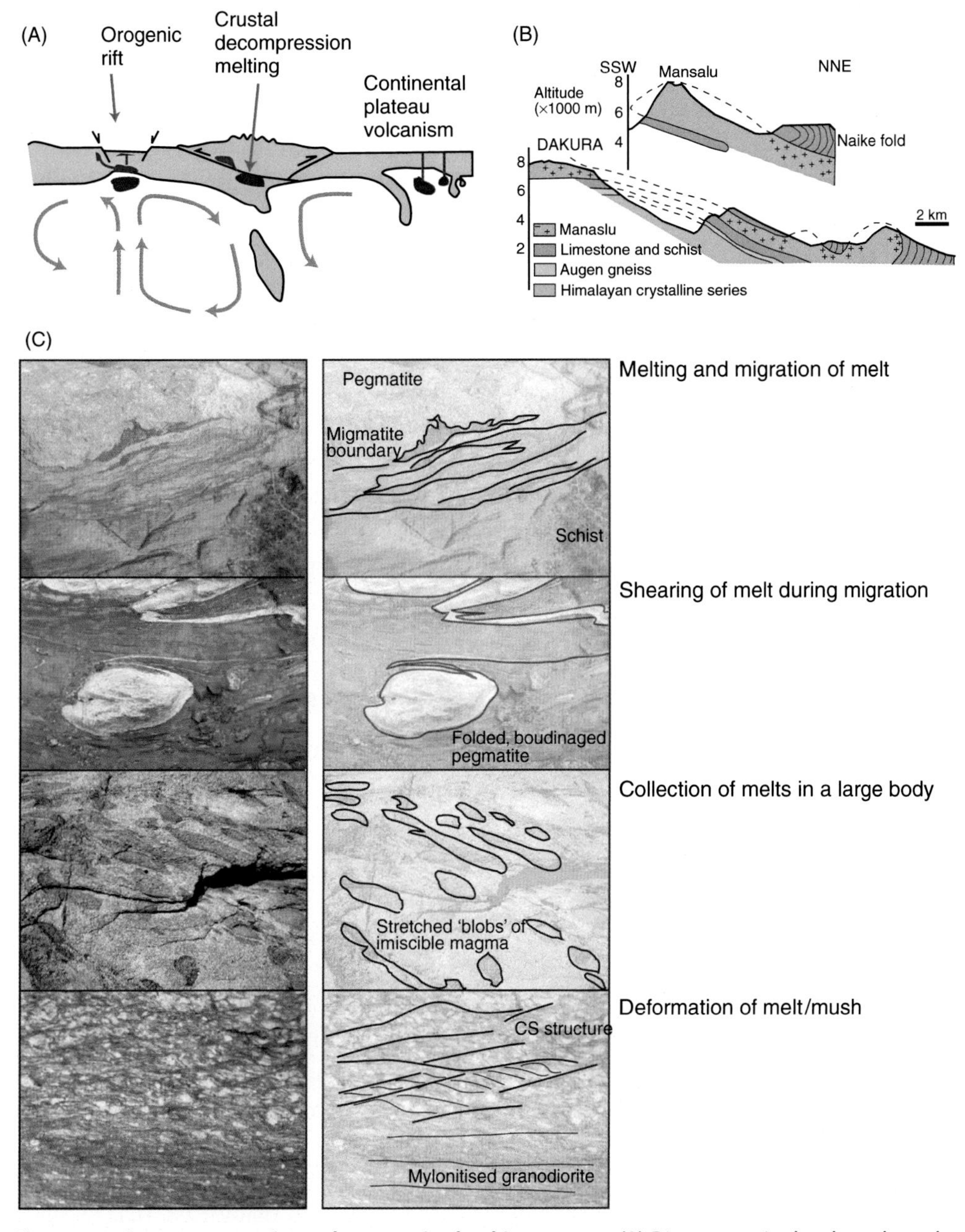

Figure 7.6 ***Convergent continental magmatic plumbing systems.*** (A) Diagrammatic sketch to show the geodynamic context of crustal melting in a continental collision, and other types of continental magma production. Crustal thickening followed by exhumation causes heating (possibly with fluids introduced) and then decompression to lead to partial melting of sedimentary rocks ('S-type' granites). These form migmatites that collect magma into ever larger bodies. Such low-viscosity zones are also zones of preferential deformation. (B) Cross-section of the Mansalu granite (Guillot et al., 1993) showing its intrusion into Himalayan orogenic sequences. (C) Images from the Cap de Creus and Roses area, Catalonia, Spain, showing the progression of migmatisation, pegmatite formation, the deformation of these in melt coeval shear zones, then the collection of liquid in the Roses granite (magma mingling and mixing), and the deformation of this partially melted mush, as deformation continued during ongoing deformation and cooling.

A good example of migmatisation and magmatic collection is found in the Cap de Creus area of Catalonia (Carreras et al., 2004) (Fig. 7.6C). Here, migmatitic melts are collected under high-pressure (sillimanite grade) conditions into pegmatite veins. The nearby Rosas Granite provides the example of a larger body composed of assembled pulses of magma. There is a very clear relationship in both cases with the tectonic strain, with early pegmatites being cut by shear zones, while later ones follow shear zones and balloon (expand) in them, but are not deformed. The pegmatites contain evidence of numerous pulses, with garnet, tourmaline and muscovite being the main accessory minerals in a quartz and feldspar groundmass.

The Rosas Granite is made up of many injections that mixed, associated with the progressive development of conjugate shear zones around the elongation axis of enclaves. Thus, deformation in both liquid and plastic states follows the same strain axes along which the pluton cooled. The mass is cut by small conjugate normal faults that also follow the same axis of compression as the ductile rocks, thus showing that the same tectonic stresses operated throughout the formation and cooling of the intrusion.

7.5 HOW DO VOLCANIC AND IGNEOUS PLUMBING SYSTEMS INTERACT WITH FAULTS, CRUSTAL RHEOLOGY AND TOPOGRAPHY?

7.5.1 The Influence of Faults on Magma Storage and Transport

Shear faults, of any type, do not tend to concentrate magma along the fault zones, as most magma tends to be transported in tensional cracks, which are not parallel to shear zones. However, magmas can also intrude by shearing the rock. Especially in the case of more viscous magmas or low-strength rocks, the creation of intrusion-related shear zones does occur (see Chapter 4). In all fault types, releasing bends, where space is opened between fault segments, can collect magma. This has already been noted in arc settings for the correlation between pull-apart structures and large magmatic systems.

Rift faulting. In rift systems, dykes are often found to open grabens above them and be associated with ground fracturing (Fig. 7.7A). Conversely, faults may be deflected around or towards magmatic systems by the topographic load of a volcano, or by the presence of a magma body (e.g. van Wyk de Vries and Merle, 1996). Volcanoes and calderas are often elongate along or oblique to rifting directions. In the case of rift-aligned calderas, such as Erta Ale (Fig. 7.7A), the strong control of rifting is clear. In many East African Silicic volcanic systems (Fig. 7.7), the caldera may either be produced by overlapping collapses, by sill intrusion oblique to the rift or by inheritance of older structures.

Detachment faults. A particular type of normal fault, detachments faults are low-angle (<20°) extensional faults. They typically form when a basin is asymmetrically extended and the asthenosphere can flow into the thinned lithosphere. A combination of decompression melting and shear melting lubricates the fault sustaining rapid motion and

(A) 1. Pre-rift, with structural grain oblique to future extension

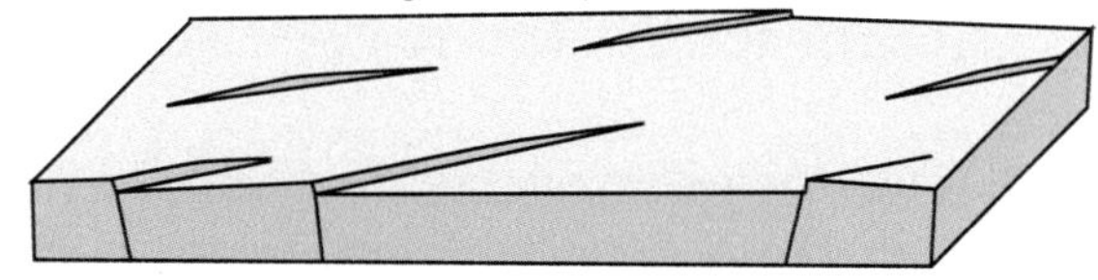

2. Rift, with large magma bodies contolled by pre-rift structure: intrusions influencing and interacting with rift faults

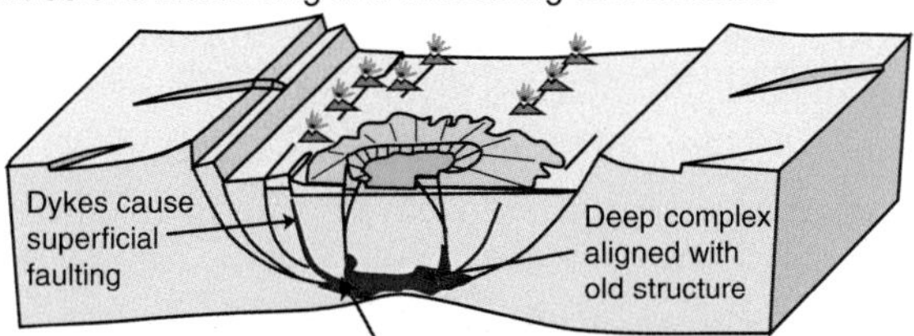

(B)

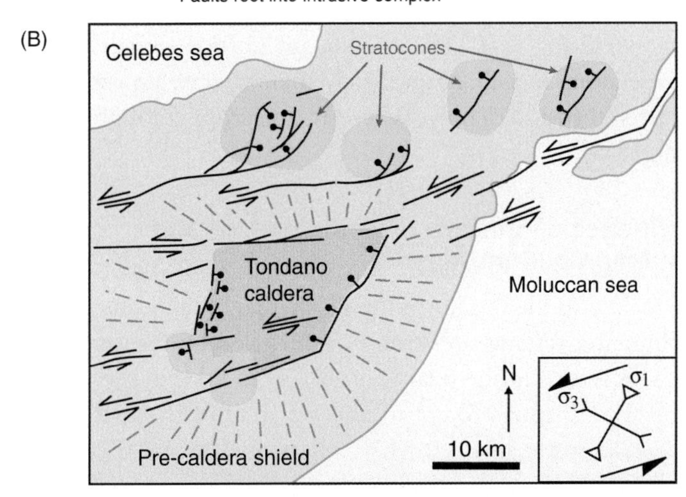

(C)

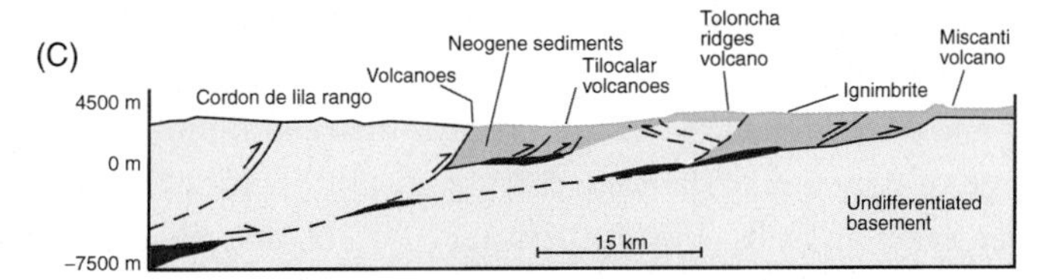

Figure 7.7 ***Magma plumbing with different tectonic regimes.*** (A) Calderas in the East African Rift are partly controlled by pre-existing deep structure (sketch modified from Robertson et al., 2016). Note that ancient structures may influence the location and shape of magma bodies, but the bodies will also affect the geometry of modern structures. (B) Magma in strike-slip zones collects in pull apart structures that can be initially generated by the load of the volcano, and the magma acting as a low-rheology zone. Example from the Sulawesi (van Wyk de Vries and Merle, 1998). (C) In compressional settings, the magma may follow thrust faults and form ramp structures. Example from the N Chilean Andes (Atacama) (Gonzales et al., 2009).

creating numerous *syn*-kinematic granite sheet intrusions (e.g. Passchier et al., 2005). This results in a portion of middle-lower crust plumbing being exposed at the surface as a core complex.

Strike-slip fault. These zones accommodate magma in pull apart structures. This is well displayed by the classic connection between granite plutons and strike-slip faults in Brittany, France, but is also seen at the surface in modern sites, such as Sumatra, Sulawesi and Central America (Fig. 7.7B). So, while transverse movement does not necessarily generate magma, it can be important for collecting and distributing it (Spacapan et al., 2016). Also, shear deformation engendered by strike-slip movement in zones of partial melt may provide a means of concentrating magma and extracting it, so transverse zones do have a role in magma generation, transport and storage.

Thrust faults. In compressional faults, magma may form sills in compressive zones and then be exploited as a flat in thrust sequences. The example of the Atacama compressional area is shown in Fig. 7.7C. The magma can intrude along the thrust plane, or the thrust can localise along the sill. Magma may then migrate up the thrust and collect in fault bend folds or thrust anticlines, where it can erupt through the crest of the fold (Gonzales et al., 2009).

7.5.2 The Influence of Shallow Crustal Stresses and Lithology on Magma Plumbing

When magma is close to the surface of the Earth, uplift is a means of creating space, such that shallow plumbing systems are strongly controlled by the way they uplift the surface (Fig. 7.8). This is seen in the formation of forced folds in the host rock (e.g. Magee et al., 2017; see also Chapters 5 and 6) and was one of the first aspects of volcanism noted by early geologists (e.g. von Buch's Craters of elevation theory: van Wyk de Vries et al., 2014).

Magma-related uplift creates a range of tectonic structures, some of which have already been described for oceanic island hot spot plumbing. In such environments, the huge volcanoes and large magma volumes produce marked uplift effects (Fig. 7.5).

The full range of structures related to intrusive uplift includes extensional fracturing and faulting over dykes and sills (Fig. 7.8), tilting of strata, marginal thrusting and strike-slip faulting. The uplift also permits lateral movement, and large sections of crust can slide away and be transported due to this process. Examples of this are the Mull volcano (Scotland), the south-west side of Etna (Italy) but such deformation can happen at all scales, and is a frequent element in monogenetic volcanoes.

Movement can also be rapid at larger scales, such as the huge heart mountain slide in Utah, USA (Mitchell et al., 2015). Such rapid displacements are found at all scales, and magma intrusion is one of the main causes of landslides on volcanoes (e.g. Bezymyanny, Mt Saint Helens; see Chapter 9).

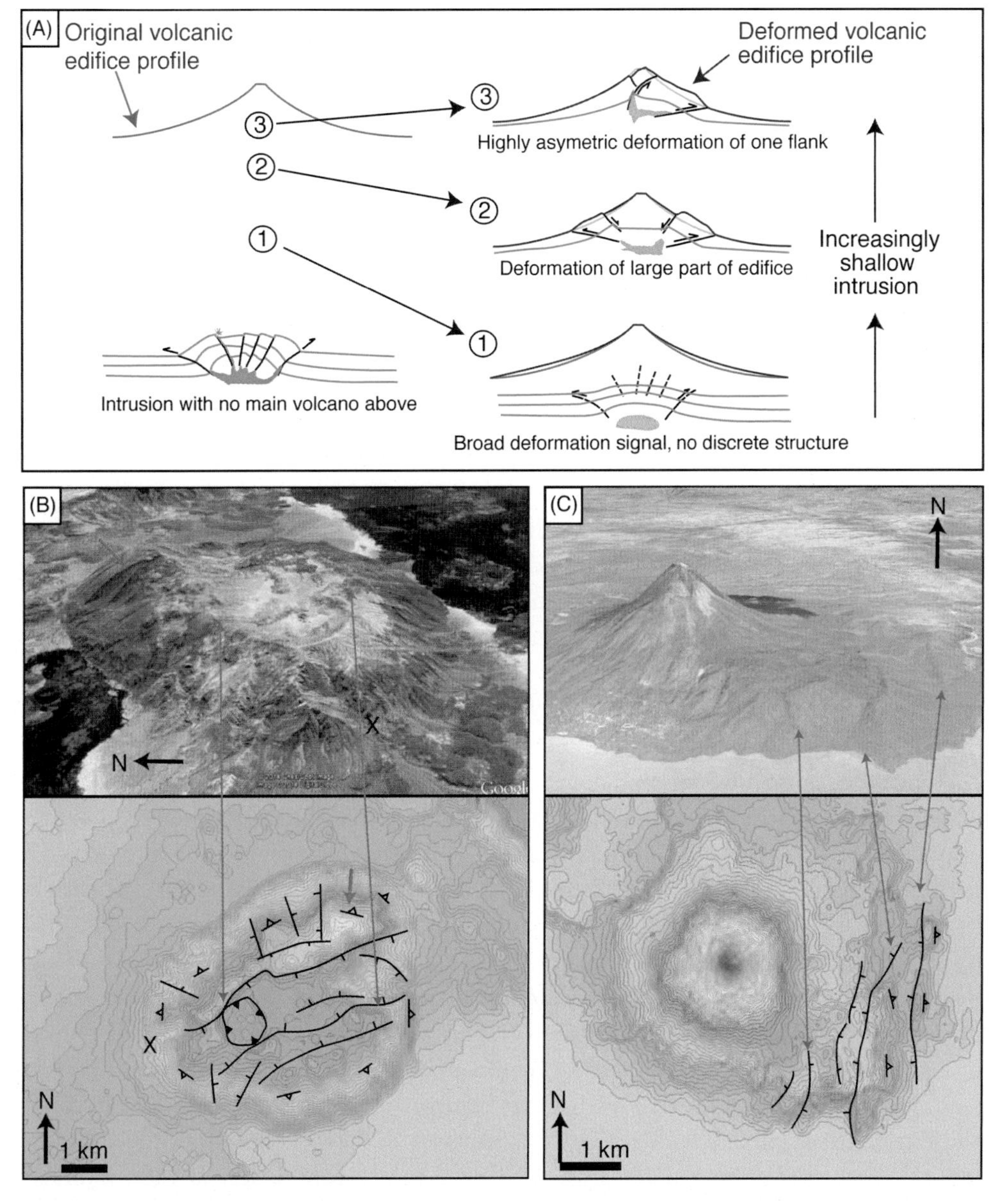

Figure 7.8 ***Shallow magma tectonics.*** (A) Sketch of the different types of tectonic uplift associated with magma intrusion, from underplating to very shallow, within edifice intrusions and intrusions with no edifice. (B) The example of Alid volcano, Eritrea, where a forced fold of 400 m height has formed due to an intrusion into proteozoic metasediments and rift sediments. Note the basaltic eruptions from the base of the dome, which has only erupted trachyte from small summit craters (Duffield et al., 1997). Blue arrows join faults in image and topographic map,' *(pink arrows mark)* the same spot in both. (C) The example of Momotombo, Nicaragua, where the east side of the volcano is uplifted by 300 m. This uplift may have occurred before the present cone had grown up, as the most recent lava flow around the faults. Blue arrows join faults in image and map.

7.5.3 The Influence of Topography on Shallow Magma Storage and Transport

Topography and gravity play important roles in creating stresses that guide magmas through the crust. They make the link between large-scale plate tectonic stresses (gravity controlled) and the local environment.

It has long been known that dyke intrusion is controlled at shallow levels by topography (e.g. experiments on rift zone formation in Hawai'i by Fiske and Jackson (1972). Observations on Etna show that dykes and eruptions often follow the margins of a valley (Valle del Bove) (Murray, 1988), thus being influenced by the topography. A similar conclusion has been reached at Stromboli (Walter and Troll, 2003; Acocella and Tibaldi, 2005) and at Calderas (Corbi et al., 2015). As described by (Maccaferri et al., 2014) and Maccaferri et al. (2015), eruptions favourably occur on the footwall, when magma is intruded near faulted rift escarpments. This illustrates the play between topographic and tectonic stresses especially well (Fig. 7.9).

Extension produced by gravity sliding or spreading also opens up pathways for magma to rise in; in contrast, flexure and sagging can compress the edifice and result in reduced magma output. These gravity tectonic processes have a major control on the evolution of magmatic plumbing. For example, at Concepcion, Nicaragua, early volcano growth resulted in a volcanic edifice that sagged into the lake basin, causing compression. The originally primitive basalts erupted were replaced by increasingly evolved rocks, and the compression eventually blocked magma ascent (Borgia and van Wyk de Vries, 2003).

Once the volcano began spreading, a rift developed and the trapped, highly evolved magma was erupted. The volcano has gone through cycles of stability and spreading and erupted a broad range of compositions reflecting the newly developed open system. The cyclic nature of the compositional variations and the stop–start nature of deformation (Saballos et al., 2014) indicate a strong interdependency between magma intrusion, gravity and regional tectonics.

In glaciated regions, ice volume can also be an important control on the volume of magma produced, erupted and stored within the plumbing system. Rapid removal of ice masses hundred to thousands of metres thick can both stimulate decompression melting at depth and allow shallow magma chambers to erupt their contents (Hardarson and Fitton, 1991; Jull and McKenzie, 1996; Slater et al., 1998). Thus, not only are volcanic and igneous plumbing systems strongly related to tectonics, they are also indirectly linked to climatic variations.

7.6 THE TECTONIC AND MAGMATIC SYSTEM

Magma plumbing systems have been shown to be an integral part of plate tectonics, and the interaction with structure occurs at all scales, from the global to the microscale. The processes that operate in each case are similar, as they all involve host-rock rheology,

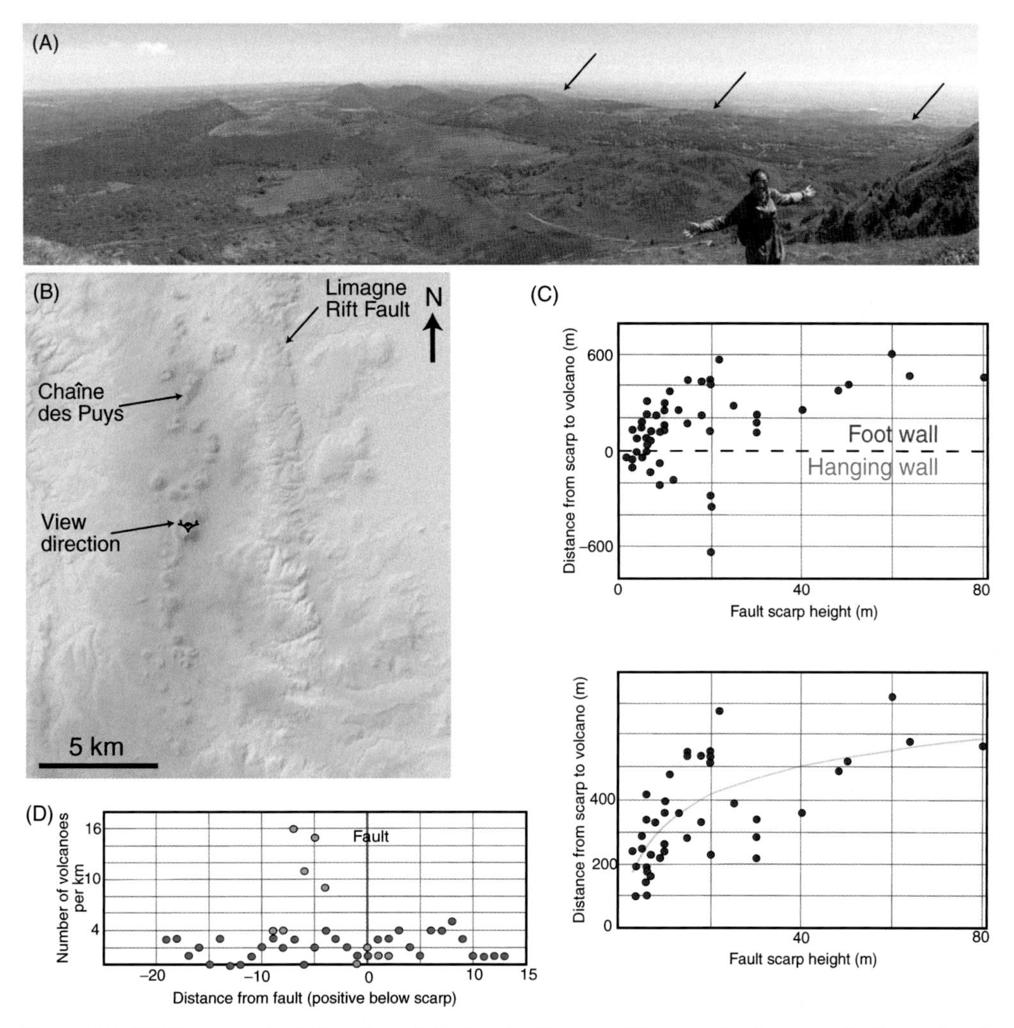

Figure 7.9 (A) Photograph of the aligned Chaîne des Puys and Limagne fault (arrows indicates fault), taken from the Puy de Dome. (B) Topographic shaded relief image of the Limagne fault and Chaîne des Puys, showing the position of the volcanoes behind the fault escarpment (eye shows view point). (C) Data from scoria cones and fault scarps in Ethiopia (Maccaferi et al., 2015), showing the tendency for volcanoes to erupt on the upper shoulder (footwall side) of the fault. (D) Distribution of volcanoes in the Limagne Rift pre- (*blue*) and post-deepening (*yellow*), showing the concentration of volcanism on the top of the escarpment.

magma rheology, density, heat, gravity and other tectonic forces. Thus, a generalised system can be built up for idealised relations between magmatic and tectonic systems. This is intended to help keep track of the importance of any individual part, irrespective of scale, on the entire system (Fig. 7.10).

(A)

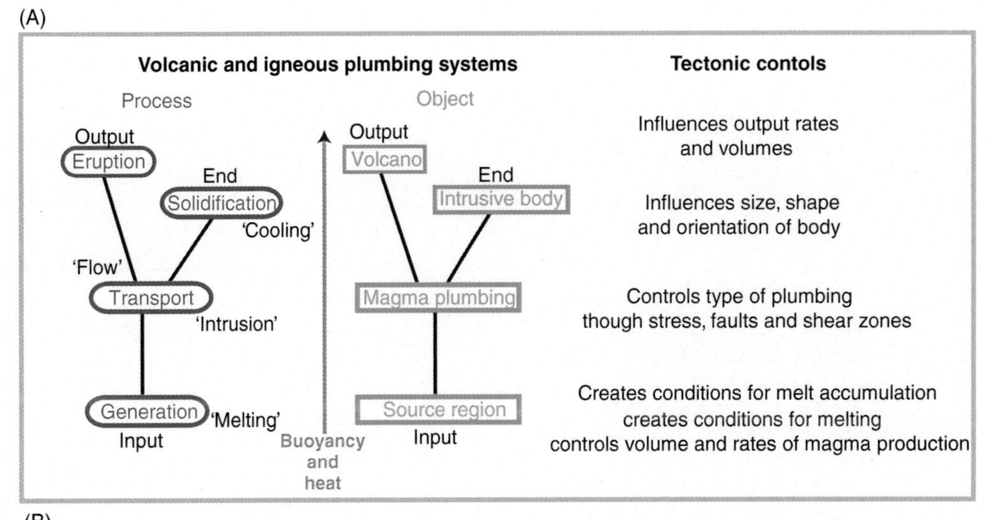

(B)

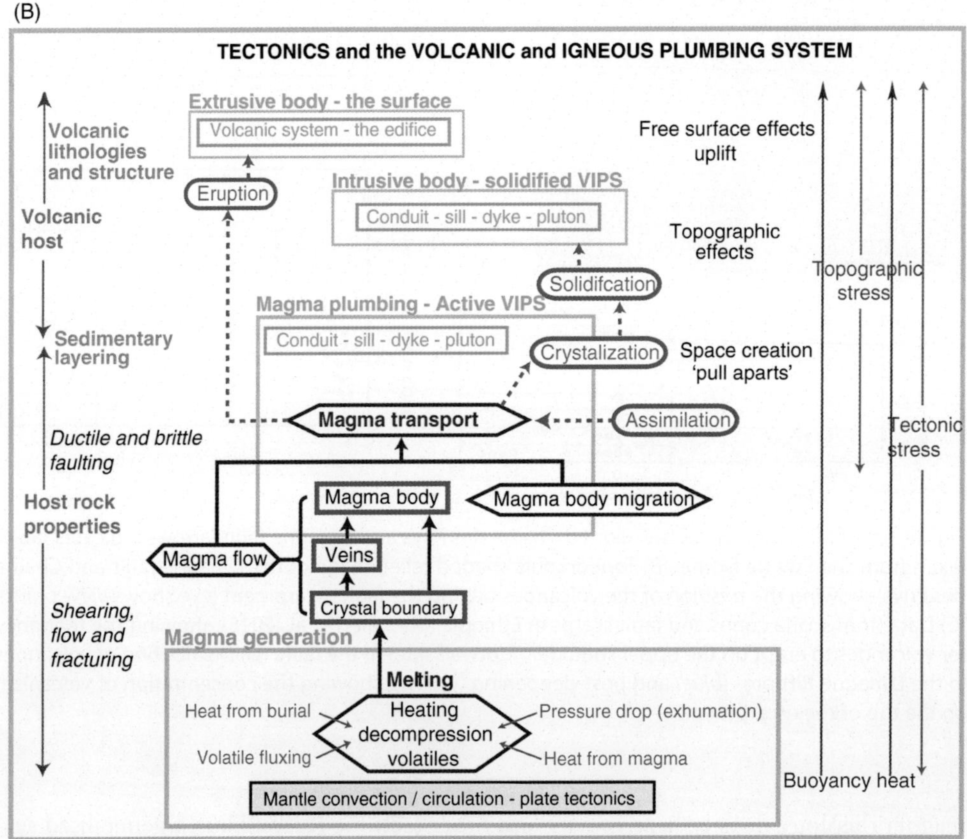

Figure 7.10 ***System diagram for the links between tectonics and volcanic and igneous plumbing systems.*** (A) A simplified system showing the main features of magma production (melting), magma collection and transport, magma migration (note flow within an intrusion is not the same as an intrusion migrating) and then magma evolution in different tectonic and lithospheric environments. (B) More full systems diagram, showing some of the many interactions within the tectonic and magmatic plumbing system.

CONCLUSIONS

Magmatic plumbing systems can be seen as a part of the plate tectonic system, where products from melting in the mantle and crust are redistributed according to temperature and density contrasts, which continuously feed back into the rest of the tectonic system.

This relation can be exemplified at the largest scale with an oceanic plate, which is almost exclusively formed of magma plumbing emplaced at the ridge. The increased load, due to the build-up of mass at the ridge, and magma pressure, associated with other large-scale forces pushes the plate away. At the other end, subducting oceanic lithosphere dehydrates and causes melting, interacting with the overlying asthenosphere and producing new magma to form an arc plumbing system. The dense slab pulls down the oceanic lithosphere (slab pull) and eventually mixes back into the rest of the mantle, to later be taken up by new ridges and hot spots.

The upper magmatic plumbing systems evolve under the joint influence of local tectonic forces, faults, topography and lithological contrasts. It influences the surface through eruptions, but also the creation of topography (bulging/sagging) and by triggering landslides and seismic events. Old extinct volcanoes, where plumbing may be exposed at the surface, provide visible evidence of these processes and valuable opportunities for public outreach.

REFERENCES

Acocella, V., Tibaldi, A., 2005. Dike propagation driven by volcano collapse: a general model tested at Stromboli. Italy Geophys. Res. Lett. 32, L08308. doi: 10.1029/2004GL022248.

Artemieva, I.M., Thybo, H., Shulgin, A., 2017. Geophysical constraints on geodynamic processes at convergent margins: a global perspective. Gondwana Res. 33, 4–23.

Borgia, A., Delaney, P.T., Denlinger, R.P., 2000. Spreading volcanoes. Annu. Rev. Earth Planet Sci 28, 539–570.

Borgia, A., van Wyk de Vries, B., 2003. Volcano-TECTONIC Spreading at Concepcion, Nicaragua. Bull. Volcanol. 65, 248–266.

Borgia, A., 1984. The dynamic basis of volcanic spreading. J. Geophys Res. 99, 17791–17804.

Carr, M.J., Feigenson, M.D., Patino, L.C., Walker, J.A., 2003. Volcanism and geochemistry in Central America: progress and problems. In: Eiler, J. (Ed.), Inside the Subduction Factory. Geophysical Monograph, 138, The American Geophysical Union, p. 153.

Carr, M.J., Rose, W.I., Stoiber, R.E., 1982. Central America. In: Thorpe, R.S. (Ed.), Andesites: Orogenic Andesites and Related Rocks. Wiley and Sons, New York, pp. 149–166.

Carreras, J., Druguet, E., Griera, A., Soldevila, J., 2004. Strain and deformation history in a syntectonic pluton. The case of the Roses granodiorite (Cap de Creus, Eastern Pyrenees). Geological Society, London, Special Publications, Vol. 224, pp. 307–319.

Corbi, F., Rivalta, E., Pinel, V., Maccaferri, F., Bagnardi, M., Acocella, V., 2015. How caldera collapse shapes the shallow emplacement and transfer of magma in active volcanoes. Earth Planet. Sci. Lett. 431, 287–293.

Corti, G., 2013. Continental rift evolution: from rift initiation to incipient break-up in the Main Ethiopian Rift, East Africa. Earth Sci. Rev. 96, 1–53.

Delcamp, A., van Wyk de Vries, B., James, M.R., Gailler, L.S., Lebas, E., 2012. Relationships between volcano gravitational spreading and magma intrusion. Bull. Volc. 74, 743–765.

Duffield, W.A., Bullen, T.D., Clynne, M.A., Fournier, R.O., Janik, C.J., Lanphere, M.A., Lowenstern, J., Smith, J.G., Giorgis, L., Kahsai, G., Mariam, K., Tesfai, T., 1997. Geothermal potential of the Alid volcanic center, Danakil Depression, Eritrea. US Geol. Surv. Open-File Rpt. 97-291, 1–62.

Einarsson, P., Brandsdóttir, B., 1980. Seismological evidence for lateral magma intrusion during the 1978 deflation of the Krafla volcano in NE Iceland. J. Geophys. 47, 160–165.

Fiske, R.S., Jackson, E.D., 1972. Orientation and growth of Hawaiian volcanic rifts. Proc. R. Soc. London 329, 299–326.

Foulger, G.R., 2010. Plates and Plumes: A Geological Controversy. Wiley-Blackwell, 328 pp.

Galland, O., Hallot, E., Cobbold, P.R., Ruffet, G., de Bremond d'Ars, J., 2007. Volcanism in a compressional Andean setting: a structural and geochronological study of Tromen volcano (Neuquén province, Argentina). Tectonics, 26. doi: 10.1029/2006TC002011.

Girard, G., van Wyk de Vries, B., 2004. The Managua Graben and Las Sierras-Masaya volcano, a case of pull-apart localization by an intrusive complex. J. Volcanol. Geotherm. Res. 144 (1–4), 37–57, 15 June 2005.

Gonzales, G., Cembrano, J., Veloso, E.E., Shyu, B.H., 2009. Coeval compressional deformation and volcanism in the central Andes, case studies from northern Chile (23°S–24°S). Tectonics 28:TC6003. doi:10.1029/2009TC002538, 20.

Guillot, S., Pecher, A., Rochette, P., Le Fort, P., 1993. The emplacement of the Manaslu granite of Central Nepal: field and magnetic susceptibility constraints. Treloar, P.J., Searle, M.P. (Eds.), Himalayan Tectonics, 74, Geological Society Special Publication, pp. 413–428.

Hall, D., Kisters, A., 2016. Epsiodic granite accumulation and extraction from the mid-crust. J. Metamorphic Geol. 34:485–500.

Hardarson, B.S., Fitton, J.G., 1991. Increased mantle melting beneath Snaefellsjokull volcano during late Pleistocene deglaciation. Nature 353 (62–64).

Jerram, D., Mountney, N., Holzförster, F., Stollhofen, H., 1999. Internal stratigraphic relationships in the Etendeka Group in the Huab Basin, NW Namibia: understanding the onset of flood volcanism. J. Geodyn. 28, 393–418.

Jull, M., McKenzie, D., 1996. The effect of deglaciation on mantle melting beneath Iceland. J. Geophys. Res. 101, 21815–21828.

Klügel, A., Longpré, M.-A., García-Cañada, L., Stix, J., 2015. Deep intrusions, lateral magma transport and related uplift at ocean island volcanoes. Earth Planet. Sci. Lett. 431, 140–149.

Korn, H., Martin, H., 1955. The Messum igneous complex in South-West Africa. Trans. Geol. Soc. South Afr. 57, 83–124.

Le Corvec, N., Sporli, B., Rowland, J., Lindsay, J., 2013. Spatial distribution and alignments of volcanic centers: clues to the formation of monogenetic volcanic fields. Earth Sci. Rev. 124, 96–114.

Maccaferri, F., Rivalta, E., Keir, D., Acocella, V., 2014. Off-rift volcanism in rift zones determined by crustal unloading. Nat. Geosci. 7 (4), 297–300.

Maccaferri, F., Acocella, V., Rivalta, E., 2015. How the differential load induced by normal fault scarps controls the distribution of monogenic volcanism. Geophys. Res. Lett. 42, 7504–7512.

Magee, C., Bastow, Jackson, C., van Wyk de Vries, A.-L., Hetherington, B., Hagos, M., Hogett, M., 2017. Structure and dynamics of surface uplift induced by incremental sill emplacement. Geology 45 (5), 431–434.

McGovern, P.J., Grosfils, E.B., Galgana, G.A., Morgan, J.K., Rumpf, M.E., Smith, J.R., Zimbleman, J.R., 2014. Lithospheric flexure and volcano basal boundary conditions: keys to the structural evolution of large volcanic edifices on the terrestrial planets. Geol. Soc. London Special Paper 401:219–237.

Mitchell, T.M., Smith, S.A.F., Anders, M.H., Di Toro, G., Nielsen, S., Cavallo, A., Beard AD, 2015. Catastrophic emplacement of giant landslides aided by thermal decomposition: Heart Mountain. Wyoming Earth Planet. Sci. Lett. 411, 199–207.

Moores, E.M., 2003. A personal history of the ophiolite concept. In: Dilek, Newcomb (Eds.), Ophiolite Concept and the Evolution of Geologic Thought. Geological Society of America Special Publication 373:17–29.

Murray, J.B., 1988. The influence of loading by lavas on the siting of volcanic eruption vents on Mt Etna. J. Volcanol. Geotherm. Res. 35, 121–139.

Nakamura, K., 1977. Volcanoes as possible indicators of tectonic stress orientation—principle and proposal. J. Volcanol. Geotherm. Res. 2, 1–16.

Passchier, C.W., Zhang, J.S., Konopasek, J., 2005. Geometric aspects of synkinematic granite intrusion into a ductile shear zone – an example from the Yunmengshan core complex, northern China. Bruhn, D., Burlini, L. (Eds.), High-Strain Zones: Structure and Physical Properties. Geological Society, 245, London, Special Publications, pp. 65–80.

Robertson, E.A.M., Biggs, J., Cashman, K.V., Floyd, M.A., Vye-Brown, C., 2016. Influence of regional tectonics and pre-existing structures on the formation of elliptical calderas in the Kenyan Rift. Wright, T.J., Ayele, A., Ferguson, D.J., Kidane, T., Vye-Brown, C. (Eds.), Magmatic Rifting and Active Volcanism, 420, Geological Society, London, Special Publications, pp. 43–67.

Saballos, J.A., Malservisi Connor, C.D., La Femina, P.c., Wetmore, P.H., 2014. Gravity and geodesy of conception volcano, Nicaragua. Geological Society of America Special Papers 498:77–88.

Saleeby, J., Ducca, M., Clemens-Knott, D., 2004. Production and loss of high-density batholithic root, southern Sierra Nevada, California. Tectonics 22, 1064. doi: 10.1029/2002TC001374.

Schilling, F.R., Partzsch, Brasse, H., Schwarz, G., 1997. Partial melting below the magmatic arc in the central Andes deduced from geoelectromagnetic field experiments and laboratory data. Phys. Earth Planet. Interiors 103:17–31.

Sigmundsson, F., 2016. New insights into magma plumbing along rift systems from detailed observation of eruptive behaviours at Axial volcano. Geophys. Res. Lett. 43 (12), 423–427.

Slater, L., Jull, M., McKenzie, D., Gronvold, K., 1998. Deglaciation effects on mantle melting under Iceland: results from the northern volcanic zone. Earth Planet. Sci. Lett. 164, 151–164.

Smith, R.B., Michael Jordan, M., Bernhard Steinberger, B., Christine, M., Puskas, C.M., Jamie Farrell, J., Waite, G.P., Husen, S., Chang, Wu-Lung, O'Connell, R., 2009. Geodynamics of the Yellowstone hotspot and mantle plume: Seismic and GPS imaging, kinematics, and mantle flow. Journal of Volcanology and Geothermal Research 188, 26–56.

Spacapan, J.B., Galland, O., Planke, S., Leanza, H.A., 2016. Control of strike-slip fault on dyke emplacement and morphology. Journal of the Geological Societydoi: 10.1144/jgs2015-166.

Sternai, P., Jolivert, L., Menant, A., Gerya, T., 2014. Driving the upper plate surface deformation by slab rollback and mantle flow. Earth Planet. Sci. Lett. 405:110–118.

van Wyk de Vries, B., 1993. Tectonics and Magma Evolution of Nicaraguan Volcanic Systems. Open University, Milton Keynes, UK.

van Wyk de Vries, B., Merle, O., 1996. The effect of volcanic constructs on rift fault patterns. Geology 24 (7), 643–646.

van Wyk de Vries, B., Merle, O., 1998. Extension induced by volcanic loading in regional strike-slip zones. Geology 26, 983–986.

van Wyk de Vries, B., Self, S., Francis, P.W., Keszthelyi, L., 2001. A gravitational spreading origin for the Socompa debris avalanche. J. Volcanol. Geotherm. Res. 105, 225–247.

van Wyk de Vries, B., Marques, A., Herrera, R., Granjas, J.L., llanes, P., Delcamp, 2014. Craters of elevation revisited: forced folds, bulges and uplift of volcanoes. Bull. Volcanol. DOI: 10.1007/s00445-014-0875-x.

Walter, T.R., Troll, V.R., 2003. Experiments on rift zone evolution in unstable volcanic edifices. J. Volcanol. Geotherm. Res. 127, 107–120.

Williams, F.M., 2016. Understanding Ethiopia: Geology and Scenery. Springer, 346 pp.

FURTHER READING

Borgia, A., 1994. The dynamic basis of volcanic spreading. J. Geophys. Res. 99, 17791–17894.

Borgia, A., Burr, J., Montero, W., Morales, L.D., Alvarado GE, 1990. Fault-propagation folds induced by gravitational failure and slumping of the Central Costa Rica volcanic range: implications for large terrestrial and Martian volcanic edifices. J. Geophys. Res. 95, 14357–14382.

Borgia, A., Delaney, P.T., Denlinger, R.P., 2000. Spreading volcanoes. Annu. Rev. Earth Planet. Sci. 28, 3409–3412.

Galland, O., 2012. Experimental modelling of ground deformation associated with shallow magma intrusions. Earth Planet. Sci. Lett. 317–318, 145–156.

Maclennan, J., 2010. Nat. Geosci. 3:229–230. doi:10.1038/ngeo833.

Roberston, E.A.M., Biggs, J., Cashman, K.V., Floyd, M.A., Vye-Brown, C., 2015. Influence of regional tectonics and pro-existing structures on the formation of eliptical calderas in the Kenyan Rift, From Wright TJ et al., Magmatic Rifting and Active Volcanism. Geological Society of London Special Publication 420:43–67.

Weinberg, R.F., 1996. Ascent mechanisms of felsic magmas: news and views. Trans. R. Soc. Edinburgh Earth Sci. 87, 95–103.

CHAPTER 8

The Petrogenesis of Magmatic Systems: Using Igneous Textures to Understand Magmatic Processes

Dougal A. Jerram[*,,†], Katherine J. Dobson[‡], Dan J. Morgan[§], Matthew J. Pankhurst[¶,††,§,‡‡]**
[*]Centre for Earth Evolution and Dynamics (CEED), University of Oslo, Norway;
[**]DougalEARTH Ltd., Solihull, UK;
[†]Queensland University of Technology, Brisbane, Queensland, Australia;
[‡]Volcanology, Earth Sciences, Durham University, UK;
[§]Institute of Geophysics and Tectonics, University of Leeds, Leeds, UK;
[¶]University of Manchester, Manchester, UK;
[††]Rutherford Appleton Laboratories, Didcot, UK;
[‡‡]Instituto Tecnológico y de Energías Renovables, Tenerife, Spain

Contents

Volcanic and Igneous Plumbing Systems. http://dx.doi.org/10.1016/B978-0-12-809749-6.00008-X
Copyright © 2018 Elsevier Inc. All rights reserved.

8.1 INTRODUCTION

The means by which magma is transported from its birthplace in the mantle to its final resting place in the crust, or as eruptive products at the Earth's surface, is not simple. From the moment a magma has collected and starts to mobilise, it will encounter scenarios of crystallisation and has potential to come into contact with rocks alien to the conditions under which the magma formed (Marsh, 1996; Davidson et al., 2007a, b; Bachmann and Bergantz, 2008). There are also key parts, or zones, in the crust where certain conditions lead to the promotion of different magmatic processes above others. Fig. 8.1 highlights some of the scenarios facing magmas as they rise from a mantle source

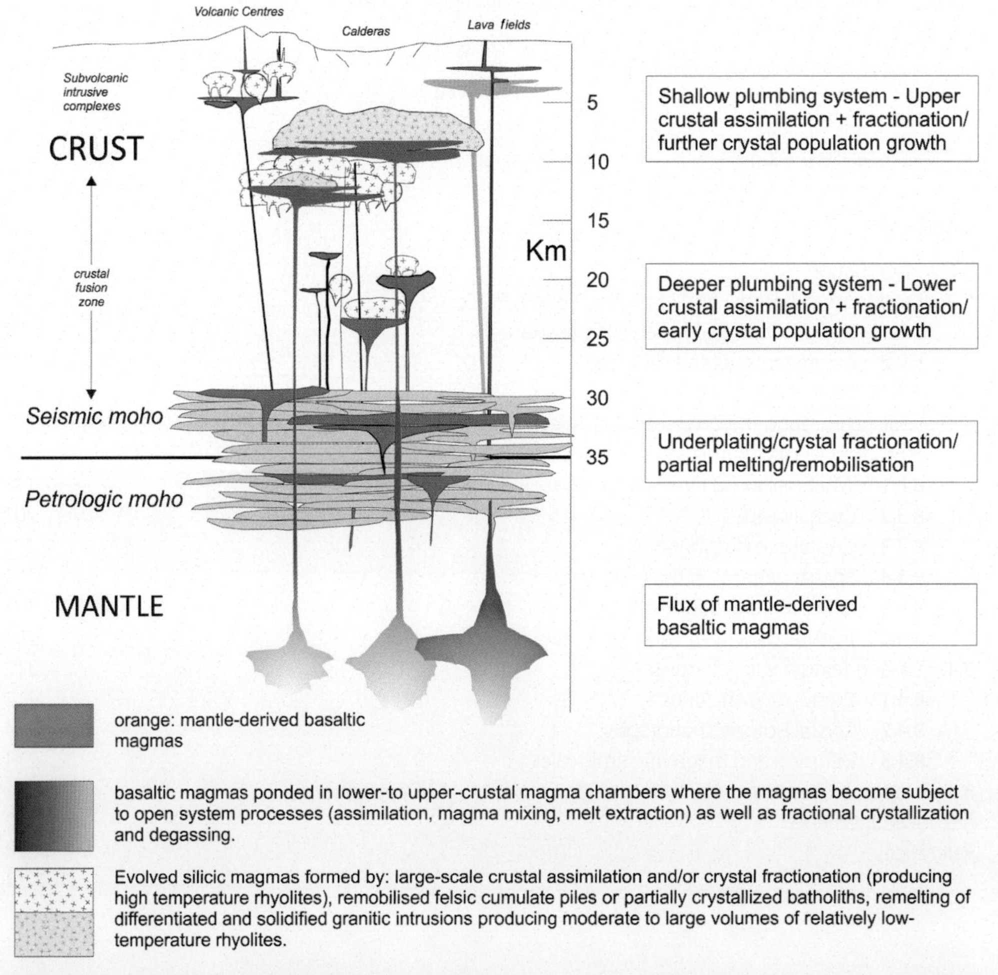

Figure 8.1 ***Schematic section through the volcanic and igneous plumbing system from the mantle to the Earth's surface.*** Adapted from Bryan et al. (2010).

area to the base of the crust and further migrate within the crust itself (Bryan et al., 2010; Jerram and Bryan, 2015). For example, the fast rise, emplacement and rapid cooling of a relatively primitive magma (e.g. as lavas or shallow small intrusions such as dykes) will potentially limit the possibilities of crustal contamination (Macdonald et al., 2010). By contrast, magmas that are able to pause within the crust for significant storage periods will undergo crystallisation, fractionation and assimilation. Furthermore, if subsequent batches of magma are introduced (see also Chapter 6), then magma mingling and mixing scenarios come into play (Martin et al., 2010).

An important aspect of the migration of magma through the crust is that at almost every stage there is some element of crystallisation taking place (Marsh, 1996). Providing that any products of this crystallisation are not lost during further ascent of the magma (e.g. through settling or floatation), crystals that are formed at these key stages within the magma petrogenesis will become part of the final rock texture. Crystals themselves can also be incorporated into relatively crystal-poor melts during the formation and migration of such melt through the magma plumbing system (Zellmer et al., 2014). The petrographic inspection of volcanic and plutonic rocks reveals a great variety of rock textures, reflecting a diverse range of histories that magma can experience (Jerram and Martin, 2008). Key information about plumbing systems and the journey that the magma has taken through the crust is often preserved in the crystals, or parts of crystals, that have been retained from the very earliest crystallisation events (Marsh, 2004; Davidson et al., 2007a).

By examining and quantifying the crystal populations within the final igneous rock, and the textural relationships between them, we can start to piece together the petrogenesis of the system and determine the characteristics of the main stages in the evolution of the magma from source to final resting place. Interrogating crystal histories, as a record of magmatic processes and conditions, is the only direct method of mapping out plumbing systems and their dynamic behaviour at the microscale. Piecing together these histories, as crystal-scale datasets reveal patterns and trends, is providing new insights to the fundamental igneous processes that have driven the physical and chemical differentiation of Earth and other rocky planetary bodies.

In this chapter, we will explore how we can shed light on magma pathways from looking closely at their preserved crystal populations. Many of the key magmatic processes, such as magma mingling and mixing, assimilation, transport and emplacement can be inferred/understood using the information preserved within the rocks. Firstly, we will briefly consider the context of the volcanic and igneous plumbing system and some of the key processes that are happening. Then, we will introduce the main components of a magmatic crystal population, as different aspects of these will be useful in constraining different processes. The rest of the chapter is set out to touch on some of the state-of-the-art methodologies of quantifying rock textures to help understand magmatic processes. This will by no means cover every aspect of the petrogenesis of the system, but rather provide a modern petrographic flavour of how it is possible to go about understanding the rock textures to better constrain magmatic processes.

8.2 KEY MAGMA CHAMBER AND PLUMBING SYSTEM PROCESSES

With the potential for the magma system to be highly complex in terms of its plumbing, it is useful to take a step back and consider a relatively simple magma chamber scenario. In the past, conceptual magma chambers were often stylised using over-simplistic balloon-type geometry with a spherical ball of magma residing within the crust (see also Chapter 1). In more recent times, and as data and understanding has grown, efforts have been developing more realistic illustrations of what have classically been called 'magma chambers' (Jerram and Bryan, 2015). When we find good examples of (fossil) magma bodies in the field, their internal structures have been dissected and can be explored, which can reveal that even the apparently most simple magma chambers must have been highly dynamic (Fig. 8.2; cf. Chapter 6) (Hunter, 1996; O'Driscoll et al., 2007a; Goodenough et al., 2008). In many areas, such is in arc settings, large zones of crystal-rich mushes may also act as important zones of transfer of crystals and melt (Marsh, 1996; Bachmann and Bergantz, 2008).

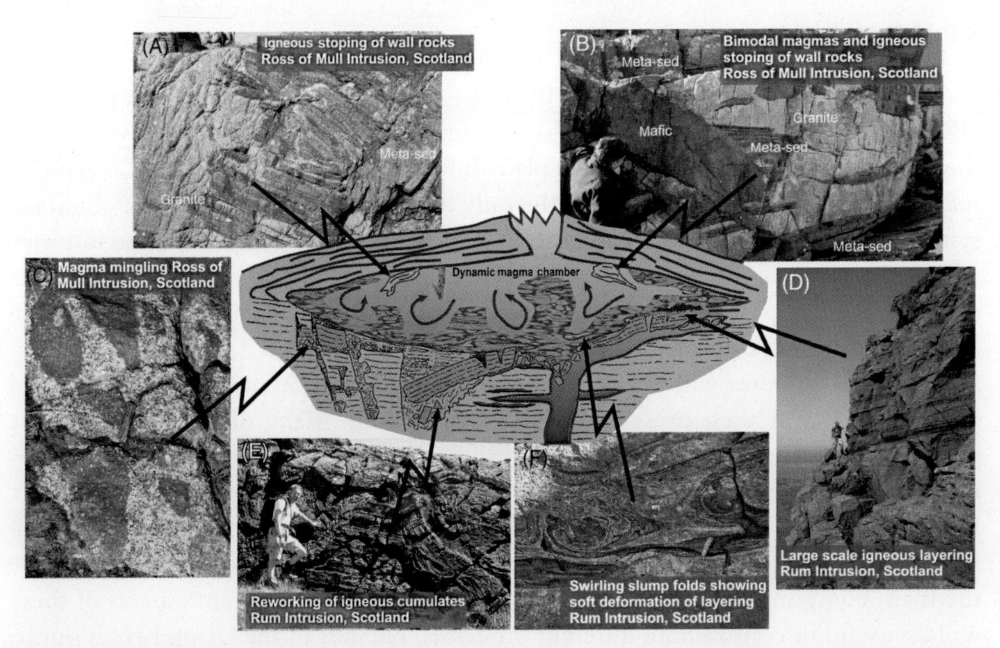

Figure 8.2 ***The dynamic magma system where new batches of magma mingle and mix with old, crystals settle to form igneous layering and magmatic stoping of roof and wall rocks (e.g. meta-sedimentary rocks) leads to crustal assimilation.***

Replenishment with new magma batches, as well as interstitial melts being expelled and mobilised to form magma reservoirs, help to set up periods of magma migration and eruption (Bachmann and Bergantz, 2003, 2004). Some of the dynamic processes which are preserved within plutonic bodies shed light on the main processes occurring within the volcanic and igneous plumbing system from source to surface. We will briefly consider three key observations from magmatic centres, which touch on the main processes that are happening.

8.2.1 Igneous Layering and Crystal Transport

Inspection of many exposed intrusive bodies reveals layered structures and banding associated with the modal distribution of the crystals that form the rock. Classic examples of these are found in intrusions such as Rum in Scotland (Fig. 8.2D), the Skaergaard intrusion in Greenland and the Bushveld intrusion in South Africa. Although such layering is more prominent within mafic intrusions, it can also be observed within many granitic bodies. What do these planar macro-textures represent in terms of igneous processes, and how might they also be recorded at the microscale?

The first thing to take from such rocks is the idea that crystals can be separated from melt. This touches on the first main process of fractional crystallisation, whereby the composition of the melt changes as certain elements partition into solid crystals, leaving behind in the melt enriched concentrations of those elements that are incompatible in that crystal's structure. Secondly, the nature of many igneous layering features, such as slumping and cross-bedding, as well as remobilisation of cumulate mushes (Fig. 8.2E) indicate that the crystals themselves can also be transported sometime after they were formed and deposited, with features indicating transport in ways analogous to those seen in sedimentary systems. If we scale this to the whole magmatic system, it suggests that crystals can be removed and transported from the environment, in which they first grew, and that this may be the rule, rather than the exception.

8.2.2 Magma Mixing and Mingling

The next, and often most striking, textural observation within large magma bodies is the juxtaposition of different magma types. This is usually observed between darker (more mafic) and lighter (more felsic) patches that are in contact with each other (Fig. 8.2C). Sometimes the contacts are sharp and angular, sometimes more rounded or intricate and sometimes they show diffuse variations between two or more different types. Some additional examples of magma contact relationships are given in Fig. 8.3.

The identification of different batches of coeval magmas interacting is hugely important when considering a magma system. It may be possible to clearly see more primitive, mafic magmas within more evolved host magmas by the identification of inclusions termed *enclaves* (Fig. 8.3A–C). Additionally, textures such as back-veining and quench breccias may also indicate multiple magmatic events with magmas of different temperatures and rheology (Fig. 8.3D). In other instances, rocks may be seen to have become hybridised; although if hybridisation is near-complete, it may not be clear in the outcrop that any magma interaction has occurred at all (Streck et al., 2007).

In this latter case, detailed interrogation of the crystal populations present can reveal previous magma mixing events that have resulted in a hybrid magma. We can see from field evidence that it is possible for crystals to be exchanged between magmas of different types (Fig. 8.3E), and magmas that are also starting to mix together (Fig. 8.3F). If this process can happen once, it is also conceivable that it may occur multiple times and in both directions within a complex magma plumbing system. If crystals continue to grow during and after such transitions, then the internal zoning, composition and even isotopic signature (Pankhurst et al., 2011) within the crystals become vitally important as a record of possible mixing events. This is where the understanding of the concepts of phenocryst, antecryst and xenocryst, and what they may reveal about magmatic histories, is of particular value.

8.2.3 Magmatic Stoping, Contact Reactions and Assimilation

The final concept to consider before we look in detail at magma textures is the interaction between magma and surrounding host rocks. Once magma has been generated and has left the immediate environment of its formation, it can start to react with any new rocks it comes into contact with. In some instances, these will be rocks that have been formed by the same magma system and, due to their similar genesis and composition will have little observable effect on the developing magma. In other instances, the magma will come into contact with host rocks of a completely different affinity. These can be everything from deep-seated crustal and/or mantle rocks, which may already be hot and ductile, to shallow sedimentary sequences within a volcanic basin.

What can initially appear to be a simple magma–host rock relationship in the field, within an intrusion or an erupted lava, may disguise that any interaction has happened at all. In other examples, it may be possible to clearly see remnants of country rock within the magma known as *xenoliths*. In larger intrusive bodies, it is possible to find large blocks of country rock (of scales up to hundreds of metres or even larger), seemingly having fallen into, and had begun to be consumed by, the magma (Fig. 8.2A and B). This process is known as *magmatic stoping*, and this is likely to be happening at one scale or another when magma pauses in reservoirs en route towards the surface (see also Chapter 2). *Stoping* and *xenoliths* provide clear evidence of magma and host rock interaction.

At a finer scale within such examples, partial melting can sometimes be observed at and around the contacts (Fig. 8.3G), and which rocks are contributing to the melt

fragments within a more felsic host, formed by magma mingling in a shallow sheet intrusion, Ardnamurcan, Scotland. (E) Large feldspar crystal being exchanged between a granitic to a more mafic magma, Elba, Italy. (F) Streaky pumice from Mt Lassen, USA. (G) Granite melt, and streaks of granitic melts forming in partially molten and ductily deforming meta-sediment block incorporated within intrusions in the Ross of Mull, Scotland. (H) Flattened and angular metasedimentary xenoliths of Jurassic shale in host granite, Cordillera Blanca, Peru (photos A, B and H courtesy of N. Petford).

Figure 8.3 ***Examples of magma interactions.*** (A) Magma mingling with 'pillows' of micro-diorite with crenulated and bulbous chilled margins within pink granite host, Jersey, Channel Islands. (B) Mafic enclaves with complex dendritic textures within granite resulting from 'viscous fingering' (disruption of the mafic fluid within a non-Newtonian granitic host), Maine, USA. (C) Syn-magmatic dyke and enclaves within a giant porphyry granite, Elba, Italy. (D) Intrusion breccia showing more mafic (dark) angular

through assimilation can be observed directly (Fig. 8.3H). Once the melt is removed from such environments and/or the direct evidence of the host rock has gone, then proxies such as isotope signatures among others will need to be used to spot the assimilation that has taken place. Here, some knowledge of the key chemical features of the various host rocks at depth is required to spot any incorporation of this material within the melt. Additionally, knowledge of the chemical signatures of different melt sources that may also be contributing to an igneous plumbing system will further help to reconcile assimilation versus magma mixing scenarios.

8.3 LOOKING THROUGH THE CRYSTAL BALL—UNDERSTANDING THE MAGMATIC PROCESSES FROM YOUR CRYSTAL POPULATION USING TEXTURAL ANALYSIS

Igneous rocks can contain complex internal structures. This is because magmas or lavas rarely arrive at their final destination without collecting a payload of crystals, the 'crystal cargo'. This crystal cargo can comprise a whole mixture of minerals, including those crystallising directly from the melt, those crystals that have been incorporated into the magma during ascent and mixed crystal populations from crystallisation events occurring as magma ascent pauses in magma chambers within a complex magma plumbing system (Wallace and Bergantz, 2002; Turner et al., 2003; Davidson et al., 2007a, b; Martin et al., 2010). In this context, the processes of assimilation, fractional crystallisation, magma mingling/mixing and the transport of crystals through the plumbing system will be imprinted on the final igneous textures and structures. In the rarer examples where a crystal-free magma is emplaced, it tends to cool with solidification–crystallisation fronts without significant development of large crystals, as seen in some sill complexes and lava sequences (Marsh, 2004). In such rare examples, whole-rock chemical information may be the only way to interrogate the magma's past. Before we consider different ways to quantify our igneous textures to help understand what magmatic processes have formed them, it is important to briefly consider the different types of crystals that make up the final crystal population.

8.3.1 What Makes up the Crystal Cargo?

As crystals form through nucleation and growth from a cooling magma, one might expect a fairly simple, or at least systematic, relationship of a magma being emplaced, cooling and crystallising to a solid. Evidence suggests, however, that magma plumbing systems can be highly complex, with multiple episodes of cooling and crystallisation, magma and crystal recycling, new magma batches mingling/mixing and complex storage zones (Figs. 8.1 and 8.2), such that the final crystal population is a mixture that contains the magma journey to its final rock. There are four main components that can be recognised: phenocrysts, xenocrysts, antecrysts and microlites (Jerram and Martin, 2008) (see Fig. 8.4). In any one case, not all four will necessarily be present, and/or it may be difficult initially to recognise such components in intrusions where significant post-emplacement growth and recrystallisation can take place.

Phenocrysts are crystallised directly from a magma, and for this reason are possibly the most intuitive to conceive. They often are shown to form comparatively large, euhedral crystals that appear, suspended in a finer grained groundmass (Gill et al., 2006). As such, phenocrysts must be in chemical equilibrium with the melt along a cooling and/or depressurisation trend toward final solidification (Fig. 8.4). They may show some fluctuation in zoning (e.g. in feldspars, XAn (Anorthite content) variations due to changes in the local crystallising conditions within the body as different minerals precipitate), but will be in isotopic equilibrium with the melt.

Xenocrysts are crystals that are foreign to the magma and to the magma system as a whole, which have been incorporated into the magma at some point during ascent by a physical process. These processes might include assimilation and stoping of country rock,

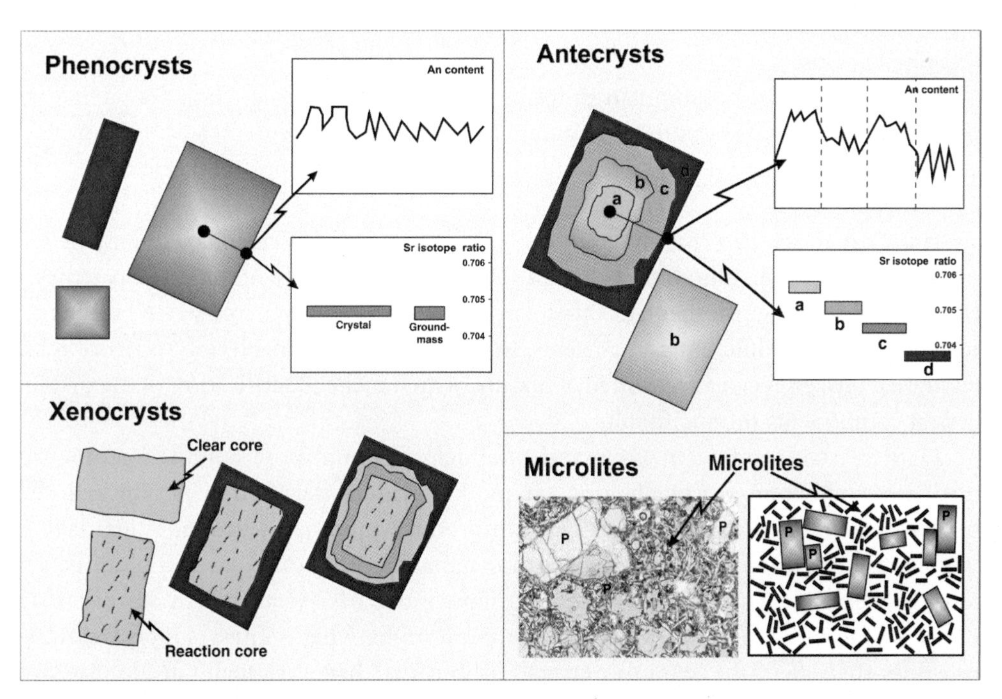

Figure 8.4 ***The four components of crystal populations, 'the crystal cargo', comprise phenocrysts, antecrysts, xenocrysts and microlites.*** Adapted from Jerram and Martin, 2008. See main text for discussions.

with individual crystals being liberated. They often contain new overgrowths or reaction rims as a consequence of interaction with the magma (Fig. 8.4B).

Antecrysts are derived from the active magma plumbing system, but have been recycled through various events (e.g. different magma storage zones, earlier crystallisation events). While they have a common genetic heritage to the magma in which they are found, their ultimate origin may extend significantly back in time and thus they record information relating to these earlier events that pre-date the formation of the carrier magma (Charlier et al., 2005; Davidson et al., 2007a; Martin et al., 2010).

In some ways analogous to the growth rings on a tree, growth patterns found in antecrysts can identify and inform our view of key stages in the evolution of the magma system as a whole (Fig. 8.4). This is often evident by studying the isotopic character of the crystals from core to rim (Davidson et al., 2007a, b). Antecrysts differ from xenocrysts in that they are a direct product of the currently active magma system. It can be difficult to separate antecrysts and xenocrysts where xenocrysts are being incorporated from older wall rocks of the volcanic edifice or solidified plutonic rocks, which may have a similar chemistry (Gill et al., 2006). Xenocrysts that are incorporated from unrelated rocks (e.g. basement rocks) are more readily identified. Equally, it can be hard to separate antecrysts and phenocrysts if the processes of crystal formation are very similar but occur widely separated in time.

Finally, and most common in eruptive products, are *microlites* (Fig. 8.4). These are commonly only a few hundred micrometres in size, typically with an acicular morphology, and form during the final degassing of the magma on eruption (Castro et al., 2003; Couch et al., 2003).

It is clear to see how the different crystal components described above may be recognised in a volcanic plumbing system. In most instances, there are textural parameters that can help distinguish them; in other cases, geochemical proxies may need to be used to help define the different components. In contrast, when considering intrusive rocks, additional processes can affect the final texture, which make identification of the original crystal components more difficult.

Here, crystals most often grow and interlock, forming more complex geometries; further crystallisation/recrystallisation can modify original sizes and shapes through textural coarsening and equilibration. This can produce modified textures, by partially or wholly resetting those formed by original magmatic processes. In layered plutonic systems, for example, cumulates represent accumulations of crystals which are often significantly modified in both texture and composition by what is known as post-cumulus processes that affect the 'cumulus crystals' (those that have originally accumulated or crystallised first).

The term *orthocumulate* describes cumulus crystals that are typically normally zoned (i.e. the rim reflects a more evolved melt composition than that of the core of the same grain) and are surrounded by new mineral phases which are nucleated from the intercumulus liquid (sometimes showing *poikilitic* relationships, where later minerals enclose grains that grew earlier). *Adcumulate* is a term that refers to texturally equilibrated examples where unzoned cumulus crystals lack any intercumulus phases. Here the rock can be made up almost entirely of one mineral phase. A *mesocumulate* texture is somewhat intermediate between that of an orthocumulate and an adcumulate: there will be some intercumulus phases present, and the cumulus crystals might show limited zoning. Such modifications of textures can be produced through a variety of processes including coarsening, creep and compaction (Hunter, 1996; Higgins, 1998).

The first stage in helping to understand and use crystal populations to get at magmatic processes is to characterise the crystal population and its context within the sample as a whole. Fundamental petrographic microscope observations should form the backbone of a study of rock textures in thin section, or at least a detailed field appraisal of mineral relationships in coarse-grained rocks from good outcrops or polished slabs. Once a number of key observations have been recorded, it is possible to quantify some of the key parameters of the texture (e.g. crystal shape, size, spatial pattern), which will not only help interpret some of the magmatic processes that were in action, but also will underpin chemical studies as part of further detailed analysis.

Characterising crystal populations has been significantly aided by analytical computer techniques, such that the size, shape and spatial distribution of a crystal population can be ascertained from a digitised 2D slice through the crystal population (Fig. 8.5). In 2D studies, the shape and size can be determined using the long- and short-axis of the crystal and its location for spatial analysis is determined by its *xy* crystal centre (Jerram et al., 2003; Higgins, 2006). As long as a statistically valid number of crystals within a sample area (>200–300) is used (Mock and Jerram, 2005; Morgan and Jerram, 2006), then such information, even from a single sample, provides a wealth of information and can act as a guide to system-scale processes. This can also be done before considering the chemical information that is stored within the constituent crystals and act as a basis to help target geochemical studies (Jerram and Davidson, 2007; Jerram and Martin, 2008).

8.3.2 Crystal Shapes

One of the most striking things about a crystal is its *morphology*. The crystal shape of simple crystal forms in 3D can be defined by the crystal's long (L), intermediate (I) and short axis (S). This can be displayed using a zing diagram (Jerram and Higgins, 2007) as a ratio where S = 1, which contains the shapes from equant to prolate(acicular) and tabular (Fig. 8.5). You can determine a 3D crystal shape from 2D crystal sections (e.g. thin section or polished/cut rock slab if the grain size is large), providing the 3D shape of all the crystals is the same using the width–length information from the 2D data (Higgins, 1994; Morgan and Jerram, 2006). To help aid this, Morgan and Jerram (2006) produced a database which can be used to statistically compare 2D shape measurements

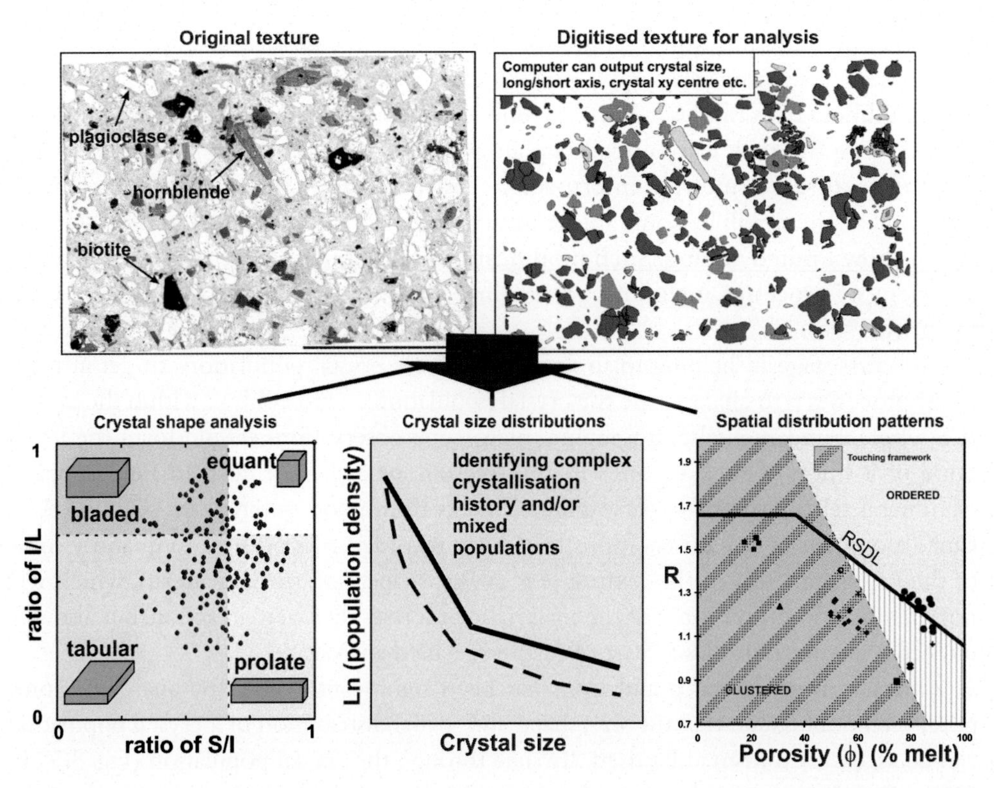

Figure 8.5 ***Textural analysis of igneous rocks has benefited from computer measurements of digitised textures. R, spatial distribution pattern; RSDL, random sphere distribution line (see Jerram and Martin, 2008 for detailed explanation).*** Data, such as crystal shape, size and spatial distribution, can now be routinely recorded from sections through igneous rocks, helping to reveal their petrogenesis.

to the best-fit 3D shape (database of sectioned reference cuboids). The estimated 3D L:I:S ratios are used to convert 2D crystal size measurements into predicted 3D crystal size distributions (CSDs; Higgins, 2006) (see section 8.3.3).

Why is shape important? If the shape can be determined accurately, this morphology helps to determine the conditions under which those crystals grew. Studies by Lofgren (1974, 1980) showed that at different degrees of undercooling, plagioclase morphology ranges in shape from acicular, to tabular, to equant. Experiments on olivine crystallisation also produce different morphologies at varying degrees of undercooling (Donaldson, 1976). Importantly, such shape parameters can be found in nature; feldspar morphologies in volcanic rocks (Mock and Jerram, 2005) and olivine cumulates in layered intrusions (O'Driscoll et al., 2007a) display the same range of shapes produced in the laboratory. It may be the case that different population components (i.e. phenocrysts, xenocrysts and antecrysts) have different morphologies originating from differing growth conditions. Studies of 3D crystal populations have shown that shape can vary significantly (Mock and Jerram, 2005), highlighting that mixed shape populations exist, and so further data sets of 3D of crystal shape may help constrain the role of crystal shape within evolving crystal populations (Jerram and Higgins, 2007).

8.3.3 Crystal Size Distributions

One of the most commonly measured textural parameters used in igneous petrology is that of crystal size (Cashman and Marsh, 1988; Marsh, 1988; Higgins, 2000). The CSD (also termed CSD analysis) is the amount of crystals of a given size within a population, recorded as the number of crystals per unit volume (a plot of population density vs. size; e.g. Fig. 8.6). The CSD of a sample can provide important information about nucleation and growth parameters (Cashman and Marsh, 1988; Marsh, 1988). The slope of the CSD plot is a function of the growth rate and residence time of the crystals. If the crystal growth rates are well constrained, timescales within magma systems can be investigated (or vice versa) (Cashman and Marsh, 1988; Higgins, 1996; Jerram et al., 2003). The shape of the CSD curve can also be used not only to identify crystallisation parameters, but may also imply physical processes, such as mixing of crystal populations or coarsening/Ostwald ripening, have affected the resultant population (Higgins, 1998; Marsh, 1998; Turner et al., 2003; Jerram et al., 2009) (Fig. 8.6C).

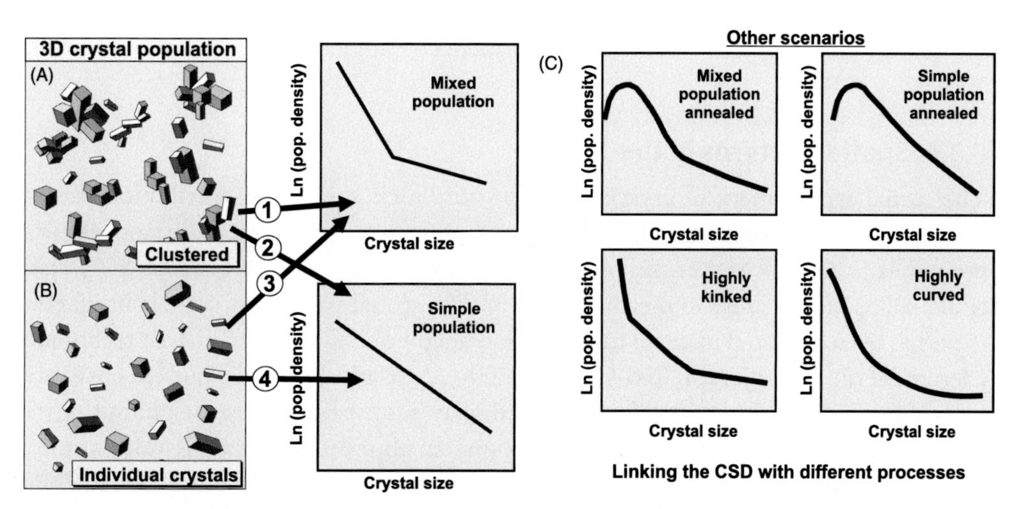

Figure 8.6 ***Example of CSD plots.*** (A) A theoretical clustered population of crystals may represent a mixed population of crystals (1) or may actually have formed from a simple population (2). (B) What may appear as a simple population of individual crystals (e.g. 4) on analysis and quantification may actually be shown to be more complex (e.g. 3). (C) Idealistic examples of how some variations in CSD plots may look. Abbreviation: *CSD*, crystal size distribution.

CSD data are commonly measured from 2D sections. The CSD itself is a calculation of number per unit volume, and so the final 3D CSD is calculated by using stereological conversions from 2D to 3D (Higgins, 2000, 2006). A handy program (*CSD corrections*) has been developed (see Higgins, 2000, 2006) which, given information on crystal shape, roundness and texture fabric, as well as raw textural data, calculates the 3D CSD of the sample. More recently, directly measured true 3D CSD measurements have been investigated, which both negates the need for 2D–3D conversions, but can also be used to test how good the conversion techniques are (Mock and Jerram, 2005; Jerram and Higgins, 2007). These studies use either serial sectioning/reconstruction techniques (Mock and Jerram, 2005) or X-ray CT analysis (Gualda, 2006; Jerram et al., 2009). There are further ways to explore CSD data using other plots (Higgins, 2002, 2006), and where examination of related samples show trends and variations.

Examples of CSDs used in igneous systems are given in Fig. 8.7. CSDs from the Emeishan province, as an example, which were used to understand the petrogenesis of giant plagioclase lava flows and shallow intrusions (Fig. 8.7A and B). These show remarkably simple profiles suggesting that the pagioclase megacrysts had grown in a static magma chamber for about hundreds to thousands of years (Cheng et al., 2014a, b). CSD patterns within a detailed section through the Poyi layered intrusion in China, as another example, show textural coarsening and crystal accumulation relationships (Fig. 8.7C and D). Specific CSD studies, such as these, where rock sequences can be compared up-section highlight CSD analysis as a very useful technique to constrain the texture development within a rock sequence (Higgins, 1998; Jerram et al., 2003; Yao et al., 2017).

8.3.4 Spatial Patterns in the Texture

The spatial arrangement of crystals or grains in a rock, the spatial distribution pattern (SDP), is another parameter that can have significance to rock texture and physical properties (e.g. porosity/permeability), crystal frameworks and melt rheology (Jerram et al., 2003; Moitra and Gonnermann, 2015). The spatial pattern of crystals in igneous rocks has been explored in detail using nearest-neighbour and cluster analysis techniques (Jerram et al., 1996; Jerram and Cheadle, 2000). A relatively simple, but effective, way to explore the spatial patterns that crystals display is to use a nearest-neighbour distribution analysis. This method plots porosity (modal abundance) versus R value, a measure of spatial arrangement based on grain centre coordinates (after Jerram et al., 1996). In this technique, the SDP pattern of randomly packed distributions of equal-sized spheres is used as a reference, as well as the SDPs of a number of known 3D textures. These provide reference points (e.g. random sphere distribution line), which allow the context of textures to be displayed in terms of a random, ordered or clustered distribution (Fig. 8.8A). It also allows an assessment of whether crystals are characterised

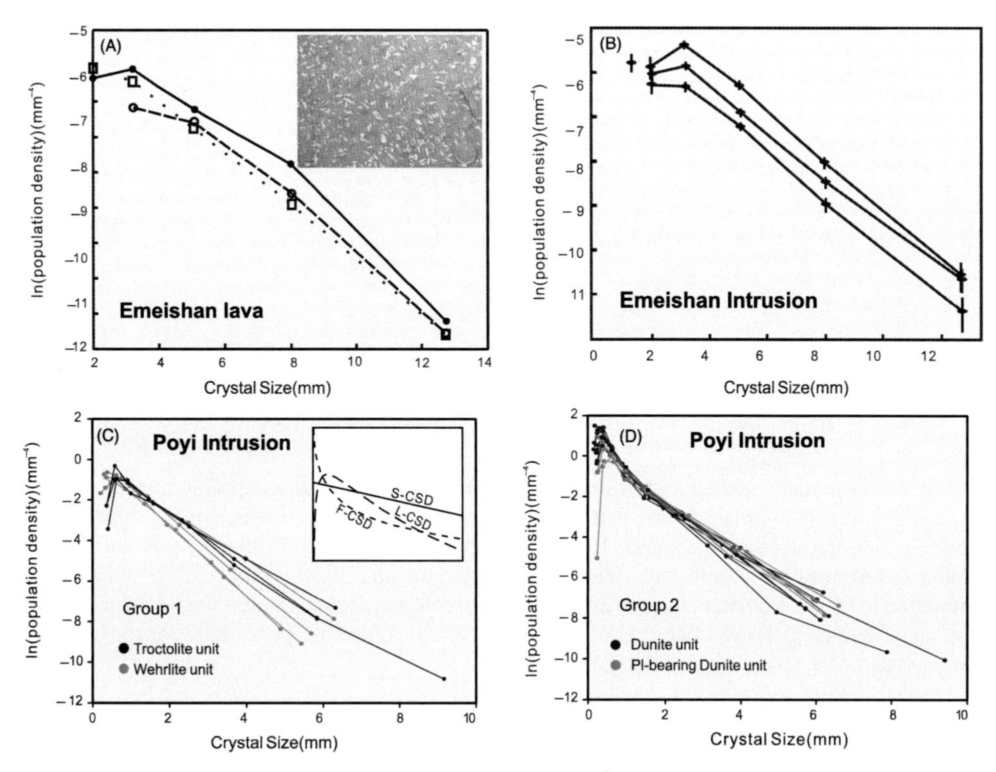

Figure 8.7 ***Examples of CSDs in igneous systems.*** (A, B) Simple CSD plots from the Emeishan province, China, showing examples from giant plagioclase basaltic rocks (inset example of giant-plagioclase lava flow) (see Cheng et al., 2014a, b). (C, D) CSD plots from olivine populations measured down an extensive borehole through the Poyi intrusion, China, with fanning and concave shapes (inset shows representative trends for: simple semi-logarithmic S-CD, logarithmic L-CSD and fragmental F-CSD) (see Yao et al., 2017 for more details).

by a touching framework or not (Jerram et al., 1996, 2003; Mock et al., 2003). This technique can be particularly useful when comparing adjacent samples in section/core (Mock et al., 2003; Yao et al., 2017), and where sample volumes are limited as in lunar and meteorite samples (Day and Taylor, 2007).

Textures in igneous rocks often show clustered crystals and crystal frameworks (Philpotts and Dickson, 2000; Jerram et al., 2003). After primary crystal frameworks are set up, the spatial patterns they form can then change with mechanical compaction (Jerram et al., 1996) and have been used to explore igneous processes in cumulates (Jerram et al., 2003; Yao et al., 2017), and during expulsion of melt from crystal mushes (Špillar and Dolejš, 2015). Yao et al. (2017), for example, showed how olivine packing and shape distributions varied within an extremely well-sampled ultramafic intrusion in China (Fig. 8.8). Again, the value of such data comes into its own when comparing within-sequence rocks, where the textural variation between samples is revealed by the textural analysis.

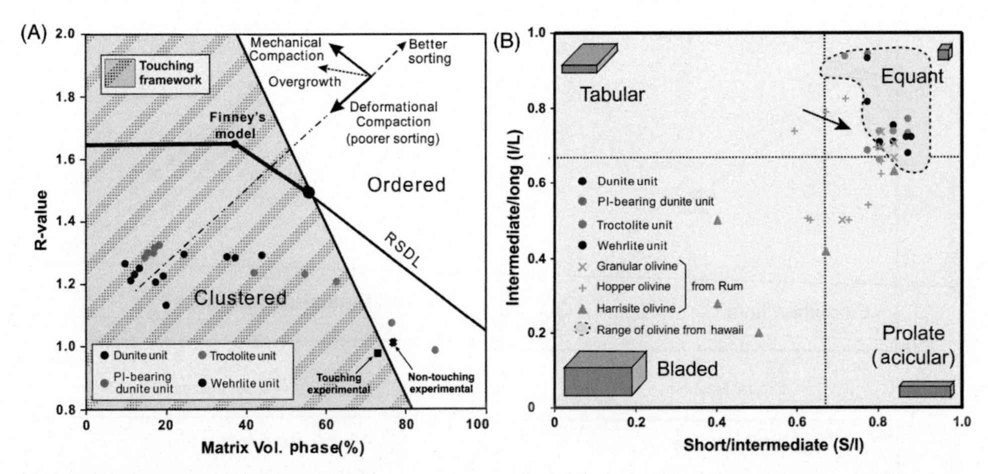

Figure 8.8 ***Example of spatial patterns and shapes from the Poyi Intrusion in China.*** (A) SDPs showing a wide range of internal packing within the intrusion. Highlighted are the zones for touching and non-touching frameworks and trends for certain processes including compaction, overgrowth and grain size sorting (after Jerram et al., 1996, 2003). (B) Zing diagram showing shape of intrusion olivines compared to ranges from Rum layered intrusion (O'Driscoll et al., 2007a), kimberlite (Moss et al., 2010), and Kilauea Volcano, Hawaii (Vinet and Higgins, 2010, 2011) (see Yao et al., 2017 for more details). Abbreviation: *SDP*, spatial distribution pattern.

8.3.5 Other Textural Data

A number of other techniques can reveal more levels of information at the grain and crystal population scales. Electron backscatter diffraction (EBSD) (Prior et al., 1999) is a technique that quantifies crystallographic orientation. This technique is often of great value to mapping out fabrics (for instance, flow indicators), or proving a lack thereof, within crystal populations, as well as help to determine grain boundaries in rocks that are almost entirely monomineralic (such as addcumulates). The application of EBSD to textures in layered igneous rocks where significant postcumulus processes have occurred is of particular use (Vukmanovic et al., 2013).

Additional methods can be employed to investigate three-dimensional fabrics within crystal populations. Orthogonal sections or grain orientation data provide the basis for the intersection method of Launeau and Cruden (1998), and the use of cumulative distribution functions by Gee et al. (2004). The preferred orientation of iron-rich minerals within rocks can be measured using anisotropy of magnetic susceptibility (AMS). In studies of igneous rocks, AMS is used to determine the shape, strength and orientation of the magnetic fabric in a sample and has been applied to intrusions to gain insights into emplacement fabrics (O'Driscoll et al., 2007b; Stevenson and Grove, 2014).

The degree of textural equilibration that has taken place, particularly in plutonic rocks (e.g. layered intrusions/crystal mushes), is also of interest since this is a record of cooling history. In this context, the angle between adjacent crystals, known as the dihedral angle, can be used (Elliott et al., 1997; Holness et al., 2005a). Here, the angle between crystals varies from ~60° in non-equilibrated textures to around ~30° in an equilibrated pore structure (Elliott et al., 1997; Holness et al., 2005a). Dihedral angle studies have been used to investigate the late-stage textural evolution and cooling of large intrusive igneous bodies (e.g. Rum, Skaergaard; Holness et al., 2007), as well as to look into the textures of clustered crystals brought up in volcanic rocks (Jerram et al., 2003; Holness et al., 2005b).

8.3.6 Textural Data from 3D X-Ray Tomography Studies

Until now, we have considered information derived from 2D sections through the rocks, where we have to extrapolate, or perform some sort of stereoscopic correction to approximate 3D information. While every textural study starts with careful 3D observation of outcrops and hand samples (which often help decide how to cut up a sample to reveal the most representative 2D perspective to analyse or perhaps bring out a certain feature), what if we could quantify the whole 3D texture and crystal population? Looking inside rocks in 3D, without damaging them, using X-ray micro-computed tomography (XMT[1]) is proving extremely useful in the geosciences (Cnudde and Boone, 2013), and especially so for textural analysis. X-ray tomography allows for visualisation and quantification of surfaces, volumes, textures and even chemical composition (Pankhurst et al., 2014). Many recent advances have been driven by the need to determine sample-specific porosity and permeability, or grain shape/size analysis in modern sedimentary petrology (Golab et al., 2010; Feali et al., 2012; Iglauer et al., 2013; Dobson et al., 2016). However, the methods are also proving indispensable in palaeontology (Monnet et al., 2009), soil science (Tracy et al., 2010), volcanology (Ersoy et al., 2010; Hughes et al., 2017), igneous petrology (Jerram et al., 2009, 2010; Pankhurst et al., 2014) and metamorphic petrology (Müller et al., 2009; Dobson et al., 2017).

So, how does it work? The way the X-rays are generated depends on if we are performing laboratory or synchrotron XMT. In almost all geoscience applications primarily for textural analysis, laboratory XMT is at least adequate and often advantageous. Here, an electron beam is fired at a metal target to generate a point source X-ray beam, which then fans out, passing through the sample before reaching a scintillator (Fig. 8.9). The scintillator converts the X-rays into light photons, which are detected by (usually) either a CMOS or CCD camera. By rotating the sample and collecting 2D *projections* or *radiographs* from different angles, a 3D image can be reconstructed using numerical algorithms. Filtered back-projection approaches are widely used and are mathematically similar to methods used to perform seismic data inversions (see Pan et al., 2009 for more information). A higher number of projections results in higher spatial accuracy; in practice, several hundred to a few thousand projections are normally used.

[1] This is also abbreviated to 'μCT' or 'XCT', but as several other computed tomography methods are also used in geosciences, we recommend the use of XMT do identify the use of X-rays.

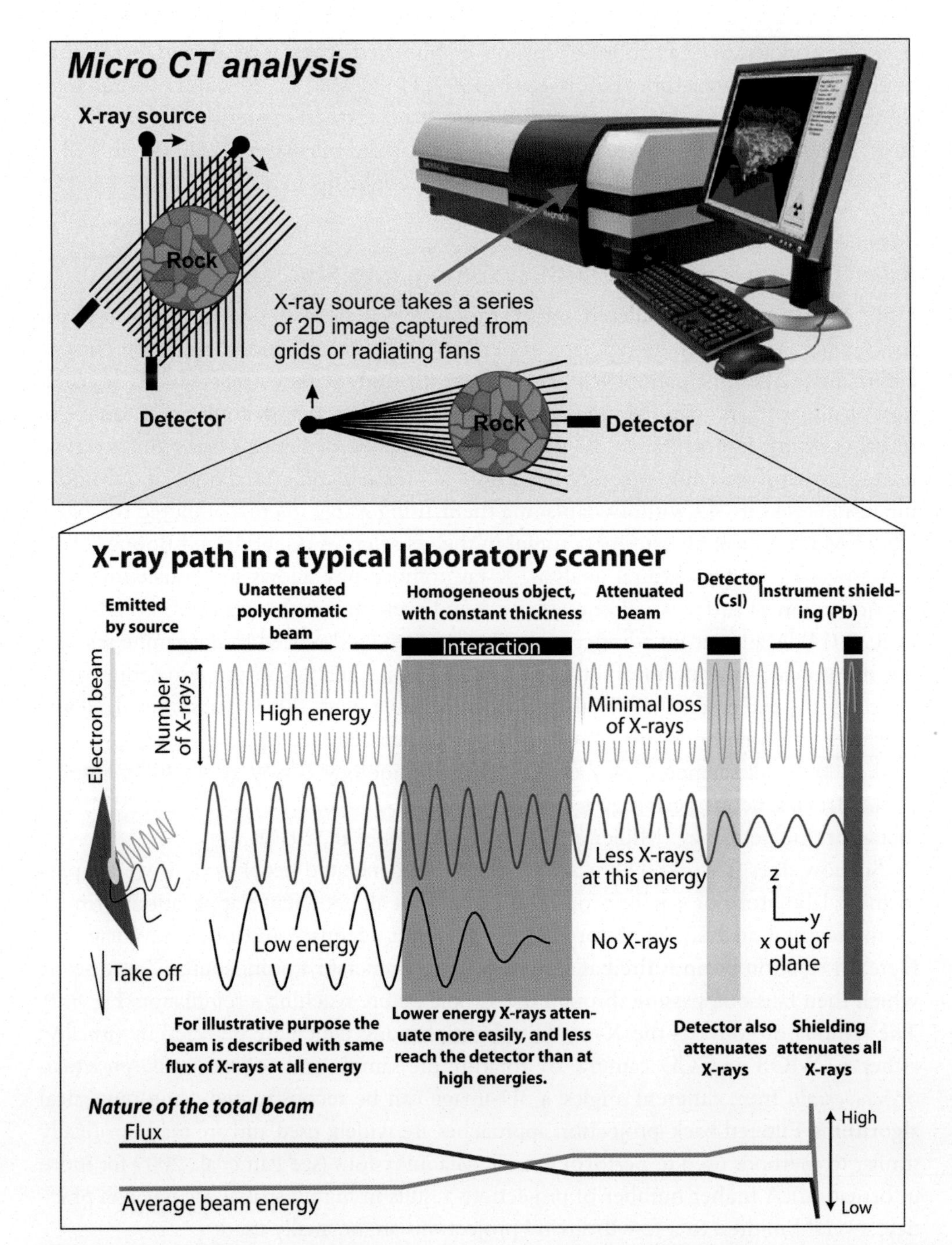

Figure 8.9 ***A small desktop ct system and how the sample and X-ray beam interact.***

Crucially, the number of photons that hit each pixel of the camera is controlled by the material(s) the beam has passed through on its path from the source to the detector, and the energy of the X-rays themselves. The higher the electron density of a material, the more it will absorb X-rays, meaning less reaches the scintillator. Different minerals often have different electron densities, and so we record a spatially variable intensity when they hit the detector (Fig. 8.9). The result is a grey-scale image volume where the greyscale value essentially corresponds to the average mineral density in the voxel (3D equivalent to a pixel) at that location. Spatial resolution of scanning systems is typically 3–30 μm (per voxel side length), for samples from 5 mm to 3 cm in diameter, and some specialised systems can image with 50-nm resolution. Like most microscopy methods, the size of the sample is inversely proportional to the image resolution. For small samples, resolution will be higher. The maximum possible sample diameter is in the region of 25 cm, but most intrusive rock samples that are larger than a few centimetres stop all X-rays with energy and flux typical of modern, commercially available, laboratory sources.

The X-ray beam in laboratory scanners contains a range of different energies (the X-ray spectra of the metal used). Each energy will be attenuated by a different amount when it passes through the minerals along the beam path. In practice, this means that more high-energy X-rays can pass through the sample than the lower-energy X-rays. This causes a phenomenon known as *beam-hardening*. As only high-energy X-rays can pass through the middle and out again, the centre appears less dense than the edges in the reconstructed image. This effect can be minimised during the 3D reconstruction process, but if not treated correctly it has the potential to risk erroneous interpretation.

Once the projections have been collected and the 3D image is reconstructed, image processing methods can be used to identify all voxels using the same basic techniques used in 2D (Fig. 8.9). In igneous rocks, this means that the distribution of each of the mineral phases can be mapped, often by highly automated methods, which supports analysis of the true geometry, orientation or other textural feature of each crystal and bubble able to be identified. Mathematically accurate quantification of a feature (crystal, bubble or fracture) can be achieved when the diameter is more than about 3 pixels (Lin et al., 2015). With intrusive igneous rocks, the crystal sizes are typically large enough that 5–10 μm pixel resolution is often sufficient.

The attenuation of any individual mineral can be calculated, but identifying a particular crystal based on X-ray attenuation alone requires it to have sufficiently different attenuation to the crystals surrounding it and that normally requires us to have some knowledge of the rock. There are some minerals (generally the more silicic ones) where this task can be extremely challenging, but with some prior (or post) microscopy or SEM analysis it is usually possible. The contrast between the rock and the vesicle, or crystal and glassy groundmass can often be more easily resolved in the scans of volcanic material (Jerram and Higgins, 2007; Hughes et al., 2017; Paredes-Mariño et al., 2017). Synchrotron XMT systems are much brighter (>10,000 times more X-rays/s) and so the incoming beam can be filtered, and images acquired using X-rays of only a single energy (monochromatic). As long as the energy used is low enough to reveal differences in attenuation between mineral phases, this is an advantage, and beam hardening is also avoided (Fig. 8.10).

Recent studies involving igneous rocks have looked at both intruded and erupted volcanic products (Fig. 8.11). High-quality 3D textural data highlight complex inter-crystalline growth of plagioclase and pyroxenes in a layered intrusion in the Dry Valleys, Antarctica (Jerram et al., 2010), as well as the role of magma mixing (Pankhurst et al., 2014; Renggli et al., 2016; Paredes-Mariño et al., 2017) and bubble growth (Pistone et al., 2015; Hughes et al., 2017; Paredes-Mariño et al., 2017) in driving eruptions. As described earlier, the need to investigate the crystal populations in erupted rocks as a window into what might be happening within the volcanic plumbing system at depth is also an important goal. This is where some of the examples of 3D textural data in the erupted products are very valuable, as this method lends a step change in data volumes able to be gathered, as well as full spatial context for in situ analysis.

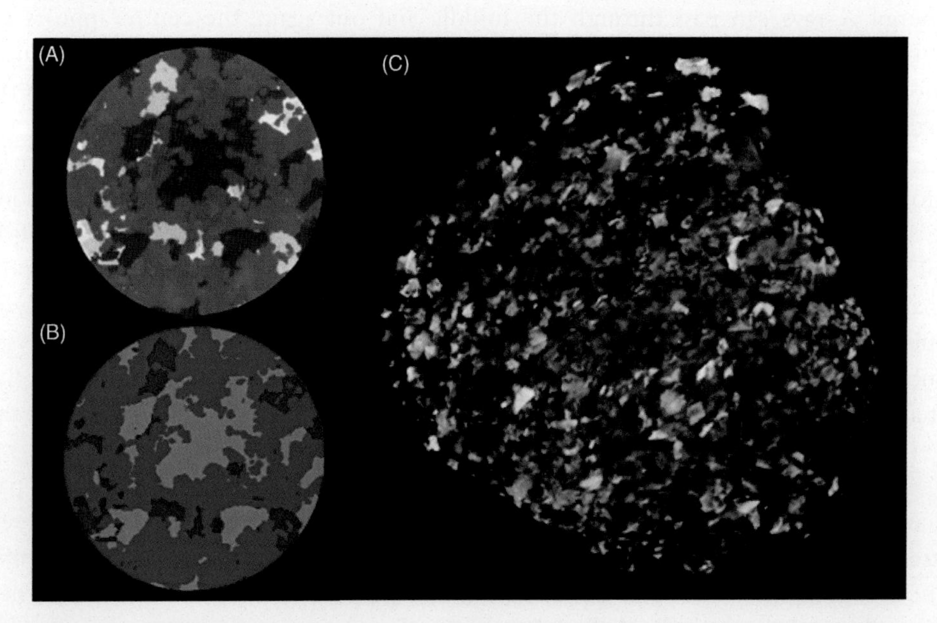

Figure 8.10 (A) A 2D slice through a 3D greyscale data set of a gabbro from Skaergaard. (B) Processed mineral map of A showing the plagioclase (pale blue), clinopyroxene (dark blue) and oxides and sulfides (*red*), and 3D render of the oxide phases (C).

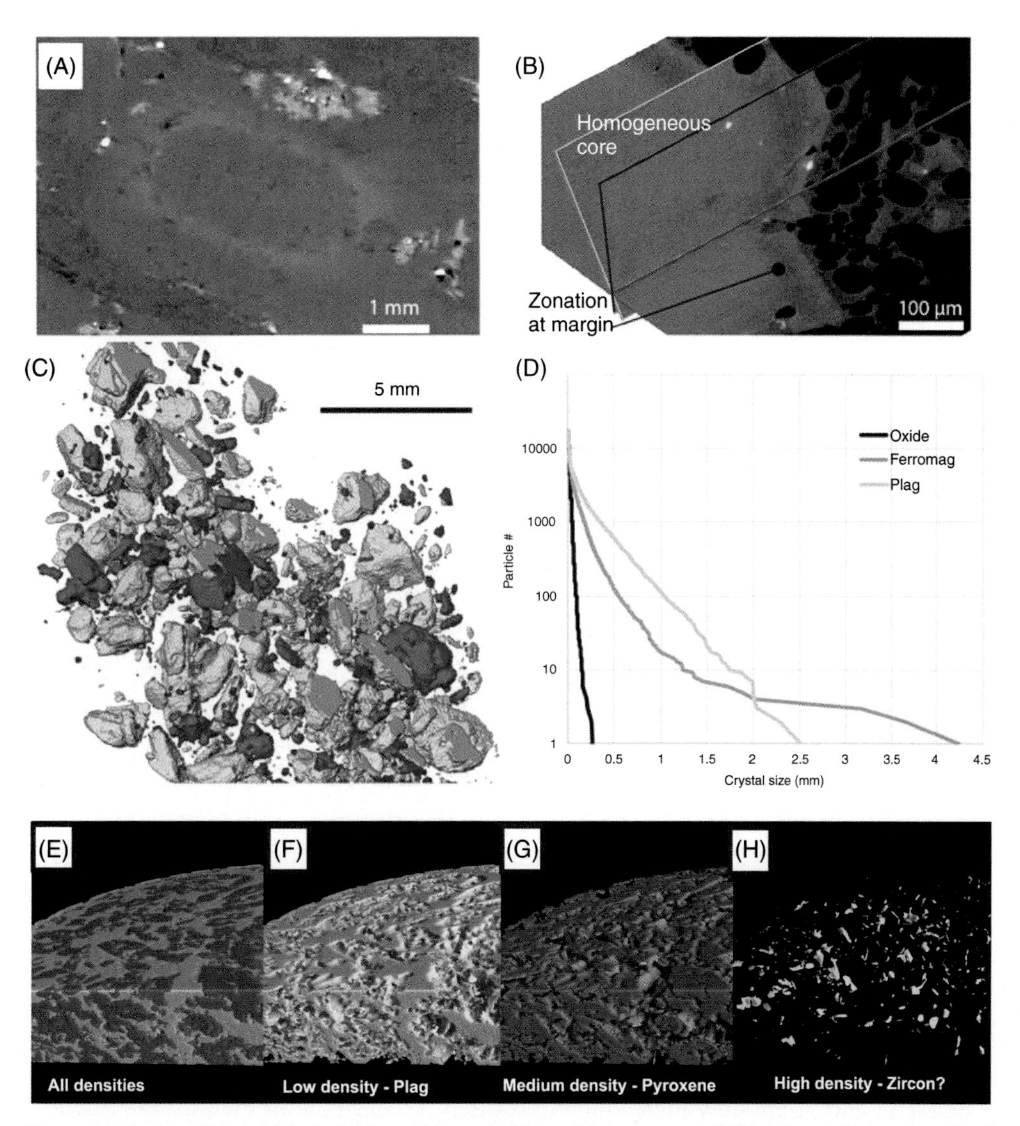

Figure 8.11 Zonation within in feldspar (A) and olivine (B) as seen in XMT data, and an example of the mineralogical 3D mapping (C) and quantitative CSD information (D) that can be extracted from the XMT datasets. (E–H) Example of multiple phase mapping from the Daias Intrusion, Dry Valleys, Antarctica (see Jerram et al., 2010).

Although somewhat in its infancy, 3D petrography is becoming increasingly accessible, not just as a way to image and measure the simple textural parameters and relationships, but also in providing a methodology to access 3D chemical data, as well as the possibility of running 4D experiments, as we will touch on in the final section of this chapter. Expanding this capacity to image and measure how crystals nucleate, grow and resorb within the magma is key to advancing the field.

8.4 LINKING TEXTURE WITH CHEMISTRY

As an igneous rock texture develops and changes, so too does the chemistry of the remaining melt. This composition is recorded by the crystals within the magma. For instance, continued crystal growth during melt composition change can result in crystals containing marked chemical zonations, which are potentially a record of many environments, and changes in those environments, during its growth history. Possible scenarios that affect a crystal's chemistry include mobilising and mixing crystals between different magma batches, multiple magmatic pulses and exchange between different, yet complexly linked, reservoirs. Therefore, a chemical view of a single crystals' journey from its origin to eruption may be highly instructive regarding the key processes of the volcanic and igneous plumbing system as a whole.

In this section, we will explore methods to quantify microchemical variations within crystals, and how these can be used to infer magmatic processes. It can also be noted that the textural quantification techniques summarised in the previous section provide essential guides to any macro- or microchemical sampling strategy.

8.4.1 Crystal Growth Zones

It may rapidly become clear from initial inspection of crystals that they contain growth zones which can be imaged optically (e.g. hand lens, petrographic microscope) or through techniques such as cathodoluminescence (CL), electron microscopy (e.g. SEM) and X-rays (e.g. EDS) (see Ginibre et al., 2007). Some examples of crystal zoning patterns are given in Fig. 8.12. Common crystals that frequently show growth zones include plagioclase, which is arguably the best studied system for zoning patterns (Ginibre et al., 2007), sanidine (Ginibre et al., 2004), quartz (Müller et al., 2003), as well as amphibole and pyroxene. The presence of the zones is evidence of changes within the magma during the crystal's growth history, such that quantification of these chemical variations can provide a window to what changes have occurred. Commonly, major and trace element analysis across the zones is used to model the melt compositions that they crystallised from. For example, trace elements and ratios in plagioclase including Fe, Mg, Ba, Ti and Sr are now increasingly measured. An example of such measurements is given in Fig. 8.12E and F (after Ginibre et al., 2007), where zoned plagioclase data from Parinacota volcano in Chile implies that at least two magmatic recharge events had been recorded by the crystals before they were erupted and preserved in the final texture. Such data helps to piece together information about magmatic plumbing systems, even from eruptive products found at the surface.

It should also be noted that within a particular phase, different crystals may show different profiles, which can indicate that they have been mixed at different times and have different individual histories of crystal exchange, magma mixing and magma storage. Additionally, the detailed investigation of zoning patterns between other mineral phases

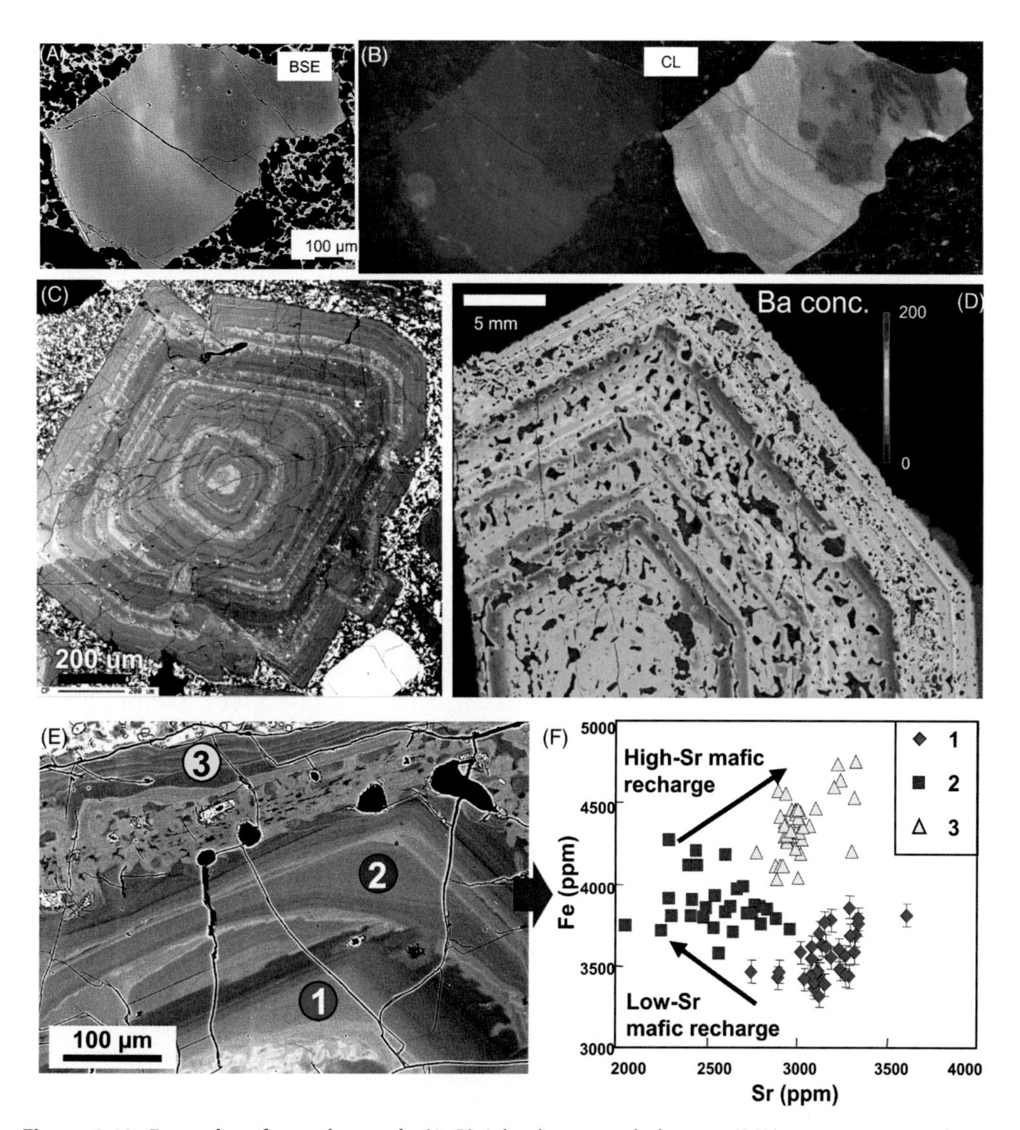

Figure 8.12 ***Examples of zoned crystals.*** (A, B) A backscattered electron (BSE) image compared to a cathodoluminescence (CL) image of a sanidine crystal from Laacher See phonolite, Germany. Only the CL image reveals complex dissolution and regrowth, whereas the BSE image highlights Ba zonation (dark core: low Ba; lighter zone: higher Ba). (C) Large zoned feldspar from Stromboli Italy (see also Fig. 8.13). (D) Ba crystal zoning in a sanidine from Taapaca, Chile (image from Gerhard Wörner). (E, F) BSE image of a zoned plagioclase crystal from Parinacota volcano (E), with Fe–Sr plot of growth zones (F) highlighting different recharge events recorded in the crystal growth history. See also Jerram and Davidson (2007), Ginibre et al. (2007) and Morgan et al. (2007).

present may also reveal different histories. So where zoned crystals are found, it is always important to inspect the textures as thoroughly as possible to fully assess the extent of the zoning preserved with a view to getting the fullest possible picture of magmatic processes.

8.4.2 Crystal Isotope Stratigraphy

As has been shown, zoning profiles preserved within crystals can sometimes be striking (Fig. 8.12) and provide some vital clues as to what has occurred within the growth history of the crystal concerned. A further development of this type of the study is the possibility to microsample crystals to obtain spatially precise geochemical and isotopic data, and in this context, it has become possible to measure isotopic signature variations within single crystals, which provides additional evidence used to support models involving different magma batches, as well as contamination events (Davidson et al., 2007a).

This type of zone quantification has become known as crystal isotope stratigraphy and has been successfully applied to Sr isotopes in plagioclase (Davidson et al., 2007b; Morgan et al., 2007). A number of magma plumbing scenarios where the isotopic zoning within a crystal will vary are highlighted in Fig. 8.13. Isotopic signatures can be due to different magma types, magmatic contamination or some combination of both. There are a few studies of intra-crystal isotopic composition from crystals within plutonic rocks (Waight et al., 2000; Tepley and Davidson, 2003; Gagnevin et al., 2005; Pankhurst et al., 2011, 2013), which have shown that significant variations in initial isotopic ratio exist and can be preserved even in the plutonic realm. This is important as it shows that although a pluton may cool and crystallise more slowly, the diffusion timescales of some isotopic systems to be fully relaxed are not exceeded (Davidson et al., 2007b).

It is important to consider such isotopic observations in the context of other chemical and textural data, in addition to field observations. Together, these multiscale observations help to interpret processes (and their timescales) involved in magma differentiation and evolution (see Davidson et al., 2007b). One example where the changes within the crystal isotope stratigraphy and the textural analysis data have been successfully linked up is at Stromboli volcano, Italy. Morgan et al. (2007) used the CSD plots and core to rim isotopic variations to produce a combined Isotope and CSD plot (termed the ICSD technique). This study showed that the change in texture (in this instance, a kink in the CSD plot) coincided with the principal variation within crystal Sr isotopic variation (Fig. 8.14).

8.4.3 Diffusion and Magmatic Timescales

Textural analysis and diffusion analysis—which summarises an array of methods exploiting the kinetics of ionic exchange and migration within crystals to determine timescales—are two highly complementary techniques. The former records the history of the crystal in terms of growth processes and successive magmatic environments; the latter allows a timescale to be placed on the history since an event is recorded in the crystal.

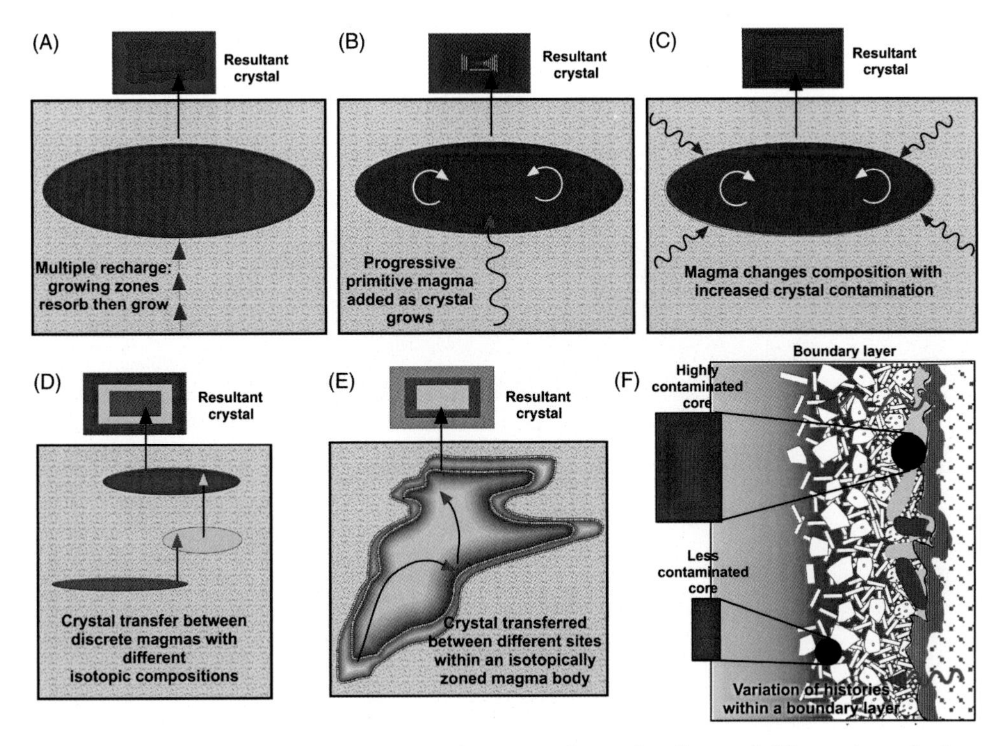

Figure 8.13 ***Different magmatic scenarios where crystal growth will record different isotopic signatures due to magma recharge and/or contamination with host rocks.*** Adapted from Davidson et al. (2007a).

If diffusional methods are used to obtain a timescale, then the use of that growth rate in interpreting the CSD will produce characteristic timescales internally consistent with the growth rate, and hence the diffusion data.

8.4.3.1 How diffusion methods work

Diffusion methods rely on some form of a compositional disjunction—a sudden step change in major- or trace-element composition, which may occur within a crystal or at its margin—often due to a change in the external environment surrounding a growing crystal (Costa and Morgan, 2010). Provided that the mineral is stable, ionic exchange will act to bring the crystal into equilibrium, homogenising effective differences. The process of homogenisation is mediated via ionic diffusion and has a characteristic and predictable behaviour. A common rule of thumb that many will have encountered is

$$x^2 \propto Dt$$

which states that distance of diffusion, squared, is proportional to diffusivity multiplied by time. This is not exact—considerations of geometry, initial condition and even how

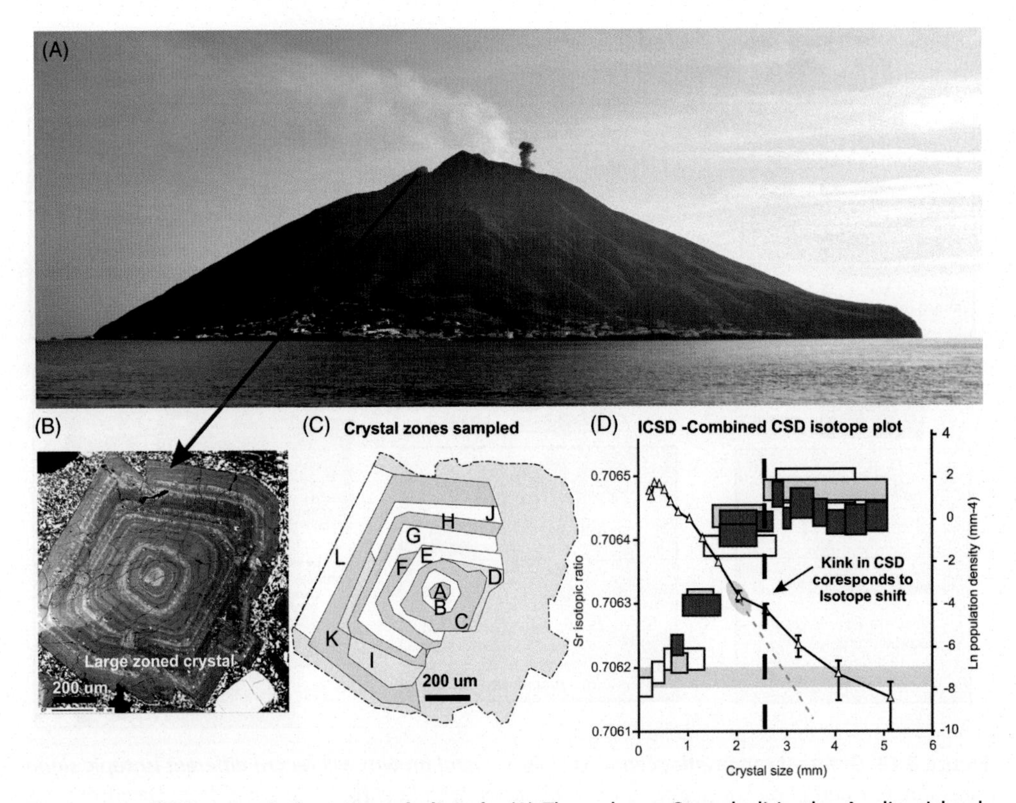

Figure 8.14 ***ICSD example from Stromboli, Italy.*** (A) The volcano Stromboli in the Aeolian Islands. (B) A crystal from sample STR46 of the Vancori period volcanic products of Stromboli imaged via SEM. (C) The locations of zones from that crystal that were individually extracted via micro-milling for isotopic determinations. (D) The isotopic data recovered. Box heights represent isotopic uncertainty, box widths represent a volume adjusted equivalency based on an extended crystal history acquired across three crystals. The data from the crystal in Fig. 13B and C is the darker of the three sets, with core on the right and rim on the left (younger crystals fall on the left of the CSD diagram). See Morgan et al. (2007) for more details.

x is measured, all have a bearing on how such a relation is implemented—but it serves to illustrate the important linkage between diffusion distance and time, and *time* is the key point here. Diffusion methods are almost universally concerned with obtaining the timescale, as this is valuable information that is both of critical use and difficult to obtain in many volcanological settings.

8.4.3.2 Governing parameters

Diffusion rates are governed by diffusivity, with units of $m^2\ s^{-1}$. Ionic diffusivities have different types, depending on the species being measured and how it travels through the crystal. Three important types are:

i) Self-diffusion, which is important for looking at isotopic exchange when an element's concentration may not change, for example, resetting oxygen isotopes in a silicate mineral.

ii) Tracer diffusion, where an element is being substituted into the lattice at trace levels, replacing something of effectively infinite concentration in relative terms; often used to study trace element substitutions in otherwise homogeneous materials.

iii) Interdiffusion, where two or more elements are equilibrating in terms of their mutual ratio. This is appropriate to many ferromagnesian minerals where Fe–Mg exchange can occur and is one of the more routine uses of diffusion methods, given the relatively high level of confidence that can be placed on Fe–Mg interdiffusion in olivine (Dohmen and Chakraborty, 2007).

Diffusivity in mineral grains can be controlled by several parameters, including temperature, pressure, crystal major element composition, oxygen fugacity and the activities or fugacities of ambient species, such as silica or water. Diffusivity is also a tensor quantity and is constrained by the symmetry of the mineral. Cubic minerals have diffusivity that is uniform in all directions, but as minerals become anisotropic, so too can the diffusion behaviour.

The most simplistic expression of a temperature-controlled diffusivity is the Arrhenian form:

$$D = D_0 e^{\frac{-\Delta H}{RT}}$$

where D is diffusivity, D_0 is a constant (the diffusion coefficient), ΔH is the activation energy of diffusion in J mol^{-1}, R the gas constant (8.314 J K^{-1} mol^{-1}) and T the absolute temperature, measured in Kelvin. This form typically gives rise to the intense temperature dependence that diffusivities show, with an order of magnitude variation in D occurring over a few tens of degrees of temperature change at magmatic conditions.

In cases where there are other dependencies, pressure typically factors into the exponential term (dependent on the equation of state), whilst composition dependences, anisotropy and ambient fugacities or activities tend to influence the D_0 term. Exact parameterisations depend on the mechanisms by which diffusion occurs in the crystal and the experimental conditions under which diffusivities were recovered. For full discussions of these effects, an introduction can be found in Costa and Morgan (2010), and more in-depth views in Costa et al. (2008) and the MSA *Reviews in Mineralogy and Geochemistry* volume 72.

8.4.3.3 What can be learned?

As noted earlier, studies typically focus on the recovery of timescales. An initial condition is forward-modelled until it most closely matches an observed profile. Models often fall into one of two types (Fig. 8.15).

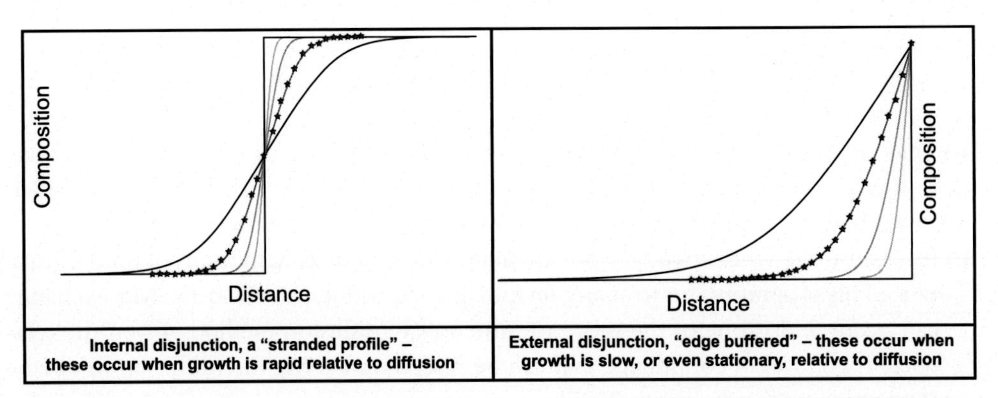

Figure 8.15 ***The two most common types of diffusion models used when modelling crystal profiles.***

There is a spectrum of models between both these end members, but the region of growth rate over which transitional behaviour exists is very small, leading measured profiles to largely look like whichever of these two end members was dominant during that part of the crystal history.

8.4.3.4 Complications and limitations

Of course, there are complications to these methods. Some of these are inherent and have briefly been mentioned.

Diffusion methods require knowledge of the magmatic temperature and are most sensitive to that variable. Thermometry of high quality, and which is appropriate to the diffusion process, is required. Typically thermometry of ±15 to 20 °C is sufficient, as long as it is appropriate to the petrologic context. For example, it would not be appropriate to use Fe–Ti oxide thermometry to model clinopyroxene, if the clinopyroxene zonation demonstrably formed at higher temperatures than the Fe–Ti oxides. Therefore, a careful consideration of thermometry, which ties together the textural information within and across the mineral phases, is key for results to be meaningful.

For certain types of diffusion, oxygen fugacity is a control and must also be constrained before modelling takes place. Dependent on the type of diffusion, parameters such as pressure, water fugacity and activity of silica in the melt must also be considered and constrained. Each additional parameter requires further measurements.

A further parameter to constrain is the crystal orientation, in order to account for anisotropy of diffusion. Such measurements are made via EBSD, and this is then combined with the diffusivity data to determine the diffusivity in the direction of the measured profile.

Sectioning angle is another parameter to consider, especially the sectioning angle across the plane of the section through the crystal, which in thin section can be measured using a universal stage. Shallow angles of a crystal edge through the thickness of the thin section can potentially produce an erroneous apparent profile.

In terms of overall limitations, there are several:

- Diffusion is typically modelled at one temperature and this may not be entirely representative.
- Unless there is a factor inhibiting diffusion, timescales cannot be longer than the modelled values, but they can be shorter if some of the profile represents changes recorded during growth.
- Diffusivities are typically constrained to between 0.2 and 0.5 log units, which is not terribly precise. This reflects directly into uncertainty on timescales and is a fundamental limitation of the technique. It is sufficient to distinguish hours from days, weeks and from months, however, and there is great potential for new experimental data to support increasing confidence in values of diffusivity.
- Regardless of absolute certainty, relative differences, such as crystal A exhibits twice as much diffusion as crystal B, will always be preserved if the geometry, temperature and composition factors have been properly considered. In this sense, the uncertainties in diffusivity, and hence time, are not stochastic by data point but should be consistently applied to all points derived from a given sample that were modelled with the same conditions.

8.4.3.5 An example deployment

A relatively simple case study is that published by Martin et al. (2008) (Fig. 8.16). They presented results on five crystals of olivine derived from mafic enclaves mobilised by the Nea Kameni (Greece) eruption of 1950. Changes in conditions, interpreted to be related to the mafic melts intruding, triggered the re-equilibration of the crystal rims towards a lower forsterite content, produced diffusion profiles of an edge-buffered type. By constraining the temperature of diffusion, oxygen fugacity, crystal orientation and using a universal stage to measure crystal face orientations in 3D, timescales of between 1 and 3 months were obtained from modelling the observed profiles. This was taken to represent the maximum timescale of mafic magma injection and residence beneath the dacite magma, and thus gives a guideline on how rapidly magma mobilisation can occur prior to future eruptions of this type. Additionally, it underscores how important magma mingling is within the igneous plumbing system.

8.4.3.6 Future developments

Developments in the field currently look towards analysing increasing numbers of crystals in order to gain better statistical certainty on results and to obtain a better handle on what results mean at the scale of the magma system. At the time of writing, papers are entering press with over 100 timescale determinations (Chamberlain et al., 2014; Kahl et al., 2015), and methods may develop to enable this to further grow. Others are looking at the possible variation of temperature over time and how this can be used to reconcile the short diffusion timescales with much longer radiometric timescales (Cooper and Kent, 2014).

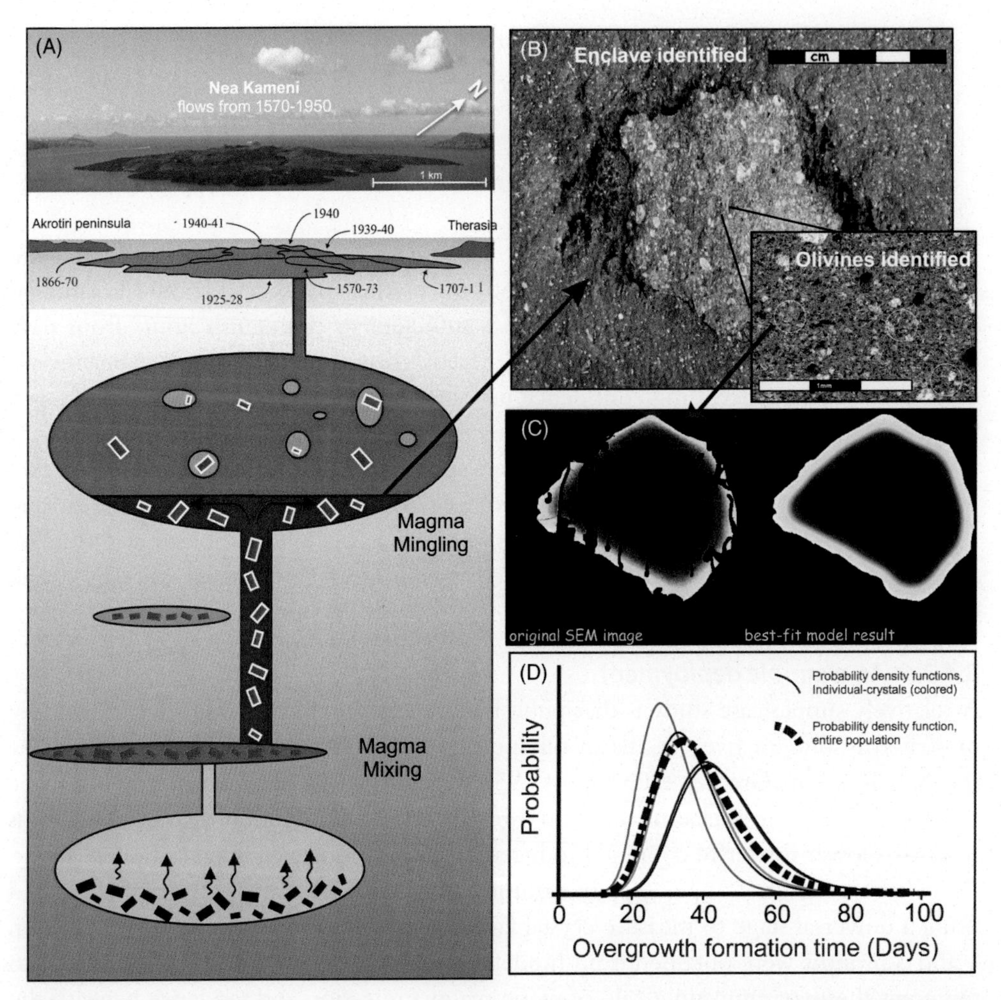

Figure 8.16 ***Using diffusion profiles to understand igneous timescales at Santorini, Greece (after Martin et al., 2008).*** (A) Nea Kemeni and a schematic representation of a possible underlying volcanic plumbing system. (B) A mafic enclave hosted in dacitic lava from Nea Kameni. (C) (left) A false-coloured BSE image of a Nea Kameni olivine, with image defects painted out. This was used as a comparison when modelling c (right) a model output attempting to reproduce the crystal data on the left, best constrained in the lower left and lower right of the crystal. (D) Five data points for Nea Kameni olivine, together with the population curve, presented as probability distributions based on diffusivity uncertainty from the diffusivity expression itself and the magmatic temperature uncertainty. More details in Martin et al. (2008).

Many different diffusion-based methods are coming forward, in some cases utilising an array of elements to remove uncertainty around initial conditions (Morgan and Blake, 2006; Till et al., 2015), or using fast-diffusing elements to investigate short-duration processes occurring in (and between) crystals and melts during eruption (Myers et al., 2016).

There are equally many cases where the methods can be turned around. Where a timescale is known or constrained from an external reference, it is possible to employ the equations to determine D. This can, due to the intense dependence of D on T, lead to a robust estimate of temperature (Morgan et al., 2013).

8.5 MAPPING CHEMISTRY IN 3D AND LOOKING FORWARD TO 4D OBSERVATIONS

As we have now seen, scientific understanding of the evolution and development of the melt and its crystal cargo has shaped the direct and indirect records preserved in that cargo. However, many of these methods are conducted on 2D thin sections, which is problematic when considering the three-dimensional nature of how crystals may relate to their neighbours and the melt. Taking these complexities into account, understanding the connectivity of the melt and the ability of that melt (and the crystal cargo) to move become important. This is especially the case when investigating the last events prior to eruption, and indeed processes that might trigger eruptions. Quantitative understanding of semi-solid magma behaviour, including melt connectivity, and its relationship to the spatial distribution of the geochemical signals preserved in the crystals is a rapidly developing field.

X-ray tomography, as introduced previously, can provide the 3D connectivity of a glassy melt phase or, more commonly for intrusive rocks, determine the last phase that precipitated from the final trapped melt and its geometry. The method can usually resolve many of the different crystal phases in the rock and can sometimes even capture the chemical zonation within crystals, but mapping the XMT grey-scale value to a specific composition is still difficult. Researchers are also working on developing more quantitative methods that use SEM-generated element maps, QEMSCAN mineral maps, EBSD (Lindgren et al., 2015; Morgavi et al., 2013a, b), or carefully calibrated standards to allow for more chemical analysis of the grey-scale XMT data. To do this, studies are characterising the beam spectra using materials (often minerals) of known thickness and purity (Davis et al., 2013; Pankhurst et al., submitted). The 2D data can be used to improve the accuracy and grey-scale resolution of the 3D reconstruction, calibrate an existing 3D data image or improve digital image segmentation processes. The advantage of the combined method is that once the XMT data has been collected, the same sample can be sliced, diced and analysed using 2D methods, and those results feed into the 3D data that has already been collected (Golab et al., 2013; Pankhurst et al., submitted).

Most of what is quantified during textural analysis of igneous rocks is the end product of what appears in many cases to be a long and complex evolution. We can start unravelling the processes that occur using chemical and petrological data and population statistics. Experimental petrology allows us to investigate the processes in more detail, as we take a suite of real or synthetic rock samples to different high-temperature and/or high-pressure conditions to observe what happens to both the textures and chemical structure of the sample. In most experimental cases, we are unable to watch the sample texture evolve: a single snap shot of the material at the end of the experiment is all that can be observed. The starting conditions can be controlled in terms of bulk chemistry, and the texture(s) present. Yet once an experimental run is commenced until when it is finished, exactly what happens to texture could not be directly observed until very recently. Using XMT methods to watch what happens in such experiments is an exciting and expanding field. Each frame of a 'live' experiment can be considered to be an individual experimental run 'frozen' in time.

The intensity of the beam available at synchrotron facilities means it is possible to collect a full high-resolution 3D image in as little as 1/20th of a second (broad-spectrum energies; Dobson et al., 2016). Even with a monochromatic beam, each 3D image can be collected in under a second, producing a real-time frame rate comparable to live-action films. Synchrotron beam line facilities often have sufficient space for fitting furnaces or other experimental equipment, which means that we can now begin to explore challenging experimental designs and simulate increasingly complex igneous processes. This work has the potential to completely revolutionise our understanding of the dynamics of magmatic behaviour, because it has the potential to capture disequilibrium processes (Fig. 8.17). The ability to image and quantify geochemical variations in 3D and to record 3D magma experiments to produce 4D datasets represent the current cutting edge of petrographic and petrological analysis. These exciting techniques as well as the more excepted and well-known ones touched on in this chapter provide the researcher with the tools to look through the crystal ball, examine their rock textures and use this information to get back to the magmatic processes that formed them.

ACKNOWLEDGEMENTS

The authors like to acknowledge the many colleagues over the years that have helped our thoughts progress, discussed and critiqued our work towards a more improved understanding, and who have generally been there in the lab, the field and remotely making Igneous petrology fun. Dougal Jerram is partly funded through a Norwegian Research Council Centres of Excellence project (project number 223272, CEED) and through DougalEARTH LTD. Some data presented were gathered using resources provided by an AXA Research Fund Fellowship to Matt Pankhurst, who also acknowledges NERC (UK) grant NE/M013561/1 for support. Katherine Dobson is supported by a NERC Independent Research Fellowship

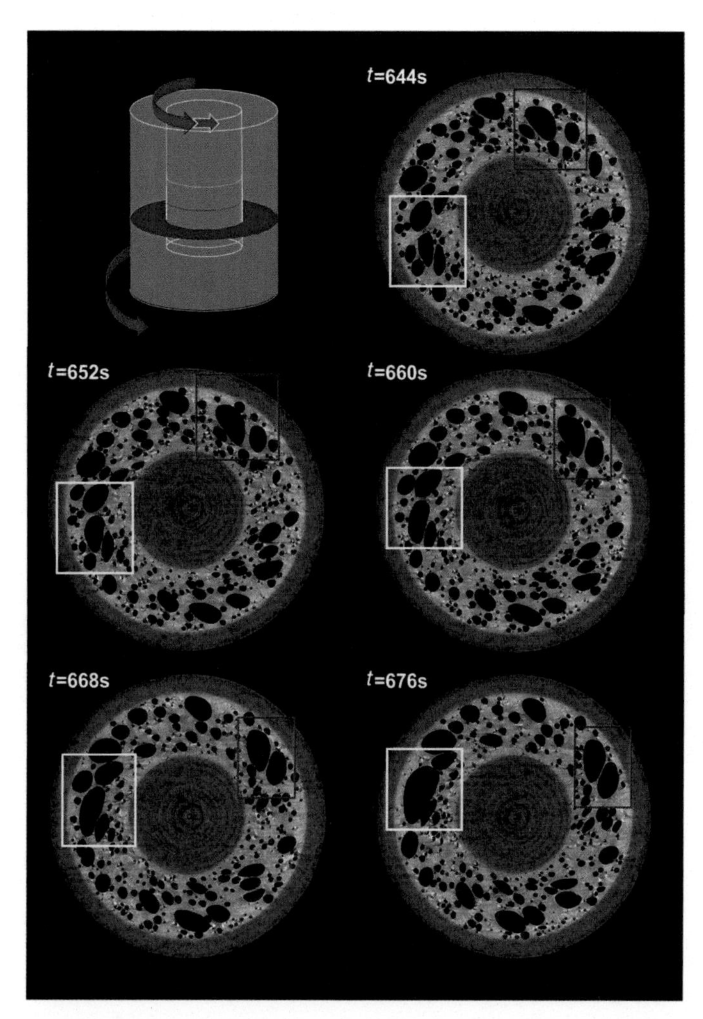

Figure 8.17 ***2D slices through a 4D dataset.*** A 3D image was acquired in 1 s with one image captured every 8 s. The data shown are from the 83–87th images collected during a rotational deformation experiment and show how bubble (black) deformation and coalescence is progressing in a synthetic magma (light grey) as it is deformed by the differential speeds of the cup and spindle (dark grey) over time.

(NE/M018687/1) and wishes to acknowledge both Diamond Light Source grants EE16040 and EE15898, and Swiss Light Source grant 20150413, that supported collection of some of the data presented. Finally we would like to dedicate this chapter to the late Jon Davidson, who helped inspire us, promoted and supported our work, and gave so much to the field of igneous petrology and volcanology.

REFERENCES

Bachmann, O., Bergantz, G.W., 2003. Rejuvenation of the Fish Canyon magma body: a window into the evolution of large-volume silicic magma systems. Geology 31 (9), 789–792.

Bachmann, O., Bergantz, G.W., 2004. On the origin of crystal-poor rhyolites: extracted from batholithic crystal mushes. J. Petrol. 45 (8), 1565–1582.

Bachmann, O., Bergantz, G.W., 2008. The magma reservoirs that feed supereruptions. Elements 4 (1), 17–21.

Bryan, S.E., Ukstins Peate, I., Peate, D.W., Self, S., Jerram, D.A., Mawby, M.R., Marsh, J.S., Miller, J.A., 2010. The largest volcanic eruptions on Earth. Earth Sci. Rev. 102 (3), 207–229, https://doi.org/10.1016/j.earscirev.2010.07.001.

Cashman, K.V., Marsh, B.D., 1988. Crystal size distribution (CSD) in rocks and the kinetics and dynamics of crystallisation. II: Makaopuhi lava lake. Contrib. Mineral. Petrol. 99, 292–305.

Castro, J.M., Cashman, K.V., Manga, M., 2003. A technique for measuring 3D crystal-size distributions of prismatic microlites in obsidian. Am. Mineral. 88, 1230–1240.

Chamberlain, K.J., Morgan, D.J., Wilson, C.J.N., 2014. Timescales of mixing and mobilisation in the Bishop Tuff magma body: perspectives from diffusion chronometry. Contrib. Mineral. Petrol. 168, 24. doi: 10.1007/s00410-014-1034-2.

Charlier, B.L.A., Wilson, C.J.N., Lowenstern, J.B., Blake, S., Van Calsteren, P.W., Davidson, J.P., 2005. Magma generation at a large, hyperactive silicic volcano (Taupo, New Zealand) revealed by U-Th and U-Pb systematics in zircons. J. Petrol. 46, 3–32.

Cheng, L.-L., Yang, Z.-F., Zeng, L., Wang, Y., Luo, Z.-H., 2014a. Giant plagioclase growth during storage of basaltic magma in Emeishan Large Igneous Province, SW China. Contrib. Mineral. Petrol. 167 (2), 1–20.

Cheng, L., Zeng, L., Ren, Z., Wang, Y., Luo, Z., 2014b. Timescale of emplacement of the Panzhihua gabbroic layered intrusion recorded in giant plagioclase at Sichuan Province, SW China. Lithos 204, 203–219.

Cnudde, V., Boone, M.N., 2013. High-resolution X-ray computed tomography in geosciences: a review of the current technology and applications. Earth Sci. Rev. 123, 1–17.

Cooper, K.M., Kent, A.J.R., 2014. Rapid remobilization of magmatic crystals kept in cold storage. Nature 506, 480–483. doi: 10.1038/nature12991.

Costa, F., Morgan, D.J., 2010. Time constraints from chemical equilibration in magmatic crystals. Dosseto, A., Turner, S.P., Van Orman, J.A. (Eds.), Timescales of Magmatic Processes: From Core to AtmosphereFirst ed. Wiley- Blackwell Publishing, Oxford.

Costa, F., Dohmen, R., Chakraborty, S., 2008. Timescales of magmatic processes from modeling the zoning patterns of crystals. Putirka, K.D., Tepley, III, F.J. (Eds.), Minerals, Inclusions and Volcanic Processes. Reviews in Mineralogy and Geochemistry, 69, Mineralogical Society of America/Geochemical Society, pp. 545–594.

Couch, S., Harford, C.L., Sparks, R.S.J., Carroll, M.R., 2003. Experimental constraints on the conditions of formation of highly calcic plagioclase microlites at the Soufrire Hills Volcano, Montserrat. J. Petrol. 44, 1455–1475.

Davidson, J.P., Morgan, D.J., Charlier, B.L.A., Harlou, R., Hora, J., 2007a. Tracing magmatic processes and timescales through mineral-scale isotopic data. Annu. Rev. Earth Planet. Sci. 35, 273–311.

Davidson, J.P., Morgan, D., Charlier, B., 2007b. Isotopic microsampling of magmatic rocks. Elements 3, 253–260.

Davis, G.R., Evershed, A.N., Mills, D., 2013. Quantitative high contrast X-ray microtomography for dental research. J. Dentist. 41 (5), 475–482.

Day, J.M.D., Taylor, L.A., 2007. On the structure of mare basalt lava flows from textural analysis of the LaPaz Icefield and Northwest Africa 032 lunar meteorites. Meteoritics Planet. Sci. 42, 3–17.

Dobson, K.J., Coban, S.B., McDonald, S.A., Walsh, J., Atwood, R., Withers, P.J., 2016. 4D imaging of sub-second dynamics in pore-scale processes using real time synchrotron X-ray tomography. Solid Earth Discussions 1–27.

Dobson, K.J., Harrison, S.T.L., Lin, Q., Ní Bhreasail, A., Fagan-Endres, M.A., Neethling, S.J., Lee, P.D., Cilliers, J.J., 2017. Insights into ferric leaching of low grade metal sulfide-containing ores in an unsaturated ore bed using X-ray computed tomography. Minerals 7(5):85.

Dohmen, R., Chakraborty, S., 2007. Fe–Mg diffusion in olivine II: point defect chemistry, change of diffusion mechanisms and a model for calculation of diffusion coefficients in natural olivine. Phys. Chem. Min. 34:409–430. DOI 10.1007/s00269-007-0158-6.

Donaldson, C.H., 1976. An experimental investigation of olivine morphology. Contrib. Mineral. Petrol. 57, 187–213.

Elliott, M.T., Cheadle, M.J., Jerram, D.A., 1997. On the identification of textural equilibrium in rocks using dihedral angle measurements. Geology 25, 355–358.

Ersoy, O., Şen, E., Aydar, E., Tatar, İ., Çelik, H.H., 2010. Surface area and volume measurements of volcanic ash particles using micro-computed tomography (micro-CT): a comparison with scanning electron microscope (SEM) stereoscopic imaging and geometric considerations. J. Volcanol. Geotherm. Res. 196 (3), 281–286.

Feali, M., Pinczewski, W., Cinar, Y., Arns, C.H., Arns, J.-Y., Francois, N., Turner, M.L., Senden, T., Knackstedt, M.A., 2012. Qualitative and quantitative analyses of the three-phase distribution of oil, water, and gas in bentheimer sandstone by use of micro-CT imaging. SPE Reservoir Eval. Eng. 15 (6), 706–711.

Gagnevin, D., Daly, S., Waight, T.E., Morgan, D., Poli, G., 2005. Pb isotopic zoning of K-feldspar megacrysts determined by laser ablation multiple-collector ICP-MS: insights into granite petrogenesis. Geochim. Cosmochim. Acta 69, 1899–1915.

Gee, J.S., Meurer, W.P., Selkin, P.A., CHeadle, M.J., 2004. Quantifying three-dimensional silicate fabrics in cumulates using cumulative distribution functions. J. Petrol. 45, 1983–2009.

Gill, J., Reagan, M., Tepley, F., Malavassi, E., 2006. Arenal Volcano, Costa Rica. Magma genesis and volcanological processes. J. Volcanol. Geotherm. Res. 157(1–3):1–8.

Ginibre, C., Wörner, G., Kronz, A., 2004. Structure and dynamics of Laacher See Magma chamber (Eifel, Germany) from major and trace element zoning in sanidine: a cathodoluminescence and electron microprobe study. J. Petrol. 45, 2197–2223.

Ginibre, C., Wörner, G., Kronz, A., 2007. Crystal zoning as an archive for magma evolution. Elements 3 (4), 261–266.

Goodenough, K., Emeleus, C.H., Jerram, D.A., Troll, V.R., 2008. Golden rum: understanding the forbidden isle. Geoscientist 18 (3), 22–24.

Golab, A.N., Knackstedt, M.A., Averdunk, H., Senden, T., Butcher, A.R., Jaime, P., 2010. 3D porosity and mineralogy characterization in tight gas sandstones. Leading Edge 29 (12), 1476–1483.

Golab, A., Ward, C.R., Permana, A., Lennox, P., Botha, P., 2013. High-resolution three-dimensional imaging of coal using microfocus X-ray computed tomography, with special reference to modes of mineral occurrence. Int. J. Coal Geol. 113, 97–108.

Gualda, G.A.R., 2006. Crystal size distributions derived from 3D datasets: sample size versus uncertainties. J. Petrol. 47, 1245–1254, https://doi.org/10.1093/petrology/egl010.

Gualda, G.A.R., Rivers, M., 2006. Quantitative 3D petrography using X-ray tomography: application to Bishop Tuff pumice clasts. J. Volcanol. Geotherm. Res. 154, 48–62. doi: 10.1016/j.jvolgeores.2005.09.019.

Higgins, M.D., 1994. Determination of crystal morphology and size from bulk measurements on thin sections: numerical modelling. Am. Mineral. 79, 113–119.

Higgins, M.D., 1996. Crystal size distribution and other quantitative textural measurements in lavas and tuffs from Egmon volcano (Mt. Taranaki), New Zealand. Bull. Volcanol. 58, 194–204.

Higgins, M.D., 1998. Origin of anorthosite by textural coarsening: quantitative measurements of a natural sequence of textural development. J. Petrol. 39, 1307–1323.

Higgins, M.D., 2000. Measurement of crystal size distributions. Am. Mineral. 85, 1105–1116.

Higgins, M.D., 2006. Quantitative Textural Measurements in Igneous and Metamorphic Petrology. Cambridge University Press, Cambridge, UK.

Holness, M.B., Cheadle, M.J., McKenzie, D., 2005a. On the use of changes in dihedral angle to decode late-stage textural evolution in cumulates. J. Petrol. 46, 1565–1583.

Holness, M.B., Martin, V.M., Pyle, D.M., 2005b. Information about open-system magma chambers derived from textures in magmatic enclaves: the Kameni Islands, Santorini, Greece. Geol. Mag. 142, 1–13.

Holness, M.B., Nielsen, T.F.D., Tegner, C., 2007. Textural maturity of cumulates: a record of chamber filling, liquidus assemblage, cooling rate and large-scale convection in mafic layered intrusions. J. Petrol. 48, 141–157.

Hughes, E.C., Neave, D.A., Dobson, K.J., Withers, P.J., Edmonds, M., 2017. How to fragment peralkaline rhyolites: observations on pumice using combined multi-scale 2D and 3D imaging. J. Volcanol. Geotherm. Res. 336:179–191. ISSN 0377-0273. doi:org/10.1016/j.jvolgeores.2017.02.020.

Hunter, R.H., 1996. Texture development in cumulate rocks. In: Cawthorn, R.G. (Ed.), Layered Intrusions: Developments in Petrology. Elsevier, pp. 77–101. doi: 10.1016/S0167-2894(96)80005-4.

Iglauer, S., Paluszny, A., Blunt, M., 2013. Simultaneous oil recovery and residual gas storage: a pore-level analysis using in situ X-ray micro-tomography. Fuel 103, 905–914.

Jerram, D.A., Bryan, S.E., 2015. Plumbing systems of shallow level intrusive complexes. In: Breitkreuz, C.H., Rocchi, S. (Eds.), Physical Geology of Shallow Magmatic Systems. Advances in Volcanology. Springer, Berlin, pp. 1–22, http://doi.org/10.1007/11157_2015_8.

Jerram, D.A., Cheadle, M.J., 2000. On the cluster analysis of grains and crystals in rocks. Am. Mineral. 85, 47–67.

Jerram, D.A., Davidson, J.P., 2007. Frontiers in textural and microgeochemical analysis. Elements 3, 235–238, https://doi.org/10.2113/gselements.3.4.235.

Jerram, D.A., Higgins, M.D., 2007. 3D analysis of rock textures: quantifying igneous microstructures. Elements 3, 239–245, https://doi.org/10.2113/gselements.3.4.239.

Jerram, D.A., Martin, V.M., 2008. Understanding crystal populations and their significance through the magma plumbing system. In: Annen, C., Zellmer, G.F. (Eds.), Dynamics of Crustal Magma Transfer, Storage and Differentiation. Geological Society, London, Special Publications, pp. 133–148. doi: 10.1016/S0167-2894(96)80005-4, https://doi.org/10.1144/SP304.7.

Jerram, D.A., Cheadle, M.J., Hunter, R.H., Elliott, M.T., 1996. The spatial distribution of grains and crystals in rocks. Contrib. Mineral. Petrol. 125, 60–74, https://doi.org/10.1007/s004100050206.

Jerram, D.A., Cheadle, M.C., Philpotts, A.R., 2003. Quantifying the building blocks of igneous rocks: are clustered crystal frameworks the foundation? J. Petrol. 44, 2033–2051, https://doi.org/10.1093/petrology/egg069.

Jerram, D.A., Mock, A., Davis, G.R., Field, M., Brown, R.J., 2009. 3D crystal size distributions: a case study on quantifying olivine populations in kimberlites. Lithos 112S, 223–235. doi: 10.1016/j.lithos.2009.05.042.

Jerram, D.A., Davis, G.R., Mock, A., Charrier, A., Marsh, B.D., 2010. Quantifying 3D crystal populations, packing and layering in shallow intrusions: a case study from the Basement Sill, Dry Valleys, Antarctica. Geosphere 6, 537–548, https://doi.org/10.1130/GES00538.1.

Kahl, M., Chakraborty, S., Pompilio, M., Costa, F., 2015. Constraints on the nature and evolution of the magma plumbing system of Mt. Etna Volcano (1991–2008) from a combined thermodynamic and kinetic modelling of the compositional record of minerals. J. Petrol. 56, 2025–2068. doi: 10.1093/petrology/egv063.

Launeau, P., Cruden, A.R., 1998. Magmatic fabric acquisition mechanisms in a syenite: results of a combined anisotropy of magnetic susceptibility and image analysis study. J. Geophys. Res. 103, 5067–5089.

Lindgren, P., Hanna, R., Lee, M., Dobson, K., Tomkinson, T., 2015. The paradox between low shock-stage and evidence for compaction in CM carbonaceous chondrites explained by multiple low-intensity impacts. Geochim. Cosmochim. Acta 148, 159–178, https://doi.org/10.1016/j.gca.2014.09.014.

Lin, Q., Neethling, S.J., Dobson, K.J., Courtois, L., Lee, P.D., 2015. Quantifying and minimising systematic and random errors in X-ray micro-tomography based volume measurements. Comput. Geosci. 77, 1–7.

Lofgren, G.E., 1974. An experimental study of plagioclase crystal morphology: isothermal crystallization. Am. J. Sci. 274, 243–273.

Lofgren, G.E., 1980. Chapter 11 Experimental studies on the dynamic crystallization of silicate melts, In: Hargraves, R.B. (Ed.), Physics of Magmatic Processes. Princeton University Press, Princeton, NJ, pp. 487–551.

Macdonald, R., Bagiński, B., Upton, B.G.J., Pinkerton, H., MacInnes, D.A., MacGillivray, J.C., 2010. The Mull Palaeogene dyke swarm: insights into the evolution of the Mull igneous centre and dyke-emplacement mechanisms. Mineral Mag. 74, 601–622.

Marsh, B.D., 1988. Crystal size distribution (CSD) in rocks and the kinetics and dynamics of crystallisation I: Theory. Contrib. Mineral. Petrol. 99, 277–291.

Marsh, B.D., 1996. Solidification fronts and magmatic evolution. Mineral Mag. 60, 5–40.

Marsh, B.D., 1998. On the interpretation of crystal size distributions in magmatic systems. J. Petrol. 39, 553–599.

Marsh, B.D., 2004. A magmatic mush column Rosetta Stone: The McMurdo Dry Valleys of Antarctica. EOS Trans. Am. Geophys. Union 85, 497.

Martin, V.M., Morgan, D.J., Jerram, D.A., Cadddick, M.J., Prior, D.J., Davidson, J.P., 2008. Bang! month scale eruption triggering at Santorini volcano. Science 321, 1178.

Martin, V.M., Davidson, J.P., Morgan, D.J., Jerram, D.A., 2010. Using the Sr isotope compositions of feldspars and glass to distinguish magma system components and dynamics. Geology 38, 539–542.

Mock, A., Jerram, D.A., 2005. Crystal size distributions (CSD) in three dimensions: insights from the 3D reconstruction of a highly porphyritic rhyolite. J. Petrol. 46 (8), 1525–1541, https://doi.org/10.1093/petrology/egi024.

Moitra, P., Gonnermann, H.M., 2015. Effects of crystal shape- and size-modality on magma rheology. Geochem. Geophys. Geosyst. 16. doi: 10.1002/2014GC005554.

Monnet, C., Zollikofer, C., Bucher, H., Goudemand, N., 2009. Three-dimensional morphometric ontogeny of mollusc shells by micro-computed tomography and geometric analysis. Palaeontol. Electron. 12 (3), 1–13.

Morgan, D.J., Blake, S., 2006. Magmatic residence times of zoned phenocrysts: introduction and application of the binary element diffusion modelling (BEDM) technique. Contrib. Mineral. Petrol. 151:58–70. DOI:10.1007/s00410-005-0045-4.

Morgan, D.J., Jerram, D.A., 2006. On estimating crystal shape for crystal size distribution analysis. J. Volcanol. Geotherm. Res. 154, 1–7, https://doi.org/10.1016/j.jvolgeores.2005.09.016>.

Mock, A., Jerram, D.A., Breitkreuz, C., 2003. Using quantitative textural analysis to understand the emplacement of shallow level rhyolitic laccoliths - a case study from the Halle volcanic complex, Germany. J. Petrol. 44 (5), 833–849. doi: 10.1093/petrology/44.5.833.

Morgan, D.J., Jerram, D.A., Chertkoff, D.G., Davidson, J.P., Pearson, D.G., Kronz, A., Nowell, G.M., 2007. Combining CSD and isotopic microanalysis: magma supply and mixing processes at Stromboli volcano, Aeolian Islands, Italy. Earth Planet. Sci. Lett. 260, 419–431, https://doi.org/10.1016/j.epsl.2007.05.037.

Morgan, D.J., Jollands, M.C., Lloyd, G.E., Banks, D.A., 2013. Using titanium-in-quartz geothermometry and geospeedometry to recover temperatures in the aureole of the Ballachulish Igneous Complex, NW Scotland. Geological Society, London, Special Publications 394:145–165. doi: 10.1144/SP394.8.

Morgavi, D., Perugini, D., De Campos, C.P., Ertel-Ingrisch, W., Dingwell, D.B., 2013a. Time evolution of chemical exchanges during mixing of rhyolitic and basaltic melts. Contrib. Mineral. Petrol. 166 (2), 615–638.

Morgavi, D., Perugini, D., De Campos, C.P., Ertl-Ingrisch, W., Lavallée, Y., Morgan, L., Dingwell, D.B., 2013b. Interactions between rhyolitic and basaltic melts unraveled by chaotic mixing experiments. Chem. Geol. 346, 199–212, 0009-2541.

Moss, S., Russell, J.K., Smith, B.H.S., Brett, R.C., 2010. Olivine crystal size distributions in kimberlite. Am. Mineral. 95, 527–536.

Müller, A., Wiedenbeck, M., van den Kerkhof, A.M., Kronz, A., Simon, K., 2003. Trace elements in quartz; a combined electron microprobe, secondary ion mass spectrometry, laser-ablation ICP-MS, and cathodoluminescence study. Eur. J. Mineral. 15 (4), 747–763.

Müller, T., Baumgartner, L.P., Foster, J.C.T., Bowman, J.R., 2009. Crystal size distribution of periclase in contact metamorphic dolomite marbles from the Southern Adamello Massif, Italy. J. Petrol. 50 (3), 451–465.

Myers, M.L., Wallace, P.J., Wilson, C.J.N., Morter, B.K., Swallow, E.J., 2016. Prolonged ascent and episodic venting of discrete magma batches at the onset of the Huckleberry Ridge supereruption, Yellowstone. Earth Planet. Sci. Lett. 451, 285–297.

O'Driscoll, B., Troll, V.R., Donaldson, C.H., Jerram, D.A., Emeleus, C.H., 2007a. An origin for harrisitic and granular olivine in the Rum Layered Suite, NW Scotland: a crystal size distribution study. J. Petrol. 48, 253–270. doi: 10.1093/petrology/egl059.

O'Driscoll, B., Hargraves, R.B., Emeleus, C.H., Troll, V.R., Donaldson, C.H., Leavy, R.J., 2007b. Magmatic lineations inferred from anisotropy of magnetic susceptibility fabrics in Units 8, 9, and 10 of the Rum Eastern Layered Series, NW Scotland. Lithos 98, 27–44.

Pan, X., Sidky, E.Y., Vannier, M., 2009. Why do commercial CT scanners still employ traditional, filtered back-projection for image reconstruction? Inverse Problems 25 (12), 123009.

Pankhurst, M.J., Vernon, R.H., Turner, S.P., Schaefer, B.F., Foden, J., 2011. Contrasting Sr and Nd isotopic behaviour during magma mingling: new insights from the Mannum A-type granite. Lithos 126, 135–146.

Pankhurst, M.J., Schaefer, B.F., Turner, S.P., Argles, T., Wade, C.E., 2013. The source of A-type magmas in two contrasting settings: U–Pb, Lu–Hf and Re–Os isotopic constraints. Chem. Geol. 351, 175–194.

Pankhurst, M.J., Dobson, K.J., Morgan, D.J., Loughlin, S.C., Thordarson, T.H., Lee, P.D., Courtois, L., 2014. Technical note monitoring the magmas fuelling volcanic eruptions in near-real-time using X-ray micro-computed tomography. J. Petrol. 55 (3), 671–684.

Pankhurst, M.J., Fowler, R., Courtois, L., Nonni, S., Zuddas, F., Davis, G.R., Atwood, R.C., Lee, P.D. submitted. Enabling three-dimensional densitometric measurements using laboratory source X-ray micro-computed tomography: SoftwareX.

Paredes-MariÃño, J., Dobson, K.J., Ortenzi, G., Kueppers, U., Morgavi, D., Petrelli, M., Hess, K.-U., Laeger, K., Porreca, M., Pimentel, A., Perugini, D., 2017. Enhancement of eruption explosivity by heterogeneous bubble nucleation triggered by magma mingling. Sci Rep 7. doi: 10.1038/s41598-017-17098-3, Article number: 16897.

Philpotts, A.R., Dickson, L.D., 2000. The formation of plagioclase chains during the convective transfer in basaltic magma. Nature 406, 59–61.

Pistone, M., Arzilli, F., Dobson, K.J., Cordonnier, B., Reusser, E., Ulmer, P., Marone, F., Whittington, A.G., Mancini, L., Fife, J.L., Blundy, J.D., 2015. Gas-driven filter pressing in magmas: insights into in-situ melt segregation from crystal mushes. Geology 43 (8), 699–702.

Prior, D.J., Boyle, A.P., Brenker, F., Cheadle, M.C., Day, A., Lopez, G., Peruzzo, L., Potts, G.J., Reddy, S., Spiess, R., Timms, N.E., Trimby, P., Wheeler, J., Zetterstrom, L., 1999. The application of electron backscatter diffraction and orientation contrast imaging in the SEM to textural problems in rocks. Am. Mineral. 84 (11–12), 1741–1759.

Renggli, C.J., Wiesmaier, S., De Campos, C.P., et al., 2016. Magma mixing induced by particle settling. Contrib. Mineral. Petrol. 171, 96, https://doi.org/10.1007/s00410-016-1305-1.

Špillar, V., Dolejš, D., 2015. Melt extraction from crystal mushes: numerical model of texture evolution and calibration of crystallinity-ordering relationships. Lithos 239:19–32. https://doi.org/10.1016/j.lithos.2015.10.001.

Stevenson, C.T.E., Grove, C., 2014. Laccolithic emplacement of the Northern Arran Granite, Scotland, based on magnetic fabric data. In: Advances in Volcanology. Springer, Berlin, Heidelberg.

Streck, M.J., Leeman, W.P., Chesley, J., 2007. High-magnesian andesite from Mount Shasta: a product of magma mixing and contamination, not a primitive mantle melt. Geology 35, 351–354.

Tepley, III, F.J., Davidson, J.P., 2003. Mineral-scale Sr-isotope constraints on magma evolution and chamber dynamics in the Rum layered intrusion, Scotland. Contrib. Mineral. Petrol. 145, 628–641.

Till, C.B., Vazquez, J.A., Boyce, J.W., 2015. Months between rejuvenation and volcanic eruption at Yellowstone caldera, Wyoming. Geology 43 (8), 695–698. doi: 10.1130/G36862.1.

Tracy, S.R., Roberts, J.A., Black, C.R., McNeill, A., Davidson, R., Mooney, S.J., 2010. The X-factor: visualizing undisturbed root architecture in soils using X-ray computed tomography. J. Exp. Bot. 61 (2), 311–313.

Turner, S.P., George, R., Jerram, D.A., Carpenter, N., Hawkesworth, C.J., 2003. Case studies of plagioclase growth and residence times in island arc lavas from Tonga and the Lesser Antilles, and a model to reconcile discordant age information. Earth Planet. Sci. Lett. 214, 279–294.

Vinet, N., Higgins, M.D., 2010. Magma solidification processes between Kilauea Volcano, Hawaii: a quantitative textural and geochemical study of the 1969–1974 Mauna Ulu lavas. J. Petrol. 51, 1297–1332.

Vinet, N., Higgins, M.D., 2011. What can crystal size distributions and olivine compositions tell us about magma solidification process in inside Kilauea Iki lava lake, Hawaii? J. Volcanol. Geotherm. Res. 208, 136–162.

Vukmanovic, Z., Barnes, S.J., Reddy, S.M., et al., 2013. Morphology and microstructure of chromite crystals in chromitites from the Merensky Reef (Bushveld Complex, South Africa). Contrib. Mineral. Petrol. 165:1031. https://doi.org/10.1007/s00410-012-0846-1.

Waight, T.E., Maas, R., Nicholls, I.A., 2000. Fingerprinting feldspar phenocrysts using crystal isotopic composition stratigraphy: implications for crystal transfer and magma mingling in S-type granites. Contrib. Mineral. Petrol. 139, 227–239.

Wallace, G.S., Bergantz, G.W., 2002. Wavelet-based correlation (WBC) of zoned crystal populations and magma mixing. Earth Planet. Sci. Lett. 202, 133–145.

Yao, Z.-S., Qin, K.-Z., Xue, S.-C., 2017. Kinetic processes for plastic deformation of olivine in the Poyi ultramafic intrusion, NW China: insights from the textural analysis of a $\sim$ 1700 m fully cored succession. Lithos 284–285 (2017), 462–476, https://doi.org/10.1016/j.lithos.2017.05.002.

Zellmer, G.F., Sakamoto, N., Iizuka, Y., Miyoshi, M., Tamura, Y., Hsieh, H.-H., Yurimoto, H., 2014. Crystal uptake into aphyric arc melts: insights from two-pyroxene pseudodecompression paths, plagioclase hygrometry, and measurement of hydrogen in olivines from mafic volcanics of SW Japan. In: Gómez-Tuena, A., Straub, S.M., Zellmer, G.F. (Eds.), Orogenic Andesites and Crustal Growth. Geological Society of London Special Publication 385(1):161–184.

FURTHER READING

Maire, E., Withers, P.J., 2014. Quantitative X-ray tomography. Int. Mater. Rev. 59, 1–43. doi: 10.1179/1743280413Y.0000000023.

Ubide, T., Galé, C., Larrea, P., Arranz, E., Lago, M., 2014. Antecrysts and their effect on rock compositions: The Cretaceous lamprophyre suite in the Catalonian Coastal Ranges (NE Spain). Lithos 206-207, 214–233, https://doi.org/10.1016/j.lithos.2014.07.029.

Jerram, D.A., Kent, A., 2006. An overview of modern trends in petrography: textural and microanalysis of igneous rocks. J. Volcanol. Geotherm. Res. 154, vii–vii10.

Holness, M.B., Vukmanovic, Z., Mariani, E., 2014. Assessing the Role of Compaction in the Formation of Adcumulates: a Microstructural Perspective. Journal of Petrology 58 (4), 643–673, https://doi.org/10.1093/petrology/egx037.

Jerram, D.A., Davidson, J.P., 2007. (Edited Elements volume). Frontiers in Textural and Microgeochemical Analysis. Elements 3 (4), 235–238, https://doi.org/10.2113/gselements.3.4.235.

CHAPTER 9

Destroying a Volcanic Edifice—Interactions Between Edifice Instabilities and the Volcanic Plumbing System

Audray Delcamp*, S. Poppe*, M. Detienne, E.M.R. Paguican†,‡**
*Vrije Universiteit Brussel, Brussels, Belgium;
**Earth and Life Institute, Université catholique de Louvain, Louvain-la-Neuve, Belgium;
†University at Buffalo, The State University of New York, Buffalo, NY, United States;
‡Earth System Sciences, Vrije Universiteit Brussel, Brussels, Belgium

Contents

9.1 INTRODUCTION

Volcanoes are complex systems with constantly evolving landscapes. Their morphology is the result of multiple, sometimes reciprocal, interactions between regional tectonics, pre-existing structures, volcano tectonics (deep and shallow), intrusive activity, the type

Volcanic and Igneous Plumbing Systems. http://dx.doi.org/10.1016/B978-0-12-809749-6.00009-1
Copyright © 2018 Elsevier Inc. All rights reserved.

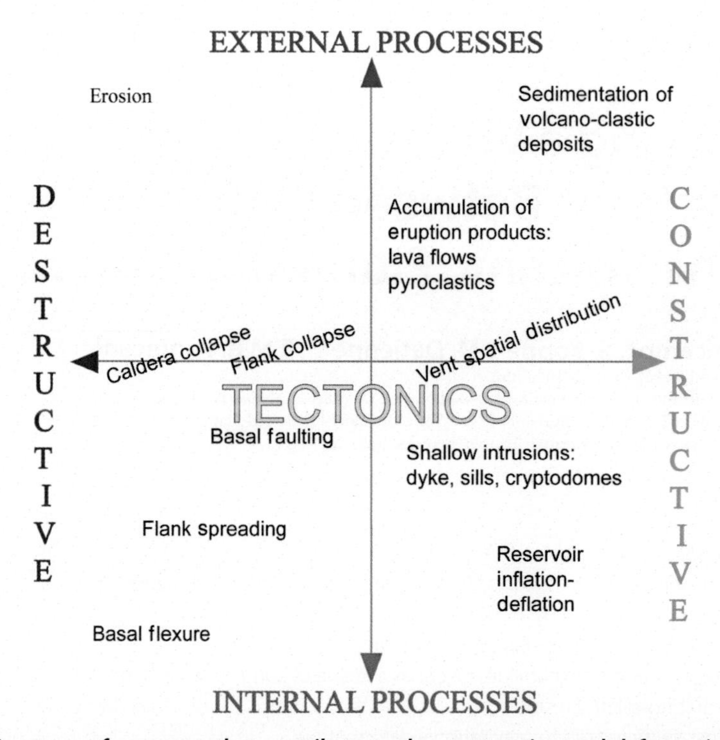

Figure 9.1 ***Spectrum of processes that contribute to the construction and deformation of volcanic edifices.*** Internal processes are directly linked to or induced by the volcanic and igneous plumbing system, whereas external processes originate from a source independent of the volcano. Additionally, constructive processes actively add volume (and height) to a volcano, whereas destructive processes remove volume from the volcanic system.

of eruptive and hydrothermal activity, erosion (see Chapters 4, 7 and 11) and gravity. All these factors, which can be grouped into internal versus external factors, play a major role in constructive and destructive processes (Fig. 9.1). This chapter focuses on the destructive processes affecting volcanoes and their consequences on the volcanic and igneous plumbing system (VIPS).

Two short-term catastrophic processes can destroy large parts of a volcanic edifice: lateral and vertical collapses. Lateral collapse occurs following long-term volcano gravitational instability and can be triggered by different mechanisms. Another form of catastrophic lateral collapse is that of a highly viscous dome, but it will not be further discussed in this chapter. Vertical volcano collapse can occur as a result of magma withdrawal from the plumbing system, creating an often circular or elliptical topographic depression referred to as collapse caldera. The 1883 Krakatoa eruption is one of the modern and spectacular examples of vertical volcano collapse with an erupted volume of ~20 km^3 (Self and Rampino, 1981). Such vertical collapses are mainly triggered by different mechanisms compared to what is discussed in this chapter, and the reader is invited to read Chapter 10 for more information on caldera volcanoes and their plumbing systems.

A major breakthrough in volcano instability and collapse knowledge came following the May 1980 lateral collapse and eruption of Mt St Helens volcano, Cascade Range, USA (Glicken, 1996). This was the first time that people extensively documented, through eyewitness accounts and pictures, a flank collapse, details of which are elaborated at the end of this chapter. Past evidence of flank collapses were also retrieved from historical accounts at other volcanoes, such as the Bandai AD 1888 collapse in Japan (Sekiya and Kikuchi, 1980), beautifully painted by the famous Inoue Tankei. Bandai volcano had successive phreatic eruptions of up to 20 times a minute ejecting ash, lapilli and blocks until the final one triggered a collapse of the northern flanks that formed a debris avalanche.

Since 1980, when volcanologists started studying the processes of volcano instability and resulting debris avalanches, it has become evident that flank collapse is a common process in the evolution of volcanic edifices. Morphological and geological traces of collapses have been identified worldwide at volcanoes in different tectonic settings, ranging from stratovolcanoes such as Mt St Helens (Glicken, 1996) to oceanic islands such as Tenerife (Ablay and Hürlimann, 2000), and small cinder volcanoes such as Lemptégy, France (Delcamp et al., 2014). Moreover, geologists documented the occurrence of multiple collapses on the same edifice, such as at Mt Kenya (Schoorl et al., 2014) and Mt. Iriga in the Philippines (Paguican et al., 2012). In such manner, more and more insights into volcano instabilities and debris avalanche emplacement have been gained from field observations, remote sensing, numerical and experimental modelling and rock mechanics. However, many questions still remain open due to the complexities of the interactions and rheology within the sliding mass. Such issues include the factors controlling the timing and geometry of the collapse and the physical processes causing the mass to slide.

Since a flank collapse unloads part of the volcanic edifice, it can drastically alter the stress field within the volcano, especially for voluminous and deep-seated collapses. This unloading has consequences on the future development and growth of the volcanic plumbing system.

In this chapter, we will review the processes of short- and long-term deformation leading to instabilities. We will then focus on the triggers and effects of flank collapse on the underlying plumbing system. We close the chapter with more detailed discussion on two well-documented examples of volcano collapse: the May 1980 Mt St Helens eruption and the Tenerife oceanic shield in the Canary Islands.

9.2 WHEN VOLCANIC EDIFICE INSTABILITIES OCCUR DUE TO SHORT-TERM DEFORMATION

9.2.1 Instability Caused by Shallow Magma Emplacement

Magma emplacement at shallow depths occurs on short time scales of hours to months, resulting in abrupt ground deformation that contributes to edifice instability (Fig. 9.2A).

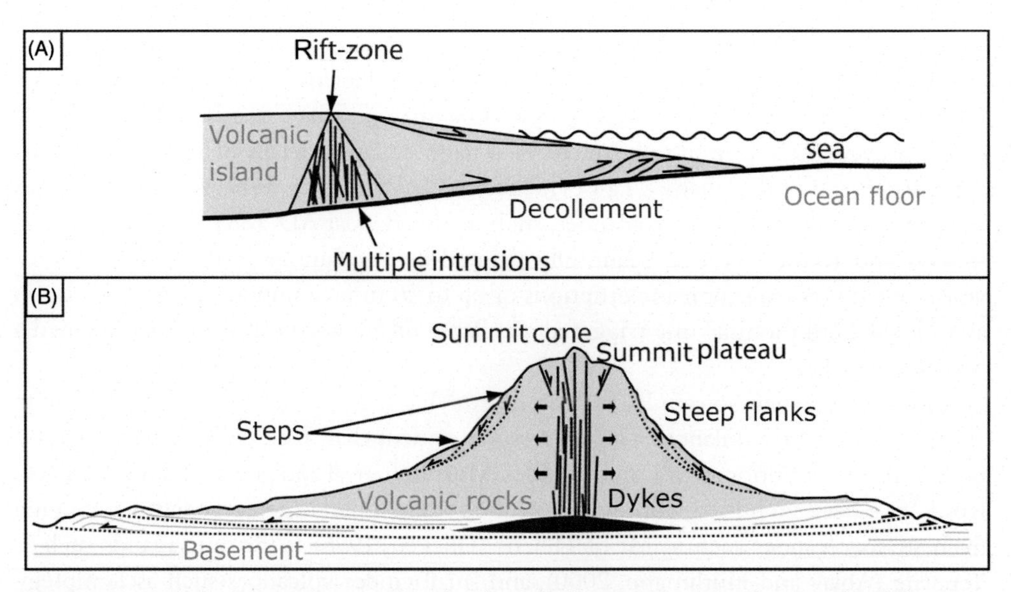

Figure 9.2 (A) Flank deformation associated with multiple intrusions in a rift zone. The deformation is accommodated by a basal decollement. *Modified from Dieterich (1988).* (B) Multiple dyke intrusions steepened and push the flank of Mt Cameroon volcano. *Modified from Mathieu et al. (2011a).*

Intrusion-related edifice deformation is observed at the rift zones of Kilauea, Hawaii, where repeated dyke emplacement over short time periods combined with gravitational relaxation of the edifice leads to seaward flank spreading (see also Chapter 4). Intrusive activity at Kilauea has been observed to increase the rate at which the Hilina slump moves towards the Pacific Ocean (8–10 cm per year; Morgan et al., 2003; Poland et al., 2017). Monitoring of surface deformation during periods of volcanic unrest using ground-based geodesy or remote sensing attributes rapid ground deformation to the emplacement of dykes, sills, laccoliths and other intrusions that can potentially lead to an eruption (Cannavò et al., 2015) (see also Chapter 11). Examples of such magma intrusion-related deformation cycles have been observed in dominantly basaltic systems not only in Kilauea (Jo et al., 2015; Poland et al., 2017) but also at Bardarbunga, Iceland (Gudmundsson et al., 2016) and at silicic volcanoes like Cordón Caulle (Castro et al., 2016). Ground deformation patterns differ depending on the type of intrusion, for example, low-viscosity magma that propagates as a dyke versus high-viscosity magma that intrudes as a laccolith or cryptodome.

Shallow ground deformation associated with the emplacement of magma can sometimes precede an eruption. In rare cases, intrusion-related ground deformation is followed by a catastrophic collapse, such as the Mt St Helens cryptodome intrusion and flank collapse (Glicken, 1996) (see end of this chapter). Intrusion-related ground deformation is often blurred by other volcano instability processes or surface features, such as vegetation, and only in a seldom case ground deformation provides an unequivocal indicator of magma intrusion, or a precursor to a volcanic eruption. At Etna volcano in Sicily, ground deformation patterns were mistakenly interpreted as related to the intrusion of a dyke, but later on appeared to have indicated the shallow failure of recently erupted and solidified lava flow packages on the volcano's steep slopes (Bonforte and Guglielmino, 2015). It is clear, however, that shallow magma emplacement disrupts and deforms volcanic edifices, alters the local stress field and thus contributes to their instability.

9.2.2 Instability Due to Ice Cap/Glacier Melting

Melting of an ice cap or glacier overlying a volcanic edifice feeds water into the edifice, contributing to hydrothermal rock alteration and high pore-water pressures (Carrasco-Núñez et al., 1993). If groundwater and alteration are focussed on certain zones within the volcanic edifice, layers or regions with low strength may be formed, providing interfaces for edifice destabilisation. In addition, ice melting removes a certain mass of ice and unloads the edifice itself, changing the gravitational stresses acting on the edifice and the underlying volcanic plumbing system.

All these processes can create instabilities within geologically short periods, either within a single edifice or on a regional scale.

9.3 WHEN VOLCANIC EDIFICE INSTABILITIES OCCUR AS A RESULT OF LONG-TERM DEFORMATION

9.3.1 Faulting-related Instability

Since the occurrence of volcanism is highly linked to tectonically active regions, it is common that volcanoes are underlain by faults (see Chapter 7). This affects the morphology and stability of the edifice as well as the potential orientation of collapse (Fig. 9.3). The effect of tectonic style on the geometry of the fractures and orientation of potential collapse has been studied using remote sensing and experimental modelling (van Wyk de Vries and Merle, 1998). Geological case studies and analogue modelling show that (1) a purely extensional or compressive regime results in collapses perpendicular to the fault; (2) a strike-slip regime results in collapses parallel, or at a low angle, to the fault and (3) a transtensional regime results in collapses similar to the ones produced by strike-slip, but the collapse scar is expected to be wider (Lagmay and Valdivia, 2006; Wooller et al., 2009; Mathieu et al., 2011b; van Wyk de Vries and Davies, 2015). To our knowledge, sector collapse in compressive regimes has only been studied through structural field work and remote sensing (Nakamura, 1977) and no experimental or numerical modelling has addressed this issue yet.

Structures induced by underlying faults, such as sigmoidal summit depressions, can create a favourable strain field and space for subsequent magma emplacement (van Wyk de Vries and Merle, 1998). The effect of regional faults and pre-existing structures on volcanoes and VIPS is further discussed in Chapter 7.

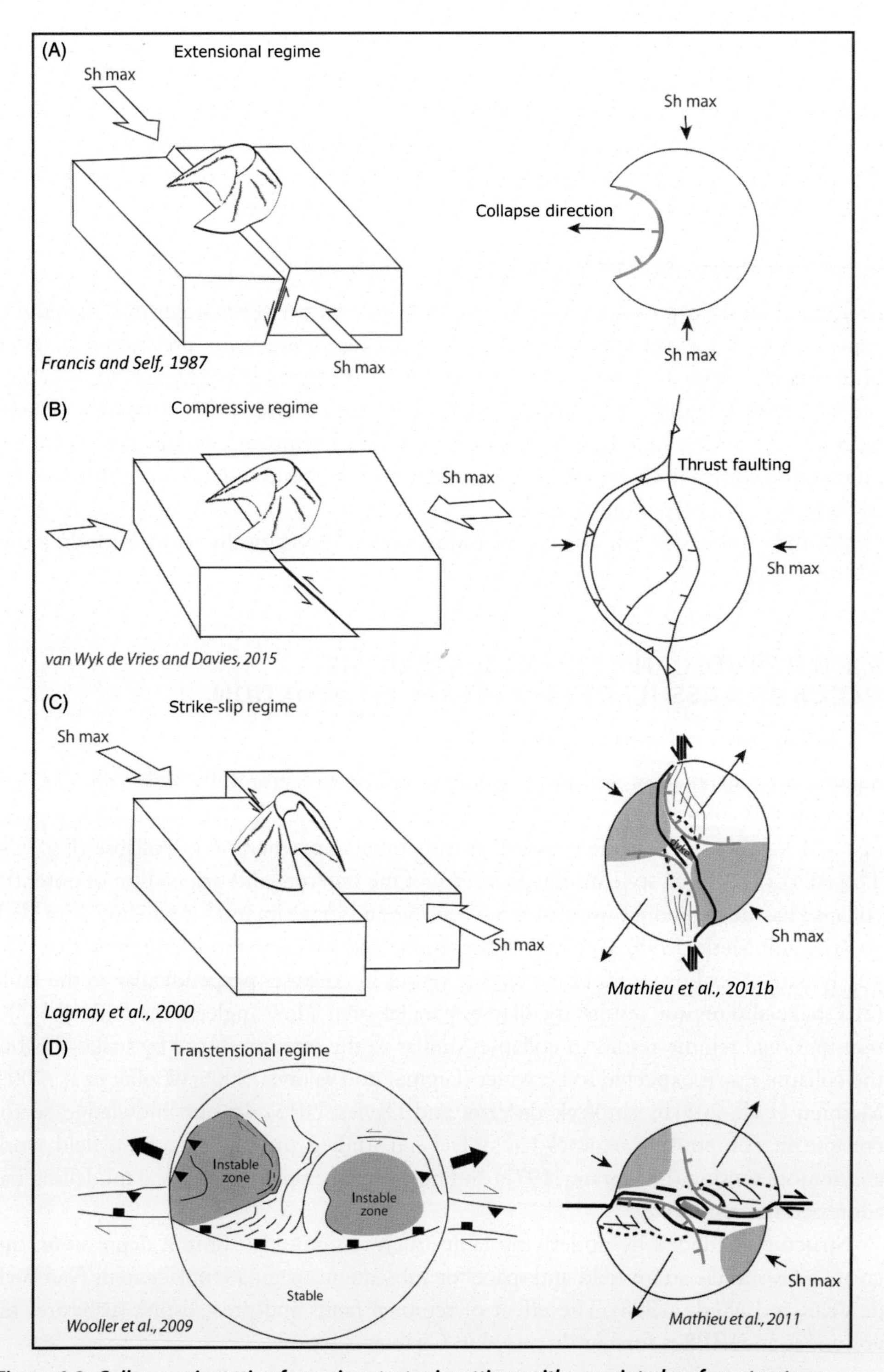

Figure 9.3 *Collapse orientation for various tectonic settings with associated surface structures.*

9.3.2 Instability Related to Gravitational Deformation

The localised load of intrusions and eruptive products causes volcanic edifices to deform under gravitational stress. Gravitational deformation is enhanced if the volcano is underlain by a weak basement or is made up of one or several weak units that may deform in a ductile way. Weak units, further referred to as low-strength layers (LSLs), include hydrothermally altered and weathered layers, salt, volcanic ashes or clays, lake sediments (Merle and Lénat, 2003; Oehler et al., 2005). Strong units deforming in a brittle way are composed of, for example, gneiss or granite in the basement and lava flows and intrusions in the volcanic edifice and its plumbing system.

The two endmember types of long-term gravitational deformation are spreading and sagging, with a transitional spectrum in between. The type of gravitational deformation in a volcanic edifice is controlled by the height of the volcano and by the ratio of the thickness of the underlying low- and high-strength basement layers (Fig. 9.4; Borgia, 1994; Byrne et al., 2013; Holohan et al., 2013).

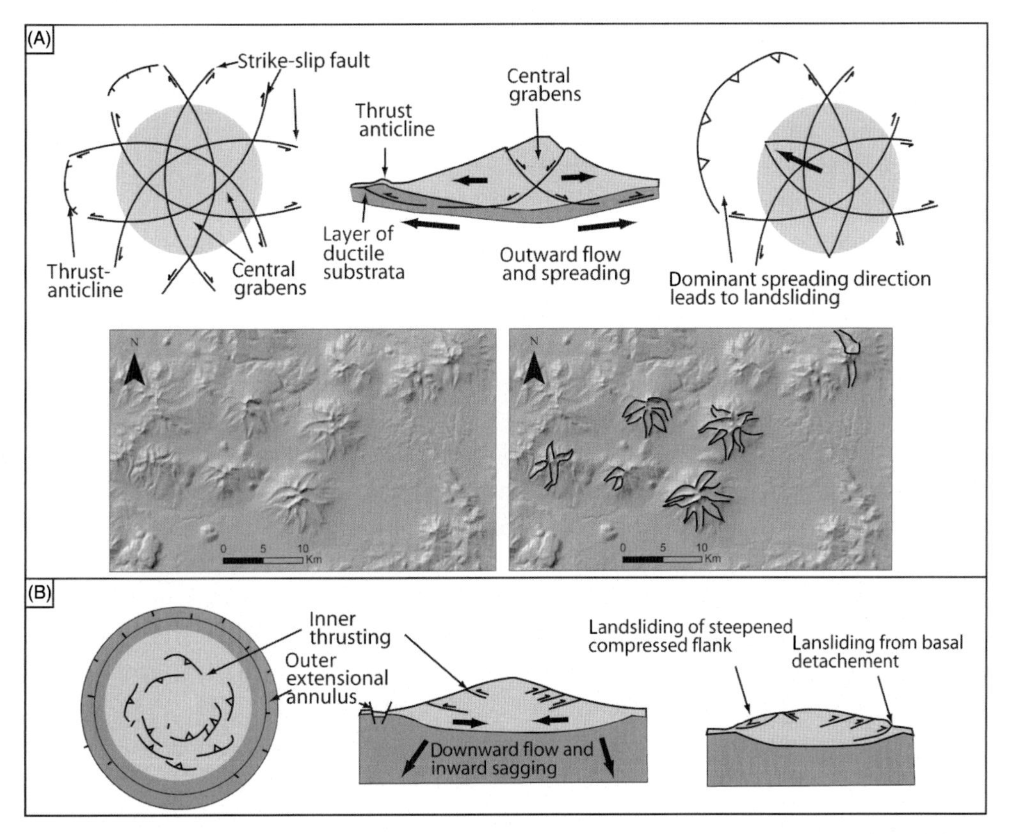

Figure 9.4 (A) Conceptual sketch of a spreading volcano and associated structures, with a hillshade image of the DEM-derived SRTM of some Bolivian volcanoes showing flower-like structures typical of spreading volcanoes. (B) Conceptual sketch of a sagging volcano and associated structures modify from van Wyk de Vries and Davies (2015).

In its simplest setting, spreading generally occurs when a volcanic edifice sits on top of a thin LSL. The edifice load applies a vertical lithostatic pressure on the LSL, which consequently progressively flows laterally, causing the edifice to spread. For a perfectly symmetrical volcano, this phenomenon results in decompression of the edifice itself and formation of summit grabens around the summit in a flower-like configuration with basal thrust and strike-slip faults accommodating the deformation (Fig. 9.4A). Examples of volcanoes that show symmetric spreading-like morphological structures include Sajama and Phasa Willk'i in Bolivia (Delcamp et al., 2008). However, the absence of long-term deformation data nor conclusive structural field evidence precludes any confirmation on the spreading origin of these structures. Spreading of the edifice is dependent on the slope, extent and location of the LSLs. For example, asymmetrical deformation of the edifice occurs when an LSL is underlying one sector of the volcano, causing spreading on that particular sector only. A classic example of this is Etna volcano in Italy (Borgia, 1994), although it remains difficult to attribute unambiguously volcano deformation to spreading alone.

Sagging, on the other hand, generally occurs when a thick LSL and thin brittle layers underlie an edifice. Due to the volcanic load, the edifice progressively sinks, or 'sags' into, and flexures, its basement. The edifice itself is increasingly compressed, resulting in terrace formation and flexure of the edifice. As of today, this deformation regime has only been identified in analogue experiments.

Spreading and sagging deformation regimes span a continuum, so it is possible to observe compressed and decompressed zones within the edifice that exhibit combined deformation structure characteristics of the two endmember styles (Byrne et al., 2013; Holohan et al., 2013). For similar volcano height, a low ratio of the thickness of high-strength versus LSLs favours sagging, while the opposite setting will favour spreading. Olympus Mons, for example, has both compressional terraces due to basement flexure (sagging) and a steep basal escarpment due to extensional spreading over a basal thrust fault (Musiol et al., 2016).

Sector and volcano spreading are dominated by extensional deformation with local compression and strike-slip movements to accommodate extension. Sagging, on the other hand, occurs dominantly as compressional deformation and is less likely to create strong instability. The presence of thrusts and basal detachments in transitional regimes with dominance of sagging can nevertheless be responsible for the formation of radial normal faults on the lowermost flanks, increasing the risk of future collapse (van Wyk de Vries and Davies, 2015; Musiol et al., 2016). Spreading is a slow process that creates extension in the edifice, which can eventually lead to landslides if the movement is accelerated by some processes such as intrusion (van Wyk de Vries and Davies, 2015). On the other hand, extension can stabilise the edifice by lowering slopes, but it can also create cliffs, that is, sub-vertical slopes, which might in turn be unstable (Oehler et al., 2005).

Long-term deformation of volcanoes is generally difficult to study through fieldwork, geophysical and geodetic monitoring techniques because of the low deformation rates. To overcome this timescale effect, the main tool used to study long-term deformation is through modelling, in particular, through analogue/laboratory experiments and numerical models. Laboratory and numerical models have revealed that gravitational deformation not only affects the volcanic edifice, but also the VIPS and eruptive activity. The spreading regime favours the emplacement of intrusion complexes including sills, dykes and other intrusions that can themselves spread and contribute to volcano instability (Delcamp et al., 2012b). Sagging, on the other hand, limits the emplacement of magma within or at the base of the volcanic edifice due to the effect of loading and compressional 'locking' of the edifice (Pinel and Jaupart, 2000). The effect of gravitational deformation and collapse on eruptive activity is outlined in detail in Section 9.5.3.

9.3.3 Instability Related to Hydrothermal Alteration

The study of debris avalanche deposits sourced from volcanic edifice collapses suggests that hydrothermal rock alteration plays a key role in promoting failure (Fig. 9.5A; Pevear, 1987; Voight et al., 2002; Salaün et al., 2011). The effect of hydrothermal alteration within a volcanic edifice is manifold (Carrasco-Núñez et al., 1993; Vallance and Scott, 1997). Firstly, hydrothermally altered rocks are generally mechanically weaker compared to fresh rocks. The weakening results from secondary minerals that form during hydrothermal alteration, such as clay minerals (intermediate argillic alteration) and amorphous silica (residual silica), that have low shear strengths (Fig. 9.5B; del Potro and Hürlimann, 2008; Pola et al., 2012; Heap et al., 2014; Mayer et al., 2016). The presence of such secondary minerals contributes to a decrease in friction and cohesion within the sliding mass, as well as between the mass and its basement. This effect accelerates the development of slide processes that commonly constitute the inception phase of the edifice failure. However, when precipitation processes prevail during alteration instead, for example, during rock silicification (Horwell et al., 2013), friction and cohesion may increase, resulting in strengthening of the rock mass (Fig. 9.5B). Altered rocks also have a tendency to deform by creep rather than by fracturing, which promotes volcano spreading. Furthermore, the formation of secondary minerals during alteration can lead to the gradual closing of fractures and pores, reducing the porosity and permeability of the altered rocks (Lowell et al., 1993; Berger and Henley, 2011; Horwell et al., 2013). A reduction of rock permeability promotes the development of high pore fluid pressures within the volcanic edifice during rainfall or magmatic gas expansion events (Day, 1996; Reid, 2004). High pore fluid pressures favour in turn pressurisation of faults, joints and more permeable layers of rock, reducing the general strength of the volcanic edifice.

Hence, the overall volcanic edifice stability is influenced by the competition between all these effects and their evolution with time, which is difficult to assess. However, in general, the impact of hydrothermal alteration on the physical and mechanical properties

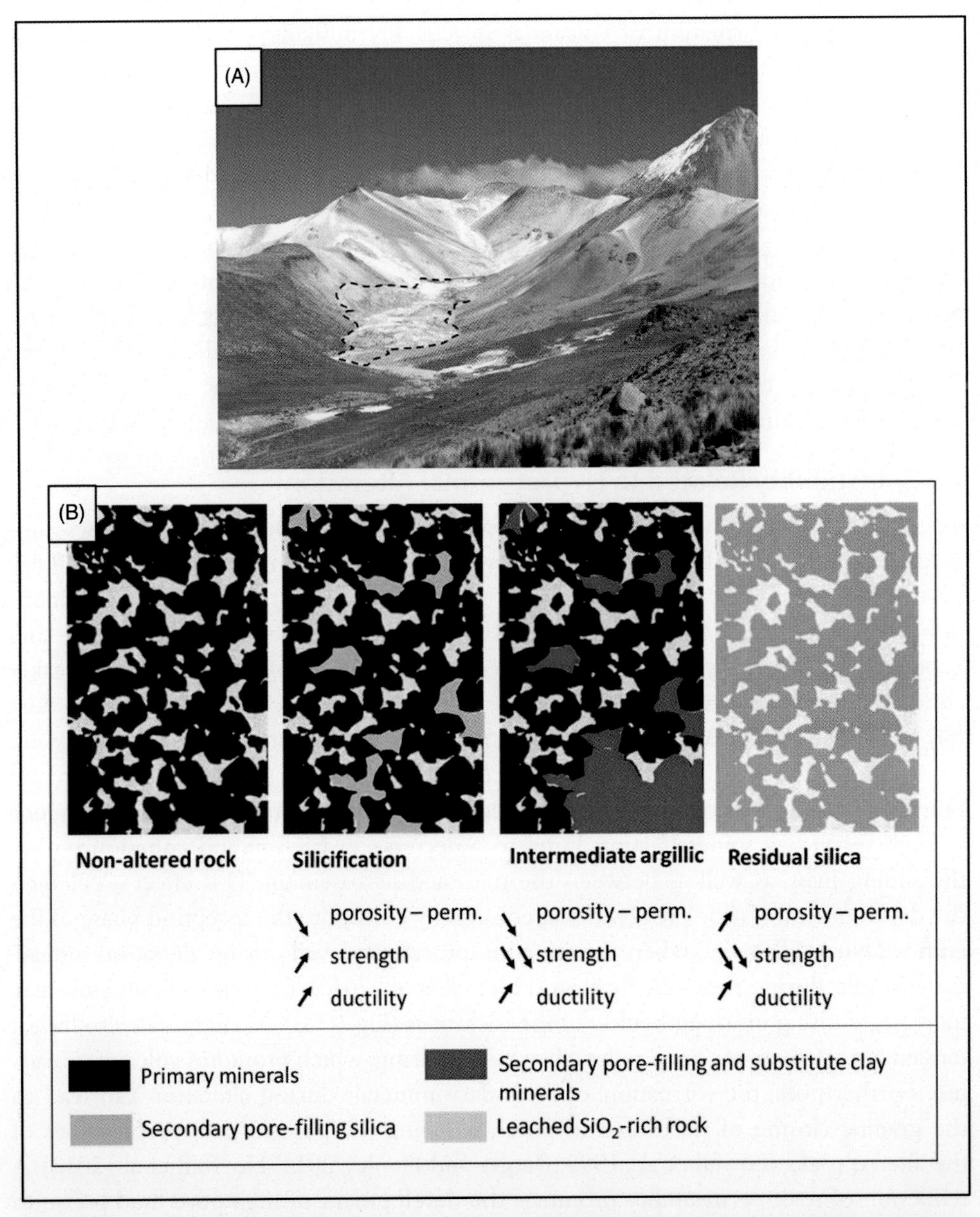

Figure 9.5 (A) Hydrothermally altered rocks of a collapsed edifice and the resulting debris avalanche deposits (dashed contours) at Tutupaca volcano, Peru (photo by Detienne, 2016). (B) Different effects of silicification, intermediate argillic alteration and leaching of rocks (residual silica) on their physical and mechanical properties: impacts on porosity and permeability (perm.), strength and ductility properties. *Arrows* indicate the evolution of altered rocks properties compared to those of an unaltered rock (upward = increase, downward = decrease) (adapted from Detienne 2016).

of rocks is most likely to lead to reduction in volcano stability. This is demonstrated by the frequent presence of altered rocks in debris avalanches, as well as by the apparent connection between altered areas and volcano flank collapses (Siebert et al., 1987; Detienne, 2016).

9.3.4 Instability Due to Long-Term Magma Emplacement

Magmatic intrusions make up a significant part of a volcanic edifice and include shallow magma reservoirs and sheet intrusions, such as discordant dykes, cone sheets (see Chapter 4) and concordant sills (see Chapter 5). These magma bodies comprise the VIPS and significantly contribute to volcano growth (Annen et al., 2001) (Chapter 4). As the volcanic plumbing system develops and grows over time scales of up to a million years, increased volumes of magma intrude into the edifice, which forces the volcano to accommodate the extra volume. Consequently, the volcano will deform by folding, uplift and breaking the surrounding layers (see also Chapters 5 and 11). The evolving volcanic plumbing system therefore strongly modifies the morphology of the volcano.

In particular, the emplacement of magma within the volcanic plumbing systems may create edifice instability in the long term by pushing the volcano flanks aside to accommodate repeated magma intrusion. This push may effectively steepen the flanks at higher-elevation slopes, as it has been proposed for the steep upper slopes of Mt. Cameroon (Fig. 9.2B; Mathieu et al., 2011a). This steepening can ultimately lead to volcano flank collapse (Elsworth and Voight, 1995). The development of rift zones through repeated dyke intrusion in basaltic volcanoes may lead to sustained flank creeping or incremental flank sliding, the dynamics of which depend on the volcano's size, intrusion return time, volume and rate. Geodetic signals at Kilauea, Etna and Piton de la Fournaise help in such way to conceive a conceptual framework of volcano flank instability (Poland et al., 2017).

Magma emplacement can also increase hydrothermal activity which, in turn, will increase pore pressure. Increasing pore pressure might again lead to volcano instability (see Section 9.3.3). On the other hand, repeated intrusion of dykes into volcanic rift zones on a gravitationally deforming volcano might impede the growth of certain types of normal faults, hence stabilising (part of) the flank of the edifice (Le Corvec and Walter, 2009) (see also Chapter 4).

Hence, the emplacement of different types of intrusions may interact with one another, and their specific configuration and timing might impede or favour volcano destabilisation.

9.3.5 Instability Due to Erosion

As volcanoes undergo dramatic topographical changes during their lifetimes (Thouret, 1999), bare and undissected, morphologically 'young-looking' volcanoes can transform into vegetated landscapes with deeply entrenched gullies and erosional valleys, as the flanks become less permeable (Ingebritsen and Scholl, 1993). The decrease in permeability alters the way water moves over and beneath the volcano surface (Jefferson et al., 2010). A combination of extremely heavy rainfall, tectonic control (fault systems and rift zones) and alternating beds of more and less resistant and permeable volcanic material (Join et al., 2005) control how an edifice erodes over time (Karátson et al., 1999).

Precipitation facilitates river incision and sediment transport by increasing water discharge. It can also trigger landslides by elevating pore pressure and accelerating the production of erodible soil by promoting chemical and physical weathering. Moreover, precipitation fosters the growth of biota, which transport sediment and break down rock (Ferrier et al., 2013).

Erosional structures, particularly single-drained or irregularly shaped erosional basins, form depending on the type of pre-existing depression such as craters or volcano-tectonic depressions, such as calderas and crater groups (Karátson et al., 1999). Amphitheatre valleys on ocean island volcanoes, basaltic shield volcanoes along the tropical–subtropical belt and subduction-related volcanoes and lava domes are formed by the parallel retreat of slope due to repeated basal seepage erosion and subsequent small landslides (Karátson and Széley, 2006) or by breaching of a primary depression from several directions, thus degrading the summit region at high rates (Karátson et al., 1999). Erosion-induced depressions in volcanic terrains are also features formed by multiple depressions of fluvially captured craters or a quasi-closed flank depression.

Importantly, along with a volcano's fresh morphology resulting from volcanic eruptions, erosion gradually degrades the surface expressions of structural volcano deformation. In contrast to large-scale sector collapses that pervasively affect a volcano's shape, traces of smaller structures such as normal faults and horst-and-graben structures induced by volcano spreading become increasingly indiscernible from purely erosive features. This highlights the usefulness of identifying structural features on remotely sensed images and fieldwork to determine whether a volcano is affected dominantly by either deformation or erosion.

9.4 WHAT TRIGGERS VOLCANO COLLAPSE? HOW DO EDIFICES FAIL?

Volcano instabilities are usually the result of long-term and repetitive processes (see Section 9.2). After having developed instabilities, a volcano can experience catastrophic collapse triggered by one or a combination of several phenomena. Earthquakes associated with fault displacements beneath the volcano, magma emplacement directly pushing or tilting the unstable flank or increasing pore pressure following an increase of hydrothermal activity can result in volcano collapse. Since the associated energy is often not enough to trigger large mass movements, volcanic eruptions alone are less likely to promote flank collapse, but might be associated with small-scale collapses.

Nevertheless, Harris et al. (2011) attributed the Abona landslide south of Tenerife to a large-scale ignimbrite eruption. Certainly, a failure plain cutting through a growing intrusion will lead to an eruption as observed for Bezymianny (Belousov et al., 1999) or Mt St Helens (Glicken, 1996). Intense rainfall or fast glacier retreat can quickly saturate superficial layers, which can turn into a volcanic landslide (Hora et al., 2007; Roberti et al., 2017).

9.4.1 Shallow Failure

A shallow failure that occurs within the volcanic edifice itself, that is, close to the summit or towards the middle flanks, will affect merely a limited volume of the edifice (i.e. <1 km^3) and will form local escarpments and flank depressions, which do not strongly modify the morphology (Fig. 9.6). Nevertheless, shallow failure may be dangerous for the population in the immediate surroundings and will contribute to a local stress field modification. Depending on the water content in the failing rock body and weather conditions during the event, shallow collapse can turn into a debris or mud flow capable of causing more damage. For example, Mt Meager, Canada, produced a shallow collapse in 2010 of 50×10^6 m^3 that quickly turned into a debris flow (Guthrie et al., 2012; Roberti et al., 2017). Another example of a shallow failure is the 1998 debris flow of Casita triggered by intense rainfall and involved a volume of 1.6×10^6 m^3 (Kerle et al., 2003).

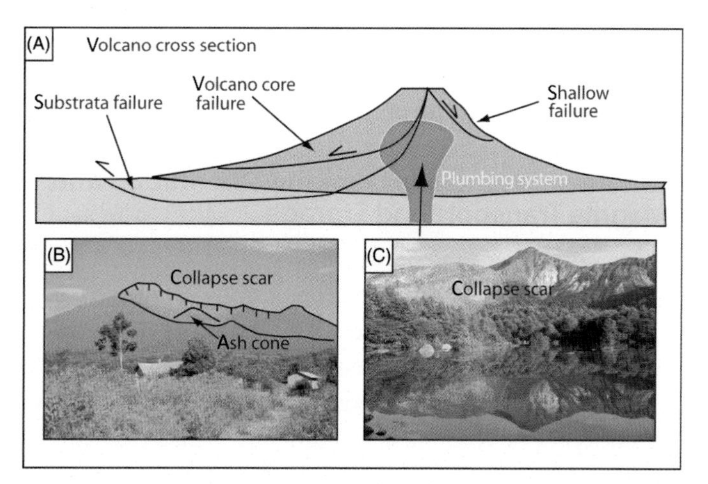

Figure 9.6 (A) Conceptual cross-section showing the location of shallow and deep failure in a volcanic edifice. (B) Scar of Meru volcano, Tanzania, with a growing ash cone within the embayment (picture: A. Delcamp). (C) Scar embayment of the 1888 collapse of Bandai San volcano, Japan (picture: A. Delcamp). The scar is cutting through the hydrothermally altered core of the volcano (white and yellow coloured rocks).

9.4.2 Deep Failure

A deep failure occurs when a deep-seated instability lies within the lower part of or below the volcanic edifice (Fig. 9.6). This is common in volcanoes with active and large hydrothermal system, or with an underlying LSL. In this case, the volcano and the substrata can catastrophically collapse altogether, resulting in a voluminous debris avalanche. This type of collapse strongly affects the stress field and volcano morphology by removing a large portion of the edifice, leaving a horse-shoe shape scar. Such failure will have a strong impact on the volcanic plumbing system. Examples of deep failures are the Socompa collapse (Chile/Argentina), the Mombacho collapse (Nicaragua) and the Bandai collapse (Japan) with collapsed volumes of 17, 1–2 and 1.5 km^3, respectively (Siebert, 1984; Francis et al., 1985; Shea et al., 2008).

Deep failure is also common in ocean island volcanoes that represent the largest volcanoes on Earth (see Chapter 7). Ocean island volcanoes rise steeply, up to 10 km from the ocean floor, where they rest on oceanic sediments (e.g. Hawaii, Canary Islands). Due to the high magma production, the flanks of these volcanoes are often over-steepened and unstable. Large horseshoe-shaped landslide scars testify to the frequent collapse of ocean island volcanoes (see Field example Tenerife). Notably, these collapses may not only lead to the destruction of large parts of the volcano and the island, but may also cause catastrophic tsunamis affecting the surrounding coasts (Hunt et al., 2013).

9.5 HOW DOES VOLCANO COLLAPSE AFFECT THE VIPS?

Volcano deformation and collapse significantly modify the orientation and magnitude of stresses within and directly below an edifice. The deep igneous plumbing system will tend to be less affected than the shallow volcanic plumbing system. Nevertheless, the stress field evolution will affect magma storage and transport.

9.5.1 Influence of Volcano Collapse on the Orientation and Structure of Magma Transport and Storage

The propagation mode of sheet intrusions (dykes, sills, cone sheets) is still a matter of debate, but their initiation and propagation are strongly controlled by the local stress field orientation and gradients (see Chapters 3–5). Stress fields occur at different scales and vary over time. Volcanoes generate their own stress field due to their gravitational force (load), the accumulation of pressurised magma bodies and the production and distribution of different types of volcanic and sedimentary materials (Kervyn et al., 2009). The volcanic stress field is usually local and restricted to the edifice and its immediate surroundings, whereas further away, regional tectonic stresses become dominant (Gudmundsson et al., 2016). Magma emplacement, collapse structure, hydrothermal alteration, erosion, gravitational deformation and other processes constantly modify the volcano stress field, which in turn influences these processes. Volcano and regional stress fields as well as pre-existing structures influence magma pathways and existing magmatic intrusions. However, it is difficult to quantify and apprehend to what extent this influence controls intrusion orientation. A few studies suggest that the regional stress field can influence the deep igneous plumbing system (up to crustal level), whereas the shallow volcanic plumbing system will be influenced by the local volcano stress field (Fiske and Jackson, 1972). The influence of regional tectonics and pre-existing structures is discussed in detail in Chapter 7. Here, we discuss the influence of volcano deformation and collapse on the shallow volcanic plumbing system.

In a symmetric and stable volcanic cone, we expect the shallow plumbing system to form radial patterns (Fig. 9.7A) (see also Chapter 4). Volcano instabilities change the stress field, which then affects subsequent intrusion orientation. One can speculate that a sagging regime impedes vertical intrusion of magma due to volcano loading and will favour sill formation while a spreading regime favours vertical sheet intrusion emplacement. Experimental modelling shows that viscous magma intrusions in a spreading volcano mimic the surface morphology and the magma intrudes into the formed graben (Fig. 9.7B). Furthermore, instability induces fractures, which, if suitably orientated, can be used by subsequent intrusions. For example, following episodes of multiple intrusion emplacements, a sector of a volcano can slowly spread and move outwards. To accommodate the local spreading, strike-slip movements will form *en-échelon* fractures along the creeping sector (Fig. 9.7C). Such space can be used by the ascending magma and the dykes will be oriented at an angle from the future scar walls as observed in, for example, Stromboli, Tahiti and Tenerife (Delcamp et al., 2012a; Hildenbrand et al., 2004; Acocella and Tibaldi, 2005). After collapse, new intrusions will tend to be oriented parallel to the collapse scar, following the un-buttressed flank, as observed in Lemptégy and Stromboli (Tibaldi, 1996; Delcamp et al., 2014). This structural configuration has also been observed through experimental modelling (Walter and Troll, 2003).

9.5.2 Post-Collapse Depressurisation and Eruption

After a shallow or deep collapse, the gravitational loading on the underlying volcanic plumbing system and the hydrothermal system suddenly decreases, which may trigger magmatic or phreato-magmatic eruptions. This was the case for, for example, Mauna Loa where depressurisation following a landslide induced phreatomagmatic explosions (Lipman et al., 1990), as well as for Mt St Helens (Glicken, 1996).

9.5.3 Influence of Volcano Collapse on Eruptive Activity and Geochemistry

Abrupt changes in the composition of erupting material after volcano collapse have been documented in many places. Often the eruption of evolved, low-density compositions prior to collapse suddenly shifts to primitive, high-density mafic

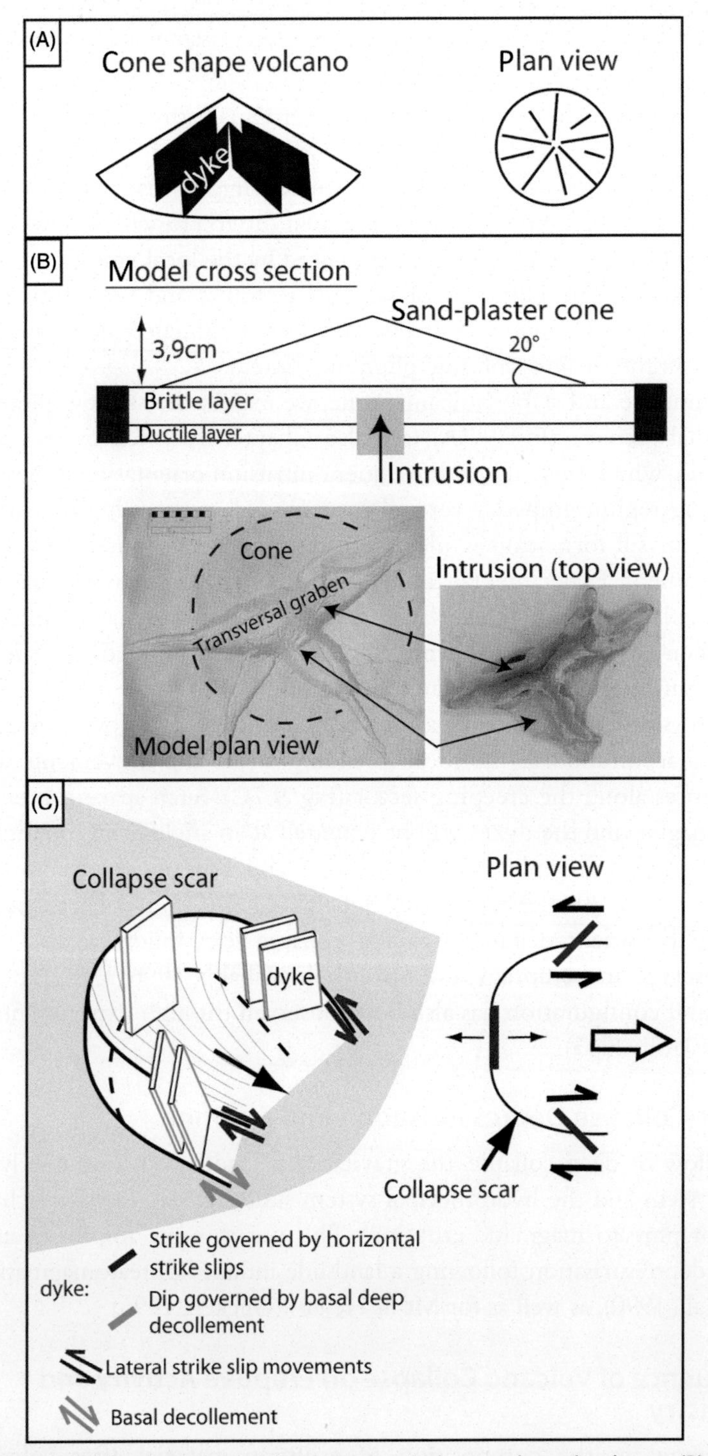

Figure 9.7 (A) Radial dyke arrangement in a symmetrical cone-shaped volcano. (B) Experimental model of a spreading volcano made of a sand-plaster cone resting on a putty silicone ductile layer. The viscous golden syrup intrudes the spreading volcano via the grabens (modified from Delcamp et al., 2012b). (C) Dyke orientation in a spreading volcano flank, as measured in the field. Dyke orientation (strike) is due to strike-slip movement along the spreading flank, whereas dyke dip is controlled by the basal decollement of the flank (Delcamp et al., 2012a).

magmas after the collapse, for example, Canary Islands (Longpré et al., 2009; Manconi et al., 2009) and Parinacota, Chile (Hora et al., 2007). The magma compositional shift is linked to the gravitational load increase as the volcanic edifice builds up (Pinel and Jaupart, 2000, 2005). The load can divert the upward propagation path of magma towards the break in slope at the volcano base (Kervyn et al., 2009), finally even impeding dense mafic magma to erupt centrally and thus favouring magma storage in shallow reservoirs. As stalled magma evolves, it becomes less dense and its buoyancy slowly increases (Chapter 8). If conditions are favourable, this less-dense magma might then erupt more easily. There is hence a strong interplay between edifice growth and magma differentiation. A collapse effectively removes load from the volcano, which decompresses the volcanic plumbing system. This decompression may result in volatile exsolution. Bubble nucleation and growth in the magma will promote a density decrease and magma remobilisation, which may lead to explosive eruptions during collapse. The removed load subsequently allows to erupt dense, mafic magma shortly after collapse, whereas it would have been trapped below the edifice in its pre-collapse loading state. Similarly, volcano spreading will favour tapping and eruption of mafic magma, whereas sagging will favour magma differentiation and eruption of low-density evolved magma (Pinel and Jaupart, 2000).

While geochemical and petrological studies have shown that magma unloading favoured decompression, resulting in an increase of partial melting (Presley et al., 1997; Hildenbrand et al., 2004), numerical modelling argues that unloading will have no effect on melt production at depth, but will rather induce pressure gradients within the deep magma plumbing system, enabling magma remobilisation as we describe above (Manconi et al., 2009). Volcano collapse thus seems to have an effect within the crust down to the upper mantle but does not affect deeper levels. Alternatively, dyke propagation requires reservoir overpressure, and a reservoir depressurisation following a collapse might impede dyke propagation and eruption (Pinel and Jaupart, 2005).

9.6 HOW FAST CAN VOLCANOES REBUILD AFTER COLLAPSE?

After collapse, most of the activity occurs within the scar embayment where volcanic load is reduced locally (Kervyn et al., 2009). Consequently, a new volcanic edifice grows within the scar embayment as at Meru volcano (Tanzania) and at Bezymianny (Russia). The rebuilding after catastrophic volcano collapse is commonly very rapid (Hora et al., 2007), which has been attributed to an increase in partial melting or rapid mafic magma remobilisation after depressurisation (Hildenbrand et al., 2004; Manconi et al., 2009). Unfortunately, few post-collapse eruptive rates have been calculated. However, at Tahiti-Nui, the eruptive rate was estimated at 5 km^3/kyr during the first 100 kyr following collapse and subsequently dropped to 0.5 km^3/kyr (Hildenbrand et al., 2004).

9.7 SUMMARY

Volcano instabilities form due to long- and/or short-term deformation, caused by, for example, magma emplacement, glacial retreat, faulting, hydrothermal alteration, and erosion. The causes of instability interact in complex ways that may either increase instability or contribute to stabilising the volcanic edifice. Long-term deformation occurs on a transitional spectrum that ranges from volcano spreading to volcano sagging, depending on the distribution and thickness of low strength layers in the edifice. Deformation changes the local volcanic stress field and produces mostly extensional structures in the spreading regime and mostly compressional structures in the sagging regime. Volcano spreading may therefore favour magma ascent and eruption, whereas sagging tends to trap magma at the base of the volcanic edifice favouring magma storage and differentiation.

Volcano deformation and destabilisation may lead to volcano failure and collapse that can occur on different scales. While shallow failure affects small volumes of the edifice only, deep failure may destroy the volcanic edifice and parts of its volcanic plumbing system. The latter can be associated with catastrophic collapses of the volcano flank and explosive eruptions. Following edifice collapse, higher rates of magmatic activity with the eruption of higher-density mafic magmas lead to rapid rebuilding of the volcanic edifice within the collapse scar.

FIELD EXAMPLES

Mt St Helens, Oregon, USA

On 18 May 1980, Mt St Helens volcano, one of several composite continental arc volcanoes of the northern Cascades experienced what would become the first catastrophic historic eruption recorded by modern monitoring equipment. After a month-long period of inflation on the north flank of the volcano, the flank bulge collapsed catastrophically, unroofing and decompressing the underlying cryptodome. The flank landslide evolved into a debris avalanche and was directly followed by a Plinian eruption with a Volcanic Explosivity Index of 5. This eruption became widely studied and stands as the textbook example of catastrophic volcano collapse.

Prior to its May 1980 events, Mt St Helens was a symmetrical cone-shaped edifice with a geological history of erupting lava flows and domes of dominantly high-viscosity dacite magma. No clear collapse scars appeared to exist in the morphology of the volcano (Fig. 9.8A). From March 1980 onwards, staff of the Cascades Volcano Observatory monitored the rapid inflation of the north flank of the volcano at rates of 1.5–2.5 m per day, forming a prominent flank bulge (Fig. 9.8B). By May 17, the movement had pushed the volcano's flank outwards and upwards by ~135 m, as was registered by angle and slope–distance measurements. Seismicity initiated at the volcano on March 20 and had its highest activity from 25 to 27 March, the day on which a first phreatic eruption took

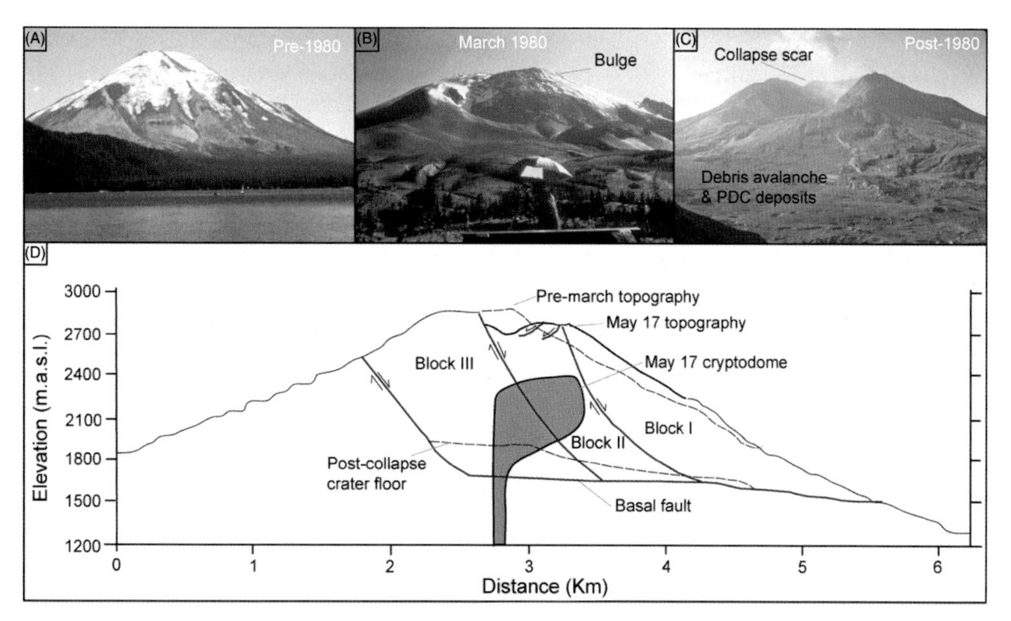

Figure 9.8 (A) Mt St Helens before the devastating May 1980 eruption seen from Lake Spirit to the north of the volcano. (B) During the spring of 1980, a cryptodome intruded into the volcano, resulting in an inflating flank bulge that is clearly visible on this iconic picture from 17 May 1980. (C) Mt St Helens a few days after the catastrophic 18 May 1980 events, which decapitated the volcano leaving a horseshoe-shaped open scar and devastated area of many km^2 where forest was knocked down and buried under debris avalanche deposits, volcanic ash and pyroclastic flows. All photo courtesy of USGS. (D) Structural sketch of the interior of Mt St Helens with the three envisioned sliding blocks and cryptodome intrusion; simplified from Glicken (1996).

place. After that, seismicity declined again. The source of the bulging was interpreted to be the shallow subterranean growth of a cryptodome, a body of high-viscous magma (see also Chapter 6). Glicken (1996) proposed that the week-long period of elevated seismicity coincided with the cryptodome intrusion clearing its path, after which it then gradually inflated within the volcano.

On 18 May, a magnitude 5.1 earthquake shook Mt St Helens. As a result, the steepened and rendered profoundly unstable north flank slid down as three retrogressive blocks, cutting through the cryptodome (Fig. 9.8D; Glicken, 1996). The slide and the consequent eruption were captured in the so-called 'Rosenquist' photo sequence (Walter, 2011). Experimental modelling with silicone putty intruding sand-plaster cones has reproduced bulges and fracturing-faulting structures similar to those observed at Mt St Helens (Donnadieu et al., 2001).

As the three blocks slid down in a landslide that incorporated ~13% of the edifice volume, they disintegrated further and turned into a devastating debris avalanche which was then constrained within the North Fork Toutle river and turned into a debris flow, running far outside the then established exclusion zone.

The slide cut deep into the cryptodome, decompressing it and giving way to a violent expansion of magmatic gas generating a highly energetic pressure wave, also called lateral blast. The subsequent eruption created an eruptive column rising up to 19 km height. The total volume of the debris avalanche deposit has been estimated at 2.5 km^3, while the total volume of the entire eruption, including the landslide, lateral blast, ash plume eruption and pyroclastic flows, has been estimated at ~4 km^3.

The eruption blasted away entire pine forests and caused significant infrastructural damage and 57 fatalities, including USGS volcanologist David A. Johnston who was tasked with observing the volcano in that fatal moment. The resulting debris avalanche and pumice deposits are well studied, thanks to the extensive field observations of Harry Glicken (1996), who later was killed by pyroclastic flows while observing the 1991 eruption of Mt. Unzen, Japan.

After the event, the volcano's summit was lowered by ~400 m and exhibited a 1-km wide horseshoe-shaped crater open towards the north (Fig. 9.8C). Such horseshoe-shaped scarps and associated debris avalanche deposits have since been identified at many volcanoes worldwide, and the documentation of the Mt St Helens May 1980 eruption has been instrumental to understanding many volcanic processes ever since. The unloading of the edifice by the lateral blast allowed the emplacement of two consecutive dacite domes within the scar embayment in 1989–1990 and 2004–2008, each time accompanied by explosive eruption magnitudes smaller compared to the 1980 cataclysm. Seismic activity continues below the volcano to this day, testifying the ongoing reloading of the magmatic system below Mt St Helens with fresh magma.

Tenerife, Canary Archipelago, Spain

Tenerife is an oceanic island that belongs to the Canary Archipelago, west of the African continent (Fig. 9.9). The island is made of three old basaltic shield volcanoes, Roque del Conde, Teno and Anaga, which built up between ~11 and 4 Ma (Thirlwall et al., 2000). After a volcanic hiatus of ~2 Ma, renewed activity formed Las Cañadas central volcano with more evolved compositions (Ancochea et al., 1990). Following a collapse at ~160–175 ka destroying Las Cañadas volcano, the Teide-Pico Viejo volcano has grown within the Las Cañadas scar (Masson et al., 2002; Boulesteix et al., 2012). Along with Teide-Pico Viejo, the recent volcanic activity has been concentrated along the rift zones of the island (Carracedo et al., 2007), with the last eruption occurring in 1909 in the northwest rift (Chinyero eruption).

Landsliding is a common process in the evolution of ocean island volcanoes as shown by the abundant collapse scars and widespread debris avalanche deposits observed offshore (Boulesteix et al., 2012; Hunt et al., 2014). In the Canaries, the volumes of the debris avalanche deposits observed through bathymetry range from 50 to 500 km^3 (Masson et al., 2002). This range highlights that mass wasting is a common phenomenon and important in shaping oceanic islands.

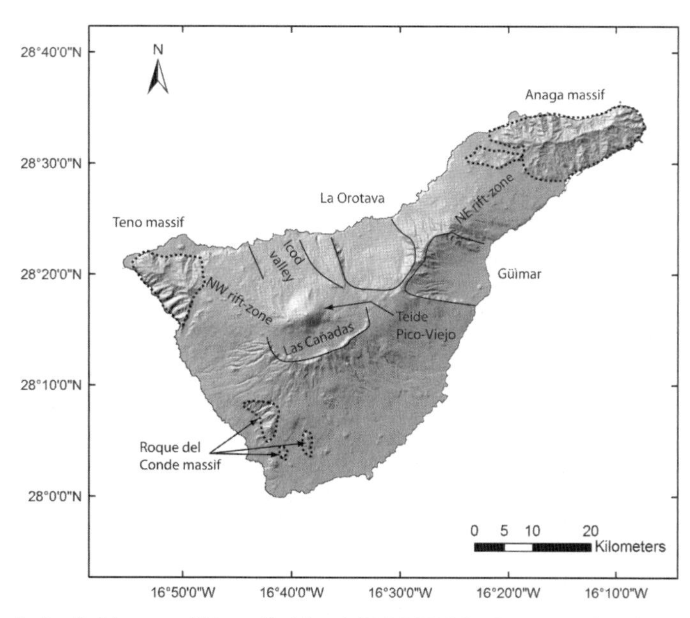

Figure 9.9 Shaded relief image of Tenerife island (DEM-SRTM), Canary Archipelago, Spain, with the main morphological and structural features highlighted.

Buried scars of previous collapse have been found in Tenerife, for example, in the Teno massif (Longpré et al., 2009), and the present geomorphology of the island is strongly marked by three horse-shoe-shaped depressions open towards the sea, called Icod, La Orotava and Güìmar (Masson et al., 2002; Hürlimann et al., 2004; Hunt et al., 2011). Debates over the origin and triggering mechanisms of these landslides exist, putting forward, for example, sea-level change, weak soil acting as detachment zone, erosion, edifice growth and seismic activity. One hypothesis concerns the role of the rift-zone orientation that would control the occurrence and constrain the orientation of these landslides (Walter and Troll, 2003) (see also Chapter 4). In the Canaries, as well as on other oceanic islands as La Réunion, widespread offshore distribution of debris avalanches surrounding the island nonetheless show that rift zones are not the only structural control that explain the abundance and orientation of collapses (Masson et al., 2002). However, intense dyke emplacement actively contributes to flank creeping, resulting in specific dyke orientation patterns (Elsworth and Day, 1999; Delcamp et al., 2012a; Chapter 4). Dyke emplacement is often associated with pore pressure increase that has a strong impact on flank stability (Voight and Elsworth, 1997). Furthermore, Tenerife has many aquifers and water reservoirs ('galerias') that could also participate to volcano deformation and destabilisation (Ecker, 1976). For the Icod scar, resulting from the Las Cañadas collapse, the lively debate concerns the mechanism of the collapse and the discussion revolves around lateral (Ancochea et al., 1990; Carracedo, 1994; Masson et al., 2002) versus vertical (caldera) collapse (Ridley, 1971; Martí et al., 2010), although it is possible to have a combination of both mechanisms. Generally, the common parameter is the steepening of the slope due to several of the factors mentioned above, such as erosion, intrusion, volcanic growth, etc. The triggering factor remains, however, a question mark and can be a combination of destabilising factors.

Mass wasting events in oceanic islands are well known to trigger tsunamis, the impact of which depends on the collapsed volume and whether the collapse occurs as a single block or as a retrogressive failure with multiple smaller blocks (Giachetti et al., 2011; Hunt et al., 2011).

Tenerife is an interesting playground to study how instability and landslides have influenced the volcanic plumbing system. Delcamp et al. (2012a) have observed the change in dyke orientation in the upper level of the northeast rift following flank creeping. Ankaramites (high-density, clinopyroxene- and olivine-rich mafic rocks) occur in the rift zone and the old shields, and Longpré et al. (2009) linked their presence with volcano unloading following major collapse that would have allowed the denser ankaramitic magma to be tapped. Boulesteix et al. (2012) similarly noticed that basaltic activity resumed after the explosive eruption following the major Icod collapse. They calculated eruptive rates of 8 km^3 kyr^{-1} for the first 10 kyr going down to ~3 km^3 kyr^{-1} over the period 161–117 ka. They proposed that after this period of Proto-Teide construction, the load of the edifice acted on the volcanic plumbing system, favouring intermittent magma storage and differentiation. This loading went along with a decline in eruptive activity and the development of a shallow magma storage system. The volcanic activity resumed at 18 ± 1 ka with the extrusion of several high-viscosity differentiated bodies at the upper level of the edifice. This example highlights the importance of collapses in the evolution of VIPS.

REFERENCES

Ablay, G., Hürlimann, M., 2000. Evolution of the north flank of Tenerife by recurrent giant landslides. J. Volcanol. Geotherm. Res. 103, 135–159. doi: 10.1016/S0377-0273(00)00220-1.

Acocella, V., Tibaldi, A., 2005. Dike propagation driven by volcano collapse: a general model tested at Stromboli, Italy. Geophys. Res. Lett. 32, 1–4. doi: 10.1029/2004GL022248.

Ancochea, E., Fuster, J., Ibarrola, E., Cendrero, A., Coello, J., Hernan, F., Cantagrel, J.M., Jamond, C., 1990. Volcanic evolution of the island of Tenerife (Canary Islands) in the light of new K-Ar data. J. Volcanol. Geotherm. Res. 44, 231–249. doi: 10.1016/0377-0273(90)90019-C.

Annen, C., Lénat, J.-F., Provost, A., 2001. The long-term growth of volcanic edifces: numerical modelling of the role of dyke intrusion and lava-flow emplacement. J. Volcanol. Geotherm. Res. 105, 263–289. doi: 10.1016/S0377-0273(00)00257-2.

Belousov, A., Belousova, M., Voight, B., 1999. Multiple edifice failures, debris avalanches and associated eruptions in the Holocene history of Shiveluch volcano, Kamchatka, Russia. Bull. Volcanol. 61, 324–342. doi: 10.1007/s004450050300.

Berger, B.R., Henley, R.W., 2011. Magmatic-vapor expansion and the formation of high-sulfidation gold deposits: structural controls on hydrothermal alteration and ore mineralization. Ore. Geol. Rev. 39, 75–90. doi: 10.1016/j.oregeorev.2010.11.004.

Bonforte, A., Guglielmino, F., 2015. Very shallow dyke intrusion and potential slope failure imaged by ground deformation: the 28 December 2014 eruption on Mount Etna. Geophys. Res. Lett. 42, 2727–2733. doi: 10.1002/2015GL063462.

Borgia, A., 1994. Dynamic basis of volcanic spreading. J. Geophys. Res. 99, 17791–17804. doi: 10.1029/94JB00578.

Boulesteix, T., Hildenbrand, A., Gillot, P.-Y., Soler, V., 2012. Eruptive response of oceanic islands to giant landslides: new insights from the geomorphologic evolution of the Teide–Pico Viejo volcanic complex (Tenerife, Canary). Geomorphology 138, 61–73. doi: 10.1016/j.geomorph.2011.08.025.

Byrne, P.K., Holohan, E.P., Kervyn, M., Van Wyk de Vries, B., Troll, V.R., Murray, J.B., 2013. A sagging-spreading continuum of large volcano structure. Geology 41, 339–342. doi: 10.1130/G33990.1.

Cannavò, F., Camacho, A.G., González, P.J., Mattia, M., Puglisi, G., Fernández, J., 2015. Real time tracking of magmatic intrusions by means of ground deformation modeling during volcanic crises. Sci. Rep. 5, 10970. doi: 10.1038/srep10970.

Carracedo, J.C., 1994. The Canary Islands: an example of structural control on the growth of large oceanic-island volcanoes. J. Volcanol. Geotherm. Res. 60, 225–241. doi: 10.1016/0377-0273(94)90053-1.

Carracedo, J.C., Badiola, E.R., Guillou, H., Paterne, M., Scaillet, S., Torrado, F.J.P., Paris, R., Fra-Paleo, U., Hansen, A., 2007. Eruptive and structural history of Teide Volcano and rift zones of Tenerife, Canary Islands. Geol. Soc. Am. Bull. 119, 1027–1051. doi: 10.1130/B26087.1.

Carrasco-Núñez, G., Vallance, J.W., Rose, W.I., 1993. A voluminous avalanche-induced lahar from Citlaltépetl volcano, Mexico: Implications for hazard assessment. J. Volcanol. Geotherm. Res. 59, 35–46. doi: 10.1016/0377-0273(93)90076-4.

Castro, J.M., Cordonnier, B., Schipper, C.I., Tuffen, H., Baumann, T.S., Feisel, Y., 2016. Rapid laccolith intrusion driven by explosive volcanic eruption. Nat. Commun. 7, 13585.

Day, S.J., 1996. Hydrothermal pore fluid pressure and the stability of porous, permeable volcanoes. Geol. Soc. London, Spec. Publ. 110, 77–93. doi: 10.1144/GSL.SP. 1996.110.01.06.

Delcamp, A., van Wyk de Vries, B., James, M.R., 2008. The influence of edifice slope and substrata on volcano spreading. J. Volcanol. Geotherm. Res. 177, 925–943.

Delcamp, A., Troll, V.R., van Wyk de Vries, B., Carracedo, J.C., Petronis, M.S., Pérez-Torrado, F.J., Deegan, F.M., 2012a. Dykes and structures of the NE rift of Tenerife, Canary Islands: a record of stabilisation and destabilisation of ocean island rift zones. Bull. Volcanol. 74, 963–980. doi: 10.1007/s00445-012-0577-1.

Delcamp, A., van de Vries, B.W., James, M.R., Gailler, L.S., Lebas, E., 2012b. Relationships between volcano gravitational spreading and magma intrusion. Bull. Volcanol. 74, 743–765. doi: 10.1007/s00445-011-0558-9.

Delcamp, A., van Wyk de Vries, B., Stéphane, P., Kervyn, M., 2014. Endogenous and exogenous growth of the monogenetic Lemptégy volcano, Chaîne des Puys, France. Geosphere 10, 998.

del Potro, R., Hürlimann, M., 2008. Geotechnical classification and characterisation of materials for stability analyses of large volcanic slopes. Eng. Geol. 98, 1–17. doi: 10.1016/j.enggeo.2007.11.007.

Detienne, M., 2016. Unravelling the role of hydrothermal alteration in volcanic flank and sector collapses using combined mineralogical, experimental, and numerical modelling studies. Université Catholique de Louvain.

Dieterich, J.H., 1988. Growth and persistence of Hawaiian volcanic rift zones. J. Geophys. Res. 93, 4258. doi: 10.1029/JB093iB05p04258.

Donnadieu, F., Merle, O., Besson, J.C., 2001. Volcanic edifice stability during cryptodome intrusion. Bull. Volcanol. 63, 61–72. doi: 10.1007/s004450000122.

Ecker, A., 1976. Groundwater behaviour in Tenerife, volcanic island (Canary Islands, Spain). J. Hydrol. 28, 73–86. doi: 10.1016/0022-1694(76)90053-6.

Elsworth, D., Day, S.J., 1999. Flank collapse triggered by intrusion: the Canarian and Cape Verde Archipelagoes. J. Volcanol. Geotherm. Res. 94, 323–340. doi: 10.1016/S0377-0273(99)00110-9.

Elsworth, D., Voight, B., 1995. Dike intrusion as a trigger for large earthquakes and the failure of volcano flanks. J. Geophys. Res. Solid Earth 100, 6005–6024. doi: 10.1029/94JB02884.

Ferrier, K.L., Huppert, K.L., Perron, J.T., 2013. Climatic control of bedrock river incision. Nature 496, 206–209. doi: 10.1038/nature11982.

Fiske, R.S., Jackson, E.D., 1972. Orientation and growth of Hawaiian volcanic rifts: the effect of regional structure and gravitational stresses. Proc. R. Soc. London A 329 (299), LP-326.

Francis, P.W., Self, S., 1987. Collapsing volcanoes. Sci. Am. 256, 72–90.

Francis, P.W., Gardeweg, M., Ramirez, C.F., Rothery, D.A., 1985. Catastrophic debris avalanche deposit of Socompa volcano, northern Chile. Geology 13, 600–603. doi: 10.1130/0091-7613(1985)13<600:CDADOS>2.0.CO;2.

Giachetti, T., Paris, R., Kelfoun, K., Pérez-Torrado, F.J., 2011. Numerical modelling of the tsunami triggered by the Güìmar debris avalanche, Tenerife (Canary Islands): comparison with field-based data. Mar. Geol. 284, 189–202. doi: 10.1016/j.margeo.2011.03.018.

Glicken, H., 1996. Rockslide-debris avalanche of may 18, 1980, Mount St. Helens volcano, Washington. Open-file Rep. (96-677), 1–5.

Gudmundsson, M.T., Jónsdóttir, K., Hooper, A., Holohan, E.P., Halldórsson, S.A., Ófeigsson, B.G., Cesca, S., Vogfjörd, K.S., Sigmundsson, F., Högnadóttir, T., Einarsson, P., Sigmarsson, O., Jarosch, A.H., Jónasson, K., Magnússon, E., Hreinsdóttir, S., Bagnardi, M., Parks, M.M., Hjörleifsdóttir, V., Pálsson, F., Walter, T.R., Schöpfer, M.P.J., Heimann, S., Reynolds, H.I., Dumont, S., Bali, E., Gudfinnsson, G.H., Dahm, T., Roberts, M.J., Hensch, M., Belart, J.M.C., Spaans, K., Jakobsson, S., Gudmundsson, G.B., Fridriksdóttir, H.M., Drouin, V., Dürig, T., AÐalgeirsdóttir, G., Riishuus, M.S., Pedersen, G.B.M., van Boeckel, T., Oddsson, B., Pfeffer, M.A., Barsotti, S., Bergsson, B., Donovan, A., Burton, M.R., Aiuppa, A., 2016. Gradual caldera collapse at Bárdarbunga volcano, Iceland, regulated by lateral magma outflow. Science (80), 353.

Guthrie, R.H., Friele, P., Allstadt, K., Roberts, N., Evans, S.G., Delaney, K.B., Roche, D., Clague, J.J., Jakob, M., 2012. The 6 August 2010 Mount Meager rock slide-debris flow, Coast Mountains, British Columbia: characteristics, dynamics, and implications for hazard and risk assessment. Nat. Hazards Earth Syst. Sci. 12, 1277–1294. doi: 10.5194/nhess-12-1277-2012.

Harris, P.D., Branney, M.J., Storey, M., 2011. Large eruption-triggered ocean-island landslide at Tenerife: onshore record and long-term effects on hazardous pyroclastic dispersal. Geology 39, 951–954. doi: 10.1130/G31994.1.

Heap, M.J., Lavallée, Y., Petrakova, L., Baud, P., Reuschlé, T., Varley, N.R., Dingwell, D.B., 2014. Microstructural controls on the physical and mechanical properties of edifice-forming andesites at Volcán de Colima, Mexico. J. Geophys. Res. Solid Earth 119, 2925–2963. doi: 10.1002/2013JB010521.

Hildenbrand, A., Gillot, P.-Y., Le Roy, I., 2004. Volcano-tectonic and geochemical evolution of an oceanic intra-plate volcano: Tahiti-Nui (French Polynesia). Earth Planet. Sci. Lett. 217, 349–365. doi: 10.1016/S0012-821X(03)00599-5.

Holohan, E., Poppe, S., Kervyn, M., Delcamp, A., Byrne, P., van Wyk de Vries, B., 2013. An experimental study of the volcano-sagging to volcano-spreading transition. In: EGU General Assembly Conference Abstracts, EGU General Assembly Conference Abstracts. p. EGU2013-1903.

Hora, J.M., Singer, B.S., Wörner, G., 2007. Volcano evolution and eruptive flux on the thick crust of the Andean Central Volcanic Zone: 40Ar/39Ar constraints from Volcán Parinacota, Chile. Bull. Geol. Soc. Am.doi: 10.1130/B25954.1.

Horwell, C.J., Williamson, B.J., Llewellin, E.W., Damby, D.E., Le Blond, J.S., 2013. The nature and formation of cristobalite at the Soufrière Hills volcano, Montserrat: implications for the petrology and stability of silicic lava domes. Bull. Volcanol. 75, 696. doi: 10.1007/s00445-013-0696-3.

Hunt, J.E., Wynn, R.B., Masson, D.G., Talling, P.J., Teagle, D.A.H., 2011. Sedimentological and geochemical evidence for multistage failure of volcanic island landslides: a case study from Icod landslide on north Tenerife, Canary Islands. Geochem. Geophys. Geosyst. 12, n/a-n/a. doi:10.1029/2011GC003740.

Hunt, J.E., Wynn, R.B., Talling, P.J., Masson, D.G., 2013. Multistage collapse of eight western Canary Island landslides in the last 1.5 Ma: sedimentological and geochemical evidence from subunits in submarine flow deposits. Geochem. Geophys. Geosyst. 14, 2159–2181. doi: 10.1002/ggge.20138.

Hunt, J.E., Talling, P.J., Clare, M.A., Jarvis, I., Wynn, R.B., 2014. Long-term (17 Ma) turbidite record of the timing and frequency of large flank collapses of the Canary Islands. Geochem. Geophys. Geosyst. 15, 3322–3345. doi: 10.1002/2014GC005232.

Hürlimann, M., Martí, J., Ledesma, A., 2004. Morphological and geological aspects related to large slope failures on oceanic islands: the huge La Orotava landslides on Tenerife, Canary Islands. Geomorphology 62, 143–158. doi: 10.1016/j.geomorph.2004.02.008.

Ingebritsen, S.E., Scholl, M.A., 1993. The hydrogeology of Kilauea volcano. Geothermics 22, 255–270. doi: 10.1016/0375-6505(93)90003-6.

Jefferson, A., Grant, G.E., Lewis, S.L., Lancaster, S.T., 2010. Coevolution of hydrology and topography on a basalt landscape in the Oregon Cascade Range, USA. Earth Surf. Process. Landforms n/a-n/a. doi:10.1002/esp.1976.

Jo, M.J., Jung, H.S., Won, J.S., Lundgren, P., 2015. Measurement of three-dimensional surface deformation by Cosmo-SkyMed X-band radar interferometry: application to the March 2011 Kamoamoa fissure eruption, Kilauea Volcano, Hawai'i. Remote Sens. Environ. 169, 176–191. doi: 10.1016/j.rse.2015.08.003.

Join, J.-L., Folio, J.-L., Robineau, B., 2005. Aquifers and groundwater within active shield volcanoes. Evolution of conceptual models in the Piton de la Fournaise volcano. J. Volcanol. Geotherm. Res. 147, 187–201. doi: 10.1016/j.jvolgeores.2005.03.013.

Karátson, D., Széley, B., 2006. Ampitheater valleys on volcanoes: characterization, evolution, surface modelling. In: Geophysical Research Abstracts. EGU06-A-09648, Vienna.

Karátson, D., Thouret, J.C., Moriya, I., Lomoschitz, A., 1999. Erosion calderas: origins, processes, structural and climatic control. Bull. Volcanol. doi: 10.1007/s004450050270.

Kerle, N., van Wyk de Vries, B., Oppenheimer, C., 2003. New insight into the factors leading to the 1998 flank collapse and lahar disaster at Casita volcano, Nicaragua. Bull. Volcanol. 65, 331–345. doi: 10.1007/s00445-002-0263-9.

Kervyn, M., Ernst, G.G.J., van Wyk de Vries, B., Mathieu, L., Jacobs, P., 2009. Volcano load control on dyke propagation and vent distribution: insights from analogue modeling. J. Geophys. Res. 114, B03401. doi: 10.1029/2008JB005653.

Lagmay, A.M.F., Valdivia, W., 2006. Regional stress influence on the opening direction of crater amphitheaters in Southeast Asian volcanoes. J. Volcanol. Geotherm. Res. 158, 139–150. doi: 10.1016/j.jvolgeores.2006.04.020.

Lagmay, A., van Wyk de Vries, B., Kerle, N., Pyle, D.M., 2000. Volcano instability induced by strike slip faulting. Bull. Volcanol. 62, 331–346.

Le Corvec, N., Walter, T.R., 2009. Volcano spreading and fault interaction influenced by rift zone intrusions: insights from analogue experiments analyzed with digital image correlation technique. J. Volcanol. Geotherm. Res. 183, 170–182. doi: 10.1016/j.jvolgeores.2009.02.006.

Lipman, P.W., Rhodes, J.M., Dalrymple, G.B., 1990. The Ninole Basalt—implications for the structural evolution of Mauna Loa volcano, Hawaii. Bull. Volcanol. doi: 10.1007/BF00680316.

Longpré, M.-A., Troll, V.R., Walter, T.R., Hansteen, T.H., 2009. Volcanic and geochemical evolution of the Teno massif, Tenerife, Canary Islands: some repercussions of giant landslides on ocean island magmatism. Geochem. Geophys. Geosyst. 10, n/a-n/a. doi:10.1029/2009GC002892.

Lowell, R.P., Van Cappellen, P., Germanovich, L.N., 1993. Silica precipitation in fractures and the evolution of permeability in hydrothermal upflow zones. Science 260, 192–194.

Manconi, A., Longpré, M.-A., Walter, T.R., Troll, V.R., Hansteen, T.H., 2009. The effects of flank collapses on volcano plumbing systems. Geology 37, 1099–1102. doi: 10.1130/G30104A.1.

Martí, J., Soriano, C., Galindo, I., Cas, R.A.F., 2010. Resolving problems with the origin of Las Cañadas caldera (Tenerife, Canary Islands): Los Roques de García Formation—Part of a major debris avalanche or an in situ, stratified, edifice-building succession? Geol. Soc. Am. Spec. Pap. 464, 113–132. doi: 10.1130/2010.2464(06).

Masson, D., Watts, A., Gee, M.J., Urgeles, R., Mitchell, N., Le Bas, T., Canals, M., 2002. Slope failures on the flanks of the western Canary Islands. Earth-Science Rev. 57, 1–35. doi: 10.1016/S0012-8252(01)00069-1.

Mathieu, L., Kervyn, M., Ernst, G.G.J., 2011a. Field evidence for flank instability, basal spreading and volcano-tectonic interactions at Mt Cameroon, West Africa. Bull. Volcanol. 73, 851–867. doi: 10.1007/s00445-011-0458-z.

Mathieu, L., van Wyk de Vries, B., Pilato, M., Troll, V.R., 2011b. The interaction between volcanoes and strike-slip, transtensional and transpressional fault zones: analogue models and natural examples. J. Struct. Geol. 33, 898–906. doi: 10.1016/j.jsg.2011.03.003.

Mayer, K., Scheu, B., Montanaro, C., Yilmaz, T.I., Isaia, R., Aßbichler, D., Dingwell, D.B., 2016. Hydrothermal alteration of surficial rocks at Solfatara (Campi Flegrei): petrophysical properties and implications for phreatic eruption processes. J. Volcanol. Geotherm. Res. 320, 128–143. doi: 10.1016/j.jvolgeores.2016.04.020.

Merle, O., Lénat, J.-F., 2003. Hybrid collapse mechanism at Piton de la Fournaise volcano, Reunion Island, Indian Ocean. J. Geophys. Res. 108, 1–11. doi: 10.1029/2002JB002014.

Morgan, J., Moore, G.F., Clague, D., 2003. Slope failure and volcanic spreading along the submarine south flank of Kilauea volcano, Hawaii. J. Geophys. 108, 2415. doi: 10.1029/2003JB002411.

Musiol, S., Holohan, E.P., Cailleau, B., Platz, T., Dumke, A., Walter, T.R., Williams, D.A., van Gasselt, S., 2016. Lithospheric flexure and gravity spreading of Olympus Mons volcano, Mars. J. Geophys. Res. Planets 121, 255–272. doi: 10.1002/2015JE004896.

Nakamura, K., 1977. Volcanoes as possible indicators of tectonic stress orientation—principle and proposal. J. Volcanol. Geotherm. Res.doi: 10.1016/0377-0273(77)90012-9.

Oehler, J.-F., van Wyk de Vries, B., Labazuy, P., 2005. Landslides and spreading of oceanic hot-spot and arc shield volcanoes on low strength layers (LSLs): an analogue modeling approach. J. Volcanol. Geotherm. Res. 144, 169–189. doi: 10.1016/j.jvolgeores.2004.11.023.

Paguican, E.M.R., van Wyk de Vries, B., Lagmay, A.M.F., 2012. Volcano-tectonic controls and emplacement kinematics of the Iriga debris avalanches (Philippines). Bull. Volcanol. 74, 2067–2081. doi: 10.1007/s00445-012-0652-7.

Pevear, D., 1987. Clay minerals in the 1980 deposits from Mount St. Helens. Clays Clay Miner. 30, 241–252. doi: 10.1346/CCMN.1982.0300401.

Pinel, V., Jaupart, C., 2000. The effect of edifice load on magma ascent beneath a volcano. Philos. Trans. R. Soc. A 358, 1515–1532.

Pinel, V., Jaupart, C., 2005. Some consequences of volcanic edifice destruction for eruption conditions. J. Volcanol. Geotherm. Res. 145, 68–80. doi: 10.1016/j.jvolgeores.2005.01.012.

Pola, A., Crosta, G., Fusi, N., Barberini, V., Norini, G., 2012. Influence of alteration on physical properties of volcanic rocks. Tectonophysics 566, 67–86. doi: 10.1016/j.tecto.2012.07.017.

Poland, M.P., Peltier, A., Bonforte, A., Puglisi, G., 2017. The spectrum of persistent volcanic flank instability: a review and proposed framework based on Kīlauea, Piton de la Fournaise, and Etna. J. Volcanol. Geotherm. Res. 339, 63–80. doi: 10.1016/j.jvolgeores.2017.05.004.

Presley, T.K., Sinton, J.M., Pringle, M., Presley, T.K., Sinton, J.M., Pringle, M., 1997. Postshield volcanism and catastrophic mass wasting of the Waianae Volcano, Oahu, Hawaii. Bull. Volcanol. 58, 597–616.

Reid, M.E., 2004. Massive collapse of volcano edifices triggered by hydrothermal pressurization. Geology 32, 373–376. doi: 10.1130/G20300.1.

Ridley, W.I., 1971. The origin of some collapse structures in the Canary Islands. Geol. Mag. 108, 477–484. doi: 10.1017/S0016756800056673.

Roberti, G., Ward, B., Van Wyk De Vries, B., Perotti, L., 2017. Structure from motion and landslides: the 2010 Mt Meager collapse from slope deformation to debris avalanche deposit mapping. Geotech. News 35, 20–24.

Salaün, A., Villemant, B., Gérard, M., Komorowski, J.-C., Michel, A., 2011. Hydrothermal alteration in andesitic volcanoes: trace element redistribution in active and ancient hydrothermal systems of Guadeloupe (Lesser Antilles). J. Geochem. Explor. 111, 59–83. doi: 10.1016/j.gexplo.2011.06.004.

Schoorl, J.M., Veldkamp, A., Claessens, L., van Gorp, W., Wijbrans, J.R., 2014. Edifice growth and collapse of the Pliocene Mt. Kenya: evidence of large scale debris avalanches on a high altitude glaciated volcano. Glob. Planet. Change 123, 44–54. doi: 10.1016/j.gloplacha.2014.10.010.

Sekiya, S., Kikuchi, Y., 1980. The eruption of Bandai-san. Tokyo Imp. Univ. Coll. Sci. J. 3 (2), 91–172.

Self, S., Rampino, M.R., 1981. The 1883 eruption of Krakatau. Nature 294, 699–704.

Shea, T., van Wyk de Vries, B., Pilato, M., 2008. Emplacement mechanisms of contrasting debris avalanches at Volcán Mombacho (Nicaragua), provided by structural and facies analysis. Bull. Volcanol. 70, 899–921. doi: 10.1007/s00445-007-0177-7.

Siebert, L., 1984. Large volcanic debris avalanches: characteristics of source areas, deposits, and associated eruptions. J. Volcanol. Geotherm. Res. 22, 163–197. doi: 10.1016/0377-0273(84)90002-7.

Siebert, L., Glicken, H., Ui, T., 1987. Volcanic hazards from Bezymianny- and Bandai-type eruptions. Bull. Volcanol. 49, 435–459. doi: 10.1007/BF01046635.

Thirlwall, M., Singer, B., Marriner, G., 2000. 39Ar–40Ar ages and geochemistry of the basaltic shield stage of Tenerife, Canary Islands, Spain. J. Volcanol. Geotherm. Res. 103, 247–297. doi: 10.1016/S0377-0273(00)00227-4.

Thouret, J.-C., 1999. Volcanic geomorphology—an overview. Earth Sci. Rev. 47, 95–131. doi: 10.1016/S0012-8252(99)00014-8.

Tibaldi, A., 1996. Mutual influence of dyking and collapses at Stromboli volcano, Italy. Geol. Soc. London, Spec. Publ. 110, 55–63. doi: 10.1144/GSL.SP. 1996.110.01.04.

Vallance, J.W., Scott, K.M., 1997. The Osceola Mudflow from Mount Rainier: sedimentology and hazard implications of a huge clay-rich debris flow. Bull. Geol. Soc. Am. 109, 143–163. doi: 10.1130/0016-7606(1997)109<0143:TOMFMR>2.3.CO;2.

van Wyk de Vries, B., Tim Davies, T., 2015. Chapter 38 – Landslides, debris avalanches, and volcanic gravitational deformation. In: Sigurdsson, Haraldur (Ed.), The Encyclopedia of Volcanoes, second edition. Academic Press, Amsterdam, pp. 665–685, doi:10.1016/B978-0-12-385938-9.00038-9.
van Wyk de Vries, B., Merle, O., 1998. Extension induced by volcanic loading in regional strike-slip zones. Geology 26, 983–986. doi: 10.1130/0091-7613(1998)026<0983:EIBVLI>2.3.CO;2.
Voight, B., Elsworth, D., 1997. Failure of volcano slopes. Géotechnique 47, 1–31. doi: 10.1680/geot.1997.47.1.1.
Voight, B., Komorowski, J.-C., Norton, G.E., Belousov, A.B., Belousova, M., Boudon, G., Francis, P.W., Franz, W., Heinrich, P., Sparks, R.S.J., Young, S.R., 2002. The 26 December (Boxing Day) 1997 sector collapse and debris avalanche at Soufrière Hills Volcano, Montserrat. Geol. Soc. London Mem. 21, 363–407. doi: 10.1144/GSL.MEM. 2002.021.01.17.
Walter, T.R., 2011. Structural architecture of the 1980 Mount St. Helens collapse: an analysis of the Rosenquist photo sequence using digital image correlation. Geology 39, 767–770. doi: 10.1130/G32198.1.
Walter, T.R., Troll, V.R., 2003. Experiments on rift zone evolution in unstable volcanic edifices. J. Volcanol. Geotherm. Res. 127, 107–120. doi: 10.1016/S0377-0273(03)00181-1.
Wooller, L., van Wyk de Vries, B., Cecchi, E., Rymer, H., 2009. Analogue models of the effect of long-term basement fault movement on volcanic edifices. Bull. Volcanol. 71, 1111–1131. doi: 10.1007/s00445-009-0289-3.

CHAPTER 10

Volcanic and Igneous Plumbing Systems of Caldera Volcanoes

Ben M. Kennedy*, Eoghan P. Holohan, John Stix†, Darren M. Gravley*, Jonathan R.J. Davidson*, Jim W. Cole*, Steffi Burchardt‡**
*University of Canterbury, Christchurch, New Zealand;
**UCD School of Earth Sciences, University College Dublin, Dublin, Ireland;
†McGill University, Montreal, QC, Canada;
‡Uppsala University, Uppsala, Sweden

Contents

10.1 INTRODUCTION

The volcanic and igneous plumbing systems associated with collapse calderas are highly complex and range dramatically in size, shape, longevity and depth. Such plumbing systems provide the heat, permeability and mineral-enriched fluids necessary for economic resources such as geothermal power and ore deposits (Stix et al., 2003). However, these systems also pose threats to society (Acocella et al., 2015). The study of magma

Volcanic and Igneous Plumbing Systems. http://dx.doi.org/10.1016/B978-0-12-809749-6.00010-8
Copyright © 2018 Elsevier Inc. All rights reserved.

storage and transport in caldera systems has thus become the focus of hundreds of studies encompassing the full range of methodologies (e.g. physical volcanology, petrology, geochemistry, geophysics, numerical modelling and analogue experiments) used by geologists. In this chapter, we focus on commonalities that make the plumbing systems of calderas distinctive.

10.2 WHAT IS A COLLAPSE CALDERA?

Collapse calderas are volcanic depressions formed by the subsidence of crustal material as magma is withdrawn from the underlying plumbing system through intrusion and/or eruption (Branney and Acocella, 2015). Crustal deformation that accommodates this subsidence can be either ductile or brittle (Kennedy et al., 2004) and involve reactivation of large tectonic structures (Gravley et al., 2007). Field (Moore and Kokelaar, 1998), numerical (Holohan et al., 2015) and experimental studies (Acocella, 2007) all offer insights into the mechanisms of subsidence, which is generally accommodated by sagging and/or ring faults. Moreover, subsidence events (Kennedy et al., 2016) and changing stress regimes (Holohan et al., 2015) provide clues into the configuration and connectivity of the magma bodies beneath, but often these are difficult to decipher.

The volumes of erupted magma linked to caldera formation—from ~0.1 to 8 km^3 for mafic caldera-forming eruptions and from ~1 to 5000 km^3 for silicic eruptions—tend to correlate with caldera size (Smith 1979). This correlation supports a strong connection between plumbing system architecture and the surface expression of the caldera. In turn, plumbing system development is dependent on the volcano's geodynamic setting, which regulates both the composition and rate of magma supply (see Chapter 7).

Caldera volcanoes commonly undergo multiple cycles of activity, each of which includes: (1) intrusion phases as the volcanic plumbing system develops, (2) caldera collapse associated with the partial or complete destruction of the plumbing system and (3) magmatic rejuvenation (Smith and Bailey, 1968). The architecture and longevity of magma storage beneath the caldera primarily control the hazard and resource potential of the volcano. It is therefore imperative to closely monitor active caldera systems to constrain these attributes (Acocella et al., 2015 and examples therein) and to interpret unrest signals, such as volcano deformation, as well as thermal, gravimetric, seismic and geochemical changes, in terms of processes in the volcanic plumbing system (see Chapter 11) and signs of impending eruption. As is evident from the geophysics of restless calderas and the complexity of intrusive geology beneath calderas (Kennedy and Stix, 2007; Kennedy et al., 2016), non-eruptive unrest events are more common than eruptions at calderas. However, life cycles of calderas can extend to 10^6 years, and understanding the when rejuvenation transitions to eruption remains a significant challenge to volcanologists.

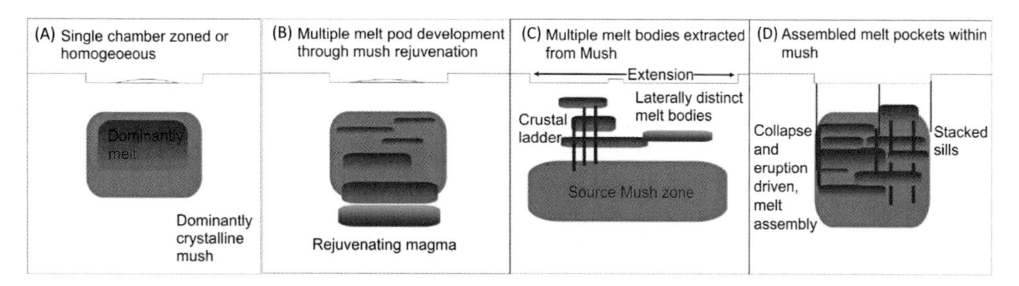

Figure. 10.1 ***Petrological models of the volcanic plumbing systems of caldera volcanoes.*** (A) Eruptive products from calderas are commonly single large-volume deposition events of chemically homogeneous or gradually zoned volcanic material, which supports a conceptual vision of a single magma chamber. (B) Other eruptive products of calderas may contain distinct but similar chemical signatures, and some cases show evidence for magmatic rejuvenation. (C) Timescales and petrological evidence also indicate that, particularly in rhyolitic systems, magma may be extracted from the mush and be stored independently in separate magma pockets before eruption, and that magma may be transported vertically or laterally during and following eruption events. (D) In some systems, evidence supports a plumbing system in which melt is stored in stacked sills in close proximity to surrounding mush. These sills can rapidly connect during eruption and subsidence.

10.3 WHAT DO ERUPTIVE PRODUCTS TELL US ABOUT THE PLUMBING SYSTEMS OF CALDERAS?

The very existence of large calderas requires large accumulations of eruptible magma (Fig. 10.1), even if the accumulation is a relatively transient phenomenon. Partially or fully crystalline magma systems may exist for 10^5 to 10^6 years (Fig. 10.1A) (cf. Reid et al., 1997; Hildreth, 2004; Bachmann et al., 2007; Kaiser et al., 2017). However, large volumes of eruptible magma can be assembled and stored at much shorter timescales of thousands of years or less (Gualda and Sutton, 2016 and references therein) through rejuvenation of a crystalline mush by intruded melt or gas (Fig. 10.1B), and the subsequent separation of interstitial melt from the mush (Bachmann and Bergantz, 2008). There is less of a consensus on the geometry of eruptible magma, whether it forms a continuous large reservoir (Fig. 10.1A) (Hildreth, 1979, 1981, 2004) or exists as separated pods (Fig. 10.1B–D), the latter being either within or vertically detached from the mush that they came from (cf. Gualda and Ghiorso, 2013; Bégué et al., 2014a,b; Cashman and Giordano, 2014; Myers et al., 2016). The timescales required to bring together such large and hazardous reservoirs of magma is of the utmost concern for quantifying unrest at calderas worldwide (e.g. the timescale of Fig. 10.1B could be comparatively long compared to Fig. 10.1D which potentially could be syn-eruptive).

In this section, we review our petrologic understanding of the geometry, depth and timescales of eruptible magma storage in both rhyolitic and basaltic caldera systems (see also Chapter 8).

10.3.1 Continuous, Compositionally Zoned Magma Chambers

The concept of a large compositionally stratified or homogeneous magma chamber (Fig. 10.1A; Williams, 1941; Lipman, 1984) has resonated with igneous petrologists for more than half a century. The concept fits nicely with the simple geometric correlation between caldera shape and the footprint of one large magma reservoir that erupts to form compositionally stratified or homogenous ignimbrites (Smith, 1979). Large stratified magma chambers were invoked to characterise several Oligocene ignimbrites in the western United States (Hildreth, 1979, 1981; Bachmann et al., 2002; Lipman, 2007; Lipman and Bachmann, 2015) or the chemically 'monotonous' crystal-rich ignimbrites of the Central Andes (De Silva, 1989; Lindsay et al., 2001; Francis et al., 1989; Wright et al., 2011). It is important to note that the large-chamber concept (Fig. 10.1A) explained geological observations for the distribution and stratigraphy of single caldera-forming ignimbrites, but it did not necessarily explain magma plumbing systems for the entire life-cycle of the caldera. For example, the life cycle of Long Valley caldera and adjacent magma systems depicts not only the interconnectivity and potential for lateral flow between magma systems, but also the transient nature of crystal-poor melt lenses or batches within a large and evolving plumbing system (Hildreth, 2004).

10.3.2 Discrete, Multiple Magma Bodies

Many caldera-forming eruptions and their associated ignimbrites require a more complex volcanic plumbing system (Eichelberger et al., 2000) than depicted in the single-large-chamber concept. The recognition of multiple magma batches, from combined geochemical and geological evidence, opened the door for new, more-complex plumbing system models (see Cashman and Giordano, 2014 for a review). Networks of distinct magma lenses or sills within a crystal mush (Cashman and Giordano, 2014), for instance, would not only allow for lateral and vertical magma migration within a mush during eruption and subsidence (i.e. syn-volcanic extraction; Fig. 10.1D), but also account for the amalgamation and storage of these discrete melt lenses prior to eruption (Fig. 10.1B and C).

Such a model is supported by many examples of exposed high-level plutonic complexes, which represent 'fossilised' versions of magmatic plumbing systems feeding caldera volcanoes (Wiebe and Collins, 1998; Barnes et al., 1990; Wiebe et al., 2002; Hawkins and Wiebe, 2004; Metcalf, 2004; Quick et al., 2009; Sewell et al., 2012). These examples indicate that volcanic plumbing systems beneath calderas are very complex, with evidence of rising melts feeding late-stage dykes, sills and laccoliths within the system (Wiebe and Collins, 1998; Bachl et al., 2001; Harper et al., 2004; Kennedy and Stix, 2007; Turnbull et al., 2010; Kennedy et al., 2016) and interaction between silicic and mafic magmas (Wiebe 1996; Snyder, 2000; Wiebe et al., 2001, 2004; Cole et al., 2014). Taken together, these eroded intrusive systems occupy several kilometres of crustal thickness, from a shallow association with eruptive deposits to the large complex magma bodies below

(Turnbull et al., 2010). In some instances, the brittle deformation at the roof of the chamber is exposed and can be correlated to layers of stoped country rock within the chamber and possible caldera-forming events (Hawkins and Wiebe, 2004). At various levels within the magma body, stacked mafic intrusions highlight fossil rheological interfaces between crystal mush and melt zones and may occur as repeated layers within or above magma bodies showing a replenishment sequence (Wiebe, 2016). Despite these detailed windows into the architecture of frozen magma bodies, the absolute depths of these eroded fossil systems are often poorly constrained.

10.3.3 Depths of Eruptible Magmas

Measured volatile concentrations in melt inclusions trapped within crystals from erupted pumice as well as other geothermobarometric methods can constrain pressures and depths of magma storage (Wallace et al., 1999; Gualda and Ghiorso, 2014). By measuring H_2O and CO_2 concentrations in melt inclusions, equilibrium pressures and depths of crystallisation can be calculated (Wallace et al., 1999). For example, at Rabaul caldera (Papua New Guinea), crystals with melt inclusions indicate entrapment of volatiles at pressures corresponding to depths of 2.5–7 km (Roggensack et al., 1996). Complementary to these methods, phase equilibria using rhyolite-MELTS geobarometry can calculate pressures, under which magmas accumulated. For example, before the eruption of the Bishop Tuff pressures range from approx. 6.5 to 7.5 km (Gualda and Ghiorso, 2014) and approx. 7 to 9 km for the Peach Springs Tuff (Pamukcu et al., 2015b). The method has also been used for several large caldera-forming and smaller caldera-sourced lava dome eruptions in the Taupo Volcanic Zone of New Zealand; the results are comparable to other geobarometers, which support a depth range of 3–5 km (Bégué et al., 2014b, 2015). Generally, depths of magma storage in caldera systems vary widely, but appear shallower for rhyolitic systems particularly in rift environments compared to basalts and intermediate magmas.

10.3.4 Timescales of Eruptible Magma Assembly and Storage

The eruptive products can also constrain timescales for developing the different components of the underlying volcanic plumbing system. From the age and zoning of the different crystals within the erupted 'cargo', the durations of periods of magma replenishment, crystallisation and accumulation can be determined.

The crystalline mushy parts of the plumbing system may persist for 10^5–10^6 years (Bachmann et al., 2007). In particular, in caldera systems dominated by monotonous intermediate magmas, the timescales for mush development are 700,000 years, as evidenced, for example, by the crystal-rich Pastos Grandes supereruption in the Central Andes (Kaiser et al., 2017). In contrast, the rejuvenation, assembly and storage of crystal-poor, highly-eruptible magmas may occur only centuries to decades prior to eruption, as was the case for the Minoan eruption at Santorini (Druitt et al., 2012) and for several eruptions in the Taupo Volcanic Zone of New Zealand (Matthews et al., 2012; Allan et al., 2013; Pamukcu et al., 2015a; Rubin et al., 2017). These remarkably short timescales overlap significantly with timeframes that humans are concerned about, and therefore the hope would be that the geophysical signals caused by such a quick assembly of large volumes of melt would be observable.

10.4 WHAT DO VOLCANO DEFORMATION AND SEISMICITY TELL US ABOUT THE PLUMBING SYSTEMS OF CALDERAS?

10.4.1 Seismic Structure of Caldera Plumbing Systems

Active caldera volcanoes commonly display signs of unrest, potentially involving movements of small to medium magma volumes within the volcanic plumbing system (Fig. 10.2; Chapter 11) (Acocella et al., 2015). Geophysical and geodetic data record

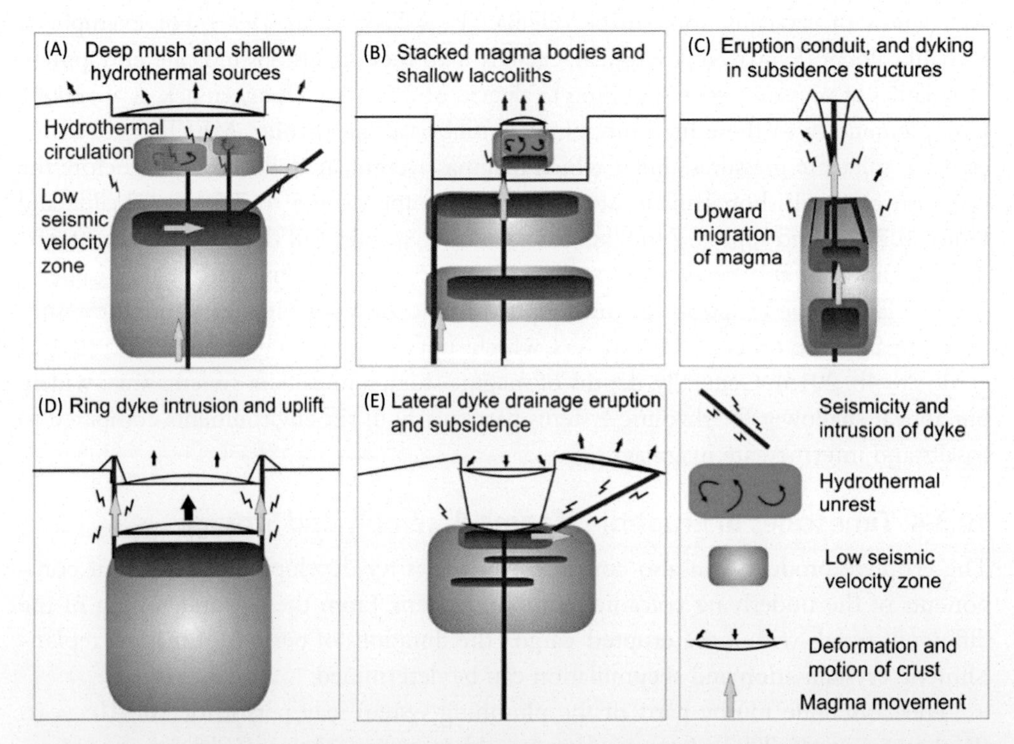

Figure. 10.2 ***Conceptual sketches of the volcanic plumbing system beneath caldera volcanoes based on geophysical and geodetic data.*** (A) Seismic low-velocity zones and broad deformation related to magma and hydrothermal activity and/or deep lateral or vertical magma movement. (B) Magma at multiple depths, local deformation driven by shallow intrusion and vertical magma transport. (C) Vertical migration of magma recorded by seismicity and facilitated by eruption and subsidence. (D) Vertical and lateral ring dyke intrusion driving seismicity and deformation. (E) Stacked sills and lateral magma withdrawal indicated by seismicity and subsidence.

ground uplift andsubsidence, as well as both vertical and lateral movement of magma. Zones of partial melt within the magma plumbing systems can be detected by combining the distribution of seismic sources and seismic wave speeds in the subsurface (Weiland et al., 1995). However, various interpretations of low-velocity zones are possible, as different combinations of melt and fluid-filled pores can have similar geophysical responses. Most interpretations involve fluids and higher-than-background geothermal gradients indicating proximity to, or a direct identification of, magma. Many restless calderas display small zones of supercritical fluids and/or partial melt at depths of 2–10 km (Fig. 10.2A) (Gudmundsson et al., 2004; Zollo et al., 2008). At depths of >8 km, the seismic large anomalies are more likely to be associated with zones of partial melt.

In the cases of large magma bodies, magma depths based on the geophysical studies are greater than those based on melt inclusion studies from prior caldera-forming eruptions (see section above). This may imply that caldera-forming magma rises from a deep long-term storage zone and accumulates at shallower depths before erupting (Fig. 10.2C).

10.4.2 Deformation Associated with Intrusive Events

Levelling, InSAR and GPS data from many active calderas have revealed episodic, time-varying, cm- to m-scale uplift and/or subsidence in the last several decades (Pierce et al., 2002; Wicks et al., 2006; Parks et al., 2015). Similar to anomalies in seismic velocity, modelling the sources of these deformation events can be interpreted as either magmatic or hydrothermal mass transfer at various depths (Fig. 10.2B).

Clearly, active caldera systems can accommodate mobile magma at shallow levels without erupting, but given the large deformations involved (>1 m) over short periods of time and the clear presence of mobile magma, the possibility of eruption must always be evaluated carefully. In some cases, the location of eruption is not the location of maximum inflation.

Smaller and more frequent deformation events occur at basaltic calderas. These events may aid our interpretation of larger or more complex deformation events at more compositionally-evolved caldera systems. Modelled InSAR solutions to deformation at the Galapagos volcanoes, for instance, suggest a range of styles and depths of intrusive events, including dykes, stacked sills and fault reactivation. For example, modelled solutions to ~5 m of inflation from InSAR data at Sierra Negra caldera in the Galapagos from 1992 to 2005 show that shallow sill/laccolith intrusion was accommodated by doming and trapdoor faulting on one side of the caldera (Chadwick et al., 2006).

In summary, active uplift at calderas indicates the presence of hydrothermal systems and small dykes and/or sills, which lie within or above deeper and larger mush bodies. Intrusive events may be recorded by geophysics as indicated from ground deformation related to emplacement or pressure change within shallow, and perhaps stacked, sources of fluid transfer. In these scenarios, it is likely that the stacked sources of deformation are linked by subsidence or regional structures that are sometimes reactivated during the deformation events.

10.4.3 Caldera Seismicity as a Manifestation of Subsidence Structures

At restless or erupting calderas, seismicity can indicate the formation or reactivation of fractures Once formed, these fractures can act as pathways that allow magma transport within the plumbing systemsand that facilitate the overall structural deformation related to uplift and/or caldera subsidence.

Large silicic caldera volcanoes may host annular, steeply dipping faults, that is, ring faults, that accommodate caldera uplift or subsidence (e.g. Rabaul). Such ring faults may be intruded and reactivated during magma transport following subsidence or during subsequent caldera cycles.

In basaltic calderas (e.g. Miyakejima, Piton de la Fournaise, Bardarbunga and Axial) where activity is recorded by seismicity, lateral withdrawal of magma through dykes from the sub-caldera magma reservoirs is an important process (Fig. 10.2E; see also Chapter 11). The subsidence is often characterised by steady exponential declines in seismicity, and the pattern of seismicity is closely linked with changes in eruption rate (Gudmundsson et al., 2016). Seismic hypocentres outline the propagation of dykes, commonly regionally controlled, from the plumbing system beneath the caldera, as well as the formation and reactivation of ring faults (Wilcock et al., 2016).

An observed incremental step-like subsidence process at several active calderas has been interpreted as a stick-slip mechanism, in which the magma reservoir is alternately pressurised and depressurised (Filson et al., 1973; Kumagai et al., 2001; Stix and Kobayashi, 2008; Michon et al., 2011). While stationary, the roof block is held in place by friction with the surrounding wall rock and by the reservoir pressure. As the reservoir is drained through the dyke, reservoir pressure decreases to a point at which gravitational forces exceed the combined magmatic and frictional forces, causing the block to subside. The subsiding block re-pressurises the reservoir, in the process also pushing magma out of the chamber into the dyke. Once the block stops moving downward, reservoir pressure and friction again hold the block in place. With further drainage, the process repeats itself.

Seismicity and asymmetric subsidence also reveal that structures beneath calderas may be highly asymmetric despite plan view symmetry. For example, at Bardarbunga (Iceland), significantly more earthquakes and greater subsidence occurred along the caldera's northern margin, indicating a trapdoor style of collapse, which is supported by an asymmetric ring-fault geometry defined by earthquake hypocentres (Gudmundsson et al., 2016). Similarly, trapdoor subsidence occurred at Fernandina in 1968, with 350 m of subsidence observed at one edge of the caldera (Simkin and Howard, 1970; Howard, 2010), although earthquakes were not sufficiently precisely located to define the ring fault seismically.

In summary, seismicity at actively deforming or erupting calderas demonstrates that there is an inherent link between magma evacuation and the development and activation of subsidence structures. The plumbing system beneath these calderas is linked by subsidence structures that are activated on different timescales and magnitudes in response to magma movement. Additionally, regional structures and stresses strongly influence both the subsidence and the lateral transport of magma (see also Chapter 7).

10.4.4 Monitoring Lateral Movement of Magma

The catastrophic eruptions that characterise many caldera-forming eruptions imply vertical movement of magma towards the surface from deeper magmatic plumbing systems (Mori et al., 1996a). This vertical magma evacuation coupled with vertical subsidence has dominated our understanding of caldera formation. However, dramatic lateral movements of magma, usually associated with basaltic caldera volcanoes, have also occurred in silicic caldera volcanoes, for example, at Katmai, Alaska (Hildreth, 1991).

Lateral magma withdrawal most commonly occurs along dykes, the lateral propagation of which is evident from surface deformation, migration of vents and/or seismicity. The direction of dyke propagation is often influenced by local stresses close to the volcano, being radially oriented, but with distance become strongly influenced by regional tectonics (Sigmundsson et al., 2015; see also Chapter 4). Lateral magma transport has been observed to occur over distances of >45 km and appears seismic in the upper, brittle crust and aseismic below (Toda et al., 2002; Sigmundsson et al., 2015), indicating that in some cases magma may also be partly sourced from below. Once the dyke is established, magma transport may be aseismic (Eibl et al., 2017). While open, the dyke-hosted conduit can allow considerable caldera subsidence over weeks to months as magma drains or erupts from the reservoir (Gudmundsson et al., 2016).

10.5 THE STRUCTURE OF CALDERA PLUMBING SYSTEMS DURING THE CALDERA CYCLE

Significant structural deformation is required during the caldera cycle (Fig. 10.3; Smith and Bailey, 1968), that is, as large magma reservoirs initiate and grow (see also Chapters 5 and 6), then empty and deflate (Druitt and Sparks, 1984), and eventually refill or regrow (Kennedy et al., 2012). The distribution of magma and the structural pathways for magma movement during the caldera cycle have been studied by using field, numerical and experimental evidence.

10.5.1 Initiation and Growth of Caldera-Forming Magma Volumes

As this chapter is focused around magma plumbing, it is important to consider the initiation and growth of a sufficiently large body of magma to form a caldera. In Smith and Bailey's (1968) classic model of caldera formation, the initial stage of magma chamber

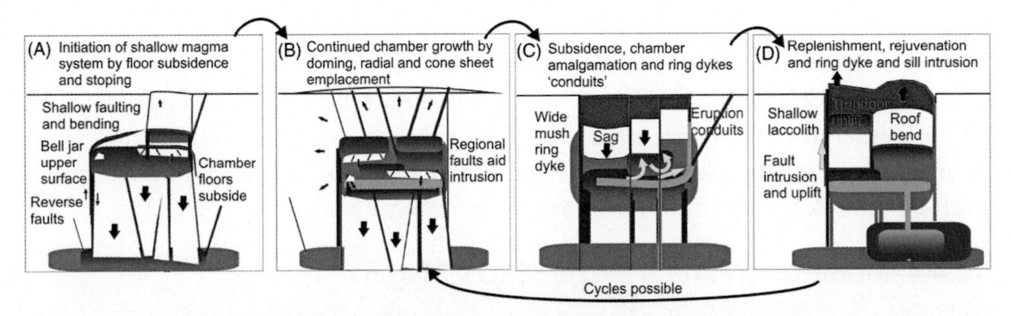

Figure. 10.3 The structural evolution of calderas: (A) Initiation of a large magma chamber via cauldron subsidence. (B) Magma plumbing prior to a caldera-forming events, intrusion aided by chamber floor subsidence, regional structures and doming. Shallow radial and concentric dykes above main area of magma. (C) The subsidence of the caldera along faults and through ductile deformation changes the plumbing system and drives magma movement, amalgamation, and eruption. (D) Re-intrusion of magma following subsidence may drive fault reactivation or ductile deformation of the crust, which is sometimes manifested as uplift at the ground surface.

emplacement is associated with uplift or 'tumescence' of the chamber roof. However, we consider that the 'space problem' associated with the huge volumes of magma involved in many caldera-forming eruptions may necessitate subsidence of the chamber floor (Fig. 10.3A). One mechanism for floor subsidence is that, as magma rises from a lower magma reservoir, 'bell-jar'-shaped reverse faults develop (see also Section 10.5.2 and reviews by Cole et al., 2005; Acocella, 2007; Marti et al., 2008). Successive sets of reverse ring faults may propagate upwards from the reservoir (Burchardt and Walter, 2010), enclosing a piston or several pistons, or zone of sagging of reservoir-roof rock, depending on the existence of regional structures and on crustal rheology. The reverse ring faults and pre-existing structures connect and detach a portion of the roof rocks (Clough et al., 1909). This mobile crust then subsides into the underlying reservoir, while magma is injected along the ring faults (i.e. ring dykes) into the opening space above the piston (Fig. 10.4; Clough et al., 1909). Hence, magma is transferred from its source reservoir and emplaced into a newly forming magma body by a process called cauldron subsidence (Clough et al., 1909).

The resulting magma body is bound by ring dykes and thus characterised by steeply outward-dipping walls and a flat to slightly bell-shaped roof, and a sub-circular to elliptical shape in map view (Yoshinobu et al., 2003; Zak and Paterson, 2006). The size of such bodies can be comparable to collapse calderas and may underlie calderas. The compositional layering of many plutons emplaced by cauldron subsidence indicates that the foundering of the central piston occurred incrementally by successive magma injections (Richey, 1927; Cargill et al., 1928; Stillman, 1970), which tend to fill the pluton from the top down (Cruden and McCaffrey, 2001). The perhaps best-exposed magma body emplaced by cauldron subsidence is the Slaufrudalur pluton in southeast Iceland, a granitic intrusion with steep walls and a flat roof (Burchardt et al., 2010, 2012). Compared

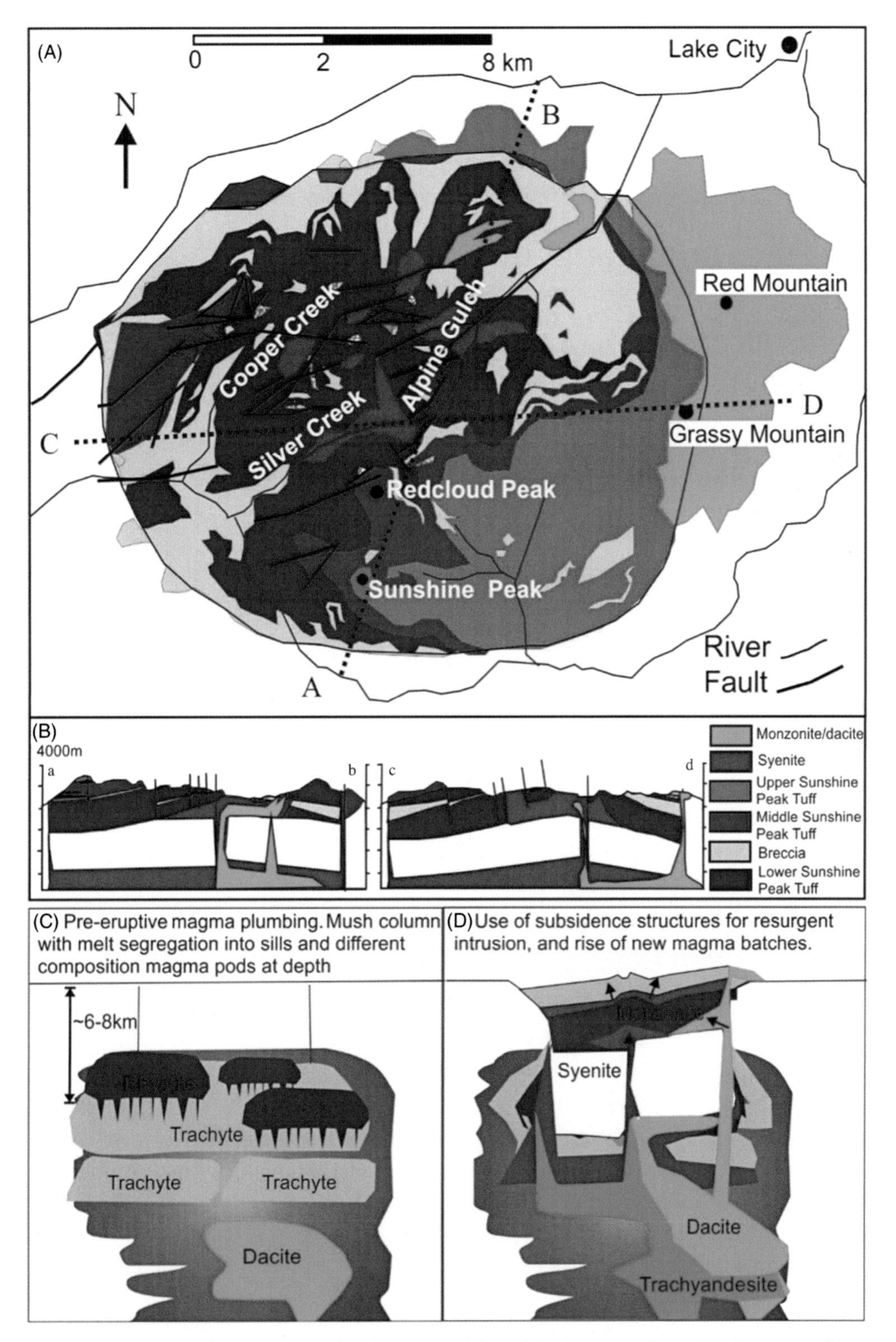

Figure. 10.4 (A) Geological map of Lake City caldera. (B) Cross-sections of Lake City caldera. (C) Schematic interpretation of pre-collapse magma plumbing system. (D) Schematic interpretation of magma plumbing after caldera collapse and during resurgence.

to the emplacement of other types of magma bodies, such as sills and laccoliths, cauldron subsidence does not strongly deform the surrounding rocks (cf. doming above laccoliths). Cauldron subsidence might explain unrest that is associated with upward-propagating seismicity, that does not lead to eruption and that is associated with surface sagging (cf. Section 10.4 and Fig. 10.2C). As magma emplacment gets shallower, roof fracturing and uplift begin to dominate (Fig. 10.3B).

There is scant detailed evidence of uplift preceding large caldera forming events (cf. Smith and Bailey, 1968; Fig. 10.3A), however, because geological exposures or structural relationships are typically overprinted by subsequent eruptive and intrusive events. An exception is the ancient Isle of Rum caldera, Scotland, where structural uplift of at least 1.5 km has occurred with a well-exposed, 8 × 12 km diameter ring fault (Emeleus et al., 1985; Holohan et al., 2009). The uplift is demonstrably pre-caldera, as the caldera infill succession oversteps uplifted and deformed country rocks within the ring fault; the equivalent country rocks outside the ring fault are essentially undeformed. The subsequent caldera collapse on Rum is evidenced by the preserved thickness of the caldera fill (>200 m) and by the syn-eruptive, down-to-centre displacement of country rocks along the ring fault system (Holohan et al., 2009).

Elsewhere, evidence for pre-collapse uplift is more subtle and often only evident from erosion and sediment transport away from the volcano, as well as from fault patterns prior to caldera collapse (Pierce et al., 2002; Troll et al., 2002). Analogue experiments (Marti et al., 1994; Acocella et al., 2000, 2001; Troll et al., 2002) show that characteristic fracturing with concentric and radial fractures can result from a complex history of both inflation and deflation of the magma plumbing system. Some basaltic calderas, such as those of the Galapagos Islands, record similar evidence for pre-subsidence inflation in the form of radial and circumferential dykes (Chadwick and Dieterich, 1995; Bagnardi et al., 2013; Chestler and Grosfils, 2013). Circumferential sheet intrusions, or cone sheets, are also well known from deeper-level exposures at ancient large caldera sites and can contribute significantly to the growth of the volcanic edifice and to surface uplift (see Chapter 4).

10.5.2 Subsidence Structures

The shape of the caldera at the surface is often used to infer the geometry and orientation of subsidence structures and of the volcanic plumbing system beneath (Smith, 1979). For example, a simple circular caldera morphology has been used to infer a single ring fault and a single cylindrical sub-caldera magma plumbing system (Lipman and Bachmann, 2015). However, a simple caldera shape at the surface may mask a more complex structure below.

Early field observations of deeply eroded calderas (e.g. Glencoe; Clough et al., 1909) and theoretical considerations of stress patterns around under-pressured magma chambers (Anderson, 1936) indicated that calderas subside along vertical to outward-inclined ring faults that also facilitate syn-collapse eruption/intrusion (see also Section 10.5.1). Subsequent observations of caldera ring faults at eroded calderas described mainly vertical (Fig. 10.3B) to inward-inclined ring faults (Lipman, 1984), a geometry favoured as the sole subsidence-accommodating structure by some numerical modelling studies (Gudmundsson et al., 1997). Analogue models (Roche et al., 2000; Kennedy et al., 2004; Acocella, 2007), field observations (Geshi et al., 2002) and more recent numerical modelling studies (Holohan et al., 2011, 2015) reveal that outward-inclined faults are coupled with peripheral inward-inclined faults that develop during caldera collapse. Moreover, analogue models show that even a single ring fault may vary in dip from inward to outward along its circumference (Roche et al., 2000; Holohan et al., 2013) and that it may or may not enclose the whole circumference of the subsided area (Kennedy et al., 2004). Such variation in ring-fault dip is seen in outcrop at Glencoe caldera (Clough et al., 1909) and is inferred from geodetic and seismic data at Tendürek volcano, Turkey, (Bathke et al., 2015) and Bardarbunga caldera, Iceland (Gudmundsson et al., 2016). With large depth-to-diameter ratios of the magma reservoir, caldera subsidence may be accommodated by several, successively-formed sets of outward- and/or inward-inclined ring faults (Roche et al., 2000; Kennedy et al., 2004; Burchardt and Walter, 2010; Holohan et al., 2011).

Another very common complexity to caldera subsidence is 'asymmetry' (Lipman, 1984). This term applies to calderas where the maximum subsidence is off-centred (Fig. 10.3B) such that the subsidence profile is asymmetric along most cross-sections of the caldera. Analogue and numerical models of caldera collapse with macroscopically symmetric starting geometries produce collapses of varying degrees of asymmetry. In these systems, as in nature, the development of fracture systems is strongly influenced by small-scale heterogeneity in the subsiding material and the shape of the roof of the magma plumbing system (Lipman, 1997). Consequently, perfect symmetry is rarely, if ever, approached. Other factors that may promote collapse asymmetry include an initially 'asymmetric' magma body (Lipman, 1997; Kennedy et al., 2004), the geometry of which may again reflect the influence of heterogeneity on various scales, as well as uneven topographic loading (Lavallée et al., 2004) and pre-existing tectonic faults (Holohan et al., 2005). One consequence of collapse asymmetry is that ring faults may show strike-slip components of displacement as well as dip-slip (Holohan et al., 2013). A second is that magma flow directions (Fig. 10.3B) along a ring fault or within a chamber during asymmetric collapse may not be simply vertically upward, but also involve a large lateral component of flow (Kennedy et al., 2008).

Caldera subsidence may also be accommodated, to a greater or lesser degree, by ductile downsagging of the rocks overlying the magma reservoir (Fig. 10.3B) (Walker, 1984; Branney and Kokelaar, 1994). Moreover, pre-existing regional faults may be reactivated to accommodate some subsidence, as well as acting as conduits for eruption. The reactivation of regional faults, as well as the generation of multiple ring faults and subsidiary fractures, may lead to a structurally complex architecture to the caldera, which may be suggestive of a non-coherent or piecemeal collapse style (Lipman, 1997).

A non-circular structural boundary of a caldera, as seen in plan view, may stem from several factors. Firstly, it may reflect regional tectonics and/or magma chamber elongation (Acocella et al., 2003, 2004; see also Chapter 7). Secondly, non-circularity and complexity of the collapse structure may be further enhanced by subsidence into several discrete magma pods (Fig. 10.3B). Indeed, many complex caldera morphologies are the product of spatially separate or overlapping caldera collapses (Gravley et al., 2007). These complex calderas may be related to multiple generations of eruptible magma bodies (Hasegawa et al., 2009; Hasegawa and Nakagawa, 2016).

10.5.3 Syn-Collapse Magma Plumbing and Conduit Systems

The location of vents before and during caldera collapse can provide additional information on the interplay between the shallow plumbing,subsidence structures and the distribution and amalgamation of sub-caldera magma bodies. Caldera-forming eruptions can occur from single vents (e.g. the younger A.D. 232 Taupo eruption; Wilson 1993), multiple vents (e.g. Huckleberry Ridge eruption at Yellowstone; Myers et al., 2016), linear vents (e.g. many Sierra Madre Occidental eruptions; Aguirre-Díaz and Labarthe-Hernendez, 2003) and ring vents (e.g. Long Valley; Wilson and Hildreth, 1997). Each of these vent types may evolve dramatically as faults develop during subsidence (Wilson and Hildreth, 1997; Holohan et al., 2008). Varied and evolving pyroclastic vent locations may imply a complex plumbing system, whereby subsiding blocks and changing stress regimes and erosive pyroclastic eruptions open and close magma pathways. As crustal blocks subside, magma can be shifted around (Folch and Marti, 1998; Kennedy et al., 2008), and previously discrete magma pods can be amalgamated (Bégué et al., 2014a; Cashman and Giordano, 2014; Kennedy et al., 2016; Myers et al., 2016).

Ancient calderas that are sufficiently well exposed reveal that vents are commonly linked to the sub-caldera reservoir via magmatic intrusions into the subsidence-related faults. Examples of ring faults intruded by a variety of magmas from the underlying magma reservoirs and with complex mingling patterns are found at Glencoe caldera (Clough et al., 1909; Kokelaar, 2007; Fig. 10.3B), Loch Bá, Isle of Mull, Scotland (Sparks, 1988), Ossipee, New Hampshire (Kennedy and Stix, 2007) and Rum caldera, Scotland (Holohan et al., 2009; Nicoll et al., 2009). One explanation for the texturally variable ring-fault intrusions at Rum, Glencoe and Ossipee is that subsidence upon a ring fault allows for a central block or blocks to subside into a mushy marginal magma zone (Marsh, 1996), trapping an area of mushy magma between the subsiding block and the chamber wall (Kennedy and Stix, 2007). Fluid dynamics experiments (Kennedy et al., 2008) have also revealed that these marginal zones of chambers are areas where vortices may be induced in magmas during subsidence, promoting the incorporation and preservation of a wide range of materials within ring dykes (Kennedy and Stix, 2007).

In some calderas, a shallow exposure level may mean that the vents are preserved only as steeply dipping dyke-like bodies of tuff with a dyke-parallel eutaxitic foliation formed by compression of fragmented magma as an eruption wanes and the dyke walls close (e.g. Sabaloka caldera, Sudan; Almond, 1977).

Finally, it is important to note that magma can migrate laterally from beneath a collapsing caldera to erupt at vents located tens of kilometres distant from the collapsing area. In this case, syn-collapse vents and conduit systems may not be found at the caldera at all. As noted above, lateral magma migration is well documented in basaltic systems (Magee et al., 2016 and references therein), with far fewer silicic examples. In some cases, the caldera volume is considerably less than the volume of magma that erupted (Hildreth, 1991; Wilson, 2001; Gravley et al., 2007; Lipman and McIntosh, 2008). This missing volume may be accounted for by volumetric expansion of the magma (Rivalta and Segall, 2008) or of the subsiding magma chamber roof (Poppe et al., 2015; Holohan et al., 2017) and/or by a hidden volume of intrusion (e.g. at Bardarbunga; Gudmundsson et al., 2016). Overall, the proclivity for lateral migration of magma into, within or out of a sub-caldera reservoir should be borne in mind when considering vent position and development.

10.5.4 Structural Uplift and Magma Intrusion Following Caldera Formation

Intrusion and structural uplift following caldera subsidence are common (see also Section 10.4.2), and if this uplift is of 10^2–10^3 m, it is termed resurgence. This may involve subsidence fault reactivation (inversion) and crustal bending associated with magma intrusion (Fig. 10.3C) (Smith and Bailey, 1968; Kennedy et al., 2012; Hildreth et al., 2017), depending on the reservoir roof geometry (Acocella et al., 2001). This relationship is clarified when erosion exposes an intertwined history of subsidence, intrusions, uplift and tectonics. However, it should be noted that uplift does not always follow caldera formation, and intrusion at many calderas may be facilitated and accommodated by regional extension (Spinks et al., 2005) or by chamber floor subsidence.

Localities that expose the solidified magma plumbing system beneath resurgent calderas provide evidence for shallow post-collapse intrusions fed by dykes. These shallow intrusions may take the form of laccoliths (Fig. 10.3C) (Kennedy et al., 2016; see also Chapter 6) or multiple stacked sills (Hildreth et al., 2017; see also Chapter 5). Their emplacement may drive variable amounts of reactivation (inversion) of regional faults or earlier subsidence-controlling faults (Browning and Gudmundsson, 2015) (Fig. 10.3C). Structural uplift may in some cases be directly linked to ring-fault intrusion (Saunders, 2001, 2004), which can in turn feed shallow intrusions within the caldera (Kennedy et al., 2016). The shallow resurgent laccoliths are typically also associated with an active hydrothermal system, which again may mask or confuse the geophysical interpretation (see Section 10.4.1). Such intrusions may in an idealised scenario feed a ring of post-collapse eruptive products, such as lava domes (e.g. at Valles caldera; Smith and Bailey, 1968). Although such rings of post-collapse vents have been used to infer the position of earlier structures, it is not a robust assumption (Walker, 1984), as a post-collapse intrusion can generate its own ring fault independently (e.g. at Grizzly Peak; Fridrich et al., 1991).

Many models have been proposed to explain caldera resurgence (Marsh, 1984; Kennedy et al., 2012). Here we propose that the drastic change in crustal structure following caldera formation, the anomalously high heat flow, and change in crustal stresses together contribute to an environment that promotes replenishment, rejuvenation and uplift associated with shallow intrusions and eruptions, and the establishment of an active hydrothermal system.

10.6 CONCLUSIONS

Our review of the volcanic and igneous plumbing system beneath calderas has drawn on petrological studies of eruptive products, geophysical studies of unrest and eruption, experimental and geological structural data as well as the theoretical framework of the caldera cycle. This review reveals some key points that we present in a conceptual view of caldera plumbing systems (Fig. 10.5). All caldera plumbing systems are distinct, but this conceptual model contains plumbing elements that are common and applicable to many sub-caldera environments.

(1) Geochemical, petrological, geophysical and numerical studies show that the state of magma beneath a caldera is transient during its life cycle. The caldera-forming magma reservoir may comprise a series of melt-rich pockets that exist within or outside of a dominantly crystalline zone of older intrusive bodies (Fig. 10.4). This reservoir system may include vertically stacked or adjacent small or large melt pods that are either isolated by fully crystalline crust or partially connected by mush that allows rapid amalgamation (Fig. 10.5). Evidence from geochemical and textural studies of large caldera-forming eruptions indicates that large bodies of interconnected melt do exist at certain times in the life cycle of the caldera. These may be chemically diverse or homogeneous, crystal-rich to crystal-poor, reside at depths of 3–10 km and with melt residence times of 10^2–10^5 years. These shallow magma bodies are fed by deeper, less evolved magma mush zones (Fig. 10.5).

(2) Studies of caldera unrest have recorded lateral transport of magma exceeding 10 km within and outside calderas along regional or subsidence-controlling structures (Fig. 10.5). As well as the more typically envisaged vertical magma ascent, this lateral transport may result in subsidence and/or inflation within the caldera, and may serve to trigger collapse in both basaltic and silicic systems. Seismicity during dyke and sill emplacement supports geological evidence for the use of planar and curviplanar dykes and sills within and outside inferred mush bodies (Fig. 10.5). Once

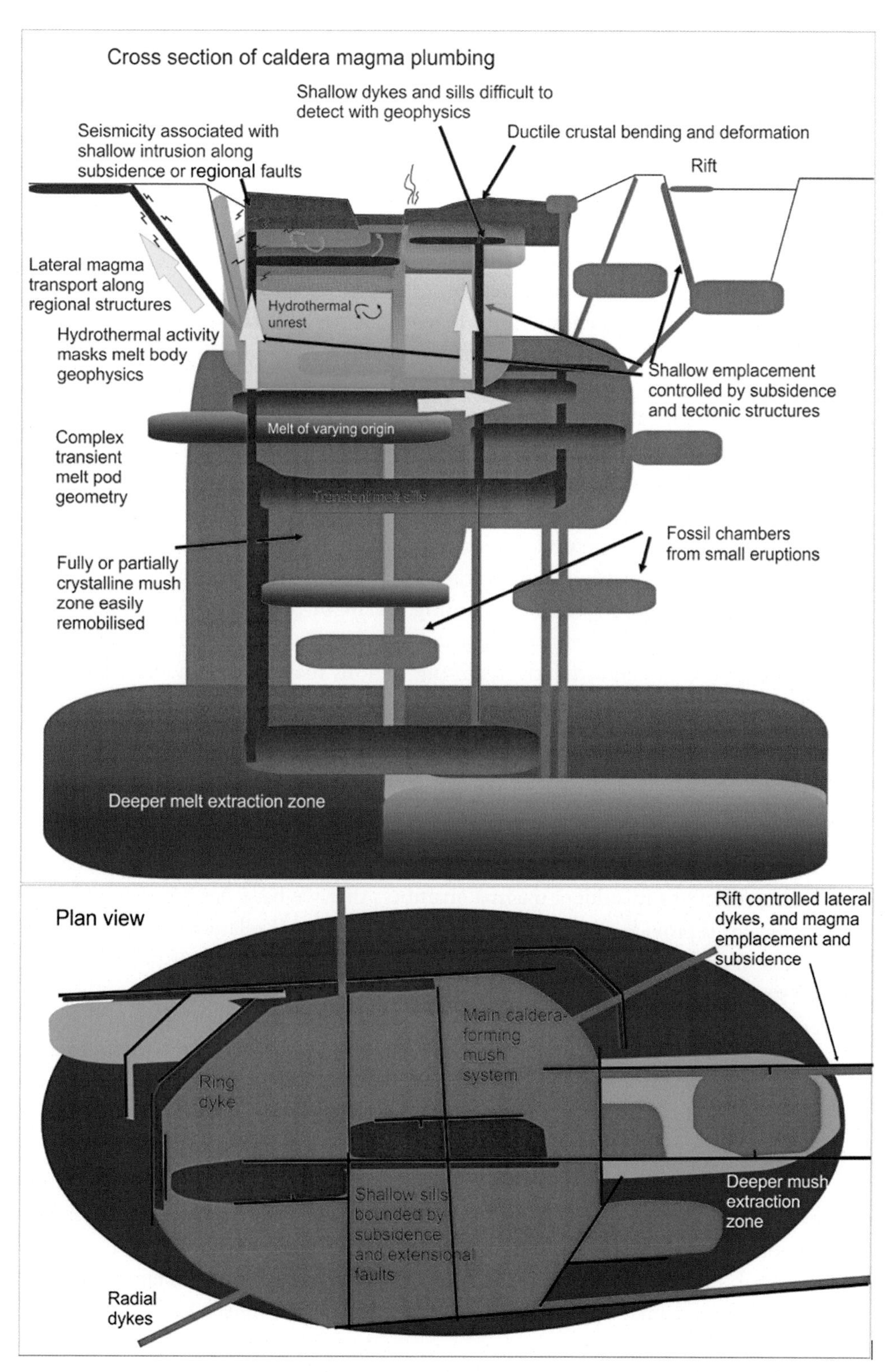

Figure. 10.5 ***Conceptual view of combined elements of volcanic and igneous plumbing systems beneath calderas.***

pathways are established, magma may be transported aseismically. Aseismic zones, S-wave anomalies and/or seismic low-velocity zones indicate the existence of small, shallow intrusions and large zones of partial melt of similar or larger dimensions than the caldera. However, these techniques struggle to identify small dykes and sills, which are a common component of sub-caldera plumbing systems.

(3) Geological and structural data from experiments and numerical models show how regional and subsidence structures can be reactivated and act as conduits for magma transport and accumulation. Planar and curviplanar subsidence structures may be re-intruded to transport magma vertically or laterally, driving deformation (Fig. 10.5). Sill-like, laccolith or lopolith intrusions are accommodated by supra-intrusion uplift or sub-intrusion subsidence of crustal blocks (Cruden, 1998). Conversely, when magma drains laterally, subsidence will occur above the source region. The classic model of the caldera cycle has implications for the transient nature of magma plumbing. Large bodies of eruptible magma may take time to accumulate and may form broad structural domes and radial dyke swarms (Fig. 10.5). However, following caldera collapse, a considerable change occurs in the stress field around the volcanic plumbing system, promoting ascent of magma along new structural pathways and resurgent uplift of the caldera floor may also occur. Smaller basaltic calderas and/or calderas in actively rifting regions may follow an alternative path, with magma intrusion and accumulation tectonically controlled and not necessarily cyclical in nature.

FIELD EXAMPLE: LAKE CITY CALDERA

Lake City caldera (Fig. 10.4A) provides a unique field example that illustrates many of the characteristics of sub-caldera magma chambers that we have described in the chapter. Specifically, it provides direct geochemical and structural relationships between the sub-caldera magma chambers and the volcanic deposits, and between the caldera and regional structure.

Geological exposure is spectacular at Lake City. Extreme vertical relief exposes >2 km vertical thickness of intracaldera volcanics and the intrusions that fed the related eruptions. Three rhyolitic to trachytic ignimbrites (300 km^3) and associated mega-breccia deposits filled the caldera. These three units show no evidence of a long time break, but do contain evidence of tapping a progressively more complex magmatic system. The first erupted Lower Sunshine Peak Tuff (Fig. 10.4B) is a relatively homogenous rhyolite. The Middle Sunshine Peak Tuff contains less-evolved trachyte evidencing the co-eruption of rhyolite and rejuvenated co-genetic trachytic crystal mush. The Upper Sunshine Peak Tuff contains evidence of several isolated magma pockets tapped together with the leftover magmas of the Middle and Lower Sunshine Peak Tuffs. This eruptive sequence points to a large magma plumbing system of pods of rhyolitic and trachytic magma (Fig. 10.4C), separated by mush zones, which catastrophically erupted and amalgamated during caldera subsidence (Hon, 1987; Kennedy et al., 2016).

Lake City sits with the larger Uncompahgre caldera and the ENE-WSW Eureka graben. The caldera is asymmetric with an ENE–WSW elongation, and faults with this orientation such as the Alpine Gulch Fault (Fig. 10.4A) were active during subsidence and resurgence. Regional tectonism is thought to have initiated in this area at around 25 Ma (Bove et al., 2001) and therefore should have been active at 23 Ma. Despite this regional structural control, the caldera has a well-defined structural boundary ring fault.

Most caldera subsidence was accommodated along a ring fracture with an additional subsidence-controlling fault parallel to regional tectonic faults. Both the ring fault and the regional structure were reactivated by intrusion following subsidence (Fig. 10.4D). The intrusions that followed subsidence provide support for the geochemical model based on the eruptive deposits with two distinct intrusion types: (1) those chemically related to the caldera-forming eruption, for example, rhyolite ring dykes, and flat-topped syenite laccoliths intruded through and at the base of the caldera filling ignimbrite; or (2) dacites ring dyke intrusions that fed lava dome eruptions and monzonite laccoliths also intruded close the base of the caldera-forming ignimbrite.

The geology and geochemistry describe an ascending magma system controlled largely by a crustal response to caldera formation and channelled along caldera structures. Prior to the caldera-forming eruption, a large magma plumbing system evolved to form rhyolitic pods and trachytic mush zones. Although not temporally constrained at Lake City, petrologic and geochemical evidence supports multiple injections and mush re-melting to develop this large volume of eruptible magma. Caldera subsidence allowed these individual pockets to erupt (Cashman and Giordano, 2014). A second magma batch rose to shallow levels along subsidence structures facilitated by crustal buoyancy changes (Kennedy et al., 2012). This second magma batch drove continued resurgent uplift, and it erupted to form lava domes. It also provided the heat for a shallow local hydrothermal system (Garden et al., 2017).

The drastic change in caldera structure, the anomalously high heat flow and change in crustal stresses all contributed to an environment that promoted replenishment, rejuvenation, shallow intrusions and the establishment of an active hydrothermal system. The post-caldera magmatism shows a range of intrusion styles that can inform interpretation of restless signals. Firstly, thin dykes exploit subsidence structures to form ring dykes. Some deformation and seismicity maybe associated with these (Saunders, 2004), but smaller vertical intrusion could be harder to detect with lower resolution geophysics. Secondly, intrusions are associated with a well-developed hydrothermal system recorded in veins, alteration zones and brecciation, particularly where subsidence-controlling faults interact (Garden et al., 2017). At active calderas, such hydrothermal activity may mask the intrusive signatures and so makgeophysical unrest difficult to assign to magmatism.

REFERENCES

Acocella,V., 2007. Understanding caldera structure and development: an overview of analogue models compared to natural calderas. Earth Sci. Rev. 85 (3), 125–160.

Acocella, V., Cifelli, F., Funiciello, R., 2000. Analogue models of collapse calderas and resurgent domes. J.Volcanol. Geotherm. Res. 104 (1), 81–96.

Acocella,V., Cifelli, F., Funiciello, R., 2001.The control of overburden thickness on resurgent domes: insights from analogue models. J.Volcanol. Geotherm. Res. 111 (1), 137–153.

Acocella,V., Korme,T., Salvini, F., Funiciello, R., 2003. Elliptic calderas in the Ethiopian Rift: control of pre-existing structures. J.Volcanol. Geotherm. Res. 119 (1), 189–203.

Acocella, V., Funiciello, R., Marotta, E., Orsi, G., De Vita, S., 2004. The role of extensional structures on experimental calderas and resurgence. J.Volcanol. Geotherm. Res. 129 (1), 199–217.

Acocella,V., Di Lorenzo, R., Newhall, C., Scandone, R., 2015. An overview of recent (1988 to 2014) caldera unrest: knowledge and perspectives. Rev. Geophys. 533, 896–955.

Aguirre-Díaz, G.J., Labarthe-Hernendez, G., 2003. Fissure ignimbrites: fissure-source origin for voluminous ignimbrites of the Sierra Madre Occidental and its relationship with Basin and Range faulting. Geology 31 (9), 773–776.

Allan, A.S., Morgan, D.J., Wilson, C.J., Millet, M.A., 2013. From mush to eruption in centuries: assembly of the super-sized Oruanui magma body. Contrib. Mineral. Petrol. 166 (1), 143–164.

Almond, D.C., 1977. The Sabaloka igneous complex, Sudan. Phil. Trans. R. Soc. London A 287 (1348), 595–633.

Anderson, E.M., 1936. The dynamics of formation of cone sheets, ring-dykes, and cauldron subsidence. Proc. R. Soc. Edinburgh 56, 128–163.

Bachl, C.A., Miller, C.F., Miller, J.S., Faulds, J.E., 2001. Construction of a pluton: evidence from an exposed cross-section of the Searchlight pluton, Eldorado Mountains, Nevada. Geol. Soc. Am. Bull. 113, 1213–1228.

Bachmann, O., Bergantz, G., 2008. The magma reservoirs that feed supereruptions. Elements 4 (1), 17–21.

Bachmann, O., Dungan, M.A., Lipman, P.W., 2002. The Fish Canyon magma body, San Juan volcanic field, Colorado: rejuvenation and eruption of an upper-crustal batholith. J. Petrol. 43 (8), 1469–1503.

Bachmann, O., Oberli, F., Dungan, M.A., Meier, M., Mundil, R., Fischer, H., 2007. 40 Ar/39 Ar and U–Pb dating of the Fish Canyon magmatic system, San Juan Volcanic field, Colorado: evidence for an extended crystallization history. Chem. Geol. 236 (1), 134–166.

Bagnardi, M., Amelung, F., Poland, M.P., 2013. A new model for the growth of basaltic shields based on deformation of Fernandina volcano, Galápagos Islands. Earth Planet. Sci. Lett. 377, 358–366.

Barnes, C.G., Allen, C.M., Hoover, J.D., Brigham, R.H., 1990. Magmatic components of a tilted plutonic system, Klamath Mountains, California. Anderson, J.L. (Ed.), The Nature and Origin of Cordilleran Magmatism, 174, Geological Society of America Memoir, pp. 331–346.

Bathke, H., Nikkhoo, M., Holohan, E.P., Walter, T.R., 2015. Insights into the 3D architecture of an active caldera ring-fault at Tendürek volcano through modeling of geodetic data. Earth Planet. Sci. Lett. 422, 157–168.

Bégué, F., Deering, C.D., Gravley, D.M., Kennedy, B.M., Chambefort, I., Gualda, G.A.R., Bachmann, O., 2014a. Extraction, storage and eruption of multiple isolated magma batches in the paired Mamaku and Ohakuri eruption, Taupo Volcanic Zone, New Zealand. J. Petrol. 55 (8), 1653–1684.

Bégué, F., Gualda, G.A., Ghiorso, M.S., Pamukcu, A.S., Kennedy, B.M., Gravley, D.M., Chambefort, I., 2014b. Phase-equilibrium geobarometers for silicic rocks based on rhyolite-MELTS. Part 2: application to Taupo Volcanic Zone rhyolites. Contrib. Mineral. Petrol. 168 (5), 1082.

Bégué, F., Gravley, D.M., Chambefort, I., Deering, C.D., Kennedy, B.M., 2015. Magmatic volatile distribution as recorded by rhyolitic melt inclusions in the Taupo Volcanic Zone, New Zealand. Geological Society, London, Special Publications 410(1):71–94.

Bove, D.J., Hon, K., Budding, K.E., Slack, J.F., Snee, L.W., Yeoman, R.A., 2001. Geochronology and geology of late Oligocene through Miocene volcanism and mineralization in the western San Juan Mountains. U.S. Geological Survey Professional Paper 1642, 30 pp.

Branney, M., Acocella, V., 2015. Calderas. In: Sigurdsson, H. (Ed.), The Encyclopedia of Volcanoes, pp. 299–319.

Branney, M.J., Kokelaar, P., 1994. Volcanotectonic faulting, soft-state deformation, and rheomorphism of tuffs during development of a piecemeal caldera, English Lake District. Geol. Soc. Am. Bull. 106 (4), 507–530.

Browning, J., Gudmundsson, A., 2015. Caldera faults capture and deflect inclined sheets: an alternative mechanism of ring dike formation. Bull. Volcanol. 77 (1), 4.

Burchardt, S., Walter, T.R., 2010. Propagation, linkage, and interaction of caldera ring-faults: comparison between analogue experiments and caldera collapse at Miyakejima, Japan, in 2000. Bull. Volcanol. 72 (3), 297–308.

Burchardt, S., Tanner, D.C., Krumbholz, M., 2010. Mode of emplacement of the Slaufrudalur Pluton, Southeast Iceland inferred from three-dimensional GPS mapping and model building. Tectonophysics 480 (1), 232–240.

Burchardt, S., Tanner, D., Krumbholz, M., 2012. The Slaufrudalur pluton, southeast Iceland—an example of shallow magma emplacement by coupled cauldron subsidence and magmatic stoping. Geol. Soc. Am. Bull. 124 (1–2), 213–227.

Cargill, H.K., Hawkes, L., Ledeboer, J.A., 1928. The major intrusions of south-eastern Iceland. Q. J. Geol. Soc. London 84, 505–539.

Cashman, K.V., Giordano, G., 2014. Calderas and magma reservoirs. J. Volcanol. Geotherm. Res. 288, 28–45.

Chadwick, W.W., Dieterich, J.H., 1995. Mechanical modeling of circumferential and radial dike intrusion on Galapagos volcanoes. J. Volcanol. Geotherm. Res. 66 (1–4), 37–52.

Chadwick, W.W., Geist, D.J., Jónsson, S., Poland, M., Johnson, D.J., Meertens, C.M., 2006. A volcano bursting at the seams: inflation, faulting, and eruption at Sierra Negra Volcano, Galápagos. Geology 34 (12), 1025–1028.

Chestler, S.R., Grosfils, E.B., 2013. Using numerical modeling to explore the origin of intrusion patterns on Fernandina volcano, Galapagos Islands, Ecuador. Geophys. Res. Lett. 40 (17), 4565–4569.

Clough, C.T., Maufe, H.B., Bailey, E.B., 1909. The cauldron-subsidence of Glencoe and associated igneous phenomena. Q. J. Geol. Soc. London 65, 1–4.

Cole, J.W., Milner, D.M., Spinks, K.D., 2005. Calderas and caldera structures. Earth Sci. Rev. 69, 1–26.

Cole, J.W., Deering, C.D., Burt, R.M., Sewell, S., Shane, P.A.R., Mathews, N.E., 2014. Okataina Volcanic centre, Taupo Volcanic Zone: a review of volcanism and synchronous pluton development in an active, dominantly silicic caldera system. Earth Sci. Rev. 128, 1–17.

Cruden, A.R., 1998. On the emplacement of tabular granites. J. Geol. Soc. London 155 (5), 853–862.

Cruden, A.R., McCaffrey, K.J.W., 2001. Growth of plutons by floor subsidence: implications for rates of emplacement, intrusion spacing and melt-extraction mechanisms. Phys. Chem. Earth Part A 26 (4–5), 303–315.

De Silva, S.L., 1989. Altiplano-Puna volcanic complex of the central Andes. Geology 17 (12), 1102–1106.

Druitt, T.H., Sparks, R.S.J., 1984. On the formation of calderas during ignimbrite eruptions. Nature 310 (5979), 679–681.

Druitt, T.H., Costa, F., Deloule, E., Dungan, M., Scaillet, B., 2012. Decadal to monthly timescales of magma transfer and reservoir growth at a caldera volcano. Nature 482 (7383), 77–80.

Eibl, E.P., Bean, C.J., Vogfjord, K.S., Ying, Y., Lokmaer, I., Moolhoff, M., Palsson, F., 2017. Tremor-rich shallow dyke formation followed by silent magma flow at Bararbunga in Iceland. Nat. Geosci.

Eichelberger, J.C., Chertkoff, D.G., Dreher, S.T., Nye, C.J., 2000. Magmas in collision: rethinking chemical zonation in silicic magmas. Geology 28 (7), 603–606.

Emeleus, C.H., Wadsworth, W.J., Smith, N.J., 1985. The early igneous and tectonic history of the Rhum Tertiary Volcanic Centre. Geol. Mag. 122 (05), 451–457.

Filson, J., Simkin, T., Leu, L.K., 1973. Seismicity of a caldera collapse: Galapagos Islands 1968. J. Geophys. Res. 78 (35), 8591–8622.

Folch, A., Marti, J., 1998. The generation of overpressure in felsic magma chambers by replenishment. Earth Planet. Sci. Lett. 163 (1), 301–314.

Francis, P.W., Sparks, R.S.J., Hawkesworth, C.J., Thorpe, R.S., Pyle, D.M., Tait, S.R., Mantovani, M.S, McDermott, F., 1989. Petrology and geochemistry of volcanic rocks of the Cerro Galan caldera, northwest Argentina. Geol. Mag. 126 (5), 515–547.

Fridrich, C.J., Smith, R.P., DeWitt, E.D., McKee, E.H., 1991. Structural, eruptive, and intrusive evolution of the Grizzly Peak caldera, Sawatch Range, Colorado. Geol. Soc. Am. Bull. 103 (9), 1160–1177.

Garden, T.O., Gravely, D.M., Kennedy, B.M., Deering, C., Chambefort, I., 2017. Controls on hydrothermal fluid flow in caldera-hosted settings: evidence from Lake City caldera, U.S.A., 2017. Geosphere 13, 6, doi:10.1130/GES01506.1.

Geshi, N., Shimano, T., Chiba, T., Nakada, S., 2002. Caldera collapse during the 2000 eruption of Miyakejima Volcano, Japan. Bull. Volcanol. 64, 55–68.

Gravley, D.M., Wilson, C.J.N., Leonard, G.S., Cole, J.W., 2007. Double trouble: paired ignimbrite eruptions and collateral subsidence in the Taupo Volcanic Zone, New Zealand. Geol. Soc. Am. Bull. 119 (1–2), 18–30.
Gualda, G.A., Ghiorso, M.S., 2013. The Bishop Tuff giant magma body: an alternative to the Standard Model. Contrib. Mineral. Petrol. 166 (3), 755–775.
Gualda, G.A., Ghiorso, M.S., 2014. Phase-equilibrium geobarometers for silicic rocks based on rhyolite-MELTS. Part 1: Principles, procedures, and evaluation of the method. Contrib. Mineral. Petrol. 168 (1), 1–17.
Gualda, G.A., Sutton, S.R., 2016. The year leading to a supereruption. PLoS ONE 11 (7), e0159200.
Gudmundsson, A., Marti, J., Turon, E., 1997. Stress fields generating ring faults in volcanoes. Geophys. Res. Lett. 24 (13), 1559–1562.
Gudmundsson, O., Finlayson, D.M., Itikarai, I., Nishimura, Y., Johnson, W.R., 2004. Seismic attenuation at Rabaul volcano, Papua New Guinea. J. Volcanol. Geotherm. Res. 130 (1), 77–92.
Gudmundsson, M.T., Jónsdóttir, K., Hooper, A., Holohan, E.P., Halldórsson, S.A., Ófeigsson, B.G., Cesca, S., Vogfjörd, K.S., Sigmundsson, F., Högnadóttir, T., Einarsson, P., 2016. Gradual caldera collapse at Bárdarbunga volcano, Iceland, regulated by lateral magma outflow. Science 353 (6296), aaf8988.
Harper, B.E., Miller, C.F., Koteas, G.C., Cates, N.L., Wiebe, R.A., Lassareschi, D.S., Cribb, J.W., 2004. Granites, dynamic magma chamber processes, and pluton construction: Aztec Wash pluton, Eldorado Mountains, Nevada, USA. Trans. R. Soc. Edinburgh 95, 277–295.
Hasegawa, T., Nakagawa, M., 2016. Large scale explosive eruptions of Akan volcano, eastern Hokkaido, Japan: a geological and petrological case study for establishing tephro-stratigraphy and-chronology around a caldera cluster. Q. Int. 397, 39–51.
Hasegawa, T., Yamamoto, A., Kamiyama, H., Nakagawa, M., 2009. Gravity structure of Akan composite caldera, eastern Hokkaido, Japan: application of lake water corrections. Earth Planets Space 61 (7), 933–938.
Hawkins, D.P., Wiebe, R.A., 2004. Discrete stoping events in granite plutons: a signature of eruptions from silicic magma chambers? Geology 32, 1021–1025.
Hildreth, W., 1979. The Bishop Tuff: evidence for the origin of compositional zonation in silicic magma chambers. Geol. Soc. Am. Special Pap. 180, 43–76.
Hildreth, W., 1981. Gradients in silicic magma chambers: implications for lithospheric magmatism. J. Geophys. Res. 86 (B11), 10153–10192.
Hildreth, W., 1991. The timing of caldera collapse at Mount Katmai in response to magma withdrawal toward Novarupta. Geophys. Res. Lett. 18 (8), 1541–1544.
Hildreth, W., 2004. Volcanological perspectives on Long Valley, Mammoth Mountain, and Mono Craters: several contiguous but discrete systems. J. Volcanol. Geotherm. Res. 136 (3), 169–198.
Hildreth, W., Fierstein, J., Calvert, A., 2017. Early postcaldera rhyolite and structural resurgence at Long Valley Caldera, California. J. Volcanol. Geotherm. Res. 335, 1–34.
Holohan, E.P., Troll, V.R., Walter, T.R., Münn, S., McDonnell, S., Shipton, Z.K., 2005. Elliptical calderas in active tectonic settings: an experimental approach. J. Volcanol. Geotherm. Res. 144 (1), 119–136.
Holohan, E.P., Troll, V.R., De Vries, B.V.W., Walsh, J.J., Walter, T.R., 2008. Unzipping Long Valley: an explanation for vent migration patterns during an elliptical ring fracture eruption. Geology 36 (4), 323–326.
Holohan, E.P., Troll, V.R., Errington, M., Donaldson, C.H., Nicoll, G.R., Emeleus, C.H., 2009. The Southern Mountains Zone, Isle of Rum, Scotland: volcanic and sedimentary processes upon an uplifted and subsided magma chamber roof. Geol. Mag. 146 (3), 400–418.
Holohan, E.P., Schöpfer, M.P.J., Walsh, J.J., 2011. Mechanical and geometric controls on the structural evolution of pit crater and caldera subsidence. J. Geophys. Res. 116 (B7).
Holohan, E.P., Walter, T.R., Schöpfer, M.P., Walsh, J.J., Wyk de Vries, B., Troll, V.R., 2013. Origins of oblique-slip faulting during caldera subsidence. J. Geophys. Res. 118 (4), 1778–1794.
Holohan, E.P., Schöpfer, M.P.J., Walsh, J.J., 2015. Stress evolution during caldera collapse. Earth Planet. Sci. Lett. 421, 139–151.
Holohan, E.P., Sudhaus, H., Walter, T.R., Schöpfer, M.P.J., Walsh, J.J., 2017. Effects of Host-rock Fracturing on Elastic-deformation Source Models of Volcano Deflation. Sci. Rep. 7, 10970.
Hon, K.A., 1987. Geologic and petrologic evolution of the Lake City caldera, San Juan Mountains, Colorado. Ph.D. thesis. University of Colorado at Boulder, 244 pp.
Howard, K.A., 2010. Caldera collapse: perspectives from comparing Galápagos volcanoes, nuclear-test sinks, sandbox models, and volcanoes on Mars. GSA Today 20 (10), 4–10.

Kaiser, J.F., de Silva, S., Schmitt, A.K., Economos, R., Sunagua, M., 2017. Million-year melt–presence in monotonous intermediate magma for a volcanic–plutonic assemblage in the Central Andes: contrasting histories of crystal-rich and crystal-poor super-sized silicic magmas. Earth Planet. Sci. Lett. 457, 73–86.
Kennedy, B., Stix, J., 2007. Magmatic processes associated with caldera collapse at Ossipee ring dyke, New Hampshire. Geol. Soc. Am. Bull. 119 (1–2), 3–17.
Kennedy, B., Stix, J., Vallance, J.W., Lavallée, Y., Longpré, M.A., 2004. Controls on caldera structure: results from analogue sandbox modeling. Geol. Soc. Am. Bull. 116 (5–6), 515–524.
Kennedy, B.M., Jellinek, A.M., Stix, J., 2008. Coupled caldera subsidence and stirring inferred from analogue models. Nat. Geosci. 1 (6), 385–389.
Kennedy, B., Wilcock, J., Stix, J., 2012. Caldera resurgence during magma replenishment and rejuvenation at Valles and Lake City calderas. Bull. Volcanol. 74 (8), 1833–1847.
Kennedy, B., Stix, J., Hon, K., Deering, C., Gelman, S., 2016. Magma storage, differentiation, and interaction at Lake City caldera, Colorado, USA. Geol. Soc. Am. Bull. 128 (5–6), 764–776.
Kokelaar, P., 2007. Friction melting, catastrophic dilation and breccia formation along caldera superfaults. J. Geol. Soc. 164 (4), 751–754.
Kumagai, H., Ohminato, T., Nakano, M., Ooi, M., Kubo, A., Inoue, H., Oikawa, J., 2001. Very-long-period seismic signals and caldera formation at Miyake Island, Japan. Science 293, 687–690.
Lavallée, Y., Stix, J., Kennedy, B., Richer, M., Longpré, M.A., 2004. Caldera subsidence in areas of variable topographic relief: results from analogue modeling. J. Volcanol. Geotherm. Res. 129 (1), 219–236.
Lindsay, J.M., Schmitt, A.K., Trumbull, R.B., De Silva, S.L., Siebel, W., Emmermann, R., 2001. Magmatic evolution of the La Pacana caldera system, Central Andes, Chile: compositional variation of two cogenetic, large-volume felsic ignimbrites. J. Petrol. 42 (3), 459–486.
Lipman, P.W., 1984. The roots of ash flow calderas in western North America: windows into the tops of granitic batholiths. J. Geophys. Res. 89 (B10), 8801–8841.
Lipman, P.W., 1997. Subsidence of ash-flow calderas: relation to caldera size and magma-chamber geometry. Bull. Volcanol. 59 (3), 198–218.
Lipman, P.W., 2007. Incremental assembly and prolonged consolidation of Cordilleran magma chambers: evidence from the Southern Rocky Mountain volcanic field. Geosphere 3 (1), 42–70.
Lipman, P.W., Bachmann, O., 2015. Ignimbrites to batholiths: integrating perspectives from geological, geophysical, and geochronological data. Geosphere 11 (3), 705–743.
Lipman, P.W., McIntosh, W.C., 2008. Eruptive and noneruptive calderas, northeastern San Juan Mountains, Colorado: where did the ignimbrites come from? Geol. Soc. Am. Bull. 120 (7–8), 771–795.
Magee, C., Muirhead, J.D., Karvelas, A., Holford, S.P., Jackson, C.A., Bastow, I.D., Schofield, N., Stevenson, C.T., McLean, C., McCarthy, W., Shtukert, O., 2016. Lateral magma flow in mafic sill complexes. Geosphere 12 (3), 809–841.
Marsh, B.D., 1984. On the mechanics of caldera resurgence. J. Geophys. Res. 89 (B10), 8245–8251.
Marsh, B.D., 1996. Solidification fronts and magmatic evolution. Mineral. Mag. 60 (1), 5–40.
Marti, J., Ablay, G.J., Redshaw, L.T., Sparks, R.S.J., 1994. Experimental studies of collapse calderas. J. Geol. Soc. 151 (6), 919–929.
Marti, J., Geyer, A., Folch, A., Gottsman, J., 2008. Chapter 6 - A Review on Collapse Caldera Modelling. Deve. Volcanol. 10, 233–283.
Matthews, N.E., Huber, C., Pyle, D.M., Smith, V.C., 2012. Timescales of magma recharge and reactivation of large silicic systems from Ti diffusion in quartz. J. Petrol. 53 (7), 1385–1416.
Metcalf, R.V., 2004. Volcanic-plutonic links, plutons as magma chambers and crust mantle interaction: a lithospheric scale view of magma systems. Trans. R. Soc. Edinburgh 95, 357–437.
Michon, L., Massin, F., Famin, V., Ferrazzini, V., Roult, G., 2011. Basaltic calderas: collapse dynamics, edifice deformation, and variations of magma withdrawal. J. Geophys. Res. 116 (B3).
Moore, I., Kokelaar, P., 1998. Tectonically controlled piecemeal caldera collapse: a case study of Glencoe volcano, Scotland. Geol. Soc. Am. Bull. 110 (11), 1448–1466.
Mori, J., Eberhart-Phillips, D., Harlow, D.H., 1996. Three-dimensional velocity structure at Mount Pinatubo: resolving magma bodies and earthquake hypocenters. Fire and Mud: Eruptions and Lahars of Mount Pinatubo, Philippines, 371–382.
Myers, M.L., Wallace, P.J., Wilson, C.J., Morter, B.K., Swallow, E.J., 2016. Prolonged ascent and episodic venting of discrete magma batches at the onset of the Huckleberry Ridge supereruption, Yellowstone. Earth Planet. Sci. Lett. 451, 285–297.

Nicoll, G.R., Holness, M.B., Troll, V.R., Donaldson, C.H., Holohan, E.P., Emeleus, C.H., Chew, D., 2009. Early mafic magmatism and crustal anatexis on the Isle of Rum: evidence from the Am Mam intrusion breccia. Geol. Mag. 146 (03), 368–381.
Pamukcu, A.S., Gualda, G.A., Bégué, F., Gravley, D.M., 2015a. Melt inclusion shapes: timekeepers of short-lived giant magma bodies. Geology 43 (11), 947–950.
Pamukcu, A.S., Gualda, G.A.R., Ghiorso, M.S., Miller, C.F., McCracken, R.G., 2015b. Phase-equilibrium geobarometers for silicic rocks based on rhyolite-MELTS-. Part3: Application to the Peach Spring Tuff (Arizona-California-Nevada, USA). Contrib. Mineral. Petrol. 169, 33.
Parks, M.M., Moore, J.D., Papanikolaou, X., Biggs, J., Mather, T.A., Pyle, D.M., Raptakis, C., Paradissis, D., Hooper, A., Parsons, B., Nomikou, P., 2015. From quiescence to unrest: 20 years of satellite geodetic measurements at Santorini volcano, Greece. J. Geophys. Res. 120 (2), 1309–1328.
Pierce, K.L., Cannon, K.P., Meyer, G.A., Trebesch, M.J., Watts, R.D., 2002. Post-glacial inflation-deflation cycles, tilting, and faulting in the Yellowstone caldera based on Yellowstone Lake shorelines. U.S. Geological Survey Open-File Report 02-0142.
Poppe, S., Holohan, E.P., Pauwels, E., Cnudde, V., Kervyn, M., 2015. Sinkholes, pit craters, and small calderas: analog models of depletion-induced collapse analyzed by computed X-ray microtomography. Geol. Soc. Am. Bull. 127 (1–2), 281–296.
Quick, J.E., Sinigoi, S., Peressini, G., Demarchi, G., Wooden, J.L., Sbisà, A., 2009. Magmatic plumbing of a large Permian caldera exposed to a depth of 25 km. Geology 37, 603–606.
Reid, M.R., Coath, C.D., Harrison, T.M., McKeegan, K.D., 1997. Prolonged residence times for the youngest rhyolites associated with Long Valley Caldera: 230 Th—238 U ion microprobe dating of young zircons. Earth Planet. Sci. Lett. 150 (1), 27–39.
Richey, J.E., 1927. The structural relations of the Mourne Granites, Northern Ireland. Q. J. Geol. Soc. London 83, 653–688.
Rivalta, E., Segall, P., 2008. Magma compressibility and the missing source for some dike intrusions. Geophys. Res. Lett. 35, L04306.
Roche, O., Druitt, T.H., Merle, O., 2000. Experimental study of caldera formation. J. Geophys. Res. 105 (B1), 395–416.
Roggensack, K., Williams, S.N., Schaefer, S.J., Parnell, Jr., R.A., 1996. Volatiles from the 1994 eruptions of Rabaul: understanding large caldera systems. Science 273 (5274), 490.
Rubin, A.E., Cooper, K.M., Till, C.B., Kent, A.J.R., Costa, F., Gravley, D.M., Deering, C., Cole, J.W., 2017. Rapid cooling and cold storage in a silicic magma reservoir recorded in individual crystals. Science 356, 1154–1156.
Saunders, S.J., 2001. The shallow plumbing system of Rabaul caldera: a partially intruded ring fault? Bull. Volcanol. 63 (6), 406–420.
Saunders, S.J., 2004. The possible contribution of circumferential fault intrusion to caldera resurgence. Bull. Volcanol. 67 (1), 57–71.
Sewell, R.J., Tang, D.L.K., Campbell, S.D.G., 2012. Volcanic–plutonic connections in a tilted nested caldera complex in Hong Kong. Geochem. Geophys. Geosyst. 13, Q01006. doi: 10.1029/2011GC003865.
Sigmundsson, F., Hooper, A., Hreinsdóttir, S., Vogfjörd, K.S., Ófeigsson, B.G., Heimisson, E.R., Dumont, S., Parks, M., Spaans, K., Gudmundsson, G.B., Drouin, V., 2015. Segmented lateral dyke growth in a rifting event at BárÐarbunga volcanic system, Iceland. Nature 517 (7533), 191–195.
Simkin, T., Howard, K.A., 1970. Caldera collapse in the Galapagos Islands. Science 169, 429–437.
Smith, R.L., 1979. Ash-flow magmatism. Geol. Soc. Am. Special Pap. 180:5–28.
Smith, R.L., Bailey, R.A., 1968. Resurgent cauldrons. Geol. Soc. Am. Mem. 116, 613–662.
Snyder, D., 2000. Thermal effects of the intrusion of basaltic magma into more silicic magma chambers, and implications for eruption triggering. Earth Planet. Sci. Lett. 175, 257–273.
Sparks, R.S.J., 1988. Petrology and geochemistry of the Loch Ba ring-dyke, Mull (N.W. Scotland): an example of the extreme differentiation of tholeiitic magmas. Contr. Mineral. Petrol. 100, 446.
Spinks, K.D., Acocella, V., Cole, J.W., Bassett, K.N., 2005. Structural control of volcanism and caldera development in the transtensional Taupo Volcanic Zone, New Zealand. J. Volcanol. Geotherm. Res. 144 (1), 7–22.
Stillman, C.J., 1970. Structure and evolution of the Northern Ring Complex, Nuanetsi Igneous Province, Rhodesia. In: Newall, G., Rast, N. (Eds.), Mechanism of Igneous Intrusion. Geol. J. Special Iss. 2:33–48.

Stix, J., Kobayashi, T., 2008. Magma dynamics and collapse mechanisms during four historic caldera-forming events. J. Geophys. Res. 113 (B9).

Stix, J., Kennedy, B., Hannington, M., Gibson, H., Fiske, R., Mueller, W., Franklin, J., 2003. Caldera-forming processes and the origin of submarine volcanogenic massive sulfide deposits. Geology 31 (4), 375–378.

Toda, S., Stein, R.S., Sagiya, T., 2002. Evidence from the AD 2000 Izu islands earthquake swarm that stressing rate governs seismicity. Nature 419, 58–61.

Troll, V.R., Walter, T.R., Schmincke, H.U., 2002. Cyclic caldera collapse: piston or piecemeal subsidence? Field and experimental evidence. Geology 30 (2), 135–138.

Turnbull, R., Weaver, S., Tulloch, A., Cole, J., Handler, M., Ireland, T., 2010. Field and geochemical constraints on mafic–felsic interactions, and processes in high-level arc magma chambers: an example from the Halfmoon Pluton, Stewart Island, New Zealand. J. Petrol. 51, 1477–1505.

Walker, G.P., 1984. Downsag calderas, ring faults, caldera sizes, and incremental caldera growth. J. Geophys. Res. 89 (B10), 8407–8416.

Wallace, P.J., Anderson, A.T., Davis, A.M., 1999. Gradients in H_2O, CO_2, and exsolved gas in a large-volume silicic magma system: interpreting the record preserved in melt inclusions from the Bishop Tuff. J. Geophys. Res. 104 (B9), 20097–20122.

Weiland, C.M., Steck, L.K., Dawson, P.B., Korneev, V.A., 1995. Nonlinear teleseismic tomography at Long Valley Caldera, using three-dimensional minimum travel time ray tracing. J. Geophys. Res. 100 (B10), 20379–20390.

Wicks, C.W., Thatcher, W., Dzurisin, D., Svarc, J., 2006. Uplift, thermal unrest and magma intrusion at Yellowstone caldera. Nature 440 (7080), 72–75.

Wiebe, 1996. Mafic-silicic layered intrusions: the role of basaltic injections on magmatic processes and the evolution of silicic magma chambers. Trans. R. Soc. Edinburgh 87, 233–242.

Wiebe, R.A., 2016. Mafic replenishments into floored silicic magma chambers. Am. Mineral. 101 (2), 297–310.

Wiebe, R.A., Collins, W.J., 1998. Depositional features and stratigraphic sections in granitic plutons: implications for the emplacement and crystallization of granitic magma. J. Struct. Geol. 20, 1273–1289.

Wiebe, R.A., Frey, H., Hawkins, D.P., 2001. Basaltic pillows mounds in the Vinalhaven intrusion, Maine. J. Volcanol. Geotherm. Res. 107, 171–184.

Wiebe, R.A., Blair, K.D., Hawkins, D.P., Sabine, C.P., 2002. Mafic injections, in situ hybridization and crystal accumulation in the Pyramid Peak Granite, California. Geol. Soc. Am. Bull. 114, 909–920.

Wiebe, R.A., Manon, M.R., Hawkins, D.P., McDonough, W.F., 2004. Latestage mafic injection and thermal rejuvenation of the Vinalhaven granite, Coastal Maine. J. Petrol. 45, 2133–2153.

Wilcock, W.S., Tolstoy, M., Waldhauser, F., Garcia, C., Tan, Y.J., Bohnenstiehl, D.R., Caplan-Auerbach, J., Dziak, R.P., Arnulf, A.F., Mann, M.E., 2016. Seismic constraints on caldera dynamics from the 2015 Axial Seamount eruption. Science 354 (6318), 1395–1399.

Williams, H., 1941. Calderas and their origin. Bull. Dept. Geol. Sci. Univ. Calif. Publ. 25:239–346.

Wilson, C.J.N., 1993. Stratigraphy, chronology, styles and dynamics of late Quaternary eruptions from Taupo volcano, New Zealand. Philos. Trans. R. Soc. London A 343 (1668), 205–306.

Wilson, C.J., 2001. The 26.5 ka Oruanui eruption, New Zealand: an introduction and overview. J. Volcanol. Geotherm. Res. 112 (1), 133–174.

Wilson, C.J., Hildreth, W., 1997. The Bishop Tuff: new insights from eruptive stratigraphy. J. Geol. 105 (4), 407–440.

Wright, H.M., Folkes, C.B., Cas, R.A., Cashman, K.V., 2011. Heterogeneous pumice populations in the 2.08-Ma Cerro Galán Ignimbrite: implications for magma recharge and ascent preceding a large-volume silicic eruption. Bull. Volcanol. 73 (10), 1513–1533.

Yoshinobu, A.S., Fowler, Jr., T.K., Patterson, S.R., Llambias, E., Tickyj, H., Sato, A.M., 2003. A view from the roof: magmatic stoping in the shallow crust, Chita pluton, Arizona. J. Struct. Geol. 25, 1037–1048.

Zak, J., Paterson, S.R., 2006. Roof and walls of the Red Mountain Creek Pluton, eastern Sierra Nevada, California (USA): implications for process zones during pluton emplacement. J. Struct. Geol. 28 (4), 575–587.

Zollo, A., Maercklin, N., Vassallo, M., Dello Iacono, D., Virieux, J., Gasparini, P., 2008. Seismic reflections reveal a massive melt layer feeding Campi Flegrei caldera. Geophys. Res. Lett. 35 (12.).

FURTHER READING

Parks, M.M., Biggs, J., England, P., Mather, T.A., Nomikou, P., Palamartchouk, K., Papanikolaou, X., Paradissis, D., Parsons, B., Pyle, D.M., Raptakis, C., 2012. Evolution of Santorini Volcano dominated by episodic and rapid fluxes of melt from depth. Nat. Geosci. 5 (10), 749–754.

Wilson, C.J.N., Charlier, B.L.A., 2009. Rapid rates of magma generation at contemporaneous magma systems, Taupo Volcano, New Zealand: insights from U–Th model-age spectra in zircons. J. Petrol., egp023.

AUTHOR BIOGRAPHY

Ben Kennedy is an Associate Professor in volcanology and geoscience education at the University of Canterbury, New Zealand. He focussed the beginning of his science career on the shallow plumbing systems of calderas; he has now defocussed onto a broad range of shallow volcanic processes and has published about 60 scientific papers. He also likes hunting crabs, wearing coloured trainers and inserting images of people riding sheep into schematic figures.

Eoghan Holohan is an associate professor of Earth Sciences at University College Dublin (UCD), Ireland. He absconded from nearby Trinity College Dublin with a PhD in geology in 2008, before hiding out in postdoctoral research positions at UCD and GFZ-Potsdam in Germany until 2016. As well as investigating the interaction of volcanic, magmatic and tectonic processes, he researches the structural and geophysical development of focused subsidence phenomena such as collapse calderas and other holes in the ground.

John Stix is Professor of Volcanology and holds the Dawson Chair in Geology at McGill University. His interests include caldera systems, magma plumbing and volatiles and gases. He is past editor of the *Bulletin of Volcanology* and associate editor of the *Encyclopedia of Volcanoes*.

Darren Gravley is a physical volcanologist at the University of Canterbury specialising in silicic caldera volcanism and magmatism, volcano-tectonics and volcanic geomorphology. His research tracks the eruptible stages of mid- to shallow-crustal magmatism through to eruption and emplacement of pyroclastic deposits associated with large to supervolcanic eruptions. Study areas include modern and ancient volcanism in New Zealand, Japan, Brazil, Italy and the United States. He also enjoys wearing coloured trainers.

Jim Cole [BSc (Spec, Leics), PhD (Wellington), FRSNZ] is currently a Professor and Research Associate at the University of Canterbury and has published nearly 200 scientific papers on volcanoes, mainly about the Taupo Volcanic Zone (TVZ), New Zealand.

Steffi Burchardt is an associate professor in structural geology at Uppsala University, Sweden. She is a volcanic plumbing system nomad, having studied plutons, dykes, cone sheets, calderas and sills, and currently on the way to approach the volcano surface. When not travelling the world to study dead volcanoes, she enjoys lifting heavy objects and baking geological cakes.

CHAPTER 11

Magma Movements in Volcanic Plumbing Systems and their Associated Ground Deformation and Seismic Patterns

Freysteinn Sigmundsson*, Michelle Parks, Rikke Pedersen*, Kristín Jónsdóttir**, Benedikt G. Ófeigsson**, Ronni Grapenthin†, Stéphanie Dumont‡, Páll Einarsson*, Vincent Drouin*, Elías Rafn Heimisson§, Ásta Rut Hjartardóttir*, Magnús Guðmundsson*, Halldór Geirsson*, Sigrún Hreinsdóttir¶, Erik Sturkell††, Andy Hooper‡‡, Þórdís Högnadóttir*, Kristín Vogfjörð**, Talfan Barnie*,**, Matthew J. Roberts****

*Nordic Volcanological Center, Institute of Earth Sciences, University of Iceland, Reykjavik, Iceland;
**Icelandic Meteorological Office, Reykjavik, Iceland;
†New Mexico Tech, Department of Earth and Environmental Science, Socorro, NM, United States;
‡Instituto Dom Luiz – University of Beira Interior, Covilhã, Portugal;
§Stanford University, Stanford, CA, United States;
¶GNS Science, Lower Hutt, New Zealand;
††University of Gothenburg, Gothenburg, Sweden;
‡‡COMET, School of Earth and Environment, University of Leeds, Leeds, United Kingdom

Contents

Volcanic and Igneous Plumbing Systems. http://dx.doi.org/10.1016/B978-0-12-809749-6.00011-X
Copyright © 2018 Elsevier Inc. All rights reserved.

11.1 INTRODUCTION

Observed ground deformation associated with volcanic unrest and eruptions is often interpreted in terms of pressure changes in sub-surface magma bodies of simple geometry associated with magma inflow/outflow (Fernández et al., 2017). The magma reservoir, a key element in a typical conceptual sketch of a volcano's interior (see Chapter 1), has been central for the interpretation of some of the data sets. In the case of volcano geodetic data, the reservoir is typically modelled as a body of homogeneous fluid experiencing pressure increase during inflow of new magma and a sudden pressure drop during an eruption or magmatic intrusion when magma flows forcefully out of a reservoir. However, recent results (Sigmundsson, 2016) suggest that this may be an oversimplification and this process is, in general, more complicated.

High-resolution spatial and temporal observations of present-day magma bodies within active volcanoes from seismology and geodetic measurements of magma transfer at the same volcano are rare. Geodetic techniques are best applied to volcanoes above sea level, whereas high-resolution seismic imaging provides better results at sea. Guided by observations at a rift-zone volcano on the ocean floor where both types of observations exist, we use the concept of a 'magma domain' (Fig. 11.1) to describe the uppermost plumbing system of volcanoes. Axial Seamount, a volcano at $\sim$1500 m depth on the Juan de Fuca mid-oceanic ridge in the Pacific Ocean, has both high-resolution seismic images of its sub-surface magmatic system and detailed results from monitoring of its most recent eruption and associated seismicity and ground deformation (Chadwick et al., 2016; Nooner and Chadwick, 2016; Wilcock et al., 2016). Observations made possible by the US Ocean Observatories Initiative make the 2015 eruption at Axial Seamount the best-monitored submarine eruption so far. Seismic imaging of the sub-surface and hence imaging of magma distribution at this volcano is superior to most other environments because of advanced analysis of extensive seismic reflection profiling at sea and the relatively simple volcanic structure. This advances our understanding when compared to earlier findings from monitored rifting episodes on land (Sigmundsson, 2016).

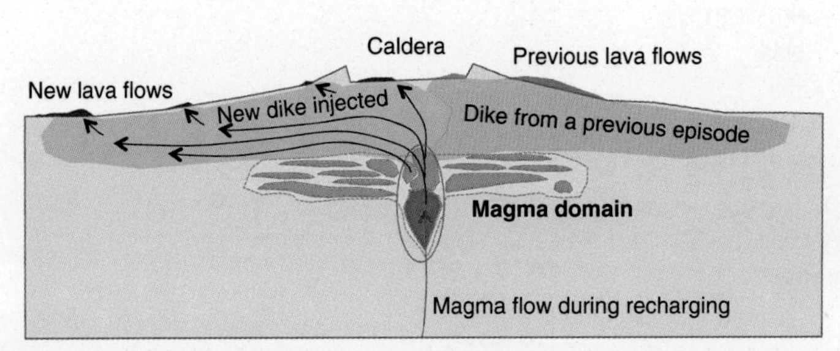

Figure 11.1 ***A general schematic model of the uppermost volcano plumbing system at a rift-zone volcano.*** The crustal volume that hosts magma at a shallow level beneath a volcano is referred to here as a magma domain (light blue hatched outline), following Sigmundsson (2016).

Geophysical magma plumbing models, in general, may benefit from incorporating more complexities, namely spatial heterogeneities in magma composition, melt content and location of major volume changes within a single magma-dominated crustal volume during eruptions. The size and shape of a magma domain can be expected to be highly variable from one volcano to another, as well as the amount and distribution of magma stored within the domain (reddish colour, Fig. 11.1). Magmas of different composition, with varying amounts of melt and crystals, can reside in pockets with variable connectivity. Ground deformation may only reflect magma inflow/outflow into relatively isolated magma bodies within the magma domain (Grapenthin et al., 2013), so to fully constrain the architecture of the magma domain, a multitude of techniques are needed (see also Chapter 1).

Here, we first describe the signatures of volcanic unrest and then focus on observations and interpretations of crustal deformation at volcanoes in terms of magma movements at depth. This requires a range of volcano geodetic measurements and different modelling approaches. We provide a number of examples from Icelandic volcanoes detailing both geodetic and seismic observations and their joint interpretations, showing how geodetic and seismic monitoring of volcanic systems can illuminate the magma domain and track magma movements. Two field examples of volcanic unrest leading to eruptions at the end of this chapter explain in detail how observed deformation relates to subsurface plumbing systems.

11.2 VOLCANIC UNREST

As magma moves to shallower levels, it usually produces characteristic surface deformation, seismicity and gas emissions. New magma arriving in volcano roots requires space causing the host rock to be displaced, which leads to surface ground deformation patterns. The resulting patterns can be interpreted in terms of models of new magmatic bodies at depth or the expansion of existing magma bodies. Arrival of new magma and associated deformation is a common precursor to eruptions, witnessed also by characteristic seismicity and volcanic gas release. These form the primary signals of volcanic unrest, together with changes in geothermal activity and surface temperature at volcanoes (Fig. 11.2). Although various signals related to volcanic unrest can be detected, it is more difficult to determine and forecast if the inflow of new magma into volcano roots will culminate in an eruption or whether magma will stall at depth. Intrusions can also form when magma flows from a pre-existing magma body residing under a volcano, commonly forming magma-filled cracks or thin sheets that may be sub-vertical (dykes; see Chapter 3), sub-horizontal (sills; see Chapter 5) or inclined (cone sheets; see Chapter 4).

Pre-eruptive processes may include magmatic intrusions or migration of magma from deep to shallow bodies. Depending on the processes involved, the rate of melt supply and depth of the magma bodies, deformation rates may vary from mm/year to

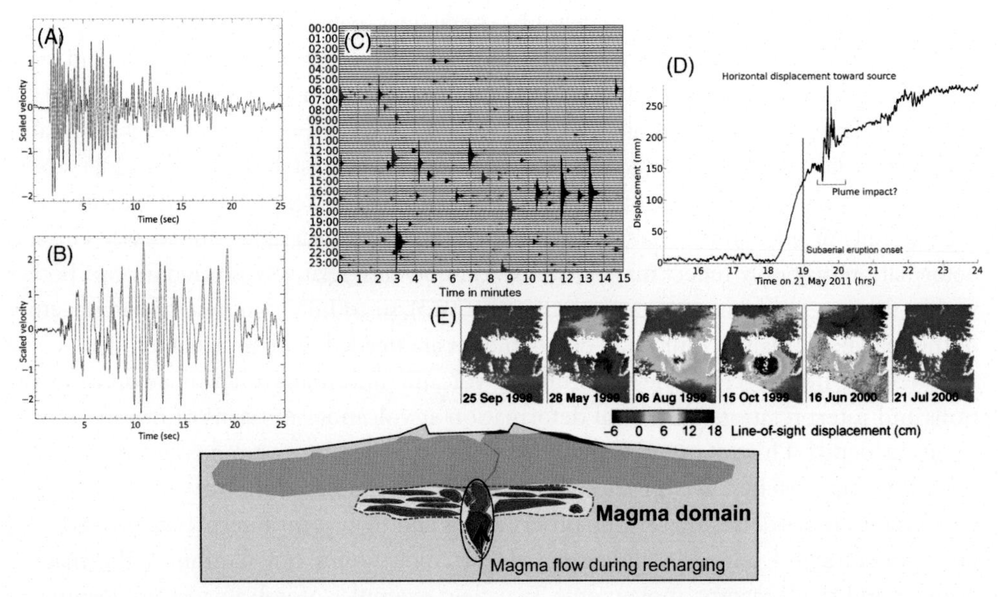

Figure 11.2 ***Seismic and geodetic signatures of volcano unrest.*** Signals can include increased numbers of both high- and low-frequency earthquakes, with examples shown in panels (A), (B) and (C). Earthquakes at Katla volcano recorded at station ALF (vertical waveform high-pass filtered above 1 Hz): (A) High-frequency event with a hypocentre inside the caldera in the top 3 km. Although shallow, it contains rather high frequencies (date: 2012-17-7 03:34:46, epicentre lat: 63.662 lon: −19.130, Ml 1.34), and (B) Low-frequency event with hypocentre west of the caldera in an area known for seasonal low-frequency seismicity, likely related to very shallow deformation (date: 2012-4-7 01:08:55, epicentre lat: 63.648, lon: −19.316, Ml 1.5). (C) Hundreds of earthquakes may occur per day in a critical unrest situation, like shown in the 'drumplot' (upper, middle) displaying 24 h of seismicity at Katla volcano on 29 September 2016 (vertical component, filtered with a highpass 2 Hz filter). (D) GPS time series of the horizontal displacement component from high rate analysis (one sample per second) at Grímsvötn volcano, revealing significant displacement prior to the onset of the 2011 eruption (Hreinsdóttir et al., 2014). (E) InSAR time series of unwrapped line-of-sight displacement (change in cm from the previous image), showing ground deformation due to an evolving intrusion in 1999 at Eyjafjallajökull (Hooper et al., 2010).

m/year, and surface displacements may be detectable on time-scales of minutes to years prior to an eruption (see, e.g. Eyjafjallajökull field example). If new magma intrudes at a shallow level prior to an eruption, then the chance of detectability is typically greater, although its detection beneath ice-covered volcanoes can be problematic, as deformation signals may not extend beyond the ice cap. If magma is injected rapidly from depth, the warning time prior to an eruption can be limited.

Magmatic intrusions often trigger small earthquakes as the melt pushes its way through the crust, related to either the migration of the magma itself or consequent changes in the local stress field. Earthquakes in volcanic environments are often classified according to their frequency content, e.g. volcano-tectonic (VT) earthquakes are associated with brittle failure of the crust, whereas long-period (LP) earthquakes are often associated with magmatic intrusions or the opening of fissures (Brandsdóttir and

Einarsson, 1992). Recent modelling work and recordings in the near field of volcanoes however suggest that the coda of shallow earthquakes gets easily trapped in shallow low velocity layers of volcanoes causing the lower frequencies to be enhanced and long ringing coda (Bean et al., 2008). Thus earthquakes, that are initially VT-brittle failure earthquakes can easily look like LP-events when recorded at some distance, even in the flanks of the volcano. Earthquakes are sometimes seen at the root of volcanoes, even at depths below the brittle-ductile boundary (in Iceland 20-30 km). These events are marked by long-periods, and often come in small swarms lasting for minutes to hours (Greenfield and White, 2015). These deep LP-events are likely to be caused by slow slip resulting in lower frequency content than VT-earthquakes. Volcanic tremor is typically observed at the onset of and during eruptions related to magma column oscillations or magma migration or dyke formation (Dawson et al., 2010; Ripepe et al., 2013; Eibl et al., 2017). The strength of the tremor varies greatly with the type of eruption. Basaltic fissure eruptions produce weak tremor whereas explosive eruptions produce strong tremor. Hypocentre locations of volcanic earthquakes, graphs of cumulative number ofVT events or cumulative seismic moment against time as well as identification of seismic tremor are employed for short-term eruption forecasting.

As melt rises within the crust, volatiles released from the magma migrate through open fracture pathways towards the surface. Pre-eruptive changes in gas emissions have been measured at many volcanoes including Usu (Japan), Redoubt (Alaska), Stromboli (Italy), Poás (Costa Rica) and El Hierro (Spain). An increase in CO_2 or He emissions or CO_2/SO_2 ratio was observed prior to some eruptions (Aiuppa et al., 2009; Allard, 2010; Padrón et al., 2013).

11.3 MEASUREMENTS OF GROUND DEFORMATION

The most important geodetic techniques used at present to map volcano deformation are two complementary space geodetic techniques. Firstly, global navigation satellite system (GNSS) geodesy, of which the global positioning system (GPS) is one example, can provide continuous displacement time series of individual points in three dimensions at a temporal resolution from fractions of a second to decades. Secondly, interferometric synthetic aperture radar (InSAR) analysis of radar images acquired by repeated satellite passes can provide a map view of change in range between ground and satellite (Sigmundsson et al., 2010a, 2015; Hooper et al., 2012). Both techniques have been used extensively at volcanoes in Iceland (Figs. 11.3 and 11.4; see Field Example 2).

GPS satellites transmit signals at two frequencies, = 1575.42 and 1227.60 MHz, referred to as L1 and L2 carriers, respectively. At least four satellites must be visible in the sky to determine the receiver's latitude, longitude, elevation and correct for the receiver´s imprecise clock. Additional satellites improve the observations. While commercial receivers determine positions from a civil ranging code on L1, geodetic receivers used in scientific applications track the carrier phase on L1 and L2 for higher-precision position estimation. The observed phase, $\mathbf{\Phi}$, scaled by the carrier wavelength, λ,

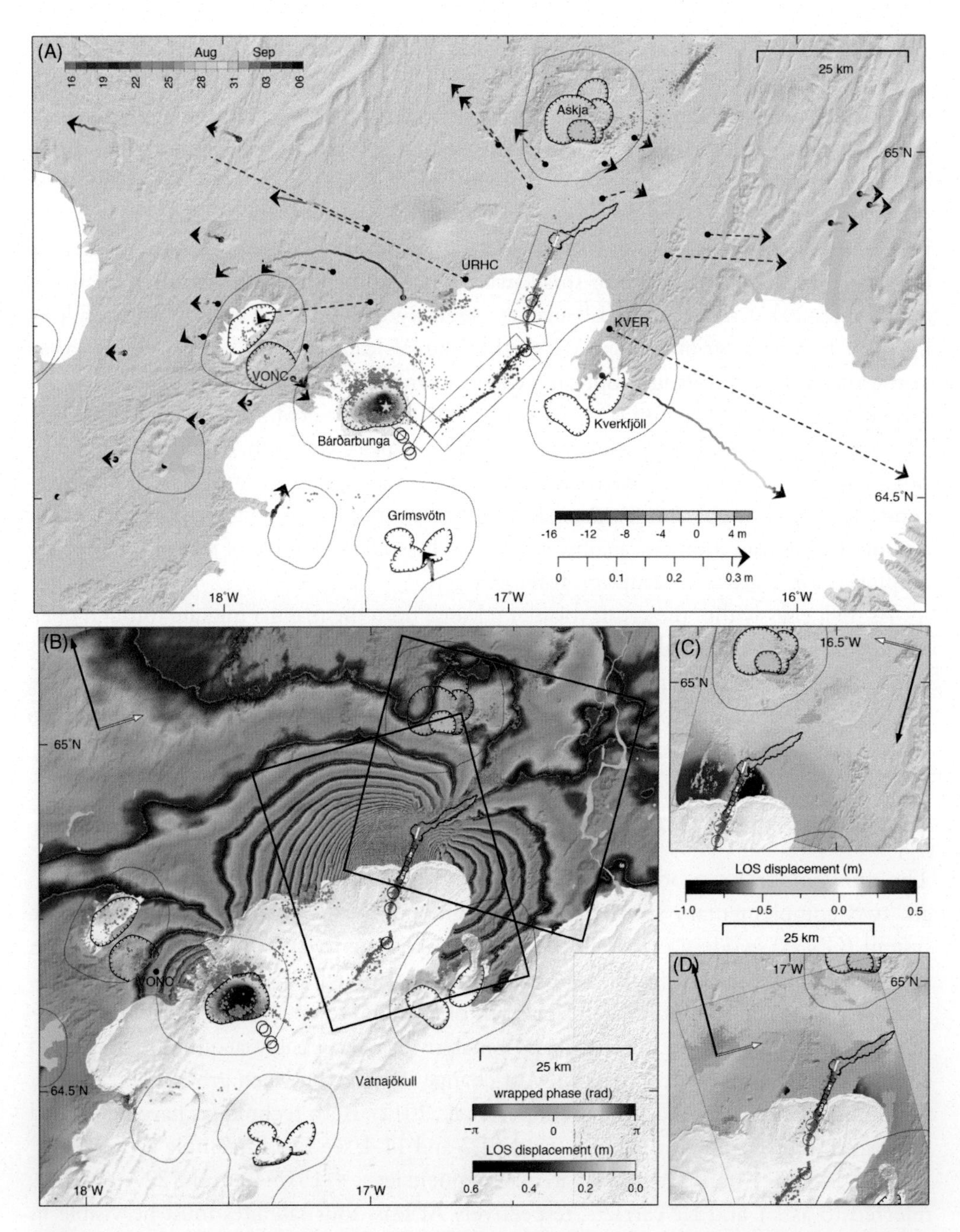

Figure 11.3 ***GPS and InSAR data (after Sigmundsson, 2015) from the beginning of the Bárðarbunga 2014–2015 unrest episode and eruption.*** (A) Earthquakes 16/08–06/09 2014 (*dots*) and horizontal ground displacements measured by GPS (*arrows*) on a map with central volcanoes (*oval outlines*), calderas (*hatched*) and northern Vatnajökull ice cap. Relatively relocated epicentres and displacements are colour-coded according to the time of occurrence, and other single earthquake locations are in

grey. Thin lines within rectangles show inferred dyke segments. The *red* shading at Bárðarbunga caldera shows subsidence up to 16 m inferred from radar profiling on 5 September at the onset of the 6-month long caldera collapse. The *star* marks the location of the magma source inferred from modelling. Also shown are ice cauldrons formed (*circles*), the outline of the lava flow mapped from a radar image on 6 September, and eruptive fissures (*white*). (B) Wrapped RADARSAT-2 interferogram spanning 08/08–01/09 2014. Shading at Bárðarbunga caldera shows unwrapped 1-day (27–28/08) COSMO-SkyMed interferogram with maximum line-of-sight (LOS) range increase of 57 cm. Also shown are earthquakes (*grey dots*), boundaries of the graben activated in the dyke distal area (*hatched lines*) and the location of interferograms in panels C and D (*boxes*). (C) Unwrapped COSMO-SkyMed interferogram spanning 13–29 08/2014. (D) Unwrapped TerraSAR-X interferogram spanning 26/07–4/09 2014. Satellite flight and viewing direction are shown with *black* and *white arrows*, respectively. LOS displacement is positive away from the satellite for all interferograms shown.

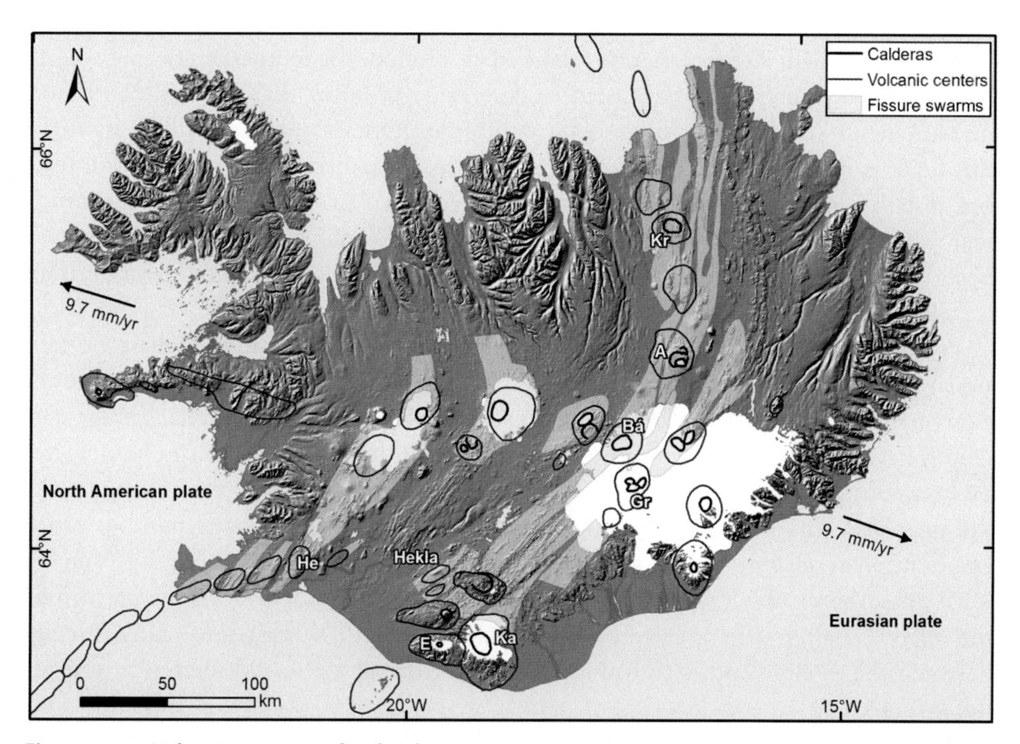

Figure 11.4 ***Volcanic systems of Iceland.*** Systems mentioned in the text are indicated: *Kr*, Krafla; *A*, Askja; *Bá*, Bárðarbunga; *Gr*, Grímsvötn; *E*, Eyjafjallajökull; *Ka*, Katla; *He*, Hengill. Volcanic centres are from Einarsson and Sæmundsson (1987), whereas fissure swarms are from Hjartardóttir et al. (2016a, b) and Einarsson and Sæmundsson (1987). Calderas are from Einarsson and Sæmundsson (1987), Gudmundsson and Högnadóttir (2007) and Magnússon et al. (2012).

determines the range (distance) between satellite and receiver r. The different frequencies of L1 and L2 allow correction of signal delay due to dispersion in the ionosphere (I_Φ). The complete phase observation model is (Misra and Enge, 2011):

$$\Phi = \lambda^{-1}\left[r - I_\Phi + T_\Phi\right] + f\left(\delta t_r - \delta t^s\right) + N + \mathrm{MP} + \varepsilon_\Phi$$

The primary sources of error, in addition to the ionosphere, are the tropospheric propagation delay, T_{Φ}, receiver and satellite clock bias, δt_r, δt^s, interference of direct and reflected signals causing the so-called multipath error, MP, as well as other, unmodeled error sources captured in ε_{Φ}. As receivers track carrier phase changes after locking onto the signal, the integer number, N, of cycles between the satellite and the receiver at the beginning of tracking is ambiguous and needs to be resolved. Careful treatment of all error sources results in precisions up to 3–4 mm in horizontal and about twice that in the vertical component (Bock and Melgar, 2016). Simultaneous tracking and analysis of multi-GNSS combining GPS, the Russian GLONASS, the European Galileo and the Chinese BeiDou systems promises positioning improvements.

The two main types of GNSS data acquisitions at volcanoes are campaign measurements, for which permanently embedded geodetic benchmarks are episodically reoccupied with geodetic equipment to determine position changes, and continuous measurements (cGPS), where geodetic equipment remains permanently installed in the field, generally telemetering observations to an analysis centre. While continuous installations that record one or more samples per second, potentially analysed in real time, are preferred to resolve the timing of rapid changes of a volcano, campaign observations can improve spatial resolution of cumulative signals even if precise timing is lacking.

Synthetic aperture radar (SAR) images record both amplitude and phase of radar signals sent by a SAR satellite and reflected back from the Earth's surface. Any ground deformation that takes place between two separate SAR acquisitions changes the path length between the ground and the satellite. The resulting phase shift can be detected by computing an interferogram, equal to the difference in the phase component of the two SAR images. While it would be impossible to infer the total number of full cycles of phase change from a single pixel (similar to the integer ambiguity N in the GPS phase observable), spatial coherence across an interferogram allows us to infer the relative change. When scaled by the signal wavelength, this gives the range change between the satellite and the ground, relative to that of a selected reference point in the image.

An originally wrapped interferogram (with fringes) shows the phase gradient. Its integration from a reference point is referred to as phase unwrapping, resulting in a value for relative phase change that includes whole phase cycles. The total phase change arises from a combination of differences in the location of the satellite overhead, topography, surface deformation, atmospheric delay and the scattering properties of the ground:

$$\Delta\varphi_{\text{int}} = \Delta\varphi_{\text{geom}} + \Delta\varphi_{\text{def}} + \Delta\varphi_{\text{atm}} + \Delta\varphi_{\text{pixel}}$$

Here $\Delta\varphi_{\text{int}}$ is the phase of the interferogram, $\Delta\varphi_{\text{geom}}$ is the combined phase contribution derived from differences in the satellite's orbital position and a perspective effect from topography, $\Delta\varphi_{def}$ results from ground deformation, $\Delta\varphi_{atm}$ is caused by atmospheric differences on the different acquisition dates. $\Delta\varphi_{pixel}$ is the contribution due to individual scatterers within a pixel.

If the scattering properties of the ground remain stable between different satellite acquisitions, contribution from $\Delta\varphi_{pixel}$ is mostly negligible. An accurate digital elevation model (DEM) and precise satellite orbits are used to correct for $\Delta\varphi_{geom}$. The $\Delta\varphi_{atm}$ term is often ignored if atmospheric delay over over the target area is small or atmospheric conditions during acquisitions are similar. However, $\Delta\varphi_{atm}$ remains a significant impediment to the measurement of small-scale deformation (<10 cm) at volcanoes, especially in tropical regions where water vapour has a considerable effect (Bekaert et al., 2015). Attempts to reduce the atmospheric phase contribution include interferogram stacking to enhance the signal-to-noise ratio, assessments of likely occurrence of stratified water vapour or use of complementary atmosphere data from optical satellites, GPS instruments or numerical weather model outputs to correct interferograms for $\Delta\varphi_{atm}$.

Once the interferogram has been corrected for non-deformation contributions to phase change, deformation $\Delta\varphi_{def}$ becomes the dominant contributor to changes in the phase, which can be converted into displacement (Δd) along the satellite's line-of-sight (LOS):

$$\Delta\varphi_{def} = -\frac{4\pi}{\lambda}\Delta d$$

where λ is the wavelength of the transmitted radar pulse. Persistent scatterer InSAR (PSI) and other advanced time-series analysis techniques can be employed to further reduce noise and generate a time series of high-resolution deformation signals (Hooper et al., 2007, 2012; Hooper, 2008). Various SAR satellites provide data to study volcanoes, and ESA's recent Sentinel mission now offers 6-day repeat acquisitions over many volcanoes.

In addition to GNSS and InSAR, ground deformation at volcanoes is often measured with borehole tiltmeters, strainmeters, repeat terrestrial radar (LiDAR) or repeated levelling surveys. Additional tools include time-lapse photogrammetry, perhaps most prominent through Gary Rosenquist's photos of the flank failure of Mt. St Helens in 1980 (Voight et al., 1981), Mimatsu diagrams that trace surface profile changes over time on paper (Minakami et al., 1951) and use of drone technology to map ground changes. Repeated levelling, when carried out on short profiles or arrays, is referred to as optical levelling tilt. This was pioneered as a volcano monitoring technique by Eysteinn Tryggvason in Iceland, beginning in 1966, mapping changes as small as 1 microradian per year (1 mm height change over a distance of 1 km). Profiles installed for this purpose at Hekla and Askja volcano (Tryggvason 1989, 1994) continue to be measured yearly.

Gravity measurements can reveal both spatial and temporal mass changes at volcanoes. Such observations can resolve if a deformation source is due to hydrothermal or magmatic activity, as the density of fluids involved is different. Relative gravity measurements are used in two different ways at volcanoes: i) to characterise the structure of volcanic plumbing systems based on a Bouguer anomaly map, for example, to decipher existence of intrusive complexes and crustal magma bodies (Gudmundsson and Högnadóttir, 2007) or ii) to monitor and quantify mass transfer over time, sometimes referred to as micro-gravity (or time lapse) observations (Carbone et al., 2017; de Zeeuw-van Dalfsen et al., 2005). Continuous gravity observations have also been carried out at a few volcanoes, including Etna and Kilauea, where rapid mass variations in relation to conduit processes have been detected.

Before the observed ground deformation can be interpreted in terms of magma plumbing and transfer, other deformation sources must also be considered (Grapenthin et al., 2010; Geirsson et al., 2012; Drouin et al., 2017). These include, in many cases, the influence of plate movements and strain accumulation at plate boundaries. In Iceland, another important contribution is glacial isostatic adjustment (GIA) due to glacier retreat since 1890. Maximum uplift velocities related to GIA exceed 30 mm/year in Iceland at present (Auriac et al., 2013), so GIA heavily influences surface deformation. Deformation induced by annual variations in snow and water loads on the surface of the Earth can be pronounced. For example, an annual peak-to-peak cycle ranging from 4 to over 20 mm is observed in the vertical component at continuous GPS time series in Iceland (Drouin et al., 2016). These effects need to be considered and corrected for before interpreting volcano deformation sources.

11.4 MODELS OF VOLCANO DEFORMATION

The use of modelling to interpret volcano deformation has recently advanced, partly because of improved computing facilities and algorithms, but also because of major advances in observing volcano deformation. The models applied to understand magma movements and associated hazards range in type. The most common models used to fit surface deformation resulting from magma inflow/outflow to/from volcano roots are simple mechanical models. These models assume that the process can be described by the emplacement of a volume of magma (of uniform properties) with idealised geometries, within a surrounding host rock most commonly approximated as a homogeneous, isotropic elastic halfspace (Fig. 11.5). Often a knife sharp discontinuity is assumed between a magmatic 'pressure source' and solid rock of uniform elastic properties. Analytical solutions for magma sources within a uniform elastic halfspace can take the form of a pressure variation in a point source (Mogi, 1958), a finite size spherical source (McTigue, 1987), an ellipsoid (Yang et al., 1988), a circular crack (sill) (Fialko et al., 2001) or thin sheets (dykes, sills) composed of one or more rectangular patches (dislocations)

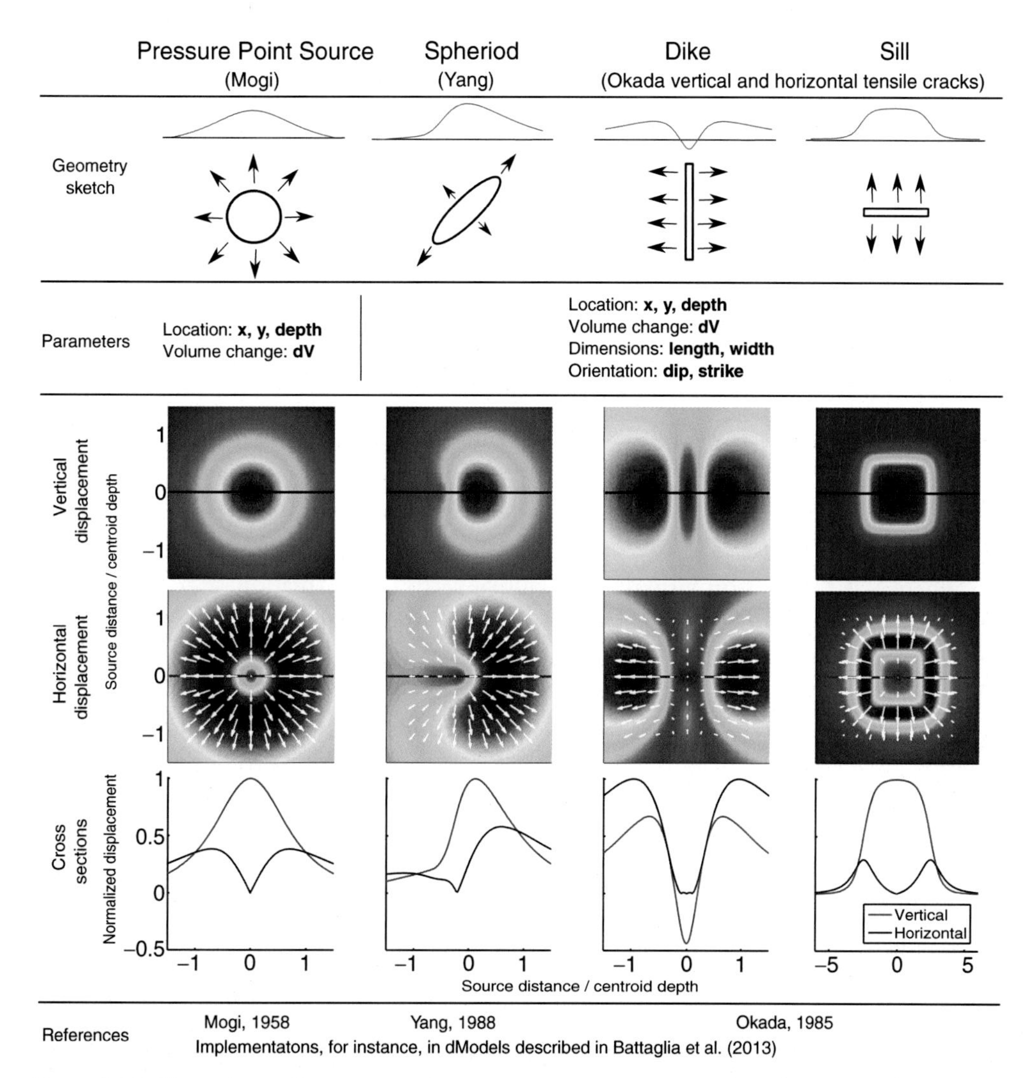

Figure 11.5 ***Illustrations of four magmatic source geometry approximations for commonly used analytical models and their predicted surface ground deformation patterns.*** Source parameter variables are listed for each model; full analytical expressions can be found in the references given at the bottom. The horizontal (*black*) and vertical (*red*) displacement fields, normalised to the maximum displacements, are given in map view, as well as for profiles transecting the displacement maxima with horizontal distances normalised to the respective source centroid depth. The plunge of the east-west striking spheroid source is 60 degrees. The dyke's ratio of centroid over burial depth of its top is 8/3. The sill's width is 5 times the centroid depth.

(Okada, 1985; Sigmundsson et al., 2010a, 2015). All these have analytical solutions that are computationally efficient to use in approaches that estimate model parameters from the data (mathematical inversion) in order to provide deformation predictions that fit the data well.

Source geometries and characteristic surface deformation patterns for the most commonly applied of these analytical models are shown in Fig. 11.5. The most simplistic model of a pressure point source (Mogi, 1958) assumes a pressure change, ΔP, or volume change, ΔV, at a depth, d, below the surface, causing radially symmetric displacements with maximum vertical and minimum horizontal motion located directly above this source. Vertical, u_z, and radial, u_r, displacements at radial distance r from the source at the surface are given by

$$u_z = \frac{(1-\nu)\Delta V}{\pi} \frac{d}{(r^2 + d^2)^{3/2}}$$

$$u_r = \frac{(1-\nu)\Delta V}{\pi} \frac{r}{(r^2 + d^2)^{3/2}}$$

where ν is the Poisson ratio. In this model, vertical deformation, u_z, dominates the amplitude over radial deformation, u_r, to about 1 source-depth radially away from the source (Fig. 11.5). The remaining models are more complex, requiring finite dimensions of length and width and can be arbitrarily striking and dipping.

Data from the 1975–1985 rifting episode at Krafla volcano in Iceland provides an example for data interpretation. Levelling observations are well reproduced with a Mogi source representing a magma body under the Krafla caldera. The volcano had about 20 inflation–deflation events. Magma accumulated in a shallow magma body during inflation causing metre-scale uplift (Fig. 11.6), which rapidly reversed when a dyke was injected into the adjacent rift zone. Levelling data show these events in a clear, sawtooth-shaped time series which well constrain the time history of volume change of a Mogi source. Such rapid deformation is generally expected to be accompanied by seismicity, and the Krafla rifting episode is no exception. However, a comparison of the volume change and the rate of earthquakes in the caldera reveals a surprising relationship (Heimisson et al., 2015). During the initial inflation period, the seismicity rate increased and displayed relatively good correlation with the volume change. After this first inflation period, the pattern changed, and the seismicity rate tended to dramatically increase only after the previous level of inflation was exceeded. This indicates that the volcano manifested what rock mechanics describes as the Kaiser stress memory effect (Kaiser, 1953). This has been reproduced in small-scale experiments with cyclic loading of rocks and other material where increases in acoustic emissions, indicating small fracture formation, are only generated when the prior load is exceeded (Lockner, 1993). Due to large stress perturbations from multiple dyke injections and eruption later on in the rifting episode, no clear relationship remains between the volume changes and seismicity rate. These data show that the Kaiser effect can occur on a kilometre-scale in

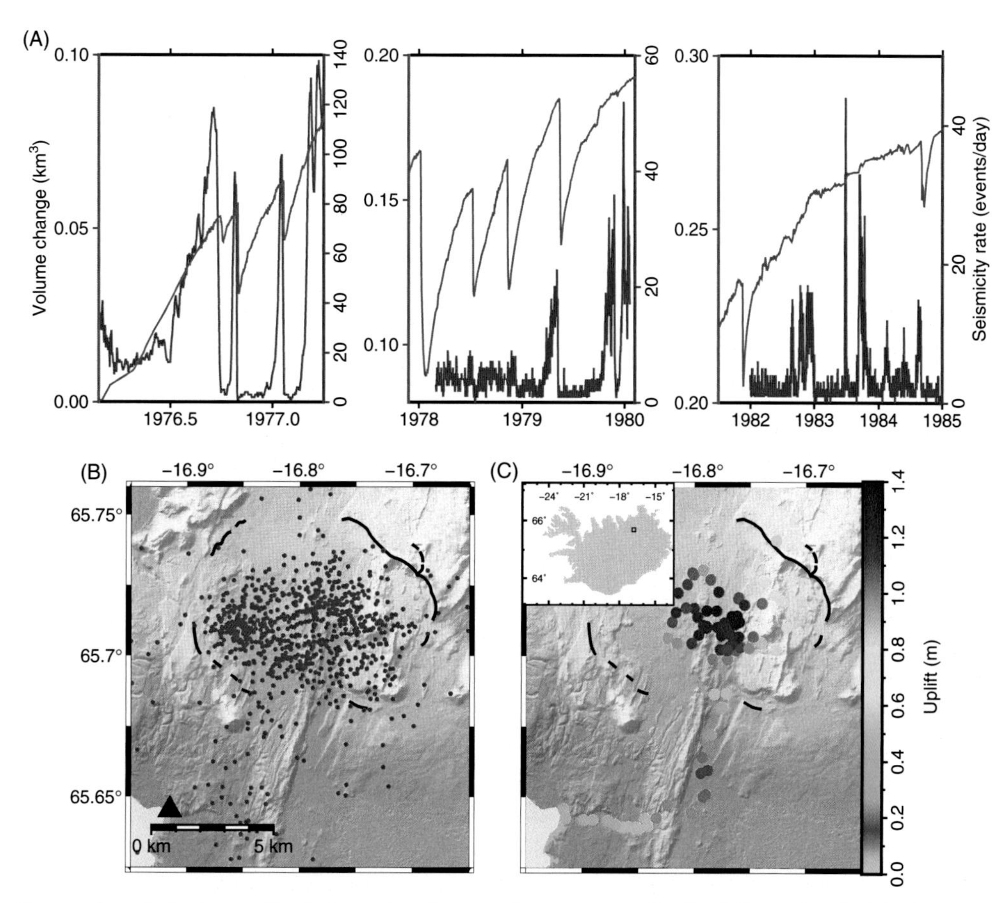

Figure 11.6 ***Inflation and inflation-related seismicity during the Krafla rifting episode.*** (A) The three time windows (left, middle and right) are the three periods where the seismicity rate from the Krafla caldera could be reliably estimated. The *blue axis* (left) and corresponding *blue curve* show cumulative volume change from Mogi source modelling of levelling data. Volume change is shown with respect to the initial deflation event. Right axis (*red*) and the *red curve* are earthquakes detected each day in the caldera. Note that in the left and middle windows, the seismicity rate dramatically increases only when volume change reaches the previous inflation level, thus demonstrating the Kaiser stress–memory effect (Heimisson et al., 2015). (B) Located inflation-related earthquakes, forming a cloud of seismicity within the caldera (shown in *black lines*). The *black triangle* marks the location of the seismic station used to measure the inflation seismicity rate at the edge of Mývatn lake (*light blue*). (C) Dots mark the location of levelling benchmarks and the colour indicates the uplift at each point. Cumulative vertical displacement is shown at 26 April 1977 with reference to the beginning of 1975. Inset shows the location of the area in Iceland. Uplift shows a mostly radially symmetric bulging of the ground. The centre of the bulge is near the centre of the cloud of earthquakes (modified from Heimisson et al., 2015).

a volcano, which has important implications for eruption monitoring and forecasting. If volcanoes inflate and deflate cyclically, the seismicity may be close to background levels, whereas a magma body may be close to failure (Heimisson et al., 2015), explaining some previously puzzling aseismic behaviour at volcanoes and emphasising the importance of multi-instrument monitoring. As, for instance, the lack of seismicity may suggest volcanic unrest to be over, when, in fact, a volcano may be rapidly inflating in preparation for another eruption.

Another example is provided by ground deformation observed over a period of almost two decades at Eyjafjallajökull volcano in south Iceland (Fig. 11.7). The deformation data can be reproduced by episodic intrusions formed over 18 years of volcanic unrest, and eventually leading to two eruptions in 2010. This activity is described in a field example at the end of this chapter.

Uncertainties on geodetic model parameters can be addressed by deriving their probability distributions utilising Bayes' theorem. This can be realized through Markov Chain Monte Carlo (MCMC) techniques that implement the Metropolis-Hastings sampling algorithm (Mosegaard and Tarantola, 1995; Menke, 2012). The process involves sampling a set of model parameters from their given a priori distributions and considering their likelihood, given the observations; the relative likelihoods of the different model parameter values are evaluated by comparing observations to model-based predictions. The final a posteriori multivariate probability density functions can be built during numerous (millions) simulations (Hooper et al., 2013; and methods section in Sigmundsson et al., 2015).

Inferring volumes of transferred magma from volcano geodesy is complicated by the simplified assumptions of the underlying models. In the case of pre-existing sources experiencing inflow/outflow, the volume change of simple analytical solutions introduced above are only equal to the volume of magma inflow/outflow if the magma is incompressible. This assumption is not realistic in most cases. In particular, as magma moves towards the surface, exsolution of volatiles due to depressurisation creates a multi-phase material containing pressurised gas bubbles that turn magma into a compressible fluid. Differences between volume decrease of shallow magma bodies as magma flows out of them and corresponding volume increases in a new dyke injected into the surrounding host rock or in eruptive material are common. For example, often the contraction volume of a shallow magma body may be only one-third or less of the volume of a dyke injected from this magma body. The effects of compressibility of magma remaining in the source magma bodies can explain this difference (Sigmundsson et al., 1992; Johnson et al., 2000; Rivalta and Segall, 2008, and references therein). The pressure decrease due to magma flowing out of a magma body results in the expansion of the remaining magma. Viscoelastic behaviour of material in a shell surrounding a spherical magma chamber can also be important (Segall, 2016) and analytical expressions for further complexities such as horizontal layering of the elastic properties (Newman et al., 2006) or viscoelastic properties (e.g., Bonafede and Ferrari, 2009) in a halfspace exist.

Holistic models of magma plumbing that often include a magma body and a conduit to the surface can help understand eruption behaviour and volcano

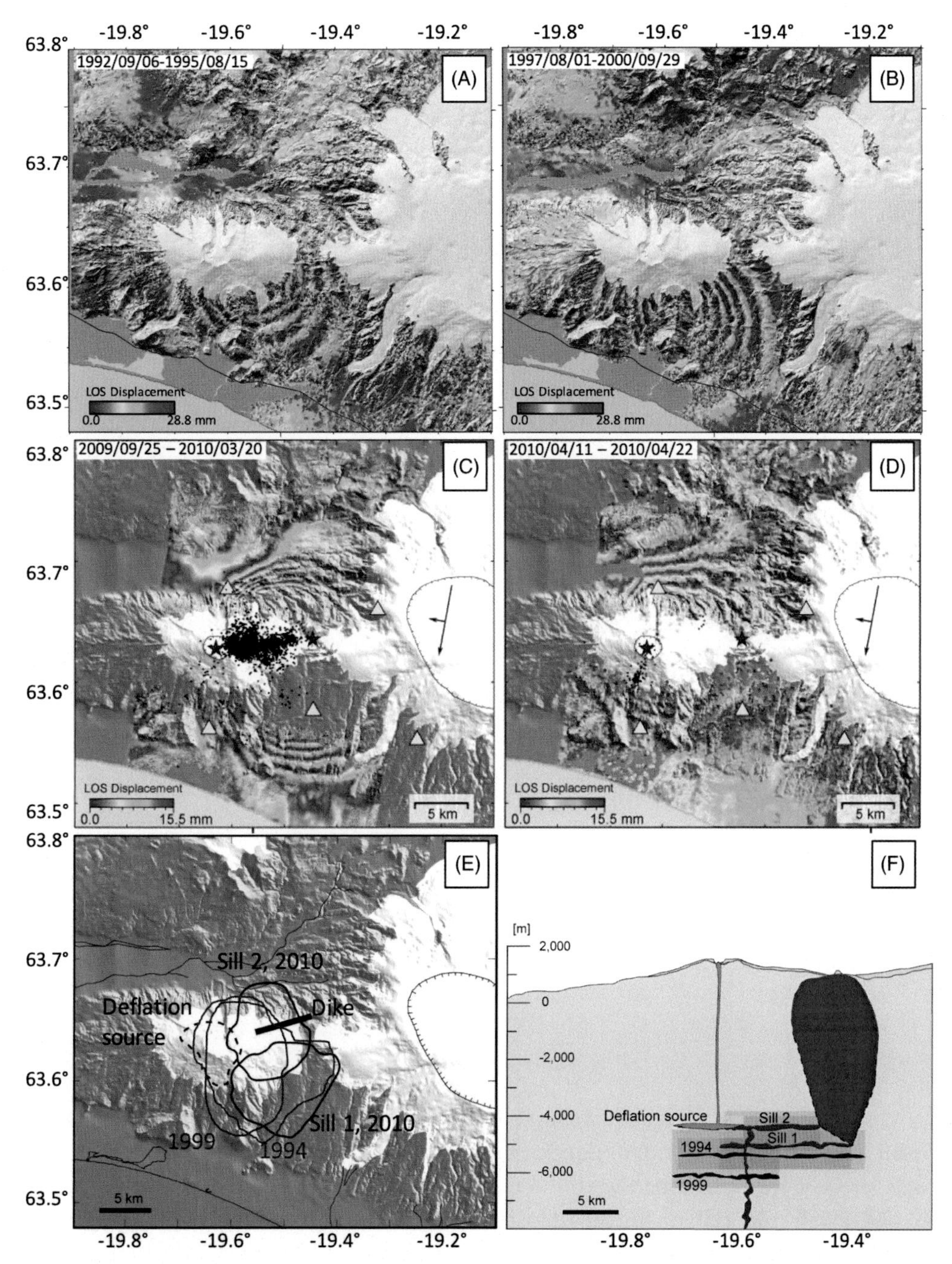

Figure 11.7 ***InSAR images illustrating surface deformation modelled as episodic sill intrusions into the Eyjafjallajökull volcano.*** Positive LOS displacement is movement away from the satellite. Note the difference in fringe size for panels A, B (28.8 mm) and C, D (15.5 mm) due to the use of different satellites (ERS and TerraSAR-X, respectively). (A) 06/09/92–15/08/95 (1994 sill intrusion); (B) 01/08/97–29/09/00 (1999 sill intrusion); (C) 25/09/09–20/03/10 (two sill intrusions and dyke formation in 2010); (D) 11/04/10–22/04/10 (initial phase of the 2010 summit eruption); (E) approximate outlines of all modelled deformation sources; (F) schematic east–west cross-section of the Eyjafjallajökull plumbing system, with all discrete modelled magmatic sources plotted at their best-fit depth (vertical exaggeration by a factor of 2). *Grey* shaded background indicates source depth uncertainties (95% confidence). Modified from Pedersen and Sigmundsson (2004, 2006) and Sigmundsson et al. (2010a,b).

dynamics. One line of progress towards improved understanding of volcanic hazards has been geophysical modelling of eruption monitoring data in the context of such models, sometimes referred to as 'physics-based' (PB) models or multi-physical models. These go beyond mechanical models of deformation sources shown in Fig. 11.5 and model jointly the dynamics of a magma body, conduit processes and eruption (Anderson and Segall, 2011, 2013; Anderson and Poland, 2016). PB models can thus be used to relate observations of deformation, gas, seismicity and eruption rates to infer important information related to magma supply and storage. This process can constrain parameters that cannot be constrained by mechanical models, such as the overall volume of magma bodies. When used together with an MCMC approach, probabilistic estimates can be derived for properties of the sub-surface plumbing system, such as depth of magma bodies, volatile content and conduit properties.

Advanced deformation modelling can be undertaken using, for example, using the numerical finite element method (FEM) (e.g., Ali, 2014) to better constrain the parameters of magma bodies and influx rates based on realistic heterogeneous layered earth models, as FEM models can be used to solve physical governing equations in arbitrarily structured domains, which are readily partitioned to account for known complexities. FEM models have many advantages in terms of including complexities in the model space (Pedersen et al., 2009; Masterlark et al., 2010, 2012, 2016). Hence, we can account for comprehensive and detailed pre-existing structural, geological and geophysical information regarding topography, sub-surface structures, background tectonic stresses and spatial variations of rheological properties. For instance, to simulate viscoelastic behaviour, a stress-dependent creep strain-rate relationship can be added to the elastic strain rates in the desired regions of the problem domain. The standard expression for creep strain rate is

$$\dot{\varepsilon} = A\sigma^n$$

where $\dot{\varepsilon}$ is the uniaxial equivalent strain rate, A is a constant that can be expanded to include a temperature dependence and thermally activated creep and σ is deviatoric stress. When combined with the elastic strain rate, the relationship is equivalent to a Maxwell material, or linear viscosity law, if $n = 1$ and A is half of the inverse of the linear viscosity. Many FEM models do not consider explicitly the temperature structure, but few valuable models of that type exist (e.g., Hickey et al., 2015). *A priori* constraints on such models are often limited. A drawback of numerical models is, however, that testing a large range of different model parameters (required for model parameter estimation) is computationally demanding and in some cases not at all feasible.

11.5 LESSONS LEARNED IN ICELAND

The volcanic regions of Iceland (Fig. 11.4) can be divided into 32 volcanic systems (see, e.g. Sigmundsson (2006), and online Catalogue of Icelandic volcanoes: http://icelandicvolcanoes.is/). Within these, there have been 22 confirmed eruptions in the last 45 years (exact number depending on the definition of what a single eruption is). Several additional small eruptions may have occurred beneath the glacier-covered volcanoes, but did not manage to melt their way through the ice and hence went unconfirmed, although inferred through seismic observations. In the past, the main methods used to monitor the state of the Icelandic volcanoes have been based on seismic observations, geodetic techniques such as GPS, tilt and InSAR observations (Sigmundsson, 2006; Sturkell et al., 2006a, b), and monitoring variations in geothermal activity, including ice cauldron development and evolution above subglacial geothermal areas. Mapping of crustal deformation and seismicity during recent Icelandic eruptions and unrest periods have provided a wealth of information concerning magma movements.

A systematic study of the behaviour of Icelandic volcanoes reveals that most of them appear to be in a state of rest. Four systems do, however, show signs of magma accumulation or increased magma pressure at the time of writing that may potentially lead to eruptions within the next few years if unrest continues; Hekla, Grímsvötn, Bárðarbunga, and Öræfajökull (unrest began in 2016/2017; not discussed here). The magma accumulation is expressed in various ways depending on the characteristics of the respective volcano. Continued accumulation will inevitably lead to an eruption, but the timing is uncertain. This depends on the rate of magma accumulation and the state of stress in the rock mass surrounding and within the magma domain, that will define how long the surrounding crust can withstand the pressure increase. Stress also governs the path of the magma when it breaks out of a magma body, whether it finds its way to the surface producing an eruption or becomes an arrested intrusion in the crust. Most of these conditions are poorly known, so an accurate prediction of these events is impossible. It is therefore important to monitor the progress of activity with geophysical instruments in order to narrow down the likely or possible scenarios.

11.5.1 Bárðarbunga

The Bárðarbunga volcano in central Iceland experienced a major unrest, lateral dyking and eruption in August 2014–February 2015 (Sigmundsson et al., 2015). The eruption was accompanied by a slow caldera collapse, a truly rare event that has not been monitored in such detail before, providing a unique opportunity for better understanding the volcanic structure and processes (Gudmundsson et al., 2016; Parks et al., 2017). The collapse was extensive as the 8×11 km^2 caldera gradually subsided and a bowl up to 65 m deep was formed, while about 1.8 km^3 of magma drained laterally along a subterranean

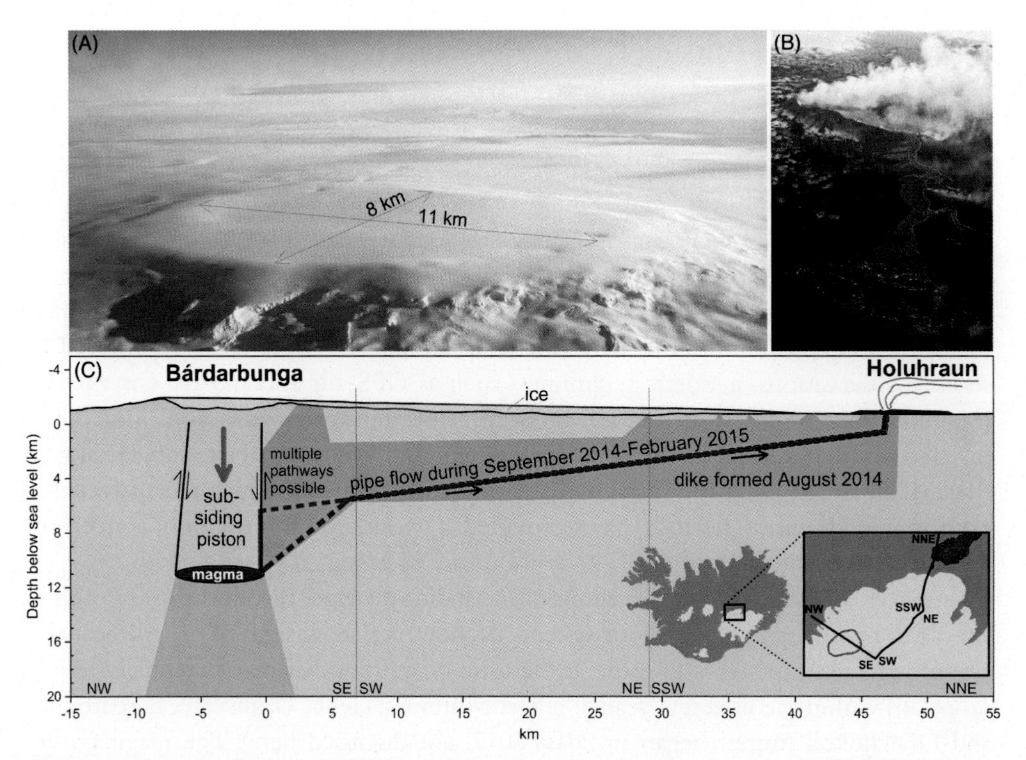

Figure 11.8 ***The Bárðarbunga caldera and the lateral magma flow path to the Holuhraun eruption site.*** (A) Aerial view (from the north) of the ice-filled Bárðarbunga caldera on 24 October 2014. (B) The effusive eruption in Holuhraun, about 40 km to the northeast of the caldera. (C) A schematic cross-section through the caldera and along the lateral subterranean flow path between a magma body and the surface. After *Gudmundsson et al. (2016).*

path, feeding an eruption 47 km northeast of the volcano (Figs. 11.8 and 11.9). This activity until the end of the eruption is described in a field example at the end this chapter.

The Bárðarbunga caldera continues to exhibit signs of unrest. After the end of the eruption, GPS deformation data display horizontal movements that are consistent with an inflation signal centred at the caldera (see GPS time series presented in the field example), but the pattern is more complicated than during the co-eruptive period. The seismicity continued to decline after the eruption, both in the far end of the dyke as well as within the caldera. In the dyke area, the activity has steadily decreased to present-day low levels. However, in September 2015, seismicity within the caldera started to increase again (Fig. 11.10). This increase was identified in terms of increased earthquake magnitudes while earthquake rate remained relatively constant. A seismic waveform correlation analysis reveals a dramatic change occurring between February and May 2015, where the first motion polarity of earthquakes reverses sign. This coincides with the ending of the caldera collapse and the eruption. Preliminary results suggest that caldera

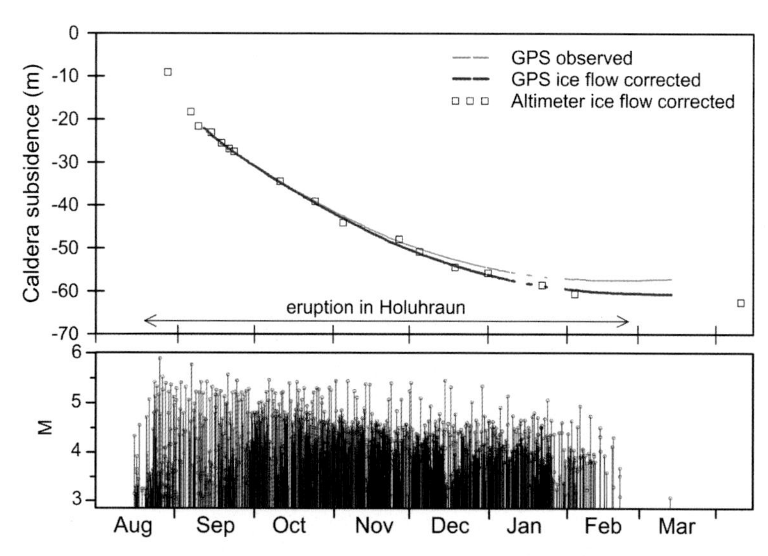

Figure 11.9 ***Caldera collapse at Bárðarbunga.*** Upper panel: Subsidence near the centre of the caldera. The two curves show the observed vertical change at a GPS station installed on the ice surface and the GPS curve corrected for ice flow, providing an estimate of the subsidence of the bedrock. Modified from Gudmundsson et al. (2016). Lower panel: Time line of seismicity during the caldera collapse as recorded by the Icelandic Meteorological Office (including 80 earthquakes between M5–M5.8 shown in *red*).

fault movements were reversed soon after the eruption ended in spring 2015 when outward movement of GPS stations around the caldera was also observed, indicating re-inflation long before any seismicity increase was detected. Over 9000 earthquakes have been recorded after the eruption at Bárðarbunga. The most intense seismicity has been located along the caldera rim in the top 5 km. Between 1 March 2015 and 20 September 2017, in total 62 earthquakes with magnitudes between M3.5 to M4.5 were recorded. The largest earthquakes (M4.5) occurred in 2017, on 2 August and 7 September.

11.5.2 Grímsvötn

The Grímsvötn subglacial volcano is the most frequently erupting volcano in Iceland, with recent eruptions in 1998, 2004 and 2011. The hours preceding eruptions are characterised by strong swarms of shallow (0–3 km deep) earthquakes (M2–3) near the southern caldera rim, but hypocentre locations are generally not well constrained (Vogfjörd et al., 2005). Activity has been increasing since the last eruption in 2011 (Fig. 11.11), with over 65% of earthquakes in the uppermost 5 km. A GPS station shows a prominent inflation/deflation cycle during and between eruptions (Sturkell et al., 2003, 2006a, b). Observations during the 2011 eruption were an improvement on earlier efforts, both because of a continuous GPS station and a tiltmeter operating close to the eruption site. The data suggest a pressure drop at a surprisingly shallow level (about 2 km depth)

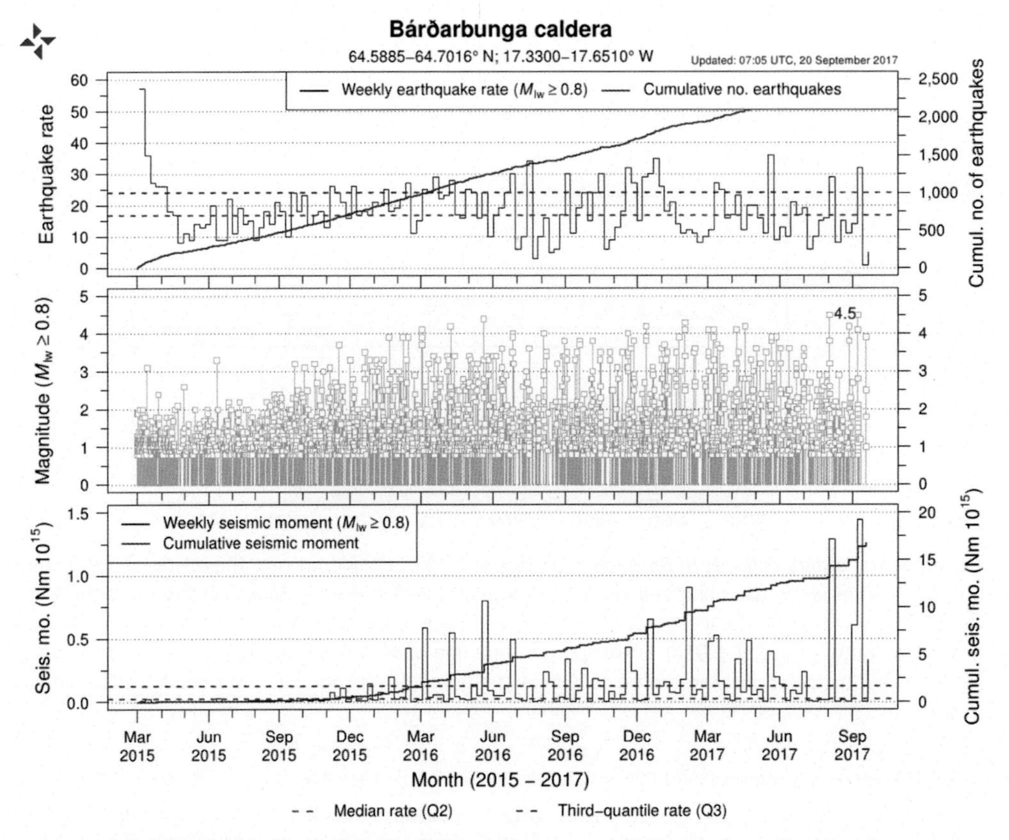

Figure 11.10 ***Seismicity at the Bárðarbunga caldera 1 March 2015–20 September 2017.*** Median monthly earthquake rate is marked with a *pink dashed line* and the third-quantile monthly rate indicated with a *blue dashed line*. Top: Earthquake rate for earthquakes larger than M0.8 (approximate magnitude of completeness). *Red line* shows the cumulative earthquake rate, histogram (grey line) shows the monthly earthquake rate. Centre: Timeline of all earthquakes used for the analysis. Bottom: *Red line* shows the cumulative seismic moment. Histogram (*grey line*) shows the monthly seismic moment.

during the eruption, in a similar location as in previous eruptions (Hreinsdóttir et al., 2014). However, Haddadi et al. (2017) utilised geobarometry of Grimsvötn eruptive products to infer average crystallisation pressure of 4 ± 1 kbar, corresponding to approximately 15 ± 5 km depth. This suggests much deeper magma storage. In the 2011 eruption, high-rate processing of the continuous GPS data indicates that 25% of the pressure drop preceded the eruption onset by about 1 h, indicating the formation of a conduit connecting the shallow reservoir to the vent at the crater rim. Lack of significant recharge (compared to the outflow rates) during the eruption enabled Hreinsdóttir et al.

GPS time series are shown in the Eurasian reference frame with seasonal variations and a 2.5 cm/yr linear term removed from the vertical to correct for uplift due to glacial isostatic movement. The effects of the 2014–2015 events at Bárðarbunga, and following changes have not been corrected for. ▶

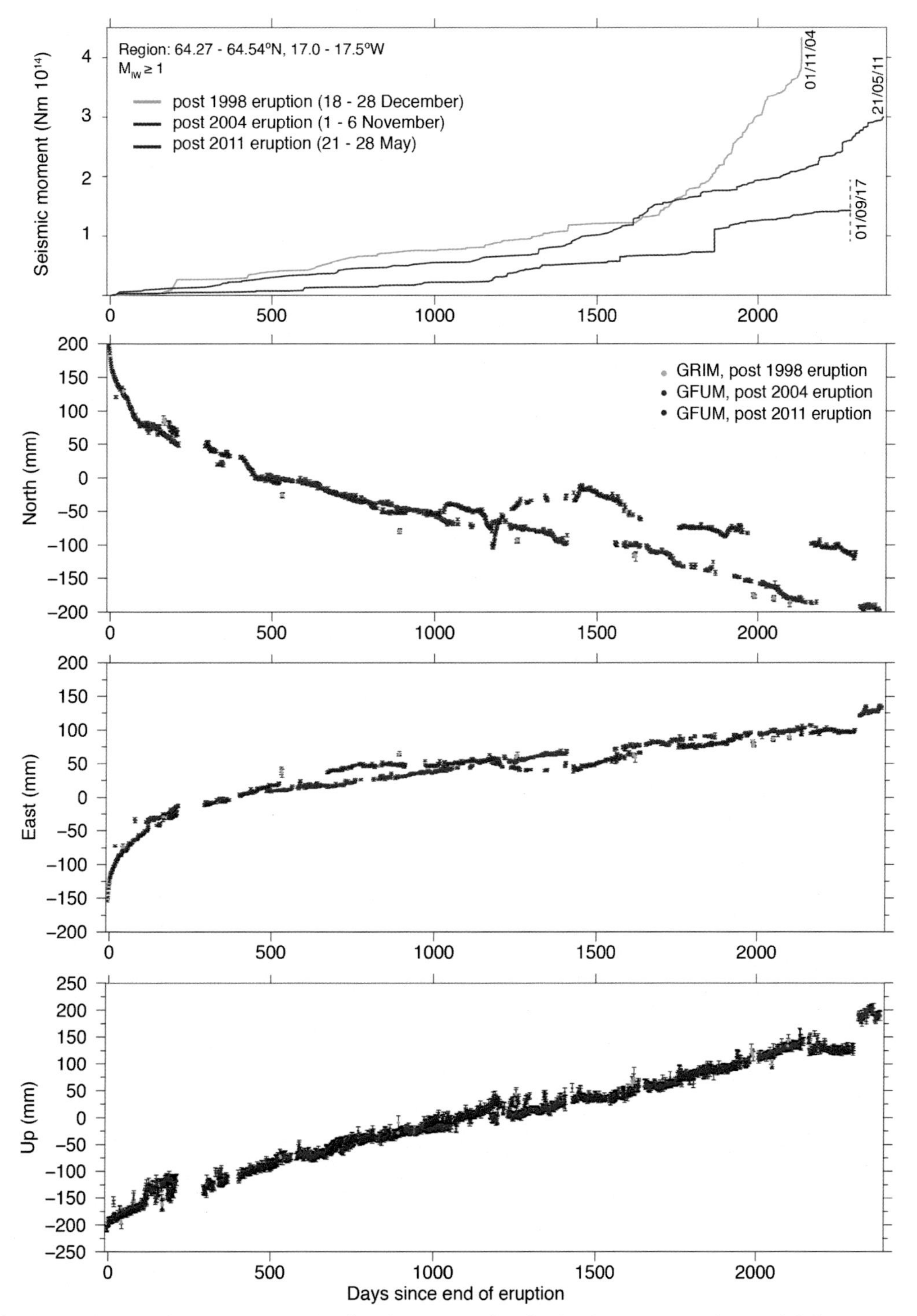

Figure 11.11 ***Grímsvötn time series showing accumulated seismic moment and crustal deformation at Mt Grímsfjall following the Grímsvotn eruptions in 1998, 2004 and 2011 up to September 2017.*** The

(2014) to link rate changes in the reservoir pressure drop to plume heights. In addition, a 10-fold difference between the DRE volume estimates of the erupted material (0.27 ± 0.07 km^3) and the geodetic model (0.027 ± 0.003 km^3) suggest a large role of magma compressibility. The geodetic data have been interpreted in terms of magma flow to one reservoir, or using two magma sources (Reverso et al., 2014).

11.5.3 Hekla

Hekla erupted in 1970, 1980–1981, 1991 and 2000 (Höskuldsson et al., 2007), and geodetic studies indicate that magma accumulates steadily in a source located between 16 and 24 km depth (Ófeigsson et al., 2011; Geirsson et al., 2012). Co-eruptive borehole strain observations (Sturkell et al., 2013) suggest that the first signs of deformation also originate from a pressure drop within a deep source, although at somewhat shallower level than the depth estimates for inter-eruptive periods (at 10-11 km depth). Only a shallow dyke (near-surface) is inferred to have formed in the 2000 eruption, separated from the underlying magma body by a conduit producing no significant deformation. Monitoring and modelling of gas release at the summit of Hekla during a quiescent state (Ilyinskaya et al., 2015) shows that the composition of the exsolved gas is substantially modified along its pathway to the surface through cooling and interaction with wall-rock and groundwater. The modification involves both significant H_2O condensation and scrubbing of S-bearing species, leading to a CO_2-dominated gas emitted at the summit in a concentrated area. These observations are in broad agreement with a preferred gas pathway through a conduit from a deep source. Recent eruptions were preceded by dense swarms of small earthquakes for 23–79 minutes but hypocentre location accuracy was rather poor (Soosalu and Einarsson, 2002; Soosalu et al., 2005). Earthquake relocation shows that between eruptions the activity is distributed in the upper 15 km, with significant clustering at 8–9 and 1–2 km depth.

The eruptions that have occurred within the last few decades have been regular and frequent: 1970, 1980–1981, 1991 and 2000. Regularity like this is rather unusual in the world of volcanoes, and in earlier times Hekla erupted only once or twice per century. The most explosive eruptions occur after a long repose period. Tilt measurements have been made at Hekla for several decades to monitor its state of inflation (Fig. 11.12). The tilt station that has shown the most consistent results is located at Næfurholt, about 11 km west of the volcano. The results indicate a magma body beneath Hekla recharges at a continuous rate between eruptions, as inferred also from GPS and InSAR. During the eruptions of 1991 and 2000, the pressure in the magma body dropped as magma flowed to the surface and the volcano deflated. Immediately after the end of the 2000 eruption, the magma body began inflating again as the volcano began preparing for the next eruption, implying a steady magma supply from depth. If this pattern continues, a new eruption will not occur until the pressure in the magma body rises beyond the previous maximum. Following the 2000 eruption, the magma pressure increased and reached the previous maximum in 2006 (Sturkell

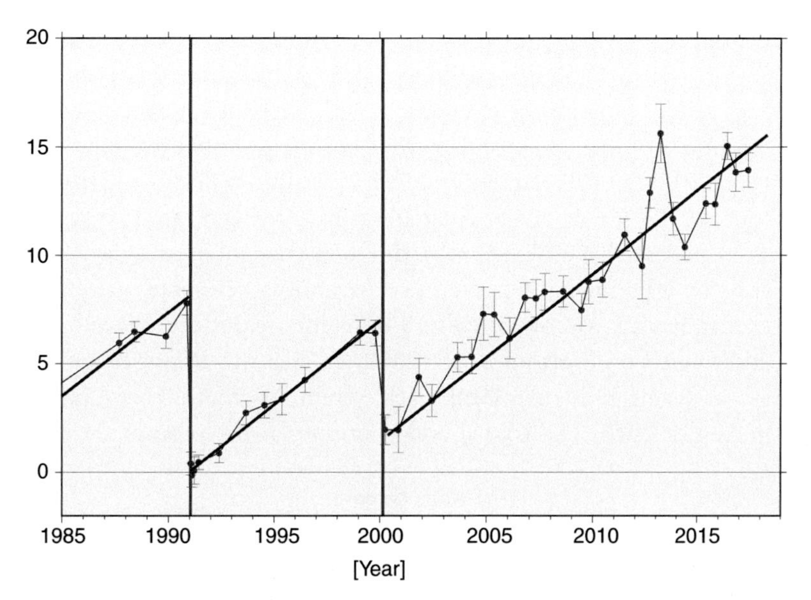

Figure 11.12 ***East component of tilt (microradians) versus time at optical levelling tilt station Næfurholt west of Hekla volcano.*** Increasing values correspond to increasing uplift towards the volcano. The curve reflects a steady increase in magma pressure beneath the volcano for the last three decades, except during the eruptions of 1991 and 2000 (red vertical lines on the graph) when the pressure decreased rapidly. *Black lines* correspond to tilt changes of 0.77 μrad/year.

et al., 2006a). This implies the pressure was sufficient at that time to initiate a new eruption if conditions otherwise remained the same in the volcano interior. Yearly measurements since then indicate continued pressure increase. Variable healing of the volcanic conduit from the Hekla magma body towards the surface in the years after an eruption may, however, influence the time interval between eruptions. If the conduit solidifies and gains strength, a longer interval of dormancy may follow despite steady magma accumulation at depth. Variations in regional stress may also have an influence on timing of eruptions.

11.5.4 Katla

Katla, partly covered by the Mýrdalsjökull ice cap, is one of the most dangerous volcanoes in Iceland. It is capable of generating large explosive eruptions with ash plume heights between 14 and 20 km (Larsen and Gudmundsson, 2016a) accompanied by major jökulhlaups (glacial outburst floods) generated through the rapid melting of large volumes of ice into water at pressures able to overcome the overburden of the ice cap. Katla's plumbing system and magma recharge mechanism are relatively poorly understood. Prior to the most recent eruption in 1918, Katla had on average two major explosive eruptions breaking through the ice per century. However, there has now been a period of quiescence of 100 years. The present long repose period may be related to the large size of the last eruption, as a positive correlation has been found at Katla between eruption magnitude and the length of the following inter-eruptive period (Eliasson et al., 2006). Alternatively, the present long repose period may relate to stress changes induced by ice retreat (Sigmundsson et al., 2010b). There are, however, suggestions that sub-glacial eruptions or intrusive events have occurred without breaking the ice cover in 1955, 1999 and 2011. The 2011 activity caused a jökulhlaup destroying the bridge across Múlakvísl, one of the glacial rivers emanating from the ice cap. It originated from a prominent ice cauldron that has been monitored for changes in sub-glacial water storage and eruptive activity. Following this event, a new, unusual seismicity pattern of repeating long-period earthquakes was observed in a different area of the volcano, on its south-east slope, interpreted by Sgattoni et al. (2016) to relate to minor magmatic injection or to changes of permeability in a local crack system. Another increase in seismic activity began in August 2016 (Fig. 11.13) and continues at the time of writing. Although

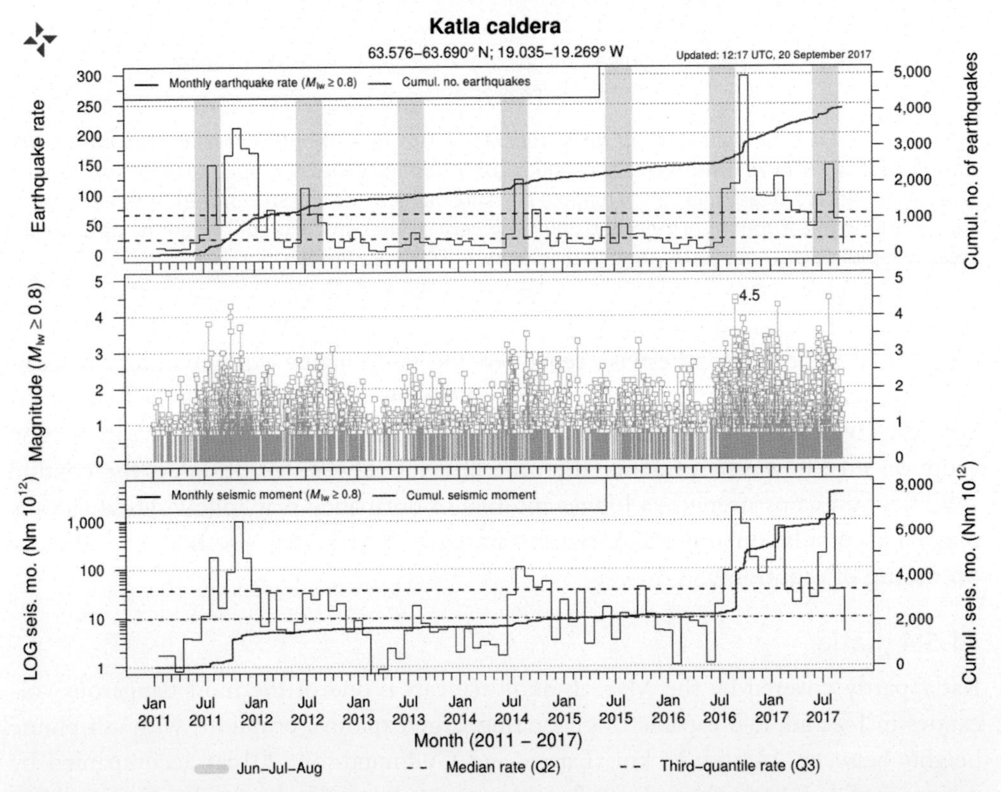

Figure 11.13 ***Seismicity in the Katla caldera from January 2011 until 20 September 2017.*** To better indicate seasonal seismicity, the months June, July and August are shaded in *blue*. Median monthly earthquake rate is marked with a *pink dashed line* and the third-quantile monthly rate indicated with a *blue dashed line*. Top: Earthquake rate for earthquakes larger than M0.8 (approximate magnitude of completeness). *Red line* shows the cumulative earthquake rate, histogram (*grey line*) shows the monthly earthquake rate. Central: Timeline of all earthquakes used for the analysis. Bottom: *Red line* shows the cumulative seismic moment. Histogram (*grey line*) shows the monthly seismic moment.

significant magma inflow during this time has not been detected with geodetic observations, the high seismic activity is interpreted as a result of conditions close to failure, indicating the volcano could erupt with a short warning time.

11.5.5 Askja

The longest time series of observed volcano deformation in Iceland is from a levelling line at Askja volcano, a nested caldera complex in North Iceland. The last major rifting episode within the Askja segment took place in 1874–1876 (Sigvaldason et al., 1992). Presumed diking events accompanied by a Plinian explosive eruption created the most recent caldera structure in the nested complex, which continued to develop for four decades. Extensive GPS, InSAR and micro-gravity (since 1988) measurements add to levelling data and seismic investigations. Deformation studies have suggested that models with two contracting Mogi sources may fit the data: one shallow (~3 km) and another at deep level (~16–21 km) (Pagli et al., 2006; Sturkell et al., 2006a, b). However, more recent deformation studies taking into consideration possible rheological variations in the sub-surface rift structure suggest that the data may be fitted by a single shallow Mogi source embedded in a rifting environment, where the presence of viscoelastic material at relatively shallow levels within the rift zone may explain the subsidence observed along the fissure system as well as within the caldera (Pedersen et al., 2009; de Zeeuw-van Dalfsen et al., 2012, 2013). Seismic tomography models from the Askja region indicate the existence of melt storage bodies in the roots of the volcanic system. Greenfield et al. (2016) found the main melt storage regions lie beneath Askja volcano, concentrated at depths of 6 km with a smaller region at 9 km. Using recorded waveforms, these authors also show that there is also likely to be a small, highly attenuating magmatic body at a shallower depth of about 2 km. The geodetic and seismic models have the potential for further joint interpretation. In contrast to the continuous subsidence, seismicity indicates that deep melt movements are taking place within the Askja rift segment as seismic activity occurs well beneath the brittle-ductile boundary (Soosalu et al., 2010). Key et al. (2011) identify three separate sites of possible magma injection from the mantle. If such deep magma movements involve only a limited volume, they may go unrecognised in surface deformation data.

11.5.6 Hengill

The Hengill triple junction is a zone of horizontal shear stress, situated where two spreading segments, the Reykjanes Peninsula Oblique Rift and the Western Volcanic Zone, intersect with the south Iceland seismic transform zone. The last eruption in the area occurred about 2000 years ago, whereas a tectonic event caused metre-scale subsidence and extension in parts of the fissure system in 1789 (Sæmundsson, 1992). A period of unusually persistent earthquake activity with roughly 80,000 recorded earthquakes (M_{max} ~ 5) occurred in the area from 1994 to 1999. The earthquake swarm coincided temporally with a 2 cm/year surface uplift rate, interpreted from InSAR data to be caused by an expanding Mogi point source at about 7 km depth (Feigl et al., 2000). The source model of the deformation corresponds to an intrusion volume of ~0.02 km^3 yielding an average intrusion rate of 0.1 m^3/s. This episode is an example of slow, deep recharge within a magmatic system, associated with high seismicity but not culminating in an eruption. Interpretation of the geodetic data shows the unrest episode caused one of the lowest increases in the stressing rate measured during recharge of an Icelandic magmatic plumbing system, but it nevertheless produced highly persistent earthquake activity. As the subtle change in the stressing rate induced by the magmatic recharge was imposed on a high background rate, a large increase in seismicity resulted (Pedersen et al., 2007).

11.6 DISCUSSION

The typical pattern of surface deformation observed at several frequently active Icelandic volcanoes (e.g. Krafla, Hekla and Grímsvötn) can be referred to as the 'volcanic deformation cycle'. In the inter-eruptive phase, variable rates of surface uplift are observed, followed by rapid subsidence during a co-eruptive phase. The subsidence reflects the sudden pressure drop due to removal of volume from the magma body within the magma domain, which is then followed by a phase of post-eruptive replenishment and uplift. This may constrain different transient surface deformation patterns according to the conditions in the magma domain immediately following an eruption, for example, to what extent the domain has been drained of magma, the stress field, the compressibility of the residing magma and whether or not it has an open connection to a deeper magma source region.

Lateral dyke propagation from a central magma chamber is characteristic of extensional tectonic settings as it is an efficient way to release centuries of stress accumulated along a plate boundary. Recent rifting events like at Bárðarbunga in August 2014 and past rifting episodes such as 1975–1984 Krafla (Iceland) and 2005–2010 Dabbahu (Afar) have all generated earthquake swarms, surface widening and faulting. Studies of these events have shown that magma can propagate laterally over 50–60 km from the spreading centre in a few days to weeks, without necessarily causing an eruption (Wright et al., 2012). Field and satellite observations reveal faulting induced by dyke emplacement that form a graben at the surface, right above the dyke top, where subsidence can reach up to 5–6 m for the largest intrusions (Ruch et al., 2016). In other instances, gradual transfer of magma towards shallower depths can interact with pre-existing faults and cause them to slip slowly, for example, at a rate of a few centimetres per month during the 2005–2010 rifting episode in Afar (Dumont et al., 2016). Magmatic–tectonic interactions and associated deformation can thus both happen rapidly and be associated with large-scale deformation, or slowly and with small fault movement (see also Chapter 7). Their study can contribute to better probe the magma domain and the volcanic plumbing system during unrest.

Geodetic studies show that some volcanoes receive a relatively steady influx of melt to crustal magma bodies during inter-eruptive periods, whereas others receive more sporadic fluxes of melt from depth. Improved understanding of magma supply and storage (including existing volumes of accumulated magma) beneath active volcanoes would significantly enhance eruption scenario modelling, aid the development of PB models and facilitate eruption forecasting. Although simple elastic deformation models have been successful for reproducing (fitting) observed deformation data for a variety of volcanoes worldwide, many geoscientists agree that the model assumptions poorly approximate volcanic systems. Additional work exploring the role of more realistic models of host rock surrounding magma bodies may therefore advance the way we interpret volcano deformation and allow for an improved understanding of volcanic processes.

11.7 SUMMARY

Interpretation of crustal deformation results together with seismicity patterns in Iceland has provided a wealth of information related to pre-, co- and post-eruptive processes including the detection of new intrusions, volumetric/pressure changes in magma bodies and estimates of melt supply rates. Future advances are expected from synergistic interpretation of geodetic and seismic observations with complementary datasets, including volcanic gas emissions and data from geochemical studies of eruptive products (e.g. geobarometry). The use of PB models may significantly enhance our understanding of volcanic plumbing systems, observed changes in volcanic behaviour and subsequently improve volcanic eruption forecasting. Conceptual models of volcanic plumbing systems constrained by a multitude of observations ideally forms the basis for interpretations of future volcanic unrest observations as well as evaluation of how a particular plumbing system may influence future eruptive activity. Such models should ideally integrate observations revealing the complexity of the magma domain and consider that different parts of this domain may be activated.

FIELD EXAMPLE: 2010 EYJAFJALLAJÖKULL ERUPTIONS AND PRECEDING 18 YEARS OF UNREST

In 2010, an explosive eruption at the ice cap that covered Eyjafjallajökull volcano in south Iceland led to the closure of large parts of Europe's airspace for several days. It was preceded by an 18-year period of intermittent volcanic unrest and magma movements, offering an unprecedented opportunity to study in detail the magmatic plumbing system of a moderately active Icelandic volcano through geodetic and seismic observations. Two eruptions spaced close in time occurred in 2010; a basaltic flank eruption preceded the evolved explosive summit eruption, with a 2-day break in the eruptive activity. Prior to that, Eyjafjallajökull had last erupted in 1821.

Eyjafjallajökull was practically aseismic during the 20-year period following the installation of a country-wide permanent seismic network. Beginning in 1991, earthquakes were detected, and periods of high activity occurred in 1994, 1996 and 1999 provided an opportunity to study the volcanic plumbing system. Ground deformation measurements (GPS, dry tilt and InSAR) showed that tens of centimetres of surface deformation accompanied the swarms at upper- to mid-crustal depths in 1994 and 1999. The deeper 1996 swarm, at lower crustal levels, caused no detectable surface deformation, but seismic analysis indicates that magma accumulated near the crust–mantle boundary (Hjaltadóttir et al., 2015). The centre of surface deformation for the 1994 and 1999 unrest periods were significantly different (Pedersen and Sigmundsson, 2004, 2006), indicating that the volcanic system was not recharging by magma flow into a single shallow magma chamber, as the centre of deformation would then be expected to remain stable. Relocations of seismic events indicated the existence of a pipe-like feature northeast of the edifice, which appeared to be feeding the two intrusions from the base of the crust. Renewed seismic unrest starting in June 2009 spurred an increase in seismic and deformation monitoring of the volcano.

The good spatial and temporal data resolution for the 2009–2010 period revealed a deformation pattern originating from a network of intrusions, as opposed to inflation of a single magmatic source (Sigmundsson et al., 2010a). In a 3-month long period, surface uplift was observed, with the centre of the surface uplift located in an area on the eastern flank of the volcano. Surface deformation was modelled as the formation of a network of sill intrusions at more than 4 km depth, which eventually led to a dyke-fed effusive flank eruption. The eruption initiated on 20 March 2010 and lasted for 3 weeks. Seismic records indicate that the dyke feeding the eruption propagated from about 4 km depth to the surface in the 4 days prior to the eruption (Tarasewicz et al., 2014). During the flank eruption, surface deformation was negligible, indicating that the net volume of drainage from shallow depth was small. The eruptive products from the flank eruption consist of olivine basalts, a primitive composition in excellent agreement with a geodetically inferred deep source region.

Following a 2-day break in eruptive activity, an explosive eruption initiated from the glacier-covered top crater on 14 April 2010. The initial explosivity of the 2-month long eruption was amplified by meltwater–magma interaction, but the erupted products were also considerably more evolved (benmoritic composition), indicating contributions from a separate magma source region (Keiding and Sigmarsson, 2012). Modelling of co-eruptive surface deformation for the second eruption confirms this, as pressure reduction in a source centrally located beneath the volcano fits the co-eruptive geodetic data (Fig. 11.7). The source was interpreted to be a pocket of evolved magma residing at about 4 km depth, which was remobilised when the primitive melts intersected the chemically evolved shallow source. Clusters of deep seismicity (10–30 km depth) occurred well into the second eruption, indicating that magma contribution occurred from progressively deeper regions throughout the eruption

(Tarasewicz et al., 2012). This theory is supported by geochemical data as the erupted products showed a gradual evolution towards more primitive products (Sigmarsson et al., 2011).

The studies of Eyjafjallajökull shows how a complex network of intrusions may facilitate transfer of primitive melts from great depth to the surface. Furthermore, the coincidental intersection of the primitive melt and a pocket of more evolved melt residing at shallow depth within this complicated plumbing system consisting of a network of intruded sheets and sills displays the unpredictable nature of moderately active volcanic systems.

FIELD EXAMPLE: LATERAL DYKING, GRADUAL CALDERA COLLAPSE OF BÁRÐARBUNGA AND THE HOLUHRAUN ERUPTION, 2014–2015

Eruptive activity of the Bárðarbunga volcanic system in the last 9000 years includes 22 verified eruptions that have occurred on the southwest fissure swarm and more than 10 eruptions on the northern fissure swarm (Larsen and Gudmundsson, 2016b). The most recent prior to 2014 were two small lava eruptions in the vicinity of Holuhraun in 1797 and 1867 (Pedersen et al., 2017 and references therein).

In 2014–2015, a major lava forming eruption occurred within the Bárðarbunga volcanic system; the eruption site was located in the Holuhraun plain north of Vatnajökull ice cap. This was the largest effusive event to take place in Iceland since the 1783–1784 Laki eruption. The eruption lasted for 6 months, produced a lava field with a volume of 1.4 ± 0.2 km^3, and a tremendous amount of gas was released into the atmosphere (11 ± 5 Mt of SO_2; Gíslason et al., 2015). The eruption was accompanied by the gradual collapse of Bárðarbunga caldera of up to 65 m (Gudmundsson et al., 2016) and intense seismicity; geochemical and geophysical observations during the eruption indicated a magma storage depth of ~10 km.

Precursory activity was identified prior to the onset of the main 2014 unrest at Bárðarbunga. Between 2012 and 2014, elevated seismicity was detected, and in mid-May 2014, three months before the main unrest started, seismicity levels increased in the Bárðarbunga caldera. Approximately 800 earthquakes of magnitudes up to M3.7 were recorded inside and around the caldera after 1 May. At the same time, a signal in line with inflation centred inside the caldera was observed at the closest cGPS station VONC (Fig. 11.14F). At the end of May, the Icelandic Meteorological Office communicated concerns about the increased seismicity in Bárðarbunga to the Civil Protection in Iceland. At the beginning of August, the joint interpretation of the seismic and cGPS data elevated concerns of impending unrest, since the data could be interpreted as increased magma pressure beneath the central volcano.

On 16 August 2014, intense seismicity started in the vicinity of Bárðarbunga caldera. During the first 24 h, a cluster of earthquakes initially propagated to the southeast of the caldera, then took a sharp 90° turn and continued to migrate to the northeast for the next

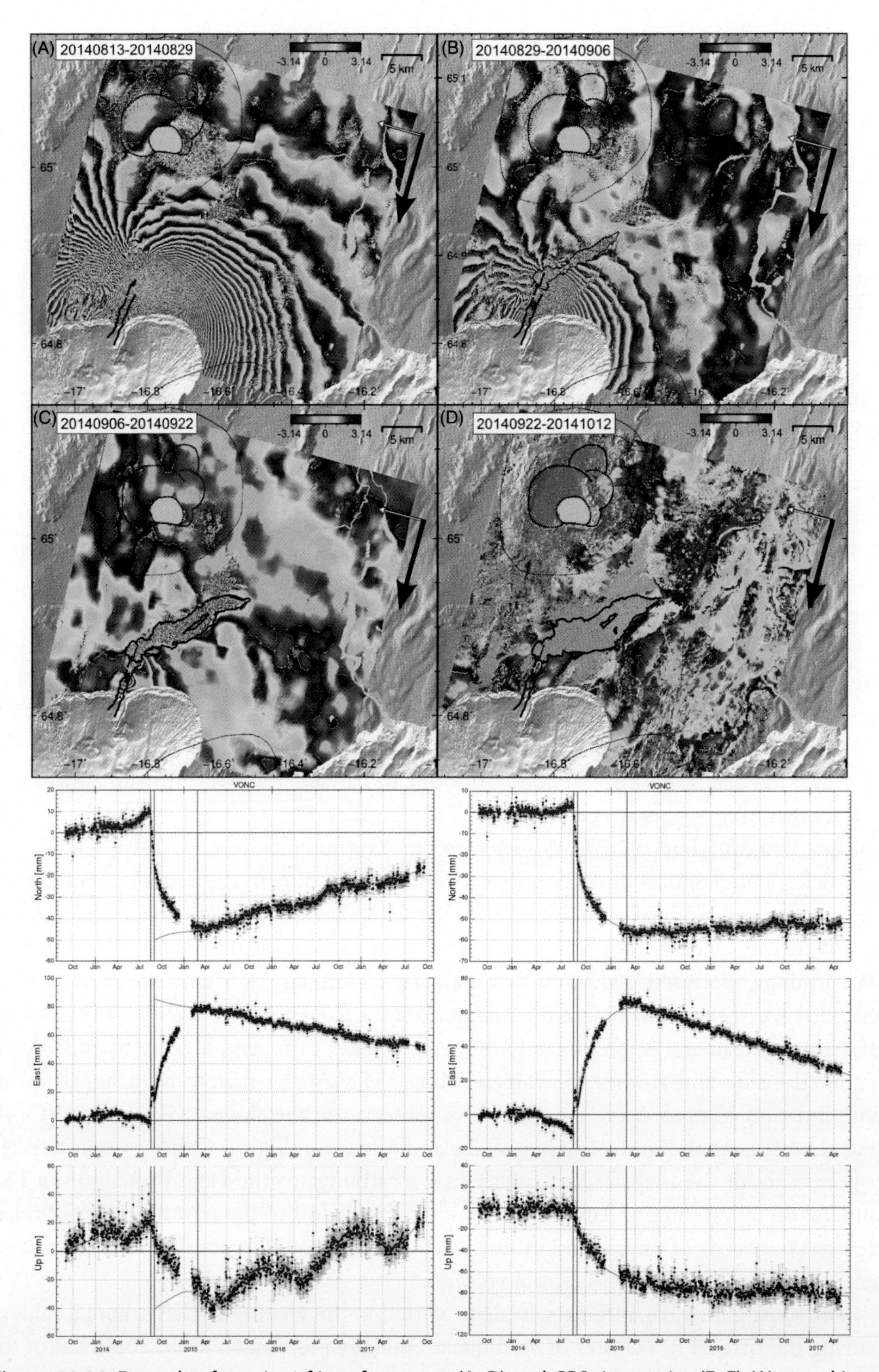

Figure 11.14 Example of a series of interferograms (A–D) and GPS time series (E, F). Wrapped interferograms formed using COSMO-SkyMed SAR satellite images show evolution of deformation induced ▶

2 weeks. Seismicity propagated laterally ~40 km beneath the ice cap, then continued another 8 km outside the glacier to the eruption site at Holuhraun. This intense seismicity (over 22,000 earthquakes) and associated deformation was representative of a segmented lateral dyke injection (Sigmundsson et al., 2015; Ágústsdóttir et al., 2016). A minor effusive eruption started at midnight on 29 August at Holuhraun lasting only a few hours. The main event began on 31 August with the opening of a 1.8-km long fissure segment (Pedersen et al., 2017) issuing spectacular lava fountains with heights reaching >100 m. This fissure remained active throughout the entire eruption, which ended on 27 February 2015. A third small eruption, 2 km south of the main fissure, occurred on 5–6 September 2014.

A combination of geodetic observations (InSAR and cGPS), radar profiling over the Bárðarbunga caldera, seismic and petrological analyses indicated that the observed deformation and seismicity were the result of melt withdrawal from a magma body beneath the Bárðarbunga caldera, slip on caldera ring faults and the transfer of magma along a 48-km long dyke which was feeding the major effusive eruption on the Holuhraun plain (Gudmundsson et al., 2016). Magma that drained from beneath the caldera was initially emplaced in a dyke and then erupted at the far end of the dyke. Good GPS and InSAR coverage of the events (Figs. 11.3 and 11.14) enabled detailed modelling (Fig. 11.15). The calculated volume change at the end of the eruption, associated with magma withdrawal from the modelled sill at ~10 km depth beneath Bárðarbunga, was -1.9 ± 0.1 and 0.7 ± 0.04 km^3 for the dyke volume (Parks et al., 2017). The dyke opening mostly occurred over a period of weeks, whereas the caldera subsidence at Bárðarbunga occurred gradually over a period of 6 months, simultaneously with the 6-month long lava flow eruption at the far end of the dyke.

◀ by the opening of a magma-filled crack: (A) 13–29 August 2014, dyke emplacement. (B) 29 August–6 September 2014, onset of the main eruption and a small eruption that initiated on 5–7 September. (C) 6–22 September 2014, encompassing the small eruption in the new graben. (D) 22 September–12 October 2014, minor deformation; and signatures of noise in the interferogram. The *black arrows* indicate the satellite flight direction, the *white* ones the satellite viewing direction. Graben border faults and outline of the lava field shown with *black lines*. The volcanic centres are shown with *dotted black lines*, and calderas with *hashed ones*. (E, F) Time series from GPS station VONC in Vonarskarð, northwest of Bárðarbunga (see Fig. 11.3 for location). *Blue vertical lines* show the onset of the Bárðarbunga rifting, *red vertical lines* mark the Holuhraun eruption period. (E) Unfiltered time series in the reference frame of the North American plate. *Blue* and *green* curves are used for later detrending process. The *green curve* shows the best-fit to annual variations for period after the Holuhraun eruption. The *blue curve* shows best-fit of the velocity prior to May 2014, hence representing base velocity estimated prior to any visible signs of inflation in Bárðarbunga. (F) The detrended time series, where annual periodic signal has been subtracted as well as the linear velocity estimated prior to May 2014. The *green line* shows a best-fit to a function of an exponential decay plus to linear component. An inflation signal is detected 3–4 months prior to Bárðarbunga rifting. At the onset of Bárðarbunga rifting event, a superposition of the formation of a dyke extending to Holuhraun and deflation can be observed. After the eruption onset, an exponential decaying deflation signal is observed along with an underling inflation, which is still ongoing in 2017 and is represented as a *green curve*.

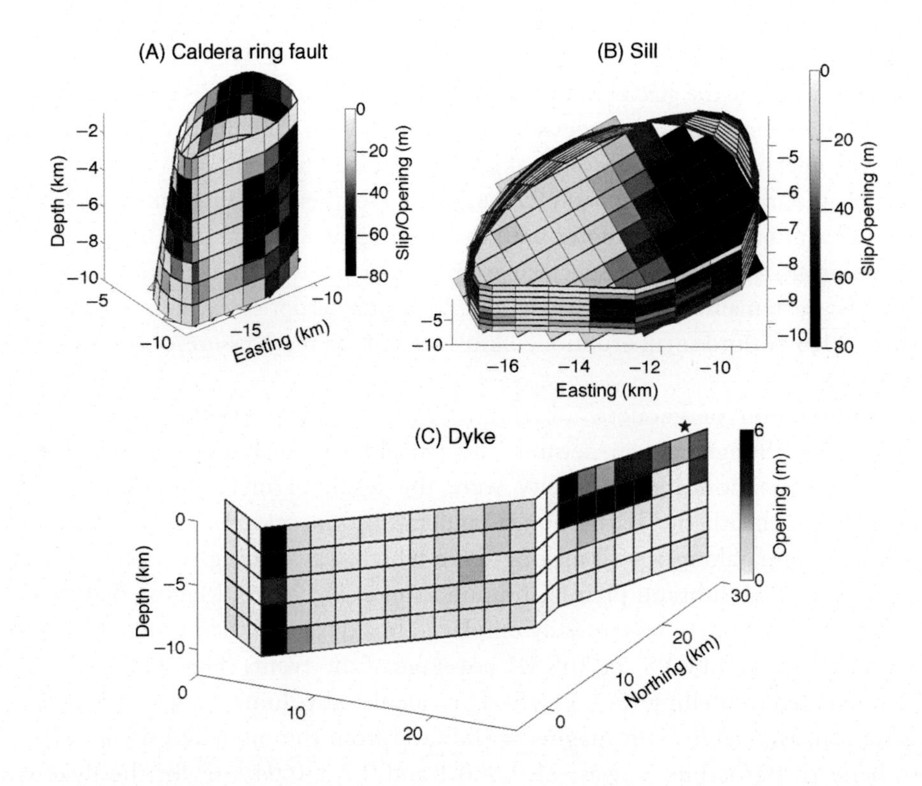

Figure 11.15 ***Bárðarbunga cumulative deformation model computed for the period 16 August 2014–10 April 2015.*** **The calculated median posterior slip on the ring fault (A), closing of the sill (B) and opening of the dyke (C). Negative slip in (A) represents downward displacement and negative opening in (B) represents closing of the sill. The *star* in (C) represents the site of the main eruption (31st August 2014 to 27th February 2015). Modified from Parks et al. (2017).**

ACKNOWLEDGEMENTS

The research and monitoring presented in this chapter has been supported by numerous projects and agencies, including: European Community's FP7 Programme Grant No. 308377 (Project FUTUREVOLC), the Icelandic Research Fund, the Research Fund at University of Iceland, crisis response funding from the Icelandic government through the Civil Protection Department of the National Commissioner of the Icelandic Police, NERC, United Kingdom and the National Science Foundation, United States (e.g., grant EAR-1464546). Space agencies and their committee on Earth Observation Satellites (CEOS) provided access to satellite data for InSAR processing through the Icelandic Volcanoes permanent supersite, a part of the initiative on Geohazards Supersites and Natural Laboratories (GSNL) by the Group on Earth Observations (GEO). An intermediate TanDEM-X digital elevation model was provided by DLR under project IDEM_GEOL0123. Field operations in remote areas have been supported by the Icelandic Coast Guard, the Landsvirkjun power company, the Iceland Glaciological Society and the Vatnajökull National Park. The authors thank the technical staff at their institutions that have made the field observations presented possible. For GPS equipment and support the authors acknowledge services provided by the UNAVCO facility with support from the National Science Foundation (NSF) and National Aeronautics and Space Administration (NASA).

REFERENCES

Ágústsdóttir, T., Woods, J., Greenfield, T., Green, R.G., White, R.S., Winder, T., Brandsdottir, B., Steinthórsson, S., Soosalu, H., 2016. Strike-slip faulting during the 2014 Bárðarbunga-Holuhraun dyke intrusion, central Iceland. Geophys. Res. Lett. 43 (4), 1495–1503.

Aiuppa, A., Federico, C., Giudice, G., Giuffrida, G., Guida, R., Gurrieri, S., Liuzzo, M., Moretti, R., Papale, P., 2009. The 2007 eruption of Stromboli volcano: insights from real-time measurement of the volcanic gas plume CO_2/SO_2 ratio. J. Volcanol. Geotherm. Res. 182 (3), 221–230.

Ali, S.T., 2014. Defmod-parallel multiphysics finite element code for modelling crustal deformation during the earthquake/rifting cycle. arXiv:1402.0429.

Allard, P., 2010. A CO_2-rich gas trigger of explosive paroxysms at Stromboli basaltic volcano, Italy. J. Volcanol. Geotherm. Res. 189 (3), 363–374.

Anderson, K.R., Poland, M.P., 2016. Bayesian estimation of magma supply, storage, and eruption rates using a multiphysical volcano model: Kīlauea Volcano, 2000–2012. Earth Planet. Sci. Lett. 447, 161–171.

Anderson, K., Segall, P., 2011. Physics-based models of ground deformation and extrusion rate at effusively erupting volcanoes. J. Geophys. Res. 116, B07204.

Anderson, K., Segall, P., 2013. Bayesian inversion of data from effusive volcanic eruptions using physics-based models: application to Mount St. Helens 2004–2008. J. Geophys. Res. Solid Earth 118 (5), 2017–2037.

Auriac, A., Spaans, K.H., Sigmundsson, F., Hooper, A., Schmidt, P., Lund, B., 2013. Iceland rising: solid Earth response to ice retreat inferred from satellite radar interferometry and viscoelastic modelling. J. Geophys. Res. Solid Earth, 118. doi: 10.1002/jgrb.50082.

Bean, C., Lokmer, I., O'Brien, G., 2008. Influence of near-surface volcanic structure on long-period seismic signals and on moment tensor inversions: Simulated examples from Mount Etna. J. Geophys. Res. 113, B08308.

Bekaert, D.P.S., Walters, R.J., Wright, T.J., Hooper, A.J., Parker, D.J., 2015. Statistical comparison of InSAR tropospheric correction techniques. Remote Sens. Environ. 170, 40–47.

Bock, Y., Melgar, D., 2016. Physical applications of GPS geodesy: a review. Rep. Prog. Phys. 79 (10), 106801.

Bonafede, M., Ferrari, C., 2009. Analytical models of deformation and residual gravity changes due to a Mogi source in a viscoelastic medium. Tectonophysics 471, 4–13.

Brandsdóttir, B., Einarsson, P., 1992. Volcanic tremor and low-frequency earthquakes in Iceland. In: Gasparini, P., Scarpa, R., Aki, K. (Eds.), Volcanic Seismology. Springer, Berlin, Heidelberg, pp. 212–222.

Carbone, D., Poland, M.P., Diament, M., Greco, F., 2017. The added value of time-variable microgravimetry to the understanding of how volcanoes work. Earth Sci. Rev. 169, 146–179.

Chadwick, Jr., W.W., Paduan, J.B., Clague, D.A., Dreyer, B.M., Merle, S.G., Bobbitt, A.M., Caress, D.W., Philip, B.T., Kelley, D.S., Nooner, S.L., 2016. Voluminous eruption from a zoned magma body after an increase in supply rate at Axial Seamount. Geophys. Res. Lett. doi: 10.1002/2016GL071327.

Dawson, P.B., Benitez, M.C., Chouet, B.A., Wilson, D., Okubo, P.G., 2010. Monitoring very-long-period seismicity at Kilauea Volcano, Hawaii. Geophys. Res. Lett. 37, 18.

de Zeeuw-van Dalfsen, E., Rymer, H., Sigmundsson, F., Sturkell, E., 2005. Net gravity decrease at Askja volcano, Iceland: constraints on processes responsible for continuous caldera deflation, 1988–2003. J. Volcanol. Geotherm. Res. 139 (3), 227–239.

de Zeeuw-van Dalfsen, E., Pedersen, R., Hooper, A., Sigmundsson, F., 2012. Subsidence of Askja caldera 2000–2009: modelling of deformation processes at an extensional plate boundary, constrained by time series InSAR analysis. J. Volcanol. Geotherm. Res., 213–214, 72–82.

de Zeeuw-van Dalfsen, E., Rymer, H., Sturkell, E., Pedersen, R., Hooper, A., Sigmundsson, F., Ófeigsson, B., 2013. Geodetic data shed light on ongoing caldera subsidence at Askja, Iceland. Bull. Volcanol. 75 (5), 1–13.

Drouin, V., Heki, K., Sigmundsson, F., Hreinsdóttir, S., Ofeigsson, B.G., 2016. Constraints on seasonal load variations and regional rigidity from continuous GPS measurements in Iceland, 1997–2014. Geophys. J. Int. 205 (3), 1843–1858.

Drouin, V., Sigmundsson, F., Ófeigsson, B.G., Hreinsdóttir, S., Sturkell, E., Einarsson, P., 2017. Deformation in the Northern Volcanic Zone of Iceland 2008–2014: an interplay of tectonic, magmatic, and glacial isostatic deformation. J. Geophys. Res. Solid Earth 122, 3158–3178.

Dumont, S., Socquet, A., Grandin, R., Doubre, C., Klinger, Y., 2016. Surface displacements on faults triggered by slow magma transfers between dyke injections in the 2005–2010 rifting episode at Dabbahu–Manda–Hararo rift (Afar, Ethiopia). Geophys. J. Int. 204 (1), 399–417.

Eibl, E.P., Bean, C.J., Vogfjörd, K.S., Ying, Y., Lokmer, I., Möllhoff, M., O'Brien, G.S., Pálsson, F., 2017. Tremor-rich shallow dyke formation followed by silent magma flow at Bárðarbunga in Iceland. Nat. Geosci. 10, 299–304.

Eliasson, J., Larsen, G., Gudmundsson, M.T., Sigmundsson, F., 2006. Probabilistic model for eruptions and associated flood events in the Katla caldera, Iceland. Comput. Geosci. 10, 179–200.

Einarsson, P., Sæmundsson, K., 1987. Earthquake epicenters 1982–1985 and volcanic systems in Iceland (map). In: Sigfússon, Þ.I. (Ed.), Í hlutarins eðli, Festschrift for Þorbjörn Sigurgeirsson. Menningarsjóður, Reykjavík.

Feigl, K.L., Gasperi, J., Sigmundsson, F., Rigo, A., 2000. Crustal deformation near Hengill volcano, Iceland 1993–1998: coupling between magmatic activity and faulting inferred from elastic modelling of satellite radar interferograms, J. Geophys. Res. 105(B11):25, 655–625, 670.

Fernández, J., Poland, M.P., Sigmundsson, F., 2017. Volcano geodesy: recent developments and future challenges. J. Volcanol. Geotherm. Res. 344, 1–12.

Fialko, Y., Khazan, Y., Simons, M., 2001. Deformation due to a pressurized horizontal circular crack in an elastic half-space, with applications to volcano geodesy. Geophys. J. Int. 146, 181–190.

Geirsson, H., LaFemina, P., Árnadóttir, T., Sturkell, E., Sigmundsson, F., Travis, M., Schmidt, P., Lund, B., Hreinsdóttir, S., Bennett, R., 2012. Volcano deformation at active plate boundaries: deep magma accumulation at Hekla volcano and plate boundary deformation in south Iceland. J. Geophys. Res. 117, B11409.

Gíslason, S.R., et al., 2015. Environmental pressure from the 2014–15 eruption of Bárðarbunga volcano, Iceland. Geochem. Perspect. Lett. 1:84–93.

Grapenthin, R., Ófeigsson, B.G., Sigmundsson, F., Sturkell, E., Hooper, A., 2010. Pressure sources versus surface loads: analyzing volcano deformation signal composition with an application to Hekla volcano, Iceland. Geophys. Res. Lett. 37 (20), L20310.

Grapenthin, R.J.T., Freymueller, J.T., Kaufman, A.M., 2013. Geodetic observations during the 2009 eruption of Redoubt Volcano, Alaska. J. Volcanol. Geotherm. Res. 259, 115–132.

Greenfield, T., White, R.S., 2015. Building icelandic igneous crust by repeated melt injections. J. Geophys. Res. Solid Earth 120, 7771–7788.

Greenfield, T., White, R.S., Roecker, S., 2016. The magmatic plumbing system of the Askja central volcano, Iceland, as imaged by seismic tomography. J. Geophys. Res. Solid Earth, 121. doi: 10.1002/2016JB013163.

Gudmundsson, M.T., Högnadóttir, T., 2007. Volcanic systems and calderas in the Vatnajökull region, central Iceland: constraints on crustal structure from gravity data. J. Geodyn. 43 (1), 153–169.

Gudmundsson, M.T., Jónsdóttir, K., Hooper, A., Holohan, E.P., Halldórsson, S.A., Ófeigsson, B.G., Cesca, S., Vogfjörd, K.S., Sigmundsson, F., et al., 2016. Gradual caldera collapse at Bárðarbunga volcano, Iceland, regulated by lateral magma outflow. Science 353(6296):aaf8988.

Haddadi, B., Sigmarsson, O., Larsen, G., 2017. Magma storage beneath Grímsvötn volcano, Iceland, constrained by clinopyroxene-melt thermobarometry and volatiles in melt inclusions and groundmass glass. J. Geophys. Res. Solid Earth, 122. doi: 10.1002/2017JB014067.

Heimisson, E.R., Einarsson, P., Sigmundsson, F., Brandsdóttir, B., 2015. Kilometer-scale Kaiser effect identified in Krafla volcano, Iceland. Geophys. Res. Lett, 42. doi: 10.1002/2015GL065680.

Hickey, J., Gottsmann, J., Mothes, P., 2015. Estimating volcanic deformation source parameters-with a finite element inversion: the 2001–2002 unrest at Cotopaxi volcano, Ecuador. J. Geophys. Res. Solid Earth 120, 1473–1486.

Hjaltadóttir, S., Vogfjörd, K.S., Hreinsdóttir, S., Slunga, R., 2015. Reawakening of a volcano: activity beneath Eyjafjallajökull volcano from 1991 to 2009. J. Volcanol. Geotherm. Res 304, 194–205.

Hjartardóttir, Á.R., Einarsson, P., Gudmundsson, M.T., Högnadóttir, Th., 2016a. Fracture movements and graben subsidence during the 2014 Bárðarbunga dyke intrusion in Iceland. J. Volcanol. Geotherm. Res 310, 242–252.

Hjartardóttir, Á.R., Einarsson, P., Magnúsdóttir, S., Björnsdóttir, Þ., Brandsdóttir, B., 2016. Fracture systems of the Northern Volcanic Rift Zone, Iceland—an onshore part of the mid-Atlantic plate boundary. In: Wright, T.J., Ayele, A., Ferguson, D.J., Kidane, T., Vye-Brown, C. (Eds.), Magmatic Rifting and Active Volcanism. The Geological Society of London, pp. 297–314.

Hooper, A., 2008. A multi-temporal InSAR methodincorporating both persistent scatterer and small baseline approaches. Geophys. Res. Lett. 35, L16302.

Hooper, A., Segall, P., Zebker, H., 2007. Persistent scatterer interferometric synthetic aperture radar for crustal deformation analysis, with application to Volcán Alcedo, Galápagos. J. Geophys. Res. Solid Earth 112 (B7).

Hooper, A., Pedersen, R., Sigmundsson, F., 2010. Constraints on magma intrusion at Eyjafjallajökull and Katla volcanoes in Iceland, from time series SAR Interferometry. In: Bean, C.J., Braiden, A.K., Lokmer, I., Martini, F., O. Brien, G.S. (Eds.), The VOLUME Project. Volcanoes: Understanding Subsurface Mass Movement, pp. 13–24, VOLUME Project Consortium, Dublin, Ireland, http://www.ucd.ie/geophysics/grouppublications/books/VOLUME%20book6.pdf.

Hooper, A., Prata, F., Sigmundsson, F., 2012. Remote sensing of volcano hazards and their precursors. Proc. IEEE 100 (10), 2908–2930.

Hooper, A., Pietrzak, J., Simons, W., Cui, H., Riva, R., Naeije, M., Terwisscha van Scheltinga, A., Schrama, E., Stelling, G., Socquet, A., 2013. Importance of horizontal seafloor motion on tsunami height for the 2011 Mw59.0 Tohoku-Oki earthquake. Earth Planet. Sci. Lett. 361, 469–479.

Höskuldsson, Á., Óskarsson, N., Pedersen, R., Grönvold, K., VogfjörÐ, K., Ólafsdóttir, R., 2007. The millennium eruption of Hekla in February 2000. Bull. Volcanol. 70 (2), 169–182.

Hreinsdóttir, S., et al., 2014. Volcanic plume height correlated with magma pressure change at Grímsvötn Volcano, Iceland. Nat. Geosci. 7, 214–218.

Ilyinskaya, E., et al., 2012. Degassing regime of Hekla volcano 2012–2013. Geochim. Cosmochim. Acta 159, 80–99.

Johnson, D.J., Sigmundsson, F., Delaney, P.T., 2000. Comment on "Volume of magma accumulation or withdrawal estimated from surface uplift or subsidence, with application to the 1960 collapse of Kilauea Volcano" by T. T. Delaney and D. F. McTigue. Bull. Volcanol. 61:491–493.

Kaiser, J., 1953. Erkenntnisse und Folgerungen aus der Messung von Geräuschen bei Zugbeanspruchung von metallischen Werkstoffen. Arch. Eisenhüttenwesen 24, 43–45.

Keiding, J.K., Sigmarsson, O., 2012. Geothermobarometry of the 2010 Eyjafjallajökull eruption. J. Geophys. Res. 117, B00C09.

Key, J., White, R.S., Soosalu, H., Jakobsdóttir, S.S., 2011. Multiple melt injection along a spreading segment at Askja, Iceland. Geophys. Res. Lett. 38, L05301. doi: 10.1029/2010GL046264.

Larsen, G., Gudmundsson, M.T., 2016a. The Katla volcanic system. In: Ilyinskaya, Larsen, Gudmundsson (Eds.), Catalogue of Icelandic Volcanoes. IMO, UI, CPD-NCIP, http://icelandicvolcanoes.is.

Larsen, G., Gudmundsson, M.T., 2016b. The Bárðarbunga volcanic system. In: Ilyinskaya, Larsen, Gudmundsson (Eds.), Catalogue of Icelandic Volcanoes. IMO, UI, CPD-NCIP, http://icelandicvolcanoes.is.

Lockner, D., 1993. The role of acoustic emission in the study of rock fracture. Int. J. Rock Mech. Min. Sci. Geomech. Abstr. 30 (7), 883–899.

Magnússon, E., Pálsson, F., Björnsson, H., GuÐmundsson, S., 2012. Removing the ice cap of Öræfajökull central volcano, SE Iceland: mapping and interpretation of bedrock topography, ice volumes, subglacial troughs and implications for hazards assessments. Jökull 62, 131–150.

Masterlark, T., Haney, M., Dickinson, H., Fournier, T., Searcy, C., 2010. Rheologic and structural controls on the deformation of Okmok volcano, Alaska: FEMS, InSAR and ambient noise tomography. J. Geophys. Res. 115, B02409.

Masterlark, T., Feigl, K.L., Haney, M.M., Stone, J., Thurber, C.H., Ronchin, E., 2012. Nonlinear estimation of geometric parameters in FEMs of volcano deformation: integrating tomography models and geodetic data for Okmok volcano, Alaska. J. Geophys. Res. 117, 17. doi: 10.1029/2011JB008811.

Masterlark, T., Donovan, T., Feigl, K.L., Haney, M., Thurber, C., Tung, S., 2016. Volcano deformation source parameters estimated from InSAR: sensitivities to uncertainties in seismic tomography. J. Geophys. Res. 121 doi: 10.1002/2015JB012656.

McTigue, D.F., 1987. Elastic stress and deformation near a finite spherical magma body: resolution of the point source paradox. J. Geophys. Res. 92, 12931–12940.

Menke, W., 2012. Geophysical Data Analysis: Discrete Inverse Theory, MATLAB Edition, Third ed. Academic Press – Elsevier Inc., San Diego, California, USA.

Minakami, T., Ishikawa, T., Yagi, K., 1951. The 1944 Eruption of Volcano Usu in Hokkaido, Japan. Bull. Volcanol. 11 (1), 45–157.

Misra, P., Enge, P., 2011. Global Positioning System: Signals, Measurements and Performance, Second ed. Ganga-Jamuna Press, Lincoln, Massachusetts, USA.

Mogi, K., 1958. Relations between the eruptions of various volcanoes and the deformations of the ground surfaces around them. Bull. Earthquake Res. Inst. Univ. Tokyo 36, 99–134.

Mosegaard, K., Tarantola, A., 1995. Monte Carlo sampling of solutions to inverse problems. J. Geophys. Res. 100, 12431–12447.

Newman, A.V., Dixon, T.H., Gourmelen, N., 2006. A four-dimensional viscoelastic deformation model for Long Valley Caldera, California, between 1995 and 2000. J. Volcanol. Geotherm. Res. 150, 244–269.

Nooner, S.L., Chadwick Jr., W.W., 2016. Inflation-predictable behavior and co-eruption deformation at Axial Seamount. Science. doi:10.1126/science.aah4666.

Ófeigsson, B.G., Hooper, A., Sigmundsson, F., Sturkell, E., Grapenthin, R., 2011. Deep magma storage at Hekla volcano, Iceland, revealed by InSAR time series analysis. J. Geophys. Res. 116, B05401.

Okada, Y., 1985. Surface deformation due to shear and tensile faults in a half-space. Bull. Seism. Soc. Am. 75, 1135–1154.

Padrón, E., et al., 2013. Diffusive helium emissions as a precursory sign of volcanic unrest. Geology 41 (5), 539–542.

Pagli, C., Sigmundsson, F., Árnadóttir, T., Einarsson, P., Sturkell, E., 2006. Deflation of the Askja volcanic system: constraints on the deformation source from combined inversion of satellite radar interferograms and GPS measurements. J. Volcanol. Geotherm. Res. 152, 97–108.

Parks, M.M., Heimisson, E.R., Sigmundsson, F., Hooper, A., et al., 2014. Evolution of deformation and stress changes during the caldera collapse and dyking at Bárðarbunga, 2014–2015: implication for triggering of seismicity at nearby Tungnafellsjökull volcano. Earth Planet Sci. Lett. 462, 212–223.

Pedersen, R., Sigmundsson, F., 2004. InSAR based sill model links spatially offset areas of deformation and seismicity for the 1994 unrest episode at Eyjafjallajökull volcano, Iceland. Geophys. Res. Lett. 31, L14610.

Pedersen, R., Sigmundsson, F., 2006. Temporal development of the 1999 intrusive episode in the Eyjafjallajökull volcano, Iceland, derived from InSAR images. Bull. Volcanol. 68, 377–393.

Pedersen, R., Sigmundsson, F., Einarsson, P., 2007. Controlling factors on earthquake swarms associated with magmatic intrusions: constraints from Iceland. J. Volcanol. Geotherm. Res. 162, 73–80.

Pedersen, R., Sigmundsson, F., Masterlark, T., 2009. Rheologic controls on inter-rifting deformation of the Northern Volcanic Zone, Iceland, Earth Planet. Sci. Lett. 281, 14–26.

Pedersen, G.B.M., et al., 2014. Lava field evolution and emplacement dynamics of the 2014–2015 basaltic fissure eruption at Holuhraun, Iceland. J. Volcanol. Geotherm. Res. 340, 155–169.

Reverso, T., Vandemeulebrouck, J., Jouanne, F., Pinel, V., Villemin, T., Sturkell, E., Bascou, P., 2014. A two-magma chamber model as a source of deformation at Grímsvötn Volcano, Iceland. J. Geophys. Res. Solid Earth 119 (6), 4666–4683.

Ripepe, M., Bonadonna, C., Folch, A., Delle Donne, D., Lacanna, G., Marchetti, E., Höskuldsson, A., 2013. Ash-plume dynamics and eruption source parameters by infrasound and thermal imagery: the 2010 Eyjafjallajökull eruption. Earth Planet. Sci. Lett. 366, 112–121.

Rivalta, E., Segall, P., 2008. Magma compressibility and the missing source for some dyke intrusions. Geophys. Res. Lett. 35, L04306.

Ruch, J., Wang, T., Wenbin, X., Hensch, M., Jónsson, S., 2016. Oblique rift opening revealed by reoccurring magma injection in central Iceland. Nat. Commun. 7, 12352.

Sæmundsson, K., 1992. Geology of the Thingvallavatn area. In: Jónasson, P.M. (Ed.), Ecology of Oligotrophic, Subarctic Thingvallavatn. OIKOS, Copenhagen, pp. 40–68.

Segall, P., 2016. Repressurization following eruption from a magma chamber with a viscoelastic aureole. J. Geophys. Res. Solid Earth 121, 8501–8522.

Sgattoni, G., Jeddi, Z., Gudmundsson, Ó., Einarsson, P., Tryggvason, A., Lund, B., Lucchi, F., 2016. Long-period seismic events with strikingly regular temporal patterns on Katla volcano's south flank (Iceland). J. Volcanol. Geotherm. Res. 324, 28–40.

Sigmarsson, O., Vlastelic, I., Andreasen, R., Bindeman, I., Devidal, J.-L., Moune, S., Keiding, J.K., Larsen, G., Höskuldsson, A., Thordarson, Th., 2011. Remobilization of silicic intrusion by mafic magmas during the 2010 Eyjafjallajökull eruption. Solid Earth 2, 271–281.

Sigmundsson, F., 2006. Iceland Geodynamics, Crustal Deformation and Divergent Plate Tectonics. Springer Verlag and Praxis Publishing, Chichester, UK, 209 pp.

Sigmundsson, F., 2016. New insights into magma plumbing along rift systems from detailed observations of eruptive behavior at Axial volcano. Geophys. Res. Lett. 43 doi: 10.1002/ 2016GL071884.

Sigmundsson, F., Einarsson, P., Bilham, R., 1992. Magma chamber deflation recorded by the global positioning system: the Hekla 1991 eruption. Geophys. Res. Lett. 19, 1483–1486.

Sigmundsson, F., Hreinsdóttir, S., Hooper, A., Árnadóttir, Th., Pedersen, R., Roberts, M.J., Óskarsson, N., Auriac, A., Decriem, J., Einarsson, P., Geirsson, H., Hensch, M., Ófeigsson, B.G., Sturkell, E., Sveinbjörnsson, H., Feigl, K.L., 2010a. Intrusion triggering of the 2010 Eyjafjallajökull explosive eruption. Nature 468, 426–430. doi: 10.1038/nature09558.

Sigmundsson, F., Pinel, V., Lund, B., Albino, F., Pagli, C., Geirsson, H., Sturkell, E., 2010b. Climate effects on volcanism: influence on magmatic systems of loading and unloading from ice mass variations with examples from Iceland, Special issue on climate forcing of geological and geomorphological hazards. Phil. Trans. R. Soc. A 368, 1–16.

Sigmundsson, F., et al., 2015. Segmented lateral dyke growth in a rifting event at Bárðarbunga volcanic system, Iceland. Nature 517, 191–195.

Sigvaldason, G.E., Annertz, K., Nielsson, M., 1992. Effect of glacier loading/deloading on volcanism: postglacial volcanic production rate of the Dyngjufjoll area, central Iceland. Bull. Volcanol. 54, 385–392.

Soosalu, H., Einarsson, P., 2002. Earthquake activity related to the 1991 eruption of the Hekla volcano, Iceland. Bull. Volcanol. 63 (8), 536–544.

Soosalu, H., Einarsson, P., Þorbjarnardóttir, B.S., 2005. Seismic activity related to the 2000 eruption of the Hekla volcano, Iceland. Bull. Volcanol. 68 (1), 21–36.

Soosalu, H., Key, J., White, R.S., Knox, C., Einarsson, P., Jakobsdóttir, S., 2010. Lower-crustal earthquakes caused by magma movement beneath Askja volcano on the north Iceland rift. Bull Volcanol. 72, 55–62.

Sturkell, E., Einarsson, P., Sigmundsson, F., Hreinsdóttir, S., Geirsson, H., 2003. Deformation of Grímsvötn volcano, Iceland: 1998 eruption and subsequent inflation. Geophys. Res. Lett. 30 (4).

Sturkell, E., Einarsson, P., Sigmundsson, F., Geirsson, H., Olafsson, H., Pedersen, R., de Zeeuw-van Dalfsen, E., Linde, A., Sacks, S.I., Stefánsson, R., 2006a. Volcano geodesy and magma dynamics in Iceland. J. Volcanol. Geotherm. Res. 150 (1), 14–34.

Sturkell, E., Sigmundsson, F., Slunga, R., 2006b. 1983–2003 decaying rate of deflation at Askja caldera: pressure decrease in an extensive magma plumbing system at a spreading plate boundary. Bull. Volcanol. 68, 727–735.

Sturkell, E., Ágústsson, K., Linde, A.T., Sacks, S.I., Einarsson, P., Sigmundsson, F., Geirsson, H., Pedersen, R., LaFemina, P.C., Ólafsson, H., 2013. New insights into volcanic activity from strain and other deformation data for the Hekla 2000 eruption. J. Volcanol. Geotherm. Res. 256, 78–86.

Tarasewicz, J., White, R.S., Woods, A.W., Brandsdóttir, B., Gudmundsson, M.T., 2012. Magma mobilization by downward-propagating decompression of the Eyjafjallajökull volcanic plumbing system. Geophys. Res. Lett. 39, L19309.

Tarasewicz, J., White, R.S., Brandsdóttir, B., Schoonman, C.M., 2014. Seismogenic magma intrusion before the 2010 eruption of Eyjafjallajökull volcano, Iceland. Geophys. J. Int. 198 (2), 906–921.

Tryggvason, E., 1989. Ground deformation in Askja, Iceland: its source and possible relation to flow of the mantle plume. J. Volcanol. Geotherm. Res. 39, 61–71.

Tryggvason, E., 1994. Observed ground deformation at Hekla, Iceland prior to and during the eruptions of 1970, 1980–1981 and 1991. J. Volcanol. Geotherm. Res. 61, 281–291.

Vogfjörd, K.S., Jakobsdóttir, S.S., Gudmundsson, G.B., Roberts, M.J., Ágústsson, K., Arason, Th., Geirsson, H., Karlsdóttir, S., Hjaltadóttir, S., Ólafsdóttir, U., Thorbjarnardóttir, B., Skaftadóttir, Th., Sturkell, E., Jónasdóttir, E.B., Hafsteinsson, G., Sveinbjörnsson, H., Stefánsson, R., Jónsson, V., Th., 2005. Forecasting and monitoring of a volcanic eruption in Iceland. EOS 86 (26), 245–248.

Voight, B., Glicken, H., Janda, R.J., Douglass, P.M., 1981. Catastrophic rockslide-avalanche of May 18. In: Lipman, P.W., Mullineaux, D.R. (Eds.), The 1980 Eruptions of Mount St. Helens. U.S. Geol. Surv. Prof. Pap. 1250, Washington, DC, pp. 347–378.

Wilcock, W.S.D., Tolstoy, M., Waldhauser, F., Garcia, C., Tan, Y.J., Bohnenstiehl, D.R., Caplan-Auerbach, J., Dziak, R.P., Arnulf, A.F., Mann, M., 2016. Seismic constraints on caldera dynamics from the 2015 Axial Seamount eruption. Science. doi:10.1126/science.aah5563.

Wright, T.J., Sigmundsson, F., Pagli, C., Belachew, M., Hamling, I.J., Brandsdóttir, B., Keir, D., Pedersen, R., Ayele, A., Ebinger, C., Einarsson, P., Lewi, E., Calais, E., 2012. Geophysical constraints on the dynamics of spreading centres from rifting episodes on land. Nat. Geosci. 5, 242–250.

Yang, X.M., Davis, P.M., Dieterich, J.H., 1988. Deformation from inflation of a dipping finite prolate spheroid in an elastic half- space as a model for volcanic stressing. J. Geophys. Res. 93, 4249–4257.

FURTHER READING

Battaglia, M., Cervelli, P.F., Murray, J.R., 2013. Modelling crustal deformation near active faults and volcanic centers—a catalog of deformation models. U.S. Geological Survey Techniques and Methods, book 13, chap. B1, 96 pp.

Dzurisin, D., 2007. Volcano Deformation, Geodetic Monitoring Techniques. Springer-Praxis, Chichester, UK, 441 pp.

Hooper, A., 2008. A multi-temporal InSAR method incorporating both persistent scatterer and small baseline approaches. Geophys. Res. Lett. 35 (16).

National Academies of Sciences, Engineering, and Medicine, 2017. Volcanic Eruptions and their Repose, Unrest, Precursors, and Timing. The National Academies Press, Washington, DC. doi: https://doi.org/10.17226/24650.

Segall, P., 2010. Earthquake and Volcano Deformation. Princeton University Press, Princeton, New Jersey, USA, 432.

CHAPTER 12

Synthesis on the State-of-the-Art and Future Directions in the Research on Volcanic and Igneous Plumbing Systems

Steffi Burchardt
Uppsala University, Uppsala, Sweden

Contents

12.1 INTRODUCTION

Our current understanding of the processes that occur within volcanic and igneous plumbing systems (VIPS) draws upon more than 300 years of observations, measurements and experiments within the fields of structural geology, igneous petrology and geochemistry, volcanology, geophysics and geodesy (see Chapter 1). During this time, these disciplines were born out of a deepening understanding of natural phenomena and the development of a large variety of measurement techniques and methods. The associated specialisation and diversification produced a very detailed, but scattered image of VIPS with contrasting perspectives instead of consensus. However, in the last decade or so, a more integrative research approach has been emerging and calling for multi-disciplinary studies of magma transport, storage, evolution and eruption. Large-scale, multi-disciplinary projects that target the complexity of volcano unrest and eruption, as well as the underlying VIPS processes have propelled research into a new era. It is in this era that VIPS studies emerge as a defined field of research with a VIPS commission of the International Association of Volcanology and Chemistry of the Earth's Interior (IAVCEI), sessions at international conferences and the first textbook on VIPS.

Volcanic and Igneous Plumbing Systems. http://dx.doi.org/10.1016/B978-0-12-809749-6.00012-1
Copyright © 2018 Elsevier Inc. All rights reserved.

This book follows the way magma takes through the Earth's crust into magma chambers and to the surface. Each chapter summarises and links knowledge from various disciplines and approaches to deliver a holistic picture of the processes that occur in VIPS. However, the chapters also highlight that there are still large gaps in our understanding of VIPS and that researchers from different disciplines may disagree on certain aspects of magma transport and storage processes. Much work remains to be done in each of the disciplines involved in VIPS studies, but, in particular, in the attempt to come to a unified picture of VIPS (Galland et al., 2013). In the following, I will outline a few common challenges and opportunities for future research, based on the questions that recurred in the chapters of this book.

12.2 STATE-OF-THE-ART AND FUTURE CHALLENGES—WHERE DO WE GO FROM HERE?

12.2.1 The Relationship Between Nature and Models

The process of research in an empirical field, such as the Earth sciences, is roughly characterised by the following steps: The researcher observes a naturally occurring phenomenon and describes her observations. To improve this description, she takes measurements of the features related to the phenomenon and analyses her data. Nature is often very systematic, so that patterns emerge from the data; and other researchers find the same patterns when they study the same phenomenon in other places or instances. Natural scientists like patterns, so they conclude that the phenomenon is generally described by the patterns they find and that there are certain processes that lead to the observed pattern. The latter conclusion is a conceptual model, that is, a generalised concept that explains a phenomenon with a process.

Examples of this research approach can be found in all chapters of this book. A specific example within the field of VIPS is the study of dyke thicknesses (see Chapter 3). Since dykes are often exposed with their thickness as the only measurable parameter and since dyke thickness is easy to measure with simple tools, many researchers have mapped dyke thicknesses in their study areas. Krumbholz et al. (2014) analysed thousands of such measurements for their fit to different statistical distributions and found that there is a very good fit with the so-called Weibull distribution. The Weibull distribution is well known in materials science and results from material failure (breaking) as a result of power-law distributed weaknesses. For rocks and dykes, this means that the weaknesses that exist in the crust (faults, joints, etc.) around a magma chamber fail at different magma chamber pressures depending on the sizes of weaknesses. As a result, the thickness of dykes injected from the same magma chamber will follow a Weibull distribution. Krumbholz et al. (2014) concluded that the host-rock properties have a major control on dyke emplacement.

Another research approach uses experiments (also called models) in the laboratory or in the computer. A researcher who wants to understand the dynamics of a certain natural phenomenon sets up an experiment to simulate, or model, the phenomenon. He chooses a number of input parameters that can be based on measured values or assumptions and that should be similar to the parameters in nature (in analogue experiments, this approach is called scaling). Once the experiment runs, the researcher observes what happens, measures and analyses the data in much the same way as described for natural phenomena above and in the hope that patterns will emerge. Compared to the study of nature, experiments have the advantage that input parameters can be controlled, that the dynamics can be observed and that the output is fully accessible. Depending on the identified patterns, the researcher also formulates a conceptual model that explains the dynamics of the natural phenomenon that he studied. However, as we do not know many of the input parameters in nature and can often not quantify them, the experiment may not actually resemble nature, which makes the concepts that emerge questionable.

An example of a model used in the study of VIPS are geodetic models that are designed to reproduce measured surface deformation caused by magma transport during volcanic unrest (see Chapter 11). The most commonly applied geodetic models use a very simple setup (Fig. 12.1): a deformation source A that represents the dyke or magma chamber, a change of pressure, volume or opening B in the source, located within a perfectly isotropic elastic half-space C that represents the Earth's crust. Testing a large number of different values for, for example, the size, arrangement and location of A and the magnitude of B, the best-fit model is identified. This model is then used to explain the process that caused the surface deformation in nature. For example, surface deformation prior to and during the Holuhraun eruption in Iceland in 2014–2015 was modelled

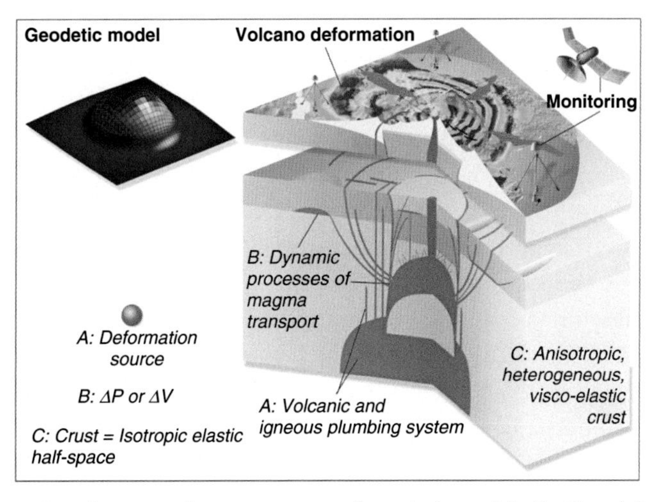

Figure 12.1 ***Comparison between the components of geodetic models (A–C) and their equivalents in nature. Note the difference in complexity.***

using an array of rectangular deformation sources that opened with time in an isotropic elastic half-space (Sigmundsson et al., 2014; Chapter 11). The best-fitting opening and arrangement of the deformation sources, together with the path of observed seismicity, were then assumed to explain the propagation path and opening of the dyke that fed the eruption. Hence, the model results are useful to explain the natural observations, although we know that some of the model assumptions are not correct.

Both research approaches are thus powerful, and their results have improved a lot since they were first employed. The main improvements in the case of natural observations are a shift from qualitative descriptions to quantitative measurements with more and more accuracy. For experimental approaches, the development of experimental techniques and increased computer capacity have led to major advances. Moreover, the increased amount and quality of data from nature implies a better quality of the model output. Additionally, the integration of data and observations from other disciplines into modelling has been a leap towards more realistic results.

However, there is one major catch here: Models are not the truth! Models are tools that explain the observations and measurements and they are concepts that help us compare our findings with those of other researchers. Our observations are always incomplete, for example, because fossil plumbing systems are not fully exposed, because in the field, we only see the solidified result of all processes that happened while the plumbing system was active, because we cannot directly observe active plumbing systems, and because we have too few exposures, measurements and monitored events. In the future, this situation will improve, but we will never have the complete picture. Therefore, we have to be careful when we use models. We should not make the mistake to force our observations to fit into a certain model. We should observe nature and measure accurately and believe in our data, but we should always be willing to take a step back and question the model, if the data does not fit.

12.2.2 The Time Scales of Magma Transport and Storage

The processes of magma transport and storage in the Earth's crust occur on time scales that range from seconds to millions of years. This scale range is a challenge for researchers and, at the same time, the key to better assess volcanic risk. Magma can be transported over large crustal distances quickly (m s^{-1}) and efficiently in dykes to either erupt at the surface or feed magmatic intrusions. The volume of magma transported is therefore a main factor controlling the eruptive volume or the amount of magma added to an intrusive body, which is why we need to understand magma supply from its source (see Chapter 2), determine the size distribution of dykes (see Chapter 3) and understand how larger magma bodies that are fed by, and can feed, dykes form (see Chapters 5 and 6).

When it comes to the formation of magma chambers, we can distinguish small magma bodies that may form through the intrusion of one continuous batch of magma on time scales of months to hundreds of years (sills and laccoliths; e.g. Castro et al., 2016) from larger magma bodies. High-precision geochronology and geochemistry as well as structural mapping have found that large magma reservoirs assemble through the addition of small magma batches ('dyke-loads'?) over time scales of thousands to millions of years (Glazner et al., 2004; Leuthold et al., 2012; Coleman et al., 2016). The size distribution of such intrusive bodies (Cruden and McCaffrey, 2001; Cruden et al., 2017), as well as field relationships of, for example, sills with multiple units, laccoliths, and plutons emplaced by incremental cauldron subsidence (Castro et al., 2016; Burchardt, 2008; Burchardt et al., 2012) indicate that large magma bodies form from small ones (see Chapters 2, 5 and 6). From numerical modelling, we learn that the emplacement of magma batches into different crustal levels is a main driver of magma differentiation, where the structure of the igneous plumbing system plays an important role (Annen et al., 2015). In addition, in order to form big and largely molten magma chambers, new magma batches need to be injected into an existing magma body frequently (Annen, 2011).

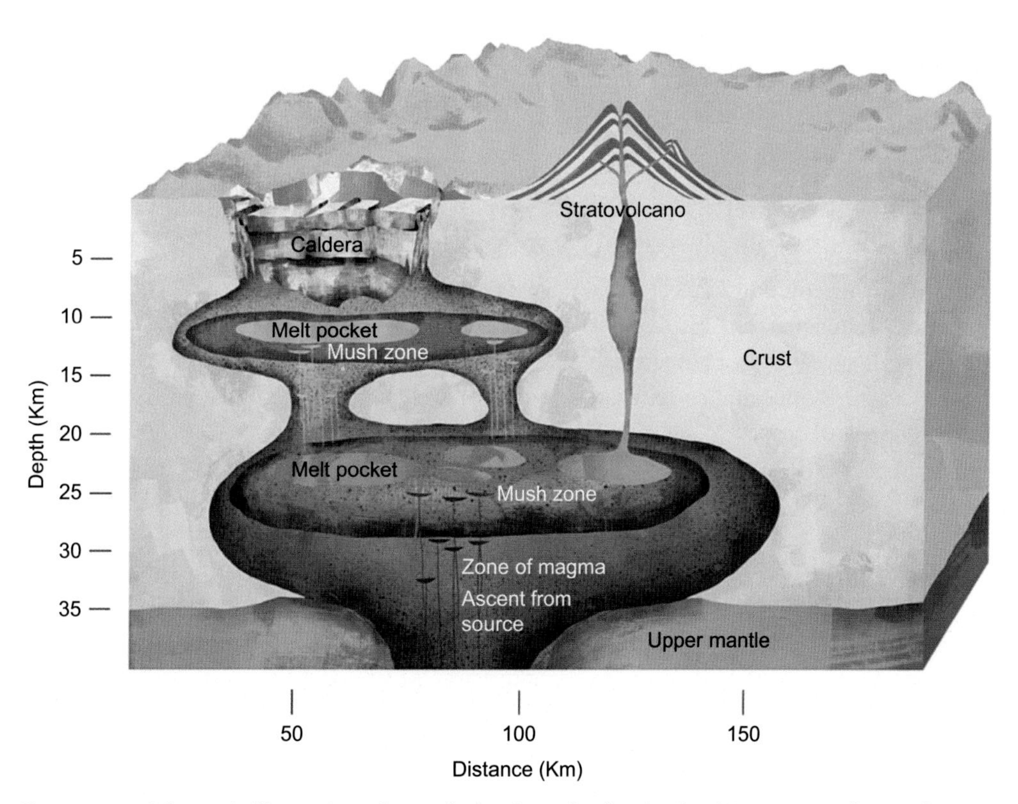

Figure 12.2 ***Schematic illustration of a mush-dominated volcanic plumbing system after Bachmann and Huber (2016).*** Note that the geometry of mush zones and magma pockets in this illustration is very schematic, while in nature they have more complex shapes that depend on the initial magma-emplacement mechanism.

Analyses of the information stored in erupted crystals (see Chapter 8) indicate that magma chambers spent a considerable amount of their lifetime as a crystal mush with much less than 50% of the reservoir volume being melt stored between crystals (Fig. 12.2; Bachmann and Bergantz, 2008; Cashman et al., 2017). Small pockets in the mush may consist of mostly melt that can be erupted (see, e.g. Chapter 10), but crystal mushes themselves are not eruptible. The VIPS of many volcanoes are probably mainly composed of mush during most of their lifetime, and this is what geophysical methods confirm for the current state of monitored volcanoes (Pritchard and Gregg, 2016).

However, the existence of large collapse calderas and of huge volumes of erupted material demonstrate that magma chambers filled with mostly melt do exist, albeit for short periods of time (cf. Chapter 10). Hence, we can assume that crystal-mush plumbing systems can be mobilised before larger eruptions (Cooper and Kent, 2014; Karakas et al., 2017). Remobilisation occurs when heat is added to the system, for example, by addition of hot, mafic melt from the mantle (Szymanowski et al., 2017), a process often referred to as recharge. Recharge that is usually fuelled by dykes occurs frequently in volcanoes and can be expressed as volcanic unrest at the surface (see Chapter 11). However, whether the added energy is sufficient for remobilisation depends on the volume and frequency of recharge and the overall structure of the crystal–mush reservoir. Hence, it is important to understand magma transport in dykes and the structure of magma chambers to unravel the time scales of formation and reactivation of volcanic plumbing systems.

12.2.3 The Interplay Between Host Rock and Magma Rheology

Another recurring theme in the chapters of this book is the interplay between the properties of the magma and its host rock. The Earth's crust consists of complex, highly heterogeneous and anisotropic materials. Depending on the temperature and pressure, as well as the structure of crustal rocks, the crust can react elastically—it goes back to its original shape—or plastically by deforming in a brittle or ductile way. During magma transport and storage, the Earth's crust is exposed to magmatic temperatures and a large range of strain rates caused by, for example, propagating sheet intrusions (m/s speed), inflating laccoliths (hundreds-of-metres/year to decade), magma rising as a diapir in the lower crust (m/thousands-of-years), which can result in different behaviour of the surrounding rocks. While the lower and middle crust are usually assumed to be plastic and deform in a ductile manner, the upper crust is often thought to be elastic at the deformation rates caused by magma transport and emplacement (e.g. Fig. 12.1). Numerical and analogue models that employ an elastic crustal rheology have produced results that strongly resemble natural phenomena, such as the shape of sheet intrusions and the associated surface deformation (e.g. Chapters 3 and 11). However, these models are still only models (see Section 12.2.1), and their results are not a proof of the general elastic behaviour of the crust.

There are still many observations that do not fit these assumptions, for example, earthquakes that indicate shear failure during dyke propagation (Sigmundsson et al., 2014) and faulting, folding and fluidisation near sill tips (Schofield et al., 2009; Spacapan et al., 2016) and that suggest that the upper crust may be viscoelastic (e.g. Chapter 5). This viscoelastic rheology may result from the interplay between the intruding magma and its host rock, leading to viscous stresses (Galland et al., 2014). This is particularly relevant for high-viscosity magma (andesitic to rhyolitic compositions) and crustal rocks with high cohesion (e.g. shale, coal). In such cases, magma emplacement may work differently compared to what is expected when an elastic behaviour of the crust is assumed (Guldstrand et al., 2017; see also Chapters 3 and 5).

Moreover, magma itself is a complex material that consists of fluid (the melt), solid (crystals) and gaseous (volatiles in bubbles) components. The amount of these components, as well as their shape, the melt composition and temperature control the magma viscosity (Caricchi et al., 2007; Giordano et al. 2008; Cordonnier et al., 2012; Pistone et al., 2012, 2013, 2016; Gonnermann and Manga, 2013). When the crystal content reaches about 60% of the magma volume, the viscosity increases dramatically, which causes the magma to stall. Its rheology is now controlled by a network of interlocked crystals, the crystal mush. This mush is not eruptible (see Section 12.2.2), but at the same time able to deform in a brittle manner (solid-state deformation).

Studies of rock textures in extrusive domes and magmatic conduits composed of high-viscosity magmas have shown that magma rheology is also influenced by strain (Chapter 4; Dingwell and Webb, 1989; Dingwell, 1997; Tuffen et al., 2003; Tuffen and Dingwell, 2005). Rock-mechanical experiments can reproduce the temperature and strain conditions within conduits and domes and conclude that the magma may change its rheology from Newtonian to non-Newtonian at high strains. This rheology change is accompanied by the magma no longer behaving as a fluid, but as a solid that is able to fracture. The exact strain, at which this rheological transition occurs, is a function of magma viscosity. Since high strains do not only occur in conduits and extrusive domes, but also during the formation of shallow crustal intrusions of these compositions, we can assume that the strain-dependent rheology change of viscous magma also affects magma emplacement (Mattsson et al., 2018). Future research will therefore reveal how magma and host-rock rheology interact during the formation of magmatic intrusions.

12.2.4 The Relationship Between VIPS, Stresses, and Pre-existing Structures

Magma transport and storage are always happening in a context. On a large scale, this context is the geodynamic setting of the plumbing system. As outlined in detail in Chapter 7, plate tectonics and magmatism are strongly linked, with magma being an essential ingredient in global geodynamics and tectonic forces being a driver of magmatism. Often, we cannot identify cause and effect in this relationship. For instance, the largest collapse calderas on Earth seem associated with strike-slip tectonics and located in pull-apart basins. At the same time, a magma chamber in the crust may be causing the localisation of a newly forming pull-apart basin (Girard and van Wyk de Vries, 2004; see also Chapters 7 and 10).

At the volcano scale, tectonic stresses are juxtaposed by more local stress fields related to the pressure within the volcanic and igneous plumbing system itself and/or to the gravitational push from the volcanic edifice. Shallow magma chambers "feel" the Earth's surface (a free surface), so that the stress field created by magma pressure in the chamber is influenced by the free surface. Propagating sheet intrusions align in this stress field and form characteristic patterns, such as the concentric pattern of cone sheets (see Chapter 4). A clear example of the extent of a volcano-scale stress field is the dyke propagation prior to the 2014–2015 Holuhraun eruption. Initially, a radial dyke was injected from the magma reservoir beneath Bárðarbunga volcano. This radial dyke propagated for a distance of about 6 km to the SE away from the volcano. There, it changed direction and subsequently propagated to the ENE, then NE and finally NNE in the direction of the Askja fissure swarm that is controlled by plate spreading, that is, the large-scale tectonic stress field (Sigmundsson et al., 2014).

Magmatic activity with eruptions and the formation of intrusive sheet swarms can, with time, lead to the growth of a volcanic edifice. The load of the edifice exerts gravitational stresses that in turn have a strong effect on magma transport paths. For instance, dykes can be forced to divert during propagation and align into volcanic rift zones (see Chapters 4, 7 and 9). The channelling of magma transport paths can, in turn, result in the formation of new magma chambers or the reactivation of existing ones. The formation of new magma chambers is an important step in the rebuilding of collapsed volcanic edifices, but can, in the long run, also lead to renewed gravitational collapse (Chapter 9).

At a local scale, magma not only interacts with host rocks of different rheologies (see Section 2.3), but it also encounters pre-existing structures, such as faults, joints and the interfaces between layers. These pre-existing structures are often "weaker" than the surrounding rocks, and from field observations, we know that magma in most cases chooses the 'path of least resistance'. More specifically, this may mean that a magmatic sheet diverts from its path and follows a bedding plane for some distance (Mathieu et al., 2015; see also Chapter 3) or that during caldera resurgence pre-existing ring faults are reactivated and intruded (see Chapter 10).

Finally, propagating magmatic sheets, such as dykes and sills, form their own, very local, stress field at the intrusion tip (see Chapters 3 and 4). The nature of this stress field controls the way magma propagates through the crust and is strongly controlled by the rheology of the host rock and the magma (see Section 2.3). For example, propagating dykes of low-viscosity magma will cause an elastic host rock to fail in tension and thus propagates as a mode-I fracture, whereas a dyke filled with high-viscosity magma may cause a viscoelastic host rock to fail in shear and thus propagates as a shear-mode fracture (see Chapter 3).

Future research will therefore unravel more details on how stress fields at different scales interact in space and time and how magmas and crustal rocks of different mechanical properties react to stresses.

12.3 CONCLUDING REMARKS

As shown in the chapters of this book and above, there is a lot we have learn about VIPS, but there is also still a lot of room for future research. Multi-disciplinary approaches and new technical developments will allow for a better conceptual and quantitative understanding of the processes that occur in VIPS, especially when we combine the strengths of our methods and overcome their weaknesses. Together with all the authors who have contributed to this book, I hope students and researchers who are interested in magma transport and storage will find this book a useful and inspiring resource and a stepping stone in their journey to study VIPS.

ACKNOWLEDGEMENTS

This book is a summary of VIPS—their various components and complex processes—that I missed as a student reading countless research articles on volcanic eruptions, dykes, plutons and magma emplacement, searching for the big picture. This book is also the resource that I missed since I became a teacher who wanted to convey my knowledge and fascination for magma transport and storage in the Earth's crust to my students. I am grateful for the opportunity to write this book and want to thank Katerina Zalina, Marisa LaFleur and the production management at Elsevier for their support. I also want to express my gratitude to all 55 authors who have contributed with their expertise to the chapters in this book, in particular, the lead authors Sandy Cruden, Janine Kavanagh, Olivier Galland, Sven Morgan, Ben van Wyk de Vries, Dougal Jerram, Audray Delcamp, Ben Kennedy and Freysteinn Sigmundsson. Working with all of you has been a pleasure and a privilege.

Moreover, I would like to thank my research group, collaborators and colleagues, in particular, Tobias Mattsson, Erika Ronchin, Christoph Hieronymus, Abigail Barker and Bjarne Almqvist, who constantly teach me new things and encourage me to question my models. Your support means a lot to me and has contributed to shape this book in many ways. I'm thankful to the members and supporters of the IAVCEI commission on VIPS. May this book help to stimulate future research and commission activities. I would also like to acknowledge the Center for Natural Hazard and Disaster Research (CNDS), the Department of Earth sciences at Uppsala University and the various research funding organisations that have supported my research on plumbing systems throughout the years.

Finally, I want to thank my family and friends in Sweden, Germany and elsewhere. I'm glad that I had you by my side when working on this book. Your support is invaluable.

REFERENCES

Annen, C., 2011. Implications of incremental emplacement of magma bodies for magma differentiation, thermal aureole dimensions and plutonism–volcanism relationships. Tectonophysics 500, 3–10.

Annen, C., Blundy, J.D., Leuthod, J., Sparks, R.S.J., 2015. Construction and evolution of igneous bodies: towards an integrated perspective of crustal magmatism. Lithos 230, 206–221.

Bachmann, O., Bergantz, G., 2008. The magma reservoirs that feed supereruptions. Elements 4, 17–21.

Bachmann, O., Huber, C., 2016. Silicic magma reservoirs in the Earth's crust. Am. Mineral. 61, 2377–2404, https://doi.org/10.2138/am-2016-5675.

Burchardt, S., 2008. New insights in the mechanics of sill emplacement provided by field observations of the Njardvik Sill, Northeast Iceland. J. Volcanol. Geotherm. Res. 173, 280–288.

Burchardt, S., Tanner, D.C., Krumbholz, M., 2012. The Slaufrudalur Pluton, Southeast Iceland – An example of shallow magma emplacement by coupled cauldron subsidence and magmatic stoping. Geol. Soc. Am. Bull. 124, 213–227, doi:10.1130/B30430.1.

Caricchi, L., Burlini, L., Ulmer, P., et al., 2007. Non-Newtonian rheology of crystal-bearing magmas and implications for magma ascent dynamics. Earth Planet. Sci. Lett. 264, 402–419. doi: 10.1016/j.epsl.2007.09.032.

Cashman, K.V., Sparks, R.S.J., Blundy, J.D., 2017. Vertically extensive and unstable magmatic systems: a unified view of igneous processes. Science 355 (6331).

Castro, J.M., Cordonnier, B., Schipper, C.I., Tuffen, H., Baumann, T.S., Feisel, Y., 2016. Rapid laccolith intrusion driven by explosive volcanic eruption. Nat. Commun. 7, 13585.

Coleman, D.S., Mills, R.D., Zimmerer, M.J., 2016. The pace of plutonism. Elements 12, 97–102.

Cordonnier, B., Caricchi, L., Pistone, M., et al., 2012. The viscous-brittle transition of crystal-bearing silicic melt: direct observation of magma rupture and healing. Geology 40, 611–614. doi: 10.1130/G3914.1.

Cooper, K.M., Kent, A.J.R., 2014. Rapid remobilization of magmatic crystals kept in cold storage. Nature 506, 480–483.

Cruden, A.R., McCaffrey, K.J.W., 2001. Growth of plutons by floor subsidence: implications for rates of emplacement, intrusion spacing and melt-extraction mechanisms. Phys. Chem. Earth Part A 26, 303–3015.

Cruden, A.R., McCaffrey, K.J.W., Bunger, A.P., 2017. Geometric scaling of tabular igneous intrusions: implications for emplacement and growth. In: Advances in Volcanology. Springer, Berlin, Heidelberg. https://doi.org/10.1007/11157_2017_1000.

Dingwell, D.B., 1997. The brittle-ductile transition in high-level granitic magmas: material constraints. J. Petrol. 38, 1635–1644. doi: 10.1093/petroj/38.12.1635.

Dingwell, D.B., Webb, S.L., 1989. Structural relaxation in silicate melts and non-Newtonian melt rheology in geologic processes. Phys. Chem. Mineral. doi: 10.1007/BF00197020.

Galland, O., Burchardt, S., Troll, V.R., 2013. Volcanic and igneous plumbing systems: state-of-the-art and future developments EOS. Trans. Am. Geophys. Union 94, 169.

Galland, O., Burchardt, S., Hallot, E., Mourgues, R., 2014. A unified model for dykes versus cone sheets in volcanic systems. J. Geophys. Res. 119, 6178–6192. doi: 10.1002/2014JB011059.

Giordano, D., Russell, J.K., Dingwell, D.B., 2008. Viscosity of magmatic liquids: a model. Earth Planet. Sci. Lett. 271, 123–134. doi: 10.1016/j.epsl.2008.03.038.

Girard, G., van Wyk de Vries, B., 2004. The Managua Graben and Las Sierras-Masaya volcano, a case of pull-apart localization by an intrusive complex. J. Volcanol. Geotherm. Res. 144, 37–57.

Glazner, A.F., Bartley, J.M., Coleman, D.S., Gray, W., Taylor, R.Z., 2004. Are plutons assembled over millions of years by amalgamation from small magma chambers? GSA Today 14, 4–11.

Gonnermann, H.M., Manga, M., 2013. Dynamics of magma ascent in the volcanic conduit. In: Fagents, S.A., Gregg, T.K.P., Lopes, R.M.C. (Eds.), Modeling Volcanic Processes. Cambridge University Press, Cambridge, pp. 55–84.

Guldstrand, F., Burchardt, S., Hallot, E., Galland, O., 2017. Dynamics of surface deformation induced by dikes and cone sheets in a cohesive coulomb brittle crust. J. Geophys. Res. 122doi: 10.1002/2017JB014346.

Karakas, O., Degruyter, W., Bachmann, O., Dufek, J., 2017. Lifetime and size of shallow magma bodies controlled by crustal-scale magmatism. Nat. Geosci. 10, 446–450. doi: 10.1038/ngeo2959.

Krumbholz, M., Hieronymus, C., Burchardt, S., Troll, V.R., Tanner, D., Friese, N., 2014. Weibull distributed dyke thickness reflects probabilistic character of host-rock strength. Nat. Commun. 5, 3272, 10/1038/ncomms4272.

Leuthold, J., Münterer, O., Baumgartnerm, L.P., Putlitz, B., Ovtcharova, M., Schaltegger, U., 2012. Time resolved construction of a bimodal laccolith (Torres del Paine, Patagonia). Earth Planet. Sci. Lett. (325–326), 85–92.

Mathieu, L., Burchardt, S., Troll, V.R., Krumbholz, M., Delcamp, A., 2015. Geological constraints on the dynamic emplacement f cone sheets—the Ardnamurchan cone-sheet swarm, NW Scotland. J. Struct. Geol. 80, 133–141. doi: 10.1016/j.jsg.2015.08.012.

Mattsson, T., Burchardt, S., Almqvist, B.S.G., Ronchin, E., 2018. Syn-emplacement fracturing in the Sandfell laccolith, eastern Iceland—implications for rhyolite intrusion growth and volcanic hazards. Front. Earth Sci. Volcanol accepted for publication.

Pistone, M., Caricchi, L., Ulmer, P., et al., 2012. Deformation experiments of bubble- and crystal-bearing magmas: rheological and microstructural analysis. J. Geophys. Res. 117doi: 10.1029/2011JB008986.

Pistone, M., Caricchi, L., Ulmer, P., et al., 2013. Rheology of volatile-bearing crystal mushes: mobilization vs. viscous death. Chem. Geol. 345, 16–39. doi: 10.1016/j.chemgeo.2013.02.007.

Pistone, M., Cordonnier, B., Ulmer, P., Caricchi, L., 2016. Rheological flow laws for multiphase magmas: an empirical approach. J. Volcanol. Geotherm. Res. 321, 158–170. doi: 10.1016/j.jvolgeores.2016.04.029.

Pritchard, M.E., Gregg, P.M., 2016. Geophysical evidence for silicic crustal melt in the continents: where, what kind, and how much? Elements 12, 121–127.

Spacapan, J.B., Galland, O., Leanza, H.A., Planke, S., 2016. Igneous sill and finger emplacement mechanism in shale-dominated formations: a field study at Cuesta del Chihuido, Neuquén Basin, Argentina. J. Geol. Soc. 174, 422–433, https://doi.org/10.1144/jgs2016-056.

Schofield, N.J., Brown, D.J., Magee, C., Stevenson, C.T., 2009. Sill morphology and comparison of brittle and non-brittle emplacement mechanisms. J. Geol. Soc. 169, 127–141, https://doi.org/10.1144/0016-76492011-078.

Sigmundsson, F., et al., 2014. Segmented lateral dyke growth in a rifting event at Bárðarbunga volcanic system, Iceland. Nature 517 (7533), 191–195.

Szymanowski, D., Wotzlaw, J.-F., Bachmann, O., Guillong, M., von Quadt, A., 2017. Protracted near-solidus storage and pre-eruptive rejuvenation of large magma reservoirs. Nat. Geosci. 10, 777–782. doi: 10.1038/ngeo3020.

Tuffen, H., Dingwell, D.B., 2005. Fault textures in volcanic conduits: evidence for seismic trigger mechanisms during silicic eruptions. Bull. Volcanol. 67, 370–387. doi: 10.1007/s00445-004-0383-5.

Tuffen, H., Dingwell, D.B., Pinkerton, H., 2003. Repeated fracture and healing of silicic magma generate flow banding and earthquakes? Geology 31, 1089. doi: 10.1130/G19777.1.

FURTHER READING

Burchardt, S., Galland, O., Studying volcanic plumbing systems and multidisciplinary approaches to a multifaceted problem. In: Karoly Nemeth (Ed.), Updates in Volcanology – From Volcano Modelling to Volcano Geology.

Cashman, K.V., Sparks, R.S.J., Blundy, J.D., 2017. Vertically extensive and unstable magmatic systems: a unified view of igneous processes, Science 355 (6331).

Coleman, D.S., Mills, R.D., Zimmerer, M.J., 2004. The pace of plutonism. Elements 32 (5), 433–436, https://doi.org/10.1130/G20220.1.

INDEX

Printed in the United States
By Bookmasters